CAMBRIDGE STUDIES IN ECOLOGY

Saltmarsh ecology

Also in the series
Hugh G. Gauch, Jr *Multivariate analysis in community ecology*
Robert Henry Peters *The ecological implications of body size*
C. S. Reynolds *The ecology of freshwater phytoplankton*
K. A. Kershaw *Physiological ecology of lichens*
Robert P. McIntosh *The background of ecology: concept and theory*
Andrew J. Beattie *The evolutionary ecology of ant–plant mutualisms*
F. I. Woodward *Climate and plant distribution*
Jeremy J. Burdon *Diseases and plant population biology*
N. G. Hairston Sr *Community ecology and salamander guilds*
Janet I. Sprent *The ecology of the nitrogen cycle*
H. Stolp *Microbial ecology: organisms, habitats and activities*
R. Norman Owen-Smith *Megaherbivores: the influence of very large body size on ecology*
John A. Wiens *The ecology of bird communities:*
Volume 1 Foundations and Patterns
Volume 2 Processes and Variations
N. G. Hairston Sr *Ecological experiments: Purpose, design and execution*
Colin Little *The terrestrial invasion: an ecophysiological approach to the origins of land animals*
R. Hengeveld *Dynamic biogeography*

Saltmarsh ecology

PAUL ADAM
University of New South Wales, Australia

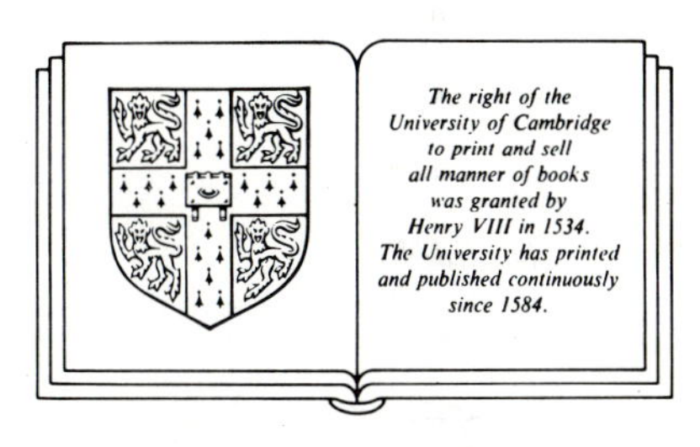

CAMBRIDGE UNIVERSITY PRESS
Cambridge
New York Port Chester
Melbourne Sydney

Published by the Press Syndicate of the University of Cambridge
The Pitt Building, Trumpington Street, Cambridge CB2 1RP
40 West 20th Street, New York NY 10011, USA
10 Stamford Road, Oakleigh, Melbourne 3166, Australia

First published 1990

Printed in Great Britain at the University Press, Cambridge

British Library cataloguing in publication data

Adam, Paul
Saltmarsh ecology.
1. Salt marshes. Ecology
I. Title
574.5′2636

Library of Congress cataloguing in publication data

Adam, Paul.
Saltmarsh ecology / Paul Adam.
p. cm. — (Cambridge studies in ecology)
Includes bibliographical references.
ISBN 0-521-24508-7
1. Tidemarsh ecology. I. Title. II. Series.
QH541.5.S24A33 1990
574.5′2636—dc20 89-22188 CIP

ISBN 0 521 24508 7

Contents

Preface

Saltmarsh has a very important place in the history of ecology. Some of the earliest field courses involved extensive study of saltmarshes; early volumes of the *New Phytologist*, *Journal of Ecology* and *Ecology* contain papers describing saltmarshes and introductory studies on physiological ecology. The ritual saltmarsh excursion is still an essential part of the curriculum in many courses, at both secondary and tertiary levels. While this educational role partly reflects the importance of tradition, it is also an acknowledgement of the enormous opportunities for demonstrating ecological phenomena provided by the saltmarsh ecosystem. In the 1960s, saltmarshes were the venue for some important studies which developed systems ecology; in the 1970s and 1980s, studies on the physiology for salt tolerance in halophytes have been one of the most active areas in the development of ecophysiology. Research is being carried out today on many aspects of the biology and ecology of saltmarsh in many parts of the world.

Despite the great number of studies on saltmarshes, there is still much we do not know. Some topics, such as the study of nutrient and energy cycles, provide great challenges and will require interdisciplinary collaboration. Others, although equally important to our understanding of the total saltmarsh resource, are more easily studied. For example, in many parts of the world, there is little documentation of the distribution of marshes or of their species composition. [Even on such a well studied coast as that of eastern North America the area of saltmarsh on Maine was underestimated by an order of magnitude for many years (Jacobson, Jacobson & Kelley 1987).] Throughout the text, I draw attention to areas of ignorance or suggest that the available evidence is inconclusive. These comments are not intended as a litany of despair but rather the very opposite. There are still numerous opportunities for further research and the prospects for future advances are exciting.

In attempting to cover such an active field, it is inevitable that an arbitrary decision has to be taken as to when to stop considering new literature, otherwise the task would be never ending. In consequence by the time the book appears, some sections may already be out of date. Such is the fate of all authors; the time taken for the majority of the text to become superseded will be a measure of the vitality of saltmarsh studies and should, therefore, be short.

My account is very heavily biased in favour of higher plants, and the longest chapter (Chapter 4) deals exclusively with plants. This reflects my own interests and background and is not to deny the existence of numerous interactions between plants, animals and microorganisms which may be important in controlling species composition and many factors in marsh development. The fauna of saltmarshes has recently been reviewed at length by Daiber (1982) who provides an introduction to the very extensive literature on this subject.

There is a tendency, in both general ecological textbooks and in more specialised literature, to regard saltmarshes as akin to roses – a saltmarsh is a saltmarsh is a saltmarsh. Generalisations about all saltmarshes are made on the basis of studies of a few sites, normally those favoured as the sites for field classes. One measure of our understanding of an ecosystem is our ability to develop general predictive models of various aspects of the system. Before this can be a profitable exercise, we must be able to recognise these features of the system about which we can generalise, without those generalisations being merely trivial. It is important therefore to be aware of the diversity of saltmarshes and to recognise those features common to all saltmarshes and those special to particular subclasses of marsh.

In Chapter 3, I have attempted an overview of one facet of this diversity, the geographical variation in the vegetation of saltmarshes. It has been my good fortune to have visited many saltmarshes, but in order to achieve a global overview, it has been necessary to rely upon accounts in the literature. While I hope that my synthesis is supportable as a broad-brush outline much more work is required to fill in the detail.

Many members of the saltmarsh biota have very wide ecological and geographical distributions. In many cases these distributions reflect the great genetic diversity of species. In Chapter 2, I have illustrated this by reference to genecological studies of a number of saltmarsh plants, although it is likely that similar diversity is shown by members of other taxonomic groups. The existence of this genetic diversity helps to explain the wide distribution of individual species and also indicates the potential available for future selection and evolution. It also highlights the inappropriateness of characterising species on the basis of studies of single populations (or even, as in some instances, single individuals).

The vegetation of saltmarshes differs markedly from that of terrestrial communities: within a saltmarsh, communities are often sharply defined. Questions may therefore be asked about how saltmarsh plants differ from those of other habitats and how, within a marsh, the characteristics of particular species relate to the habitat of the communities in which they grow. The physiological ecology of halophytes has long attracted interest but in the last few decades has been a particularly active field of endeavour. The physiological basis of salt tolerance is now sufficiently understood to explain the exclusion of non-halophytes from saltmarshes. Explanations for the segregation of species within a marsh (which is likely to reflect traits other than salt tolerance) are, as yet, not readily apparent. There is great potential for ecologists and physiologists to formulate hypotheses, which can be tested, to explain the physiological basis of community composition. In Chapter 4, physiological studies on saltmarsh species are reviewed in an ecological context.

Not only is our understanding of the physiological segregation of communities poor but the maintenance of communities has also been poorly studied, although to predict future development of marshes, and their response to environmental change, will demand better knowledge of the population ecology of individual species. Chapter 5 provides a brief review of aspects of the population ecology of halophytes. Although a great deal is known about the germination of seeds of halophytes under laboratory conditions, very little is known about the behaviour of species in the field.

Aspects of the functioning of saltmarshes as ecosystems have been the subject of many studies. However, the majority of these studies have been on the very distinctive *Spartina alterniflora* marshes of the eastern USA and few investigations have been made of other marsh types. It is still unclear whether generalisations about productivity and nutrient and energy exchanges can be made for *S. alterniflora* marshes, let alone whether results from *Spartina* marshes can be extrapolated to other marsh types. The ecology of *S. alterniflora*-dominated marshes has been comprehensively reviewed by Pomeroy & Wiegert (1981).

If we are to make rational management and planning decisions about coastal resources, it is clearly essential that we know more about saltmarsh productivity and exchanges between saltmarshes and adjacent waters.

Saltmarshes have a long history of exploitation, modification and destruction. In recent years, there has been concern in many western countries at the pace of wetlands loss, and regulations to protect saltmarshes have been introduced. Nevertheless, the threats from numerous 'minor' incursions, habitat degradation by pollution, and the possibility of imminent global sea level rise as a consequence of the 'Greenhouse Effect' suggest there are no grounds for complacency about the adequacy of existing conservation measures. The conservation of saltmarsh has been discussed at length by Daiber (1986) and Chapter 7 provides only a brief introduction to issues relevant to saltmarsh management.

My attempts to understand saltmarshes have been guided and assisted by many colleagues – the international fraternity of saltmarsh researchers is marked by its friendliness and enthusiasm. I am grateful to all those who have discussed and debated their ideas with me, given up their time to show me field sites or who have extracted me from the clutches of glutinous mud. Especial thanks are due to Alan Gray who first introduced me to the study of saltmarshes (and *Puccinellia maritima*), Tony Smith-White whose championing of *Sporobolus virginicus* provided the opportunities to visit much of the Australian coast and the late Derek Ranwell, who, while never being totally convinced about the merits of phytosociology, was unstinting in his advice and encouragement.

No book can be prepared in isolation and I am very grateful to all those who have helped with the task. The final version of the manuscript was entered onto the word processor by Joan Ratcliffe and Alaric Fisher, who now know more about saltmarshes than they ever wanted to; their contribution was invaluable. Alaric Fisher also prepared the majority of the figures. Robert King, Gunter Kirst and Bob Vickery read sections of the manuscript and made many useful suggestions; however, any responsibility for errors and omissions is mine alone.

Thanks are due to the institutions which have provided facilities during my explorations of saltmarshes: the Botany School and Emmanuel College, Cambridge; University College and the Botany Department, Durham, and the Botany School, University of New South Wales.

The invitation to write this book came from John Birks who has also read the manuscript in its entirety; his editorial advice was very useful. I owe John a debt of gratitude for his help, advice and friendship over many years; a debt for which this response to his invitation is inadequate recompense.

P.A.

Abbreviations for frequently-occurring species names

For species whose names appear frequently in the text, generic names are abbreviated as in the list below. For species referred to infrequently normal conventions apply.

Agrostis stolonifera	*A. stolonifera*
Armeria maritima	*Arm. maritima*
Artemisia maritima	*Art. maritima*
Aster tripolium	*A. tripolium*
Avicennia marina	*Av. marina*
Distichlis distichophylla (= Agropyron pungens)	*D. distichophylla*
Distichlis spicata	*D. spicata*
Elymus pycnanthus	*E. pycnanthus*
Festuca rubra	*F. rubra*
Glaux maritima	*G. maritima*
Halimione portulacoides	*H. portulacoides*
Juncus gerardi	*J. gerardi*
Juncus kraussii	*J. kraussii*
Juncus maritimus	*J. maritimus*
Juncus roemerianus	*J. roemerianus*
Leptochloa fusca (= Diplachne fusca)	*L. fusca*
Limonium vulgare	*L. vulgare*
Mesembryanthemum crystallinum	*M. crystallinum*
Plantago maritima	*Pl. maritima*
Plantago coronopus	*Pl. coronopus*
Puccinellia maritima	*Pu. maritima*
(for other *Puccinellia* species *Pu.* is adopted as the generic abbreviation)	

Salicornia europaea	*S. europaea*
Salicornia pacifica	*S. pacifica*
Salicornia virginica	*S. virginica*
Spartina alterniflora	*S. alterniflora*
Spartina anglica	*S. anglica*
Spartina foliosa	*S. foliosa*
Spartina maritima	*S. maritima*
Spartina patens	*S. patens*
Sporobolus virginicus	*Sp. virginicus*
Suaeda maritima	*Su. maritima*

(for other *Suaeda* species *Su* is adopted as the generic abbreviation)

1

General features of saltmarshes and their environment

Introduction

Coastal saltmarshes may be defined as areas, vegetated by herbs, grasses or low shrubs, bordering saline water bodies. Although such areas are exposed to the air for the majority of the time, they are subjected to periodic flooding as a result of fluctuations (tidal or non-tidal) in the level of the adjacent water body.

Distinction is drawn between saltmarsh and two other vegetation types found in similar habitats. Seagrass beds are for the most part permanently submerged but in some localities are found on mud or sandflats exposed at low water during spring tide periods. The characteristic floristic composition and structure of these intertidal stands is sufficient to maintain a distinction from any areas conventionally regarded as saltmarsh. Mangroves differ from saltmarsh in being dominated by trees. Where mangroves are well developed, forming forests 30 m high, the distinction between mangrove and saltmarsh is clear. However, there are areas where the differences are less striking. In southern Australia, at the southern limit of mangrove distribution, *Avicennia marina* forms a low shrubland, which can be shorter in stature than *Sclerostegia arbuscula* on adjacent saltmarsh. In such cases the distinction is on the basis of floristics and convention. *Avicennia*, which at lower latitudes is a medium-sized tree, is regarded as always being a mangrove, while *Sclerostegia* is never accorded mangrove status and is conventionally accounted as a member of the saltmarsh flora.

The definition of coastal saltmarsh stresses that the flooding waters are saline – not necessarily seawater but more saline than freshwater. Coastal saltmarsh is thus a subset of the wider category, tidal marsh (Daiber 1986). In many estuaries there is a continuum between coastal saltmarsh and marshlands in the upper estuary subject to tidal but non-saline flooding. Freshwater tidal marshes have, in comparison with saltmarshes, been a neglected habitat for study but Odum (1988) has presented a review of the ecology of the two marsh types. For most purposes in this book an arbitrary cut-off is taken at the limit of saline incursions.

There are also extensive non-coastal areas with saline soils. These often have floras and vegetation with marked similarities to that of coastal marshes but are not discussed in any detail in this book. In several parts of the world, species regarded as characteristic of intertidal marshland are also found on seacliffs where vegetation similar to that of saltmarshes may be found in very exposed localities subject to frequent drenching with salt spray.

Notwithstanding the focus established by the definition of coastal saltmarsh, the adaptations of organisms to salinity and flooding are similar regardless of habitat. In Chapter 4, where adaptation is discussed in detail, some examples are taken from both mangroves and inland saline habitats.

Coastal saltmarsh occupies the interface between land and sea. Its environment has some features of both land and sea and its biota, both flora and fauna, has both marine and terrestrial elements. Nevertheless, the organisms which are essential for the recognition of saltmarsh are the vascular plants which are terrestrial in origin. Saltmarsh is best regarded as a highly modified terrestrial ecosystem.

In this introductory chapter, those features of the environment by which saltmarsh differs from fully terrestrial ecosystems are discussed. In many ways, the dominant factor is the water level fluctuations, which, with the exception of microtidal areas, such as the inner Baltic Sea, are tidal in origin. Tides control (in conjunction with other factors) soil salinity and degree of waterlogging. Tides also carry sediment into marshes.

Saltmarshes are highly dynamic environments subject to erosion, accretion and progradation. While the study of these processes is the province of coastal geomorphologists, it is important for ecologists to be aware of the agents of change in coastal marshes and how these might be studied.

The environmental factors which characterise saltmarsh are universal. At a regional and local level, other factors influence the biota and allow the recognition of particular types of marsh as discussed in Chapter 3. However, the environmental similarity of all saltmarshes is responsible for some general features of the nature and distribution of the biota (particularly the flora) and these are discussed in this chapter.

Tides

The environmental feature which distinguishes coastal saltmarshes from terrestrial habitats is tidal submergence.

Tides are periodic sea level changes generated by the gravitational effects of sun and moon on the waters of the oceans. There are three distinct types of tidal submergence pattern found on different coasts of the world (Fig. 1.1). The regime which characterises north European coasts is semi-diurnal with two high waters of approximately equal height occurring *c.* 12.5 h apart. In a mixed tide regime the difference in height of the two high waters occurring each day is considerable (Fig. 1.1b). Under the diurnal regime there is a single high water each day.

The distribution of land masses and the shape of the sea bed cause the sea to be divided into a series of tidal basins, each with its own characteristics. Within each such basin, the tidal wave rotates about a node, the amphidromic point, at which point the tidal amplitude is zero.

The maximum tidal amplitude, or the tidal range, varies according to distance from the amphidromic point and local coastal morphology. The tidal range may be greatly amplified within funnel-shaped bays or estuaries (for example, the Bristol Channel and Bay of Fundy), although in the upper reaches of estuaries a considerable reduction in range occurs. In estuaries or bays with very restricted mouths the tidal range throughout may be very much less than on nearby open coasts. (The 'double tide' characteristic of the Solent region (Fig. 1.2) is the product of modification of a semi-diurnal regime by local topography.)

The tidal amplitude varies temporally at a number of different scales. The largest tidal forces are experienced when sun, moon and earth are aligned, which occurs one and a half days after new and full moons, and produces tides of large amplitude known as spring tides. Seven days after spring tides are experienced, the forces are at a minimum and tides of smaller amplitude, neap tides, result. The amplitude of spring tides varies over an annual cycle with particularly high tides around the solstices and equinoxes. At a longer time scale, of 18.6 years, caused by variation in solar and lunar declination, there is a further cycle of variation in tidal amplitude (see Kennish 1986).

The four superimposed cycles affecting tidal amplitude can be accommodated in the calculation of tide tables which provide predictions of tidal heights. The actual tide experienced may be very different from that predicted; in addition to astronomical forces, tide height is strongly influenced by meteorological conditions. Changes in atmospheric pressure alter mean sea level [an increase in pressure of 1 mm Hg reduces sea level by 13 mm (1 millibar equivalent to a change in level of 9.75 mm)]. Persistent onshore winds will raise effective tide heights, whereas offshore winds will lower them. Under certain weather conditions it is possible for very considerable increases in sea level (storm surges) to occur in shallow seas; such surges have been responsible for considerable loss of life. Over longer periods, changes in relative land and sea levels cause alterations in the extent

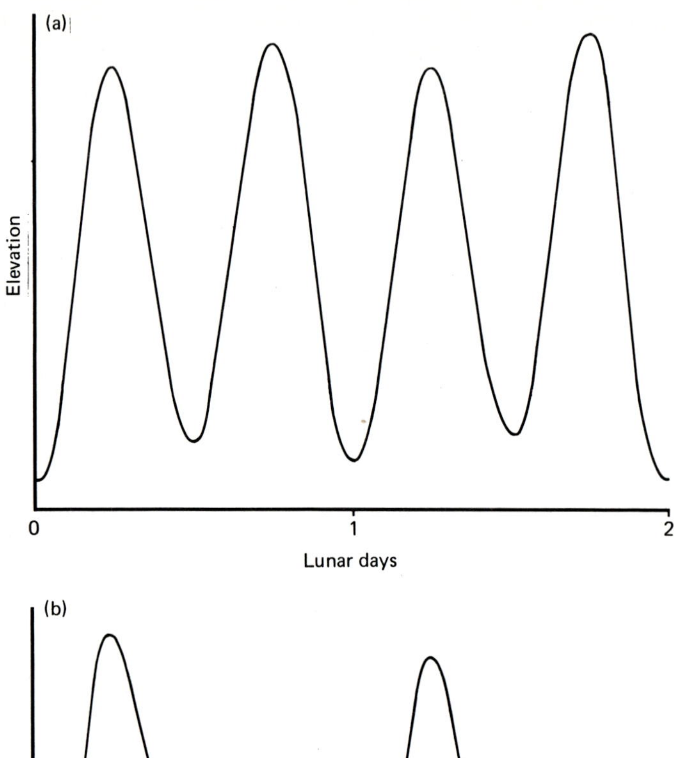
(a)
Elevation
0
1
2
Lunar days

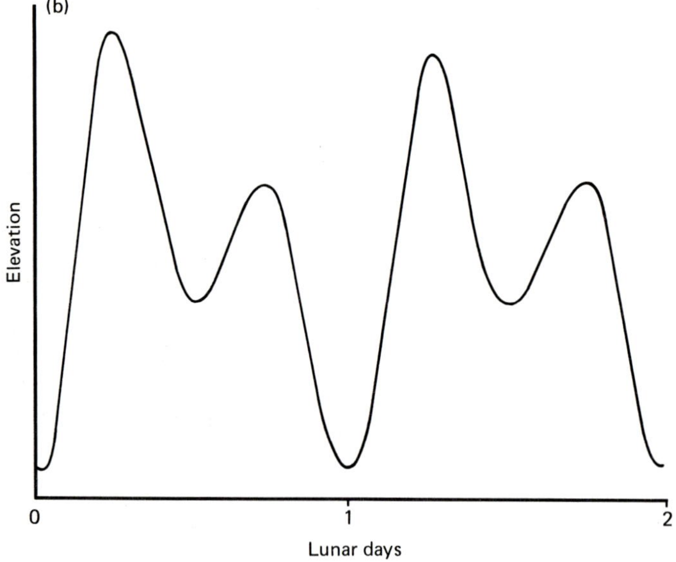
(b)
Elevation
0
1
2
Lunar days

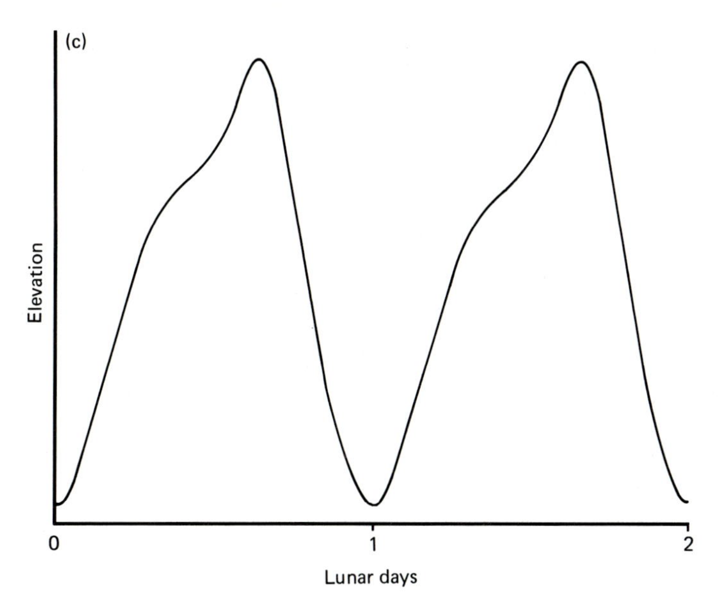

Fig. 1.1 Tidal curves for the three tidal regimes: a, semi-diurnal; b, mixed; c, diurnal.

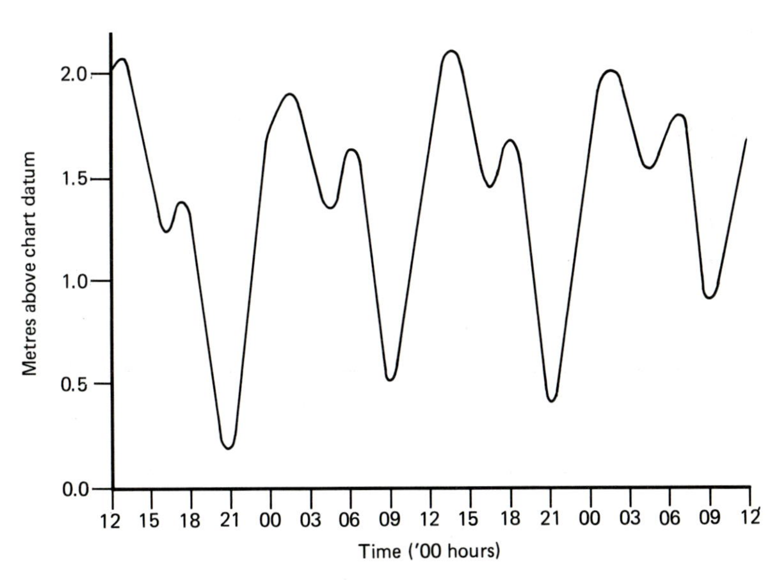

Fig. 1.2 Spring tide curves for Poole Bridge (Poole Harbour, Dorset, UK), showing the 'double tide' characteristic of the Solent region. (Redrawn from Gray 1985*b*.)

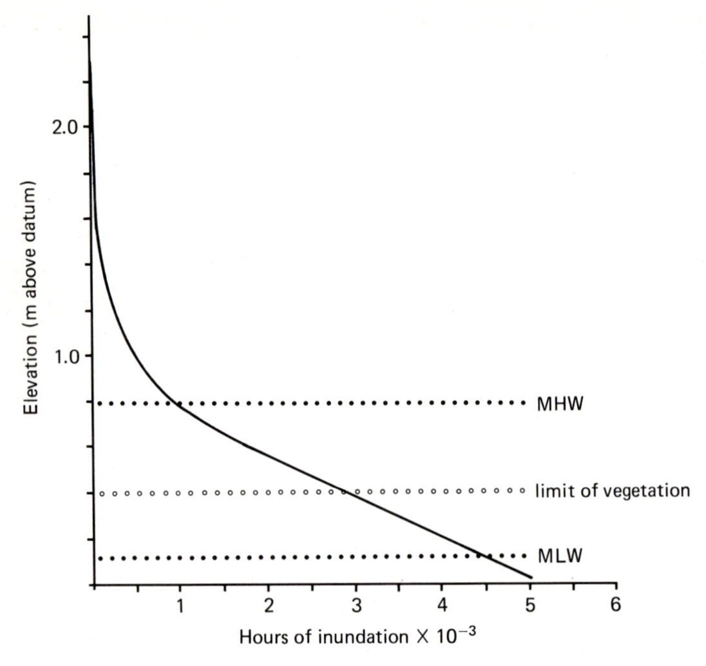

Fig. 1.3 Tidal submergence curve (expressed as hours of inundation per year) for Esbjerg, Denmark, calculated from data for 1900–1962. (Redrawn from Jensen, Henriksen & Rasmussen 1985.)

of the terrain exposed to tidal action. Where changes in relative sea level result in considerable change in the shape of the coastline, there may be changes in local tidal range. In regions of small tidal range and with predictable climatic conditions, water level fluctuations affecting coastal marshes may be essentially seasonal. This is the case in the Baltic (see Ericson & Wallentinus 1979) and also in south-western Australia (Hodgkin 1978). A review of tidal behaviour in estuaries is presented by Kennish (1986).

Regardless of the particular nature of the tidal regime, the number of submergences and the length of time for which submergence occurs will decrease with increasing elevation on the shore. The submergence curve (either in terms of number of tides or hours of submergence) is sigmoidal in shape (Fig. 1.3).

Ecologists working on rocky shores have paid particular attention to the shape of the submergence curve and there has been considerable debate as to the existence or otherwise of so called 'critical tidal levels' (CTLs). There are two different concepts of CTL current in the literature. Colman (1933) suggested that the annual submergence curve was not smoothly sigmoidal, but that there were sections of the curve where the number of submergences (or, conversely, the period of exposure to air) changed more rapidly with height, and the levels at which this change in the gradient of the curve occurred could be termed 'critical levels'. Subsequent to Colman's suggestion, critical levels were recognised on the submergence curves for several localities. Underwood (1978) recalculated the submergence curve for Devonport, the station used by Colman, and argued that it was smoothly sigmoidal with no evidence of critical levels. Hartnoll & Hawkins (1982) computed submergence curves for a number of British stations on the basis of recorded tide heights rather than predictions. The shapes of the curves differed surprisingly between locations. For some stations, critical levels could be identified, while at other sites the curves were smoother with no obvious breaks in the slope. Whether these critical levels are of biological significance still requires evaluation.

The annual submergence curve presents a synoptic view of the tidal flooding regime in which information on the timing and duration of individual flooding events is lost. Doty (1946) discussed CTLs in terms of exposure and submergence events defined both by timing and tidal height. Swinbanks (1982) analysed various tidal curves and identified a number of CTLs using these criteria. He emphasised the importance of the various temporal cycles which affect tidal amplitude in determining particular CTLs. Variation in the timing of tidal flooding is likely to be more important for many species than relatively small changes in the total incidence of flooding.

There is considerable variation in the seasonal incidence of tidal inundation at different levels on a saltmarsh (Fig. 1.4; Jensen, Henriksen & Rasmussen 1985; Clarke & Hannon 1967; Jefferies 1977*a*,*b*; Eleuterius & Eleuterius 1979). *A priori* it seems likely that these various submergence patterns may be factors of considerable importance in the biology of saltmarsh species. Notwithstanding the effects of the 18.6-year cycle, the annual pattern of submergences is highly predictable. Jefferies (1977*a*), Jefferies & Perkins (1977) and Jefferies, Davy & Rudmik (1979) have argued that the interaction of tidal flooding and climate in East Anglia leads to a predictable occurrence of a hypersaline soil solution in the upper marsh and that upper marsh populations of several species have genetically determined growth rhythms which reduce growth during the hypersaline period (see pp. 266–268). It is possible also that the reproductive phenology of species (particularly timing of flowering and of seed release) may be adapted to particular tidal events.

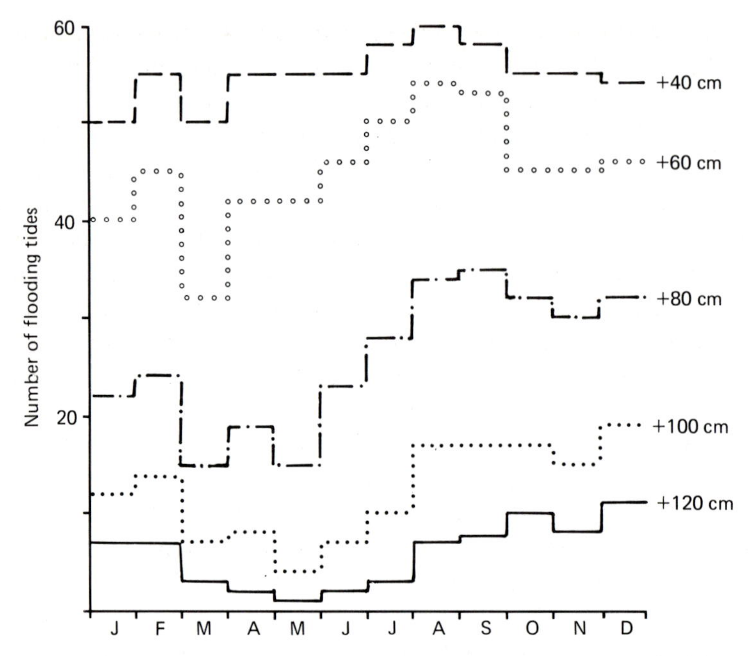

Fig. 1.4 Variation in the incidence of flooding tides over a period of a year (calculated from data for 1945–1965) at different levels on the shore near Esbjerg. Note the low number of tides flooding the upper-marsh in late spring and early summer. (Redrawn from Jensen *et al.* 1985.)

While the potential for tidal events to influence many aspects of the biology of saltmarsh species (both animal and plant) is readily apparent, there are few detailed investigations of tidal effects on species' behaviour. Similarly, there are very few studies which have determined the elevation of community boundaries so that it is not possible to compare species or community distribution between sites. Data on the consistency or otherwise of the elevation range of species relative to tidal flooding would allow hypotheses on the effect of tides on species distribution to be proposed and tested.

Although many tidal phenomena are predictable, interactions between meteorological and tidal events can produce extreme environmental conditions which are unpredictable in their occurrence and which may impose severe perturbations on ecosystems. It is easy to invoke such events to explain distribution patterns of organisms. However, it is more difficult to investigate such hypotheses. Bleakney (1972) documented the effects of very high tides (and their subsequent extreme low water) in both summer and winter in the Bay of Fundy and suggested that the interaction between tides and climate may 'impose a sudden bias on the annual natural selection forces that effect the various phenotypes and genotypes within that broad spectrum of individual variations which we term a species'.

Although Bleakney (1972) stressed the unpredictability of tidal/climatic interactions, his analysis also showed the importance of the 18.6-year cycle in tidal amplitude on the occurrence of extreme tidal events. From the point of view of individual organisms with short life spans, tidal extremes may, indeed, be unpredictable. Nevertheless, Swinbanks (1982) suggests that the distributions of populations of such species may still be related to CTLs generated by long-term tidal cycles. Many long-lived saltmarsh plants may live through several long-term tidal cycles, and must be tolerant of the great diversity of conditions which this entails.

Swinbanks (1982) suggests that the longer-term tidal cycles periodically provide colonisation opportunities over several years for the expansion of seagrass beds and, based on data of Wells & Coleman (1981), mangrove stands. It would be of interest to monitor the extent of low marsh vegetation to see whether there is evidence for episodic expansion which can be related (possibly causally) to tidal cycles.

The existence of long-term tidal cycles complicates the determination of the upper limits of saltmarshes. There will be areas of the upper marsh which are only flooded at intervals of several years. During the intervening period, colonisation by glycophytes (species with their major distribution in non-saltmarsh habitats) may occur, some of which will be killed by subsequent flooding tides but others, after the initial establishment phase, may persist. The great diversity in species composition of the extreme upper saltmarsh in northern Europe can be partly related to land use patterns but may also reflect the vagaries of invasion by glycophytes and their subsequent winnowing by tidal flooding.

Detailed studies of the palynology and stratigraphy of a Long Island saltmarsh indicate that alternation in the upper marsh between Gramineae and Cyperaceae domination occurred 2–3 times between 1835 and 1930 (Clark 1986). These changes coincide with the 18.6-year tidal cycle, the expansion of Cyperaceae occurring when tidal range was at its minimum.

Although exceptional stands of vegetation may endure almost continuous flooding during part of the year (for example, *Spartina anglica* in Poole Harbour may be flooded for up to 23.5 h per day during the November neap tides – Hubbard 1969 – and *Spartina alterniflora* in Mississippi at its lower limit may be flooded for c. 7645 h per annum – 87% of the year – Eleuterius & Eleuterius 1979), much saltmarsh vegetation is exposed to the air for the majority of the time (see Fig. 1.3) – the habitat is essentially a terrestrial one, occasionally flooded, rather than a submarine one occasionally exposed to the air.

The effects of tidal flooding will differ according to the nature of the organisms concerned. Organisms of terrestrial origin are mainly active when exposed; organisms of aquatic origin are active when submerged; and, while some examples are found in the upper marsh, the majority are restricted to the lower marsh or to local aquatic habitats such as pans and creeks.

Ranwell (1972) makes the distinction between submergence and emergence marshes – submergence marshes are those subject to frequent flooding and in which the biota contains those species adapted to, or tolerant of, submergence. [A similar distinction was drawn by Chapman (1934, 1960a), and Steers (1964), while Redfield (1972) distinguished between intertidal and high marsh.] Emergence marshes are less frequently flooded and the environmental conditions acting on the biota are determined during the periods of exposure to the air. In the submergence marsh, conditions of soil moisture and salinity are relatively constant; in the emergence marsh, both soil moisture and soil salinity may be highly variable [see, for example, Jefferies (1977*a*,*b*) and Jefferies & Perkins (1977)] and change in response not only to tidal flooding but also to the climate.

While the concept is appealing, determining the elevation at which the transition between emergence and submergence conditions occurs is difficult (Ranwell 1972). Chapman (1960*a*) suggested the change occurs at that level on the shore subject to about 360 flooding tides a year. This may be an appropriate distinction in south-east England, although Ranwell (1972) pointed out that the data in Chapman (1960*a*) did not show a sharp disjunction at this level, but is not necessarily so elsewhere. Because of the importance of the interaction between tidal flooding and climate in the development of emergence conditions, the transitional level may well be modulated by regional climatic conditions and, thus, may be at a different level on the wetter west coast of Britain than on the drier south-east and is likely to be at very different relative levels on temperate and mediterranean region saltmarshes.

Unfortunately, there are very few marshes where the distribution limits of species and communities have been determined accurately in relation to standardised datum levels. Such data as do exist suggest that, while it may be possible to define absolute limits for species in terms of numbers and length of flooding tides, the distribution of species varies considerably between sites.

After one of the most detailed local studies of elevation ranges of species,

Johnson & York (1915) concluded that 'the vertical range of a littoral plant is exactly proportional to the range of the tide'. Adams (1963) suggested that 'The mean elevation of occurrence above MSL divided by one-half the mean tide range of the area concerned is a characteristic constant for each of the various saltmarsh species'. Such claims have encouraged a belief in the constancy of species' behaviour in respect to tidal levels and, in the absence of independent data, the use of species' limits to estimate the position of various levels on the shore.

Nevertheless, the inversion of species zonation patterns in upper estuarine sites is a well documented phenomenon (Gillham 1957*a*; Beeftink 1966, 1985*b*; Ranwell 1972; Eleuterius & Eleuterius 1979), suggesting that relationships between species distributions and tidal levels may not hold even within single estuaries, let alone between regions.

Comparison of the accounts given by Gray (1972) and Chapman (1934) indicate that saltmarshes in Morecambe Bay are restricted to much higher levels on the shore than those on the Norfolk coast, and that the elevation range of many individual species is similarly truncated. The tidal range in Morecambe Bay is considerably greater than that along the Norfolk coast and it is possible that there is an inverse relationship between the proportion of the tidal range occupied by saltmarsh and the tidal range. There is a slight indication of this in the data from the east coast of the United States assembled by Adams (1963). Frey & Basan (1985) suggest that around the USA coast there is variation in marsh levels which can be related to the type of tidal regime.

The nature of the sediment may influence the elevation at which colonisation is possible. Randerson (1979) showed that around the Wash, UK, the elevation of the pioneer zone was correlated with sand content of the sediment (Fig. 1.5). On the sandiest substrates, the lowest marsh zones were at a considerably higher elevation than on silt. Nutrient content of the sands was lower than that of finer sediments and Randerson (1979) suggested that low nutrient status impaired plant growth and thus colonisation potential. This hypothesis would also explain the truncation of saltmarshes on the sandy sediments of Morecambe Bay compared with the silts of North Norfolk.

Beeftink (1965, 1977*a*) has suggested that inversion of zonation patterns may be correlated with tidal range, with communities dominated by the shrub *Halimione portulacoides* being found at lower elevations than the *Pu. maritima* dominated grasslands in Europe when the tidal range exceeds 8 m. Whether tidal range is the major factor involved is uncertain. At a continental scale, there is an increasing tendency for *Halimione* to occur lower on the

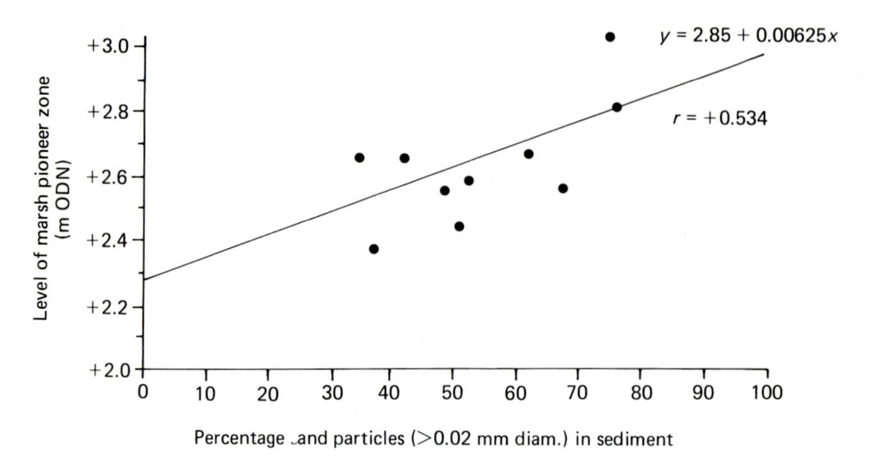

Fig. 1.5 Elevation (in metres above Ordnance Datum, Newlyn) of the lowest vegetation zone around the Wash, UK, plotted against sediment sand content. (Redrawn from Randerson 1979.)

shore at lower latitudes while, at a local scale, there appears to be a substrate effect, with low marsh Halimionetum in south-east England being found on coarse, shell-rich sediment (Adam 1976*a*), in comparison to the major midmarsh occurrences on silt.

Not only is there considerable variation in the proportion of the intertidal zone occupied by saltmarsh, but also the relative position of species in the zonations may vary between sites, even in the seaward estuarine zone (Adam 1981*a*).

Until more data are available from a wide range of sites, it will not be possible to determine how the effects of tides are modified by other factors (which may include genetic differences in species between sites), or whether any generalisations about the distribution of any particular species relative to the frequency and duration of tidal flooding can be sustained.

The length of daylight for which a marsh is submerged will vary according to the time of high tide. For marshes in temperate latitudes during the summer there will be greater loss of daylight when high tides occur at 0600 and 1800 than at 1200 and 2400 hours. During midwinter the converse may be true. In a semi-diurnal tidal regime the timing of spring tides is a characteristic of a particular localitity. Thus, in Morecambe Bay in north-west England, high spring tides occur around midnight/midday; on the Norfolk coast they are around 0600 and 1800 hours. Whether there are systematic differences between sites in the tidal ranges of species according to the length of submergence in daylight remains to be investigated, despite

Johnson & York (1915) having drawn attention to the problem in one of the earliest detailed studies of species distributions relative to tidal levels. Tutin (1942) showed that the upper limit of the seagrass *Zostera marina* in Britain could be correlated with the time of low water on spring tides, being at lower levels where low water occurred in the middle of the day. Intolerance of drying out is probably the major factor controlling this relationship.

The limits of saline incursion and tidal influence in an estuary are not synonymous, unless some artificial constraint such as a weir separates the freshwater river from its estuary. Normally, however, the influence of the tide extends further upstream than that of salinity. The freshwater tidal reach of the estuary may be fringed by extensive marshes (Zonneveld 1960; Heckmann 1986; Odum 1978; Good, Whigham & Simpson 1978), but these freshwater tidal marshes are not considered in any detail here. Nevertheless, such marshes may be important components of the overall nutrient and energy cycling processes within the total estuary (Odum 1978; Good *et al.*, 1978; Heckman, 1986). Frequently, the tidal amplitude in the freshwater reach is much reduced compared with that in more seaward sites, but this is not necessarily so – the freshwater tidal section of the Elbe estuary experiences a mean tidal range of over 3 m (Heckman 1986). The position of the limit of saline influence in an estuary is not constant, as the degree of fluctuation will reflect temporal variation in river discharge. The limit of saline incursion in the Hawkesbury River in New South Wales, for example, may move upstream several kilometres during drought periods – but, in addition, is affected by changes in morphology of the river bed, resulting from commercial sand extraction (SPCC 1983).

Within the saline-influenced section of the estuary, the patterns of distribution of salinity vary with estuary type and the relative extent of marine and riverine influence. Depending on the particular combination of factors operating in an estuary, the estuarine waters may be predominantly mixed and brackish with a gradient in average salinity along the estuary, or fresh and saline water masses may be discrete and stratified (see Fig. 1.6, and reviews of estuarine hydrology by Dyer 1979, Officer 1976, and Kennish 1986).

Salinity of saltmarsh soils

As the elevation of the marsh surface increases, the number of flooding tides decreases. While the flux of salt into the marsh will also decrease with elevation, the salinity of the interstitial soil water does not display any simple or consistent relationship with elevation.

In the lower saltmarsh, with frequent immersion throughout the year,

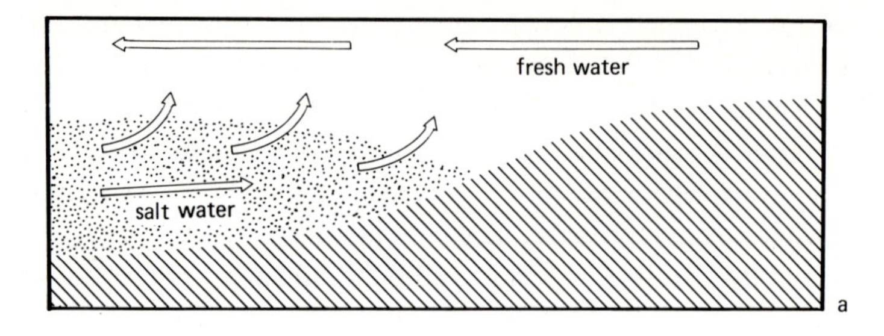

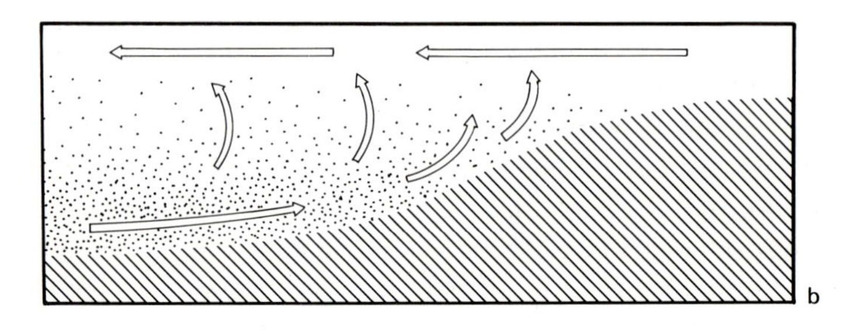

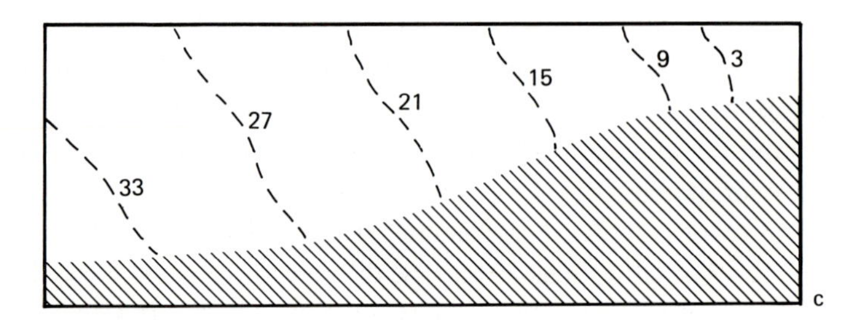

Fig. 1.6 Salinity distributions in estuaries: a, salt wedge estuary; b, partially mixed estuary; c, mixed, or vertically homogeneous estuary (isohalines in parts per thousand). After various sources.

the soil salinity is relatively constant and rarely exceeds that of the flooding water. At higher elevations, there is an interaction between the influence of flooding and climate, leading to greater variability in soil salinity. Between periods of tidal flooding, rainfall will reduce soil salinity. During drier periods, evapotranspiration will increase soil salinity, sometimes to the point that a salt crust may form on the soil surface. Even when the flooding water is brackish, the soil salinity can reach high values, exceeding that of seawater (Beeftink 1965, 1977*a*).

Even under temperate conditions with relatively high rainfall, the interaction between patterns of tidal flooding and seasonal changes in climate may result in predictable periods of high soil salinities in the upper saltmarsh (Jefferies 1977*b*; Jefferies & Perkins 1977). On arid tropical and subtropical coasts, long periods of hypersalinity may prevent the establishment of perennial vegetation, leading to extensive salt flats (Thom, Wright & Coleman 1975) (Figs. 1.7, 1.8).

The position of the zone where the highest salinities are experienced will vary from site to site, being influenced by rainfall patterns and ground and surface water inputs into the upper marsh. While greater variability and highest maxima are most frequently found at relatively high elevations, there are regions where climatic and hydrological conditions combine to give consistently low salinities throughout a marsh (Ewing & Kershaw 1986).

In upper estuaries where the flooding tide is brackish, the low marsh zones may show low soil salinities with higher salinities only experienced in the upper marsh. Inversion of zonation, recorded in many estuaries (Gillham 1957*a*; Beeftink 1965, 1966, 1977*a*; Ranwell 1972; Adam 1976*a*), where species found in mid and low marsh communities in the seaward portion of an estuary are restricted to higher elevations in the upper estuary may reflect this salinity pattern. *Phragmites australis*, *Scirpus* (*Bulboschoenus*) spp. and *Schoenoplectus* spp., which characteristically provide the low marsh fringe in brackish water in many parts of the world, are restricted in the lower estuary to sites where salinity conditions are permanently ameliorated.

Effects of flooding

Saltmarsh soils are frequently waterlogged and anaerobic. After tidal immersion many areas of saltmarsh drain slowly, primarily because local topography is rarely conducive to rapid drainage, but also because of the low hydraulic conductivity of many saltmarsh soils (Clarke & Hannon 1967, 1969).

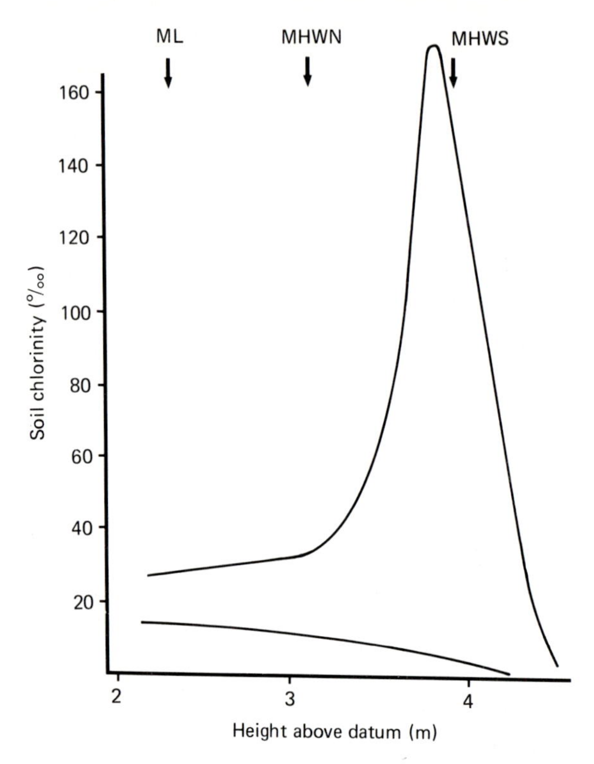

Fig. 1.7 Range of variation in soil chlorinity with elevation at Gladstone, Queensland, a subtropical region, over the period 1975–1980. (Based on a figure in Hutchings & Saenger 1987.)

The physiochemical environment of waterlogged soils is very different from that of freely draining soils and is reviewed by Ponnamperuma (1972), Gambrell & Patrick (1978) and Armstrong (1982).

If a soil is depleted in oxygen and the diffusion rate of oxygen is inadequate to maintain aerobic respiration, various soil bacteria populations develop which utilise electron acceptors other than oxygen for respiratory oxidation. In consequence, a range of substances, both inorganic and organic, are converted to a reduced state, and this is reflected by a lowering of the oxidation–reduction (redox) potential of the soil.

The redox potential of a soil is a measure of its tendency to receive or supply electrons and is controlled by the types and proportions of the oxidising and reducing substances it contains (Armstrong 1982). There is a sequential series of reductions as the redox potential is lowered; nitrate is reduced to ammonia at a higher potential than reduction of manganic ions

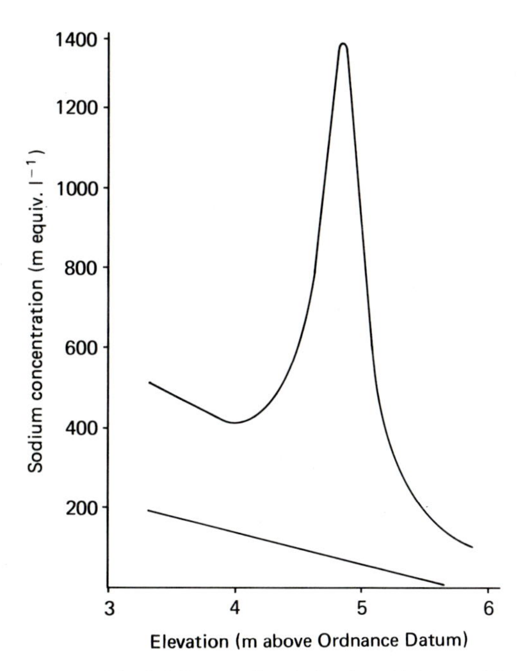

Fig. 1.8 Range of variation in soil water sodium concentration with elevation at various sites around Morecambe Bay, north west England, between 1972 and 1974. (Adam, unpublished data.)

to the manganous state, which in turn occurs at a higher potential than the ferric–ferrous ion transition.

At lower redox potentials still, sulphate-reducing bacteria, mainly of the genus *Desulphovibrio*, utilise sulphate as their terminal electron acceptor in respiration. (*Desulphovibrio* shows optimal activity at pHs close to neutrality – such conditions prevail in many saltmarshes.) In most non-saline soils, the majority of the sulphide so formed reacts with metal ions (such as iron or manganese) to form insoluble inorganic sulphides. Howarth (1979) has reported rapid production of pyrite (FeS_2) in saltmarsh sediments. However, sulphate is present in abundance in seawater so that sulphide may be produced in excess of any capacity of the soil to provide metal ions for the formation of insoluble sulphides. Sulphide in solution (either as the acid sulphide ion HS^-, or as dissolved hydrogen sulphide gas) can reach high concentrations in saltmarsh soils. Ingold & Havill (1984) demonstrated considerable fine-scale heterogeneity of sulphide concentration in saltmarsh soils. They suggest that the activity of sulphate-reducing bacteria may be limited by the availability of usable carbon sources; thus, the distribution of sulphide reflects the local release of suitable organic substrates during the decomposition of plant material (most probably algal mats).

Anaerobic microbial respiration may liberate a wide range of organic compounds. In many freshwater wetlands, methane is a major product, but methanogenic bacteria are inhibited by sulphate (Winfrey 1984; Kristjannson & Schönheit 1983), so that methanogenesis is less frequent in saltmarsh soils. In anaerobic soils the majority of biologically available nitrogen is in the form of ammonium ions; some ammonia is the product of microbial reduction of nitrate, but the majority is formed during the anaerobic degradation of organic matter (Armstrong 1982).

As the oxygen concentration in waterlogged soil declines, the carbon dioxide concentration increases. However, Armstrong (1982) argues that rarely, if ever, are lethal concentrations for plant growth reached.

Chapman (1938) suggested that during tidal flooding an aerated layer persisted in the soil. However, studies by Clarke & Hannon (1969) and Armstrong *et al.* (1985) demonstrated that such an aerated layer is not a general feature of saltmarsh soils; if an aerated layer is to be found then it is localised or transient in its occurrence.

There are relatively few measurements of the redox potentials in saltmarshes and a number of the measurements in the literature are of dubious value because of technical problems with the procedures used (Armstrong 1979; Armstrong *et al.* 1985).

Armstrong *et al.* (1985) measured redox potentials at a number of soil depths in a range of communities over a season in a saltmarsh in the Humber estuary.

Three basic aeration patterns emerged from the study. In the low-marsh *S. anglica* dominated community, reducing conditions persisted throughout the summer over most of the soil profile. Brief periods of oxidation occurred during neap tide periods down to 5 cm depth, but there was no change at greater depths (Fig. 1.9). At higher levels in the marsh, in a *Pu. maritima* dominated community, the soil remained reducing at depth below 20 cm, but above this depth became strongly oxidising during neap tide periods, with rapid falls in redox potential following tidal flooding (Fig. 1.10). At higher elevations in a community termed by Armstrong *et al.* (1985) a General Salt Marsh [corresponding to the *Limonium – Armeria* nodum in the Puccinellietum maritimae of Adam (1981*a*)], the soil was well oxidised, even at depth, for much of the summer, with only brief anaerobic phases following the highest spring tides (Fig. 1.11).

Armstrong *et al.*'s (1985) data demonstrate the rapidity and magnitude

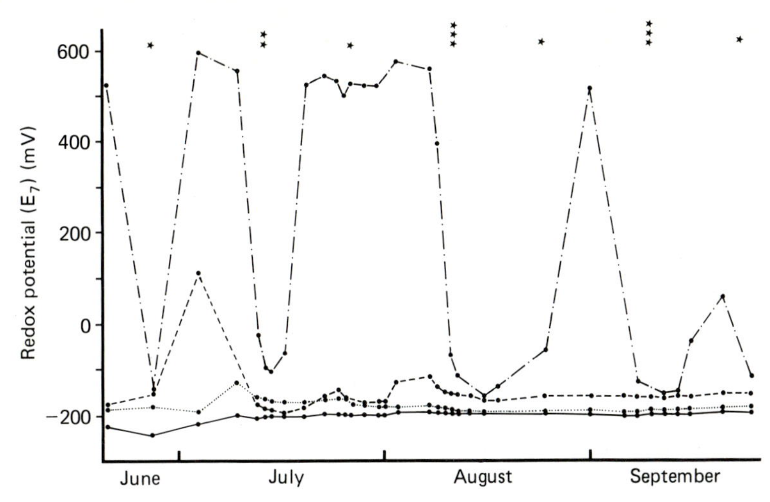

Fig. 1.9 Fluctuations in soil redox potentials between June and September 1979 in a *Spartina* sward at Welwick Marsh, Humber estuary, UK (● – · – ● at 1.5 cm depth, ● – – – ● at 5 cm, ●· · · ·● at 10 cm, ● — ● at 30 cm). (★, low spring tides) ($\leq$6.9 m above chart datum), (★★, high spring tides) ($\geq$7.1 m above chart datum), (★★★, very high spring tides) ($\geq$7.3 m above chart datum). (Redrawn from Armstrong *et al.* 1985.)

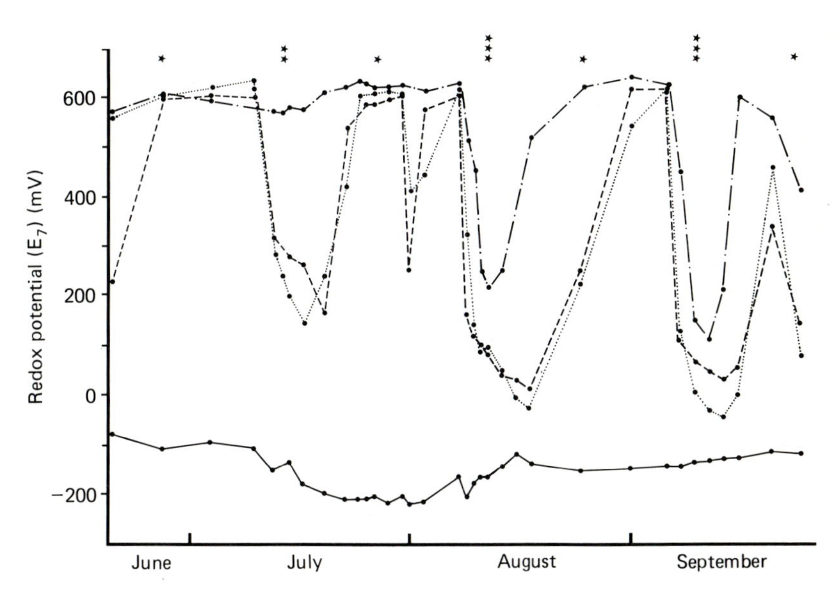

Fig. 1.10 Fluctuations in soil redox potentials between June and September 1979 in a *Puccinellia maritima* dominated sward. Symbols as for Fig 1.9. (Redrawn from Armstrong *et al.* 1985.)

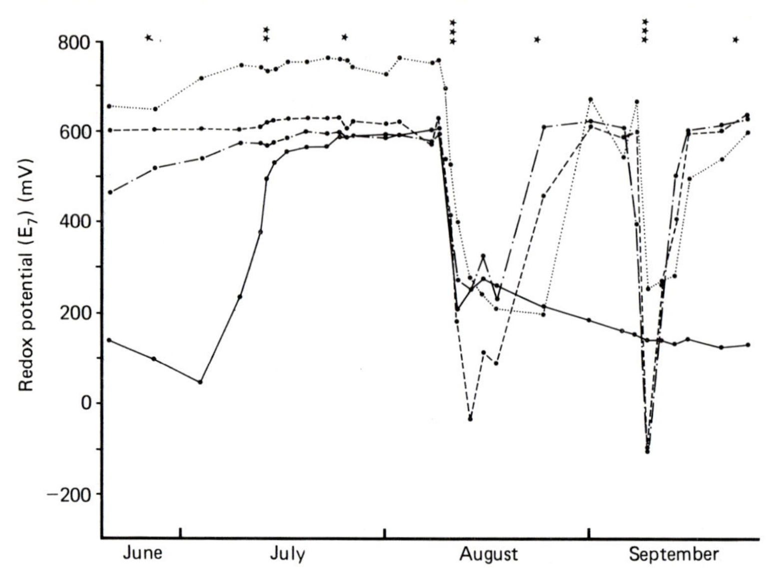

Fig. 1.11 Fluctuations in soil redox potential in the general saltmarsh community at Welwick Marsh for June to September 1979. Symbols as in Fig 1.9. (Redrawn from Armstrong *et al.* 1985.)

of the fluctuations in the redox potential of saltmarsh soils. The lowering of potential on tidal flooding and the subsequent rise after the flooding cycle indicate that removal and return of oxygen is, in itself, the major factor in the redox fluctuations (Armstrong *et al.* 1985). The reduction in potential occurs very rapidly on flooding, but recovery, although still rapid, must be delayed until some time after the end of the flooding spring tide cycle (see, for example, Fig. 1.10).

In well drained sites, such as creek banks, tidal flooding may have a flushing effect, preventing the build up of high salinities and replenishing nutrients (Smart 1982). In more poorly drained sites, the effect of flooding on soil aeration is likely to outweigh any ameliorating effects on salinity and nutrient status.

The dynamic nature of the saltmarsh environment

Sedimentation and marsh development

Vertical growth of marsh surfaces

The tidal waters flooding saltmarshes are normally turbid. As the tide floods the marsh, the vegetation acts as a baffle allowing the sediment to settle out. Under this model, the sedimentation rate would be expected to be inversely proportional to elevation (Fig. 1.12). The model is supported,

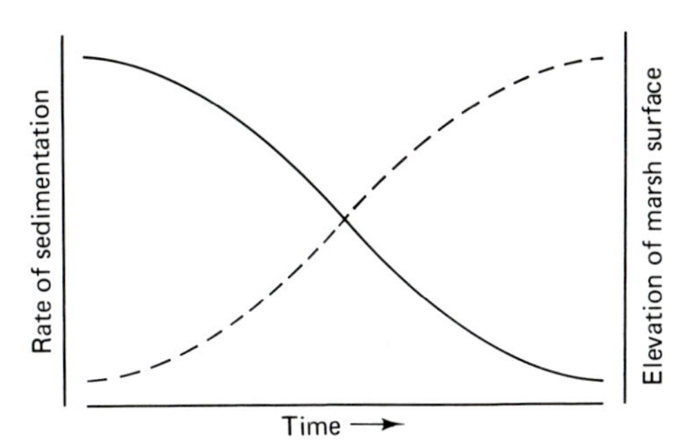

Fig. 1.12 Hypothetical model of the relationship between elevation (- - -) and rate of sedimentation (—).

in outline at least, by a number of observations of sedimentation rates at different marshes (see *inter alia* Richards 1934; Steers 1938, 1948, 1964; Ranwell 1964*a*; Randerson 1979). The rate of sedimentation in the upper-marsh declines not only because of fewer flooding tides, but also because much sediment would be trapped in the lower marsh before reaching the higher levels (Randerson 1979). As the flooding tide reaches the marsh surface, not only by general flow from the seaward but also by overbank flow from creeks, this model would also predict the formation of creek levees (Steers 1938, 1948). It would also be predicted that coarser sediment would settle out first so that the lower marsh and creek levees might be expected to have a higher proportion of sand than the upper marsh where active sedimentation would involve predominantly silt and clay fractions.

Detailed studies, however, suggest a number of modifications to this simple model.

The highest sedimentation rates are not necessarily at the lowest level of the marsh as the model might predict, but at the lowest level of continuous vegetation (Richards 1934; Randerson 1979) – this supports the basic model in that it suggests that the baffle effect of continuous vegetation is an important control on the settling out of sediment. As an additional extension to the model, it could be suggested that the attenuation of sedimentation rate would be influenced not simply by elevation but also by the lateral extent of closed vegetation. A decline in sedimentation rate with increasing distance inland from the marsh edge was recorded by Richards (1934).

Stumpf (1983) has shown for a saltmarsh in Delaware that sedimentation of fine sediments in the upper marsh is greater than would be predicted if the main process was simply the consequence of the stilling of the floodwaters and settling of sediment (Fig. 1.13), during the short period of slackwater experienced at the upper limit of the tide. Stumpf (1983) suggests that the

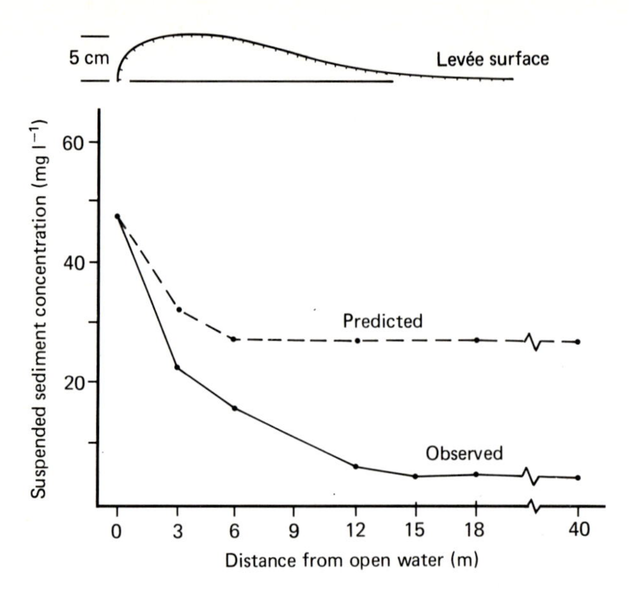

Fig. 1.13 Floodwater sediment concentrations over the marsh surface. Loss of suspended sediment from the water column is greater than predicted by a simple model of sedimentation. Data of Stumpf (1983) from a Delaware marsh. (Redrawn from Stumpf 1983.)

additional sediment is initially impacted onto the vegetation. The sediment could then reach the soil surface in several ways – by being washed off by rain, in the faecal pellets of gastropods grazing on the vegetation, or by the collapse onto the marsh surface of dead stems and leaves.

The assessment of patterns of accretion is influenced by the time scale of sampling. Long-term averages suggest a relatively simple inverse relationship between accretion and elevation. Repeated measurements at short intervals tend to reveal a more complicated picture with rates of accretion varying over time and, in the lower marsh, accretion being interrupted by phases of erosion (Fig. 1.14). Such patterns may be influenced by variation in the availability of sediment, variation in the trapping and stabilising capacity of the vegetation (for example, decline in the above ground biomass in winter, and, in the case of low marsh zones dominated by annual species, the total absence of cover in winter), and seasonal variation in erosive forces (for example, the winter erosion in the lower marsh zone recorded by Richard 1978 – see Fig. 1.14 – is largely due to ice scour). Short-term measures of sedimentation may provide high estimates of longer-term accretion of the marsh surface as they do not allow for settlement and consolidation.

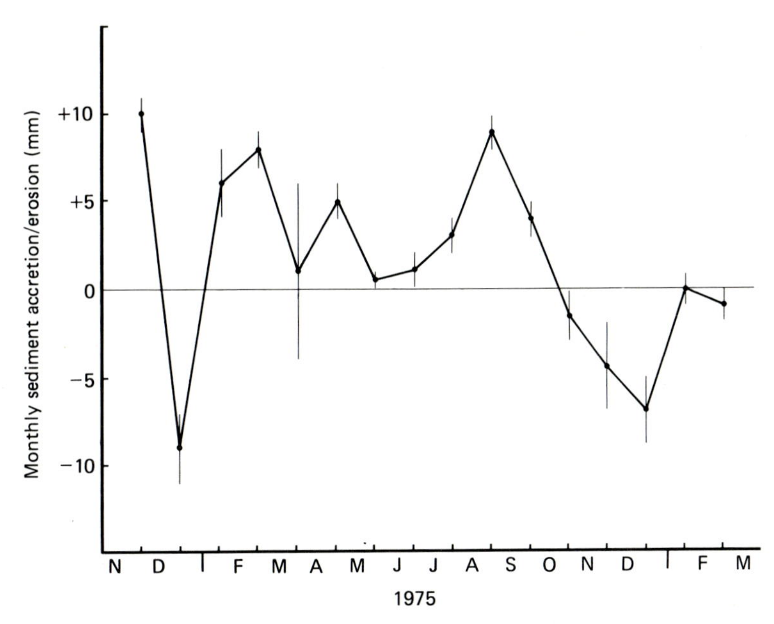

Fig. 1.14 Variation in monthly sediment accretion (or erosion) in a low-marsh *Spartina alterniflora* community. (Redrawn from Richard 1978.)

In addition to variation in sediment input between seasons, there are often substantial variations between years (Harrison & Bloom 1977). The most likely cause of this variation is the incidence of storms (Harrison & Bloom 1977). Storms may be responsible for major erosion (particularly in the lower marsh – Ranwell 1964*a*; Richard 1978), but may also be a means of increasing sedimentation in the upper marsh (Stumpf 1983). Stumpf (1983) indicates that the occurrence of coarse silt and sand in upper marsh sediments is incompatible with a 'normal' tidal input model and can only be explained by influx of these fractions during storms. Stumpf (1983) argues that storms may have greater influence on marshes experiencing micro- and meso-tidal regimes than under macro-tidal conditons, but there are few data to test this hypothesis. Schubel & Hirschberg (1978) showed that more than 50% of the sediment deposited in the upper Chesapeake Bay between 1905 and 1975 was contributed during only two events – a major flood in March 1936 and Hurricane Agnes in June 1972.

Where saltmarshes occur behind barrier dune systems, storms breaching the dunes may deposit substantial layers of sand across marshes (Chapman, 1960*a*).

Sources of sediment

The ultimate source of non-biogenic sediment deposited on saltmarshes by the tides is erosion of terrestrial rocks. Saltmarsh sediments often include calcium carbonate fragments from molluscan shells – either from animals living in the marsh or from offshore mudflats. In some case, carbonate may constitute a high proportion of the sediment. The material in suspension in estuarine waters can usefully be divided into two components: that entering the estuary from the seaward and that of more immediate terrestrial origin. The balance between the two will vary enormously between estuaries, depending *inter alia* upon the size of the available sedimentary masses, catchment size and management, and run-off. In rivers building emergent deltas [for example the Mississippi or Tigris-Euphrates (Shepard 1977)] major inputs are from river borne sediment, reflecting erosion in catchments. In other estuaries there is, at the present time, little input of river borne sediment and marine sediment predominates. In some instances, marine and river borne material may occur as discrete masses within the same estuarine system (Fig. 1.15). In some estuaries, there may be no net input of sediment, but relocation of material within the system may be reflected by localised erosion and accretion of marshes.

The balance between the different sources may vary between adjacent rivers and between regions. The movement of large amounts of material in rivers requires high flows and erodible catchments. In many parts of the world, catchment clearance following European settlement had led to major erosion and consequent documented extension of saltmarsh or mangroves (Thorogood 1985). In Britain, the major phases of forest clearance may have been accompanied by increased sediment inputs in estuaries, but at the present time catchment erosion is a relatively minor phenomenon and reworking of marine deposits is the more likely proximal source of sediment for marsh accretion (except in a few special cases where industrial activity promotes inputs – as in the Fal estuary where very rapid marsh accretion is a consequence of china clay extraction – Ranwell 1974). Much of the coastal sediment around Britain is glacially deposited material and is, thus, variable in composition. Along the New South Wales coast (unglaciated for many millions of years) the majority of offshore sediment is sand and, in most sedimentary compartments, there is currently no net onshore movement (Chapman *et al.* 1982). Although most river systems in NSW have comparatively small catchments and by world standards, low discharges, extensive forest clearance over the last 200 years has exposed highly erodible soils so that the major active net sediment inputs are river borne.

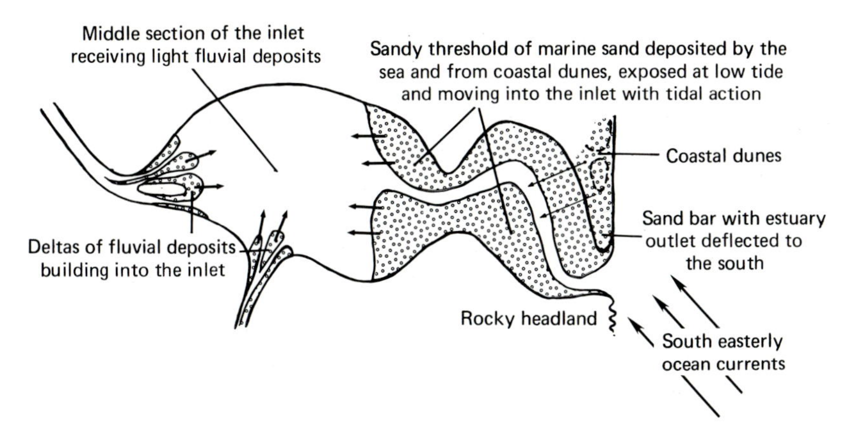

Fig. 1.15 Schematic representation of variation in sediment source within an estuary on the New South Wales south coast. (Redrawn from Milledge 1980.)

Frequently, there is a substantial difference between riverine and marine sediments – marine sediments being well sorted and predominantly sandy, while riverine sediment is less well sorted, but with a higher proportion of fine fractions. The physiochemical regime in saltmarsh soils is largely a reflection of sediment type, so the balance between marine and riverine sources in sediment input might be expected to have considerable influence on the vegetation. Where the marine sediment is itself of comparatively recent glacial origin, the effect may be less marked.

Although the majority of sedimentary input into marshes is tide borne, aeolian supplies may be significant in some circumstances. In regions with macro- or meso- tidal regimes, extensive sand flats may be exposed seaward of the marsh at low tide and, with strong winds, appreciable amounts of sand may be blown into the lower marsh. Where marsh abuts dunes, some wind borne sand may settle in the upper marsh zone.

Recorded organic matter levels of saltmarsh soils are very variable, but are frequently low, although often increasing with elevation (Packham & Liddle 1970; Gray & Bunce 1972). The major inputs of organic matter are from the *in situ* vegetation, although some material may also be tide borne. Rarely, peat eroded from the catchment may be redeposited in saltmarshes. The sediment in saltmarshes in eastern North America is frequently referred to as peat but its organic content is often below 30%, although in some cases the peat epithet is better deserved with organic matter around 50% (Nixon 1980).

Where sea level is rising relative to the land, previous terrestrial peat deposits may become intertidal and support saltmarsh vegetation. In western Ireland, for example, saltmarsh vegetation may grow on a thin veneer of

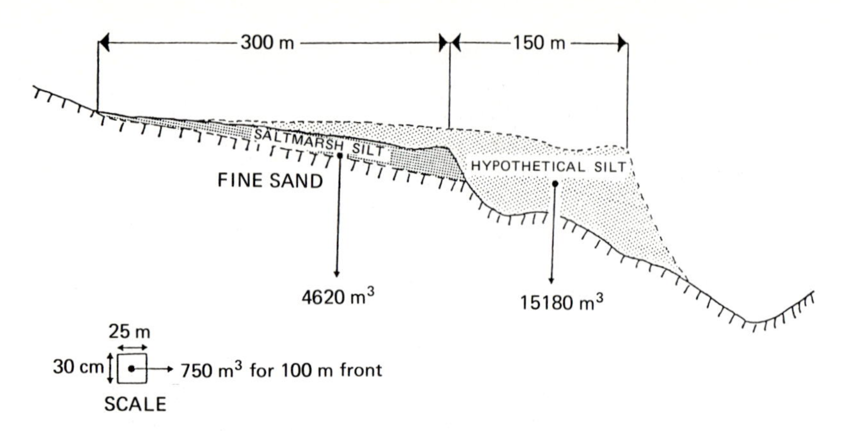

Fig. 1.16 As a marsh progrades, the volume of silt required to raise the marsh level to a given elevation increases. (From Jeffrey 1984.)

inorganic sediment over deep blanket peat deposits which themselves overwhelmed the earlier forest vegetation.

The floristically distinctive marshes of western Scotland (Adam 1981*a*) often have very organic soils which sometimes can be categorised as peat. These sediments are not overwhelmed terrestrial deposits and their origin is obscure.

On stable or slowly sinking coasts, accumulation of organic matter in the upper marsh theoretically provides a mechanism by which succession from saltmarsh to fully terrestrial vegetation could occur. Convincing demonstrations that this does occur are lacking, whereas transitions to terrestrial vegetation on rising coasts – as a result of isostatic uplift or tectonic activity, or following sedimentation caused by clearance of immediately adjacent steep catchments – are well attested.

Horizontal growth of saltmarshes

As well as accreting vertically, saltmarshes have the capacity to extend seawards (progradation). The rates of horizontal spread can be rapid; for example, marshes near the mouth of the River Nene in the Wash extended at an average of 50 m y^{-1} for about ten years in the later nineteenth century (Kestner 1962).

Extension of marshes requires colonisation of mud or sandflats, which implies a source of propagules (either from existing marsh to landward or from adjacent areas) and relatively stable conditions with low wave energy and tidal scour.

As vertical accretion occurs and the marsh front moves seawards, the

Fig. 1.17 Channel erosion of a marsh front at Roudsea, north west England.

surface profile becomes an inclined plane – initially steepening in gradient and then becoming gentler as sedimentation in the upper marsh declines. Assuming an ample supply of sediment, seaward growth of the marsh would cease when the erosive forces become too strong for successful colonisation – theoretically, continued accretion could then produce a uniformly flat marsh with an abrupt cliff at the seaward edge. Jeffrey (1984) has pointed out that as a saltmarsh grows the volume of sediment required for a given extension increases (Fig. 1.16). Unless there is an increase in the supply of sediment the growth rate of marshes is thus likely to decline with increasing size.

In reality, there is unlikely to be a continuous phase of marsh growth. The position of major channels in estuaries is often highly dynamic and movement of ten, or even hundreds, of metres in a few tide cycles can occur (Kestner & Inglis 1956; Inglis & Kestner 1958). When a channel moves into a marsh, erosion occurs producing an erosion cliff (Fig. 1.17). As the marsh edge recedes, some of the sediment eroding from the cliff face is deposited on the remaining marsh, increasing the elevation of the marsh edge (analogous to the formation of a levee along a creek). When the channel recedes, new marsh develops to seawards. Many marshes may have undergone several such erosion and extension cycles, leaving a series of old cliff edges separated by saucer-shaped terraces (see Jakobsen 1964; Beeftink 1977*a*). These cliffed marshes are characteristic of north-west England (Chapman 1960*b*; Gray 1972) – cliffs are much rarer in south-eastern England, although Greensmith & Tucker (1966) described a cycle of erosion, cliff formation and new marsh formation in Essex.

Gray (1972) has documented the growth of saltmarshes in Morecambe Bay (Fig. 1.18). Partial stabilisation of the major river channels following the construction of railway viaducts produced a long period of sustained marsh growth (which more than replaced the area of marsh reclaimed during the railway construction in 1857). In the early 1980s, however, there has been extensive erosion of marshes on the eastern side of the Bay. While the pattern of marsh erosion and growth in many estuaries has been altered by construction of training walls and constraints on the movement of channels, it is likely that, even before human intervention, marshes were subject to cycles of expansion and erosion. [Haynes & Dobson (1969) suggest that there may be a long-term periodicity to movements of estuarine channels.] As well as erosion by channel changes, wave erosion during major storms may be an important factor reshaping marsh morphology.

Harmsworth & Long (1986) have documented sustained losses of marsh area over a 21-y period at Dengie in Essex (despite local marsh growth in some sections). This could be part of a long-term erosion/accretion cycle or it could be a one-way process. The Essex coastline is sinking relative to sea level and given that the upper edge of the marsh is now artificially fixed by a seawall, preventing landward movement of the marsh, Harmsworth & Long (1986) suggest there is the possibility of complete eradication of the saltmarsh. [Harmsworth & Long (1986) point out that seawalls, by altering tidal flow patterns, may accelerate marsh erosion.] They also suggest that loss, through disease, of seagrass beds seaward of the marsh may have exposed the saltmarsh to greater wave energy, promoting erosion.

The role of biota in marsh development

The initial establishment of vascular plants requires relatively stable conditions. Coles (1979) has shown that microalgae (particularly diatoms) which secrete mucus may promote sedimentation and stabilisation of mudflat surfaces. These algae are grazed by gastropods and, at high tide, by fish. Reduction in grazing may lead to increased sedimentation (Coles 1979). It is possible that algal stabilisation of lower marsh sediments is required to facilitate subsequent establishment of vascular plants.

Once vascular plants are established, then the vegetation performs two major roles – trapping sediment and stabilising accreted material. (In addition, plants contribute organic matter to the soil.) The models developed by Randerson (1979) and tested with data from North Norfolk and the Wash suggest that, while both roles are performed, the stabilising role is the more important.

There are widely held views that saltmarsh (and mangroves) are landbuilders – the seaward extension of saltmarsh promotes accretion and further habitat development. However, the requirement for low energy conditions for marsh establishment suggests that it is more appropriate to view saltmarsh as taking advantage of sites where sediment accumulation is already occurring and it is the ability of plants to stabilise this sediment against subsequent erosion which determines whether a marsh develops.

Saltmarshes are also claimed to protect shorelines from erosion. Any such ability is limited. Once established, a marsh can disappear very rapidly if there is a change in conditions resulting in exposure to increased erosive forces. Nevertheless, the binding of sediment by plant roots does confer some resistance to erosion (van Eerdt 1985) and differences between species in root distribution and tensile strength affect erodibility.

The perception that saltmarshes played a role in shoreline protection was related to the view that saltmarsh was a plentiful, relatively unimportant resource, so that erosion of substantial areas of saltmarsh occurred with little notice being taken of it. In major storms considerable erosion of marshland may have taken place but such episodes were rare. In the interval between storms, marsh recovery of the marshland was possible. However, if much of the marshland is reclaimed, the area left to be 'sacrificed' to storms is small and recovery after damage uncertain. In these circumstances erosion of embankments and damage to the reclaimed land may occur after comparatively minor storms.

The role of animals in marsh development is complex, but poorly studied. Given the relationship between plant biomass and net accretion (Randerson 1979), reduction in biomass would be expected to affect accretion rates. Utilisation of marshes by domestic livestock reduces aboveground biomass and also results in a vegetation structure which may be less effective in trapping and retaining sediment. When grazing was introduced as a factor into Randerson's (1979) model, it resulted in a change in species composition. However, the effect of domestic stock on marsh development does not appear to have been studied in the field.

Saltmarshes may, on a seasonal basis, be subject to very heavy grazing and trampling pressure from geese and ducks (Jefferies, Jensen & Abraham 1979; Smith & Odum 1981; Smith 1983; Roberts & Robertson 1986). The

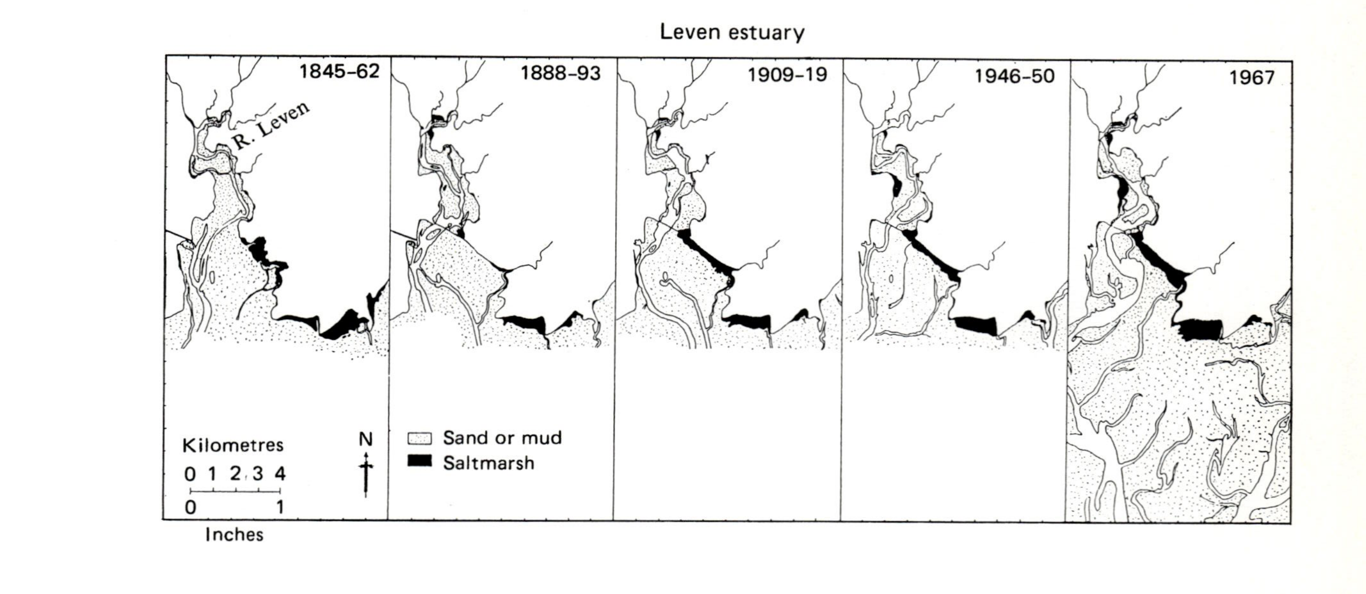
Leven estuary
1845-62
1888-93
1909-19
1946-50
1967
R. Leven
Kilometres
0 1 2 3 4
0 1
Inches
N
Sand or mud
Saltmarsh

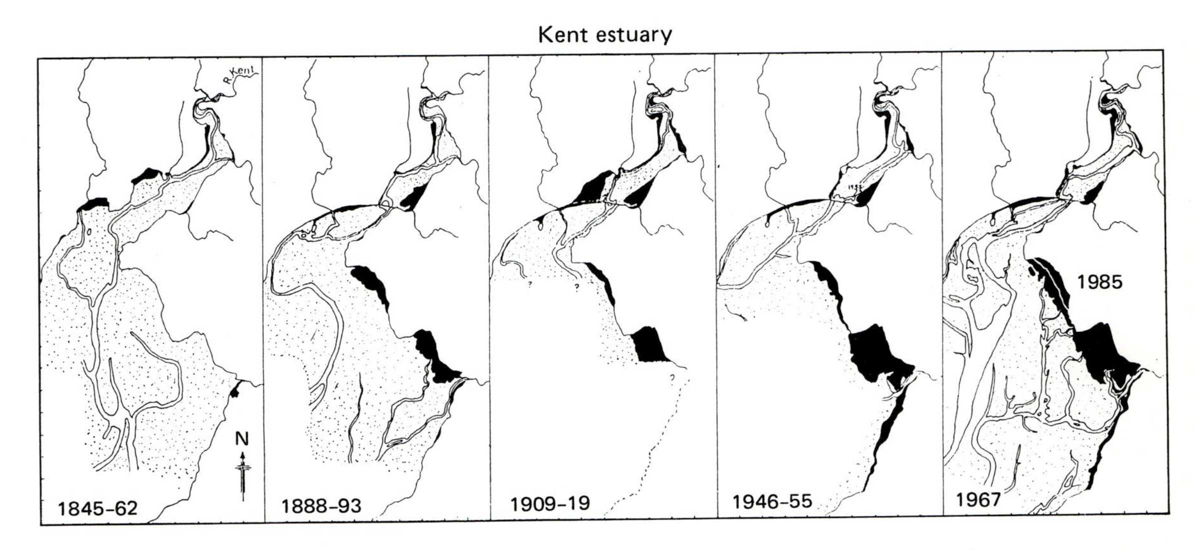

Fig. 1.18 Changes in the extent of saltmarshes around the Leven and Kent estuaries, Morecambe Bay; data derived from various Ordnance Survey maps and local maps. (From Gray 1972; 1985 edge of Silverdale marsh drawn from field observation.)

bare patches which can be created by birds may act as the nucleus for pan formation or more extensive erosion.

Invertebrates may be very important 'processors' of saltmarsh sediment. The fauna of mudflats and in creeks includes many deposit and filter feeders. The sediment input to the marsh may contain a large proportion of faecal pellets from this fauna (Frey & Basan 1985). Not only is the particle size of the sediment altered, but also its chemistry – the faeces and pseudofaeces represent concentrated inputs of available nitrogen and phosphorus (Long & Mason 1983). However, while this may be important to some marshes, it is not equally so in all cases; mussel beds such as those described by Kuenzler (1961*a*,*b*) and Frey & Basan (1985) are absent from many marshes. Within the saltmarsh itself, the burrowing activities of the fauna may have considerable effects on soil aeration and chemistry, but the impact of this upon the growth and distribution of the flora has only been studied in relation to *S. alterniflora* (Bertness 1985).

The mudflat and marsh molluscan fauna, after death, provide carbonate particles which are incorporated within the marsh sediment.

Heterogeneity of marsh sediments

The basic model of marsh accretion implies that marsh sediment characteristics will vary both horizontally across the marsh and vertically, with finer sediment being deposited at higher elevation than sand. This pattern may be modified by storm events (pp. 22–23) but has been widely demonstrated (Long & Mason 1983). At a smaller spatial scale, similar changes occur between creek levees of relative coarse sediment and the finer-grained material in inter-creek basins.

Changes in soil physical and chemical characteristics resulting from this pattern of sediment distribution are considerable and might be expected to have an influence on the vegetation. Gray & Bunce (1972) demonstrated large changes in soil chemistry between high and low marshes in Morecambe Bay, and the various plant communities they recognised had distinctive soil conditions.

Given the influence of tidal flooding on soil conditions, it might be anticipated that at a given elevation the soil environment would be relatively constant. However, a number of studies have shown spatial heterogeneity within zones (Ranwell *et al.* 1964; Adam 1976*a*; Vestergaard 1982). Heterogeneity at this fine scale may reflect many factors. The activities of animals and of roots may create locally oxidised sediments, changing the availability and form of several elements. Selective uptake of ions by roots may alter both concentrations and composition within the rhizosphere – the very low hydraulic conductivities of saltmarsh soils (Clarke & Hannon

Fig. 1.19 An oblique aerial photograph of the Keyhaven saltmarsh (Hampshire, UK) showing numerous creeks and pans. (From Steers 1960.)

1967, 1969) mean that, even following tidal flooding, such heterogeneity may persist. The distribution of plants in the pioneer zone may promote localised sedimentation (as illustrated by the hummock formation by *Pu. maritima* pictured in Yapp, Johns & Jones 1917). The effects of such differential sedimentation in the early stages of marsh development may still be reflected in the heterogeneity of the mature marsh. The importance of the variation in soil properties at this spatial scale to the biota is unknown. As many saltmarsh plants exhibit clonal growth, a single individual may occupy an area encompassing considerable variation in soil properties. However, for sediment-dwelling fauna, this variation may be a significant factor in controlling local distribution patterns.

Pans and creeks

Pans (small pools, normally with abrupt rather than gently shelving shorelines) and creeks are a characteristic feature of many saltmarshes. There is, however, very considerable variation in the form and density of these features, both within and between sites.

Pans and creeks may occupy a large part of the area of a marsh (see Figs. 1.19, 1.20), but in some regions, creeks and pans are almost completely

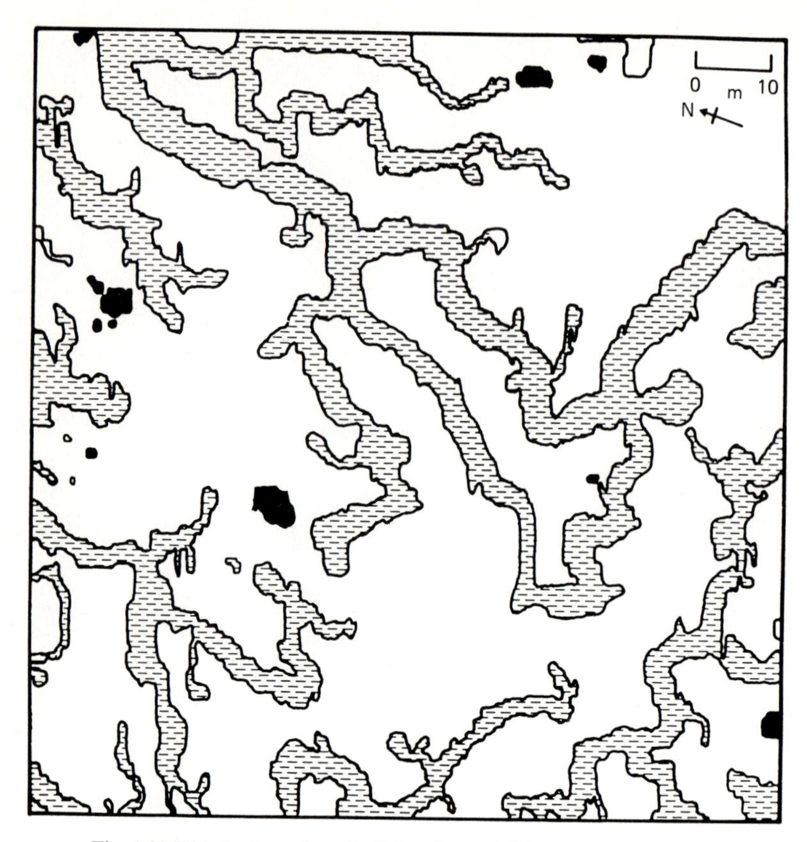

Fig. 1.20 Distribution of creeks [creek symbol] and pans [pan symbol] on a small area of saltmarsh at Colne Point, Essex, UK. (Redrawn from Hussey & Long 1982.)

absent, as in the case in many saltmarshes in southern Australia (Adam, Wilson & Huntley 1985) and the vast marshes of Hudson Bay (Glooschenko 1980*a*). High pan density may occur on marshes with few creeks, as in many upper marshes in north west England; high creek density may be associated with a paucity of pans, as in many marshes in south east England (Abd. Aziz & Nedwell 1986*a*).

These topographic features create a series of habitats in themselves, but also influence the surrounding marsh, being responsible for local gradients in environmental conditions, often cutting across the overall marsh zonation patterns. These local gradients are reflected in the mosaic of plant communities within each zone (Fig. 1.21).

Even within a single marsh, pans vary considerably in their form – differing in area, extent, shape, orientation of long axis, depth and permanence of water body.

The origin and development of pans have been the subject of some controversy. Yapp, Johns & Jones (1917) suggested that pans could have several origins (see also Steers 1964). In the initial stages of colonisation and marsh development, vegetation cover is not continuous. If the vegetated patches accrete vertically as well as extending laterally, then unvegetated areas may become enclosed to form pans. Once formed, the pan is likely to remain unvegetated because of the lack of drainage. Such a pan was categorised as a 'primary' pan by Yapp, Johns & Jones (1917). In later stages of marsh development, it was suggested that creeks may cut back into such pans and, after drainage, the pan site may be rapidly colonised by vegetation.

Other pans are elongated in outline and can be related, particularly on inspection of aerial photographs, to former creeks. Creeks can become blocked in several ways: by overgrowth of vegetation, by damming following undercutting of creek banks, or accumulation of inwashed material (Selisker & Gallagher 1983).

Yapp, Johns & Jones (1917) also recognised 'secondary' pans. The horizontal extension of marshes is rarely a continuous process, but is interrupted by a phase of erosion when a cliff forms. Below the cliff there is often an accumulation of eroded blocks of the old marsh surface (see Fig. 1.17). At the end of the erosion phase, marsh growth resumes and a new marsh terrace is formed below the cliff. As the vegetation consolidates, pans are formed as in primary marsh, although in this case the nuclei around which vegetation develops may be the eroded blocks rather than new seedlings. The marsh developing after the erosive phase was termed secondary marsh and any pans, secondary pans. However, most marshes displaying microcliff features may have gone through several cycles of erosion and accretion and any distinction between primary and secondary pans may be difficult to make by simple inspection.

On the basis of the Yapp, Johns & Jones (1917) model, it would be predicted that, while the number of channel pans would increase with increasing age (elevation) of a marsh surface, the density of other pans (whether primary or secondary) would remain constant or decline (as a result of capture by creeks and subsequent colonisation by vegetation). The model does not provide a mechanism by which pans can form in established vegetation other than following large-scale erosion. However, it is frequently the case that pan density is highest in the upper zones of the marsh. This is particularly obvious in south-eastern England (see Pethick 1974, 1984), where the highest pan density occurs in the mid to upper marsh, a zone characterised by dominance of herbs rather than grasses (Adam 1981*a*).

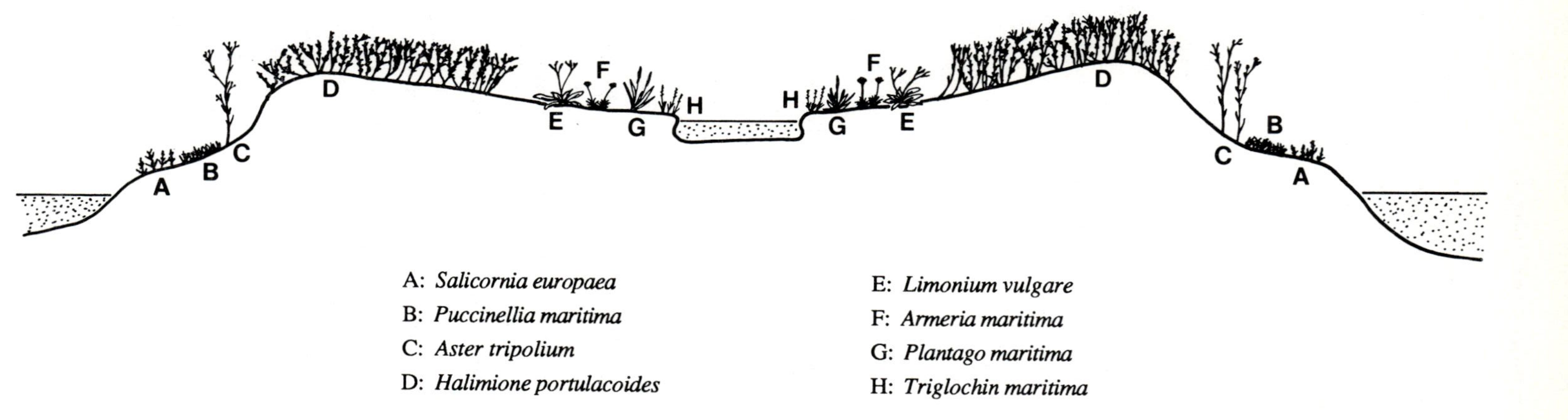

Fig. 1.21 Variation in vegetation associated with local variation in microtopography on a north European saltmarsh.

It is not clear how these pans are formed. Various agencies may be responsible for the death of vegetation: smothering by drift litter, ice action at higher latitudes, and the activity of animals (see, for example, Jefferies, Jensen & Abraham 1979; Roberts & Robertson 1986). Hypersalinity persisting for long periods may also be responsible for death of vegetation. However, once vegetation has been killed, a mechanism by which the protopan becomes deepened is required. It is possible that once the protective vegetation cover is removed, tidal scour and wave action may cause the pan to be excavated, while in some regions continued ice action may be important (Roberts & Robertson 1986).

Smith (1979; see also Kesel & Smith 1978) has proposed that pans may be formed from below as a result of undermining of marsh surfaces by piping. It is suggested that pipes may be initiated by periods of desiccation of the marsh surface when deep cracks may form. Even under the high rainfall conditions experienced in the Outer Hebrides, Smith (1979) argues that this mechanism is a major cause of pan formation. This model has the advantage that it would promote pan formation in high-level marshes where pan densities may, indeed, be much higher than at lower elevations. However, further work is required to confirm that piping may be a widespread agency in pan formation. [The development of desiccation cracks on a saltmarsh, under a temperate oceanic climate, has also been described by Carr & Blackley (1986*a*,*b*). While these authors do not relate these cracks to pan formation, they do point out that they provide pathways for the movement of water and other materials into the subsurface sediments.]

It seems probable that the relative importance of different mechanisms of pan formation will vary between sites, being influenced by local physiographic conditions, vegetation types, climate and the availability of appropriate agencies for initiating local vegetation death. The failure to confirm hypotheses derived from the Yapp, Johns & Jones (1917) model, while disproving its general applicability, does not rule it out as being relevant under certain circumstances (Pethick 1974).

The primary colonist in the Dovey estuary at the time of Yapp, Johns & Jones (1917) study was *Pu. maritima*, which on a sandy substrate tends to produce low hummocks separated by bare areas which may develop into 'primary' pans. The lower marsh communities in south-east England, however, tend to form a continuous vegetation cover, broken only by creeks, without obvious potential pan sites. It would be of interest to re-study pan distribution in the Dovey estuary. For the last few decades the primary colonist in the estuary has been *S. anglica* and the vegetation cover in the lower marsh is almost continuous. At the same time, the accumulation of litter, largely produced by *Spartina*, on the upper marsh terraces is much increased.

The creek systems of saltmarshes are important for the interchange of materials between marshes and adjacent estuaries and, in addition, may provide an essential habitat for various animals. Nevertheless, the distribution and abundance of creeks vary considerably between marshes and, at some sites, creeks are virtually absent. Creek density and tortuosity are, at least in part, related to the sediment type and tidal range. Thus, saltmarshes in south-east England, with fine silt sediment and moderate tidal range have very complex creek systems, while marshes on the west coast on sandy sediment and with large tidal range have much simpler 'herringbone' systems.

The major elements of a creek system may be inherited from the mudflats preceding marsh development and most of the system is established in the early stages of marsh colonisation. Although the immediate impression of many creeks is of instability, with signs of bank erosion and channel movement, most such changes are local in occurrence and produce only temporary fluctuations. Over the long term the remarkable feature of much of most creek systems is stability – individual creeks can be traced on charts or aerial photographs virtually unchanged over decades. Garofalo (1980) suggests that most channel migrations occur as a result of storms rather than in consequence of continuously active processes.

As the general elevation of the marsh surface increases with age, so the differential between marsh and creek bank increases. Substantial levees may be produced, often supporting different communities from the surrounding marsh. The levees, developed as a result of greater sediment capture close to the creek, are better drained and have more aerated soil conditions that the adjacent marsh. As the levees develop, the drainage conditions in the inter-creek basins may deteriorate. At the highest levels in the marsh, where the input of inorganic sediment is low, the differential elevation of the levees may be reduced. In sites exposed to ice action, ice scour may partially level levees each year (Ogden 1981).

Drift litter

Tides redistribute large amounts of debris on saltmarshes (Fig. 1.22). This material may be autochthonous or allochthonous, including both material of terrestrial origin carried into estuaries by rivers and material from marine sources. The majority of this material is plant matter, but animal material may also be included after storms when large quantities of mudflat fauna may be washed ashore. Even on shores remote from

Fig. 1.22 Drift litter at the upper edge of the saltmarsh, Bull Island, Dublin.

habitation, a component of the drift litter will be of human origin; close to major cities, a major portion of the material may consist of the products of industrial society.

Beeftink (1965, 1966, 1977*a*) distinguished two types of plant litter – debris of higher plants and larger brown algae concentrated along spring tide marks, and other algal material spread over a diffuse zone in the lower marsh.

Both sorts of litter play an important, if poorly quantified, role in the nutrient economy of saltmarshes. Driftlines also provide a specialised habitat supporting a range of communities. In northern Europe, the most widespread plant communities on litter are characterised by annual nitrophiles (Nordhagen 1940; Beeftink 1965, 1966, 1977*a*; Adam 1981*a*). In most cases, redistribution of material during winter storms prevents the development of longer-lived communities. However, drift deposited above the normal high spring tide mark by exceptional storms may persist for several years and exhibit a succession from communities dominated by annuals to a more permanent cover. The dynamics of litter deposits have been poorly studied. The availability of potential litter will vary seasonally while the distribution of deposits will be influenced by variation in tide heights and the incidence of storms. Storms may move large quantities of litter (Hackney & Bishop 1981) and short-term measurements of litter movement may be poor predictors of longer-term budgets. However, the very unpredictability of storms means that it is difficult to carry out research on the topic.

While most drift litter accumulates in the upper marsh, large deposits may, at least transiently, be found at any level in the marsh. The impact of the deposition of drift litter on the underlying vegetation has been the subject of much speculation, but few detailed studies. While it has been suggested many times that such litter deposits may initiate pan formation, the evidence is inconclusive (Hartman, Caswell & Valiela 1983; see also pp. 35–37). Litter accumulation may damage the existing vegetation and may result in death of extensive patches which remain as bare ground when the litter is redistributed. In most cases, however, these areas are rapidly recolonised and, while there may be a persistent slight depression, litter does not appear to be a major initiator of deeper pans. Hartman *et al.* (1983) demonstrated that different communities in marshes on the east coast of the USA differed in their susceptibility to damage by litter; it is probable that this finding would be applicable in other regions.

Mineralisation of drift litter and incorporation of organic matter from litter in the soil may be important processes in the development of saltmarsh soils and the nutrient economy of the whole ecosystem. However, in many marsh systems, the greatest bulk of material eventually accumulates in a narrow zone of the high marsh. The rates of breakdown, the role (and, indeed, the composition) of the often abundant invertebrate fauna (Rammert 1981) in breakdown processes, and the fate of mineralised nutrients (do they accumulate under the litter zone, are they redistributed to the marsh, or are they leached directly to the estuary after spring tides?) remain poorly studied topics.

The quantities of litter involved vary considerably between marshes, but can be very considerable. Human modification of estuarine processes may have considerable impact upon patterns of litter accumulation. Heckman (1986) described how, in the freshwater tidal portion of the Elbe estuary, litter which formerly was spread over the floodplain by winter storm tides now accumulates (up to 2 m deep) at the foot of the seawall, from where it is cleared by snow ploughs and burnt. Heckman (1986) calculated that, in terms of loss of a fertiliser effect, by which nutrients in the litter were returned to the agricultural lands of the flood plain, and from loss of water quality improvement resulting from uptake of nutrients from the estuary by vegetation, elimination of tidal marsh following seawall construction was responsible in the Elbe for a loss, at current prices, of more than 20×10^6 DM per annum from effects on phosphorus and nitrogen budgets alone.

The constituents of the drift litter vary very much between sites. On very heavily grazed marshes, the drift may be a mixture of terrestrial-derived material and marine algae; on marshes with extensive *Spartina* spp. stands or reedbeds most of the litter may be of marsh origin; at other sites, extensive offshore seagrass beds provide the bulk of the litter. Whether the nature of the litter influences the plant communities which develop on it in any consistent way is uncertain. Various physical and chemical properties (such as mean fragment size and degree of lignification) vary between litter types and it might be anticipated that this would influence rates of mineralisation. In much of northern Europe, the spread of *S. anglica* may have increased the quantity of litter and will also have altered its composition. Beeftink (1975) suggested that the greater uniformity of drift deposits since the advent of *Spartina* may have been accompanied by a reduction in diversity of strandline vegetation.

If plant and animal debris tends to accumulate in particular parts of a marsh, then so will other tide-borne material. Litter deposits are thus likely to be sinks for various pollutants, although the consequences of this are virtually unknown.

Measurement of changes in marshes

Vertical accretion

Rates of accretion in saltmarshes have been estimated using a variety of techniques. The most widely used method is to employ a marker horizon such as coal, brickdust or aluminium glitter (Richards 1934; Harrison & Bloom 1977; Richard 1978). After the marker layer is laid on the marsh surface, the accretion of sediment above the layer is measured over a period – most such studies have been relatively short term, but some have involved periods of a decade or more (Harrison & Bloom 1977; Steers 1977). In some cases, a natural marker, such as a layer of sand associated with a particular storm, allows assessment of accretion rates over large areas of a marsh. In recent years, the use of silica flour cores rather than marker layers has gained popularity (Kestner 1979; Coles 1979).

For short-term studies, stakes have been used to monitor changes in marsh elevation (for example see Ranwall 1964*a*; Hubbard & Stebbings 1968). However, if tidal current velocities are high, scour around the stake may prevent any useful measurement of surface changes.

Average sedimentation rates over longer periods can be estimated from radiometric dating methods – for example, using ^{210}Pb (Armentano & Woodwell 1975). Where the marsh sediment contains appreciable organic matter, or where rising sea level has resulted in terrestrial organic matter being entombed by saltmarsh, then ^{14}C dating can be used (provided appropriate measures are taken to avoid problems caused by the presence of carbonate shell fragments).

A potential source of error in the use of both marker horizon technique and radiometric techniques is that bioturbation may profoundly disturb the sediment stratigraphy. Frey & Bassan (1985) document the extent of bioturbation in some American east coast marshes. However, the degree of bioturbation appears to be inversely correlated with latitude and, in northern Europe and on the northwest coast of the USA, bioturbation is not a major factor (Seliskar & Gallagher 1893).

In some marshes, the sediment is finely laminated – this is readily apparent in erosion cliffs (Fig. 1.23). The surface of each lamina represents a temporary hiatus in sedimentation; but whether this occurs at regular intervals, such that the laminations could be used to date marsh development, is not known. Kellerhals & Murray (1969) report the development of varve-like stratification in high level tidal flats of the Fraser River Delta, British Columbia. In the summer, the flats are covered with a dense algal mat (predominantly blue green algae, but also including the chlorophytes *Enteromorpha* and *Rhizoclonium*). This algal layer traps and stabilises sediment. In winter, sand deposited during storms covers the algae, resulting in an annual stratification of alternating sandy and more organic layers. Whether a similar mechanism is involved in the development of laminated sediments within saltmarsh is unknown.

Lateral movement

The investigation of the evolution of coastlines over the post glacial period is essentially the province of geomorphologists – much of the story is built up from interpretation of core stratigraphy (see, for example, Steers 1977; Roy 1984). Such studies have important implications for ecologists as they provide evidence of the transience of many habitats and the consequent repeated opportunities for the establishment of new communities (see Roy 1984).

Ecologists are most likely to be interested in the most recent phases of marsh development. These can be studied by comparison of charts and maps and, more recently, aerial photographs and satellite imagery. In some instances, changes are sufficiently rapid that monitoring of quadrats and transects is a feasible method of investigating marsh development.

There have been many studies of marsh development from map evidence (see *inter alia* Steers 1964, 1977; Marker 1967; Gray 1972). Care has to be taken with interpretation of both hydrographic charts and terrestrial maps as, in many instances, only a general indication of the presence of marsh is

Fig. 1.23 Laminated sediment revealed in an eroding marsh cliff, Roudsea, north west England.

given and boundaries are not accurately delimited. Even where the marsh edge has been surveyed, it may not always be revised on later editions so that monitoring changes over time may be difficult. Despite these limitations, there is a vast amount of cartographic data available for the study of marsh history.

On terraced marshes, the various cliffs represent past erosion episodes. Often these cliffs are not indicated on maps, but sometimes other documentary records allow their origin to be dated (for example, the uppermost terrace on West Plain marsh in Morecambe Bay represents that area of marsh remaining after major erosion caused by movement of the channel of the River Leven in 1828 – Gray & Adam 1974). Much of the readily identifiable historical documentation for saltmarshes refers to reclamation – such information is valuable as it allows assessment of the rate of accretion of new marshland. Nevertheless, careful interpretation of local documents reveals much about unreclaimed marshes (see Parkinson 1980, 1985).

Aerial photography provides the opportunity of studying marsh development in greater detail – as well as changes in position of the marsh front, changes in creek and pan distribution can be studied. In some circumstances, even the growth of individual clones (of species such as *Juncus* or

Spartina) could be monitored. With appropriate ground truthing and use of suitable film, changes in community composition may be detected. While there have been many studies utilising aerial photography, there are very few sites for which repeated compatible imagery is available.

Recent technological developments in remote sensing permit reflectance of particular spectral bands to be measured. These data may allow estimation of above ground biomass – Hardisky *et al.* (1986) review studies on saltmarsh vegetation and suggest that these techniques have considerable potential.

Satellite imagery is very valuable in providing a synoptic overview of wetland resources. However, it does not offer sufficient resolution to map features such as creeks and pans or community boundaries. While the newer systems permit resolution at a scale of 20 m (Hardisky *et al.* 1986), this is still inadequate for studies of individual sites, although data with much finer resolution should soon become more readily available.

Sea-level change

Within the context of the last two million years of the earth's history, the present-day experiences unusual environmental conditions. The period has been one of widespread glacial conditions at high and mid latitudes, interspersed by comparatively brief warmer interludes, such as the current Holocene interglacial. During the cold periods, there was a global lowering of sea level by as much as 100–150 m. The present coastline has only been in its current position for approximately the last 6000 years, since the end of the phase of rapid rise in sea level since the last glaciation.

Little is known about the detailed geomorphology of the coastline at the time of the lowest sea-level periods (see Kennish 1986), but estuaries would have been small (Schubel & Hirschberg 1978) and it is likely that the areas available to support intertidal saltmarsh were also small. During the very rapid rise in sea level between 15 000 and 6000 years ago (on average 1 m/century), the rate of change in sea level was probably too great for extensive marsh development. In the last 6000 years, global sea level has been relatively stable, within metres of its current position. However, for particular localities, considerable changes in relative land–sea levels have occurred as a result of isostatic and tectonic movements of the land.

Schubel & Hirschberg (1978) argue that estuaries reached their peak development in number and size around 3000–5000 B.P. Since that time, geomorphological evolution and, in recent centuries, human activity, have reduced the area of estuaries considerably (Schubel & Hirschberg 1978; Nichols & Biggs 1985; Roy 1984). The rate of evolution varies considerably and estuaries at various stages in their development can be found in close proximity (Roy 1984). Nevertheless, in geological terms, all estuaries must be regarded as ephemeral features destined to be infilled with sediment. For most estuaries, their remaining lifespan may be estimated to be in the order of a few tens of thousands of years (Schubel & Hirschberg 1982); although, for some, the future is of more rapid disappearance. It is estimated, for example, that the estuary of the Chang Jiang (Yangtze), one of the world's major rivers, has a remaining life of only 1000 years (Schubel & Hirschberg 1982).

While the last 5000 years have seen a continuing reduction in the area of estuarine waterbodies, this is not necessarily the case when coastal wetlands are considered. The infilling of estuaries with sediment provides the opportunity for saltmarsh development and extensive marshes may occur in later stages of estuarine evolution (Roy 1984; Nichols & Biggs 1985). However, even if the effects of humans are discounted, marsh growth has not been continuous. At the local scale, changes in marsh area will have been controlled by channel changes (see p. 28), the balance between sedimentation rates, relatively small eustatic changes in sea level and isostatic changes, and tectonic events. While tectonic activity is restricted in both time and space, it has had major impacts on some extensive saltmarshes, such as those on the southern Alaska coast affected by the 1964 earthquake (Ovenshine & Bartsch-Winkler, 1978; Crow, 1971).

Some marsh systems appear to have been relatively stable for hundreds, or even several thousand years. For others, extensive growth or reduction over decades can be documented. Very extensive intertidal wetlands may have had a transient existence in some estuaries. Woodroffe, Thom & Chappell (1985) have documented the existence, for about 500 years (between 6500 and 7000 years ago), of vast mangrove stands on what are now the freshwater wetlands of the Alligator Rivers region of northern Australia.

Along the New England coast (north-eastern USA), accretion rates in saltmarshes over the last few thousand years have kept pace with increases in relative sea level. This has allowed saltmarshes to extend both landward, over previous terrestrial or freshwater wetland communities, and seaward onto mudflats (Niering & Warren 1980). In the Gulf of Bothnia, continuing isostatic recovery results in former coastal marsh becoming fully terrestrial (Ericson & Wallentinus 1979). In Hudson Bay, isostatic uplift at an estimated 1.0–1.5 m/100 y (Hunter 1970) is a major influence on saltmarsh development.

Intertidal wetland distribution and extent have, thus, been subjected to continuous change throughout the Holocene. During the Pleistocene, extensive intertidal wetlands may only have been present for a small fraction of the period during the later part of the interglacial interludes.

The implications of this pattern of ever-changing wetlands are considerable. Any hypotheses about the importance of saltmarsh productivity to adjacent estuarine and coastal waters, or about the role of coastal wetlands in geochemical cycles must take cogniscence of the great changes in marsh area that have occurred over short time periods and of the fact that for most of the last 2 million years total saltmarsh area may well have been much less than at present. Within the wider context of the whole estuary, Schubel & Hirschberg (1978) have questioned the application of the concept of 'estuarine dependence' to various species and suggest that: 'The biological importance of estuaries to the survival of organisms and to the maintenance of the total marine ecosystem appears to have been exaggerated'. Schubel & Hirschberg (1978) postulate that many supposed estuarine-dependent species persisted in low salinity open waters at times when estuaries were rare features. While this may be so, the fluctuations in saltmarsh area must, nevertheless, have had considerable impact on estuarine ecosystem functioning.

The floristic composition of saltmarshes at times of lower sea level is unknown. Although the simplest hypothesis would be that, as sea level fell, saltmarsh migrated seaward and, thus, remained essentially unchanged in composition – this is unlikely. Sea-level change is the consequence of other environmental changes which may have considerably influenced saltmarsh vegetation. At high and mid latitudes, the climate was cooler and marshes may have been exposed to greater ice scouring. The seasonality and volume of freshwater run-off would differ from the present, leading to different salinity regimes in estuarine and coastal waters. The sedimentary regimes during glacial periods would also differ from those of the present. Overall, environmental conditions of saltmarshes during glacial periods are likely to have been different from those prevailing today.

Terrestrial plant communities, even when their current distribution is closely correlated with particular environmental conditions, have not migrated as entities around the landscape through the climatic changes of the Pleistocene. The response of taxa to environmental change has been individualistic (as documented by Huntley & Birks 1983); while some current species' assemblages may have occurred in the past, other groupings are unique to brief periods of time.

In the case of saltmarshes, it might be suggested that, *a priori*, species' assemblages may be more likely to be recurrent in time than those of terrestrial vegetation. The halophytic element in floras is normally very small, and over the Pleistocene, at least, there is no evidence either for extensive diminution or augmentation. This being so, it could be suggested that opportunities for development of new assemblages would be limited. Nevertheless, many colours can be produced from a limited palate and, under different environmental conditions, even a slight shift in habitat preferences could result in species combinations unlike any of those of today.

It is argued in Chapter 3 that the geographical distribution of major saltmarsh types is correlated with climatic factors. If, for purposes of argument, it is accepted that this correlation, at least partly, reflects causality, then species' assemblages of high and mid latitude marshes might not have been able to survive simply by moving seaward; rather, considerable migration to lower latitudes may have been required. While the very wide distribution of many saltmarsh species may be taken as evidence of their dispersal ability, little is known about potential rates of spread or of differential rates of migration between species. However, if survival of saltmarsh species demanded extensive migration, then, by analogy to the terrestrial situation, it might be suggested that community integrity is unlikely to have been maintained.

Unfortunately, direct evidence as to the nature of interglacial coastal saltmarshes is unlikely to be forthcoming. Any sedimentary record of such marshes will have been submerged by the rising sea level long ago.

If it is assumed that no angiosperm is obligately halophilous (see Chapter 4), then survival of saltmarsh species may not have demanded that they follow the moving shoreline. It is conceivable that the saltmarsh and mangroves of previous high sea level times were 'stranded' by the falling sea level and survived along abandoned shorelines. This is, however, unlikely – when saltmarshes are removed from direct marine influence by human activity some halophytic species may persist for long periods, but the overall species composition changes very rapidly. In addition, climatic conditions, at least at higher latitudes, would have been inimical to the survival of the whole species assemblage at inland localities. There is no convincing evidence for the long-term survival of 'stranded' former intertidal wetlands. The inland stand of the mangrove *Av. marina* in north western Western Australia, claimed to mark a relict Tertiary shoreline (Beard 1967), may more plausibly be explained as having developed from propagules discarded by Aboriginals. [It is known that Aboriginal groups travelled to the coast to collect mangrove fruit.]

Although the survival of communities at the site of old shorelines is unlikely, individual species may have been able to occupy a range of inland habitats. There is evidence for several saltmarsh species having wider non-coastal distributions during glacial times (see Bell 1969; Adam 1977*a*; Beeftink 1985*b*). While some of these occurrences may have been at inland saline sites (Bell 1969), others may indicate no more than the presence of relatively open, unshaded, vegetation. Nevertheless, climatic conditions in such vegetation would be very different from those on the coast – with far greater diurnal and seasonal variations. On the basis of their present behaviour, it is difficult to conceive of all the halophytic species being able to survive at inland localities, even in the absence of competition.

The history of the majority of halophytic species is likely to remain obscure but in the absence of data, it is, nevertheless, possible to speculate upon the history of saltmarshes – it is also possible to speculate upon future changes. Two conflicting scenarios may be advanced. In one, the onset of the next glaciation will be marked by a lowering of sea level and migration and re-assortment of the saltmarsh biota – in the other, a substantial rise in sea level occurs in consequence of global warming as a result of a greenhouse effect caused by increasing concentrations of atmospheric carbon dioxide, methane, nitrous oxide and chlorofluorocarbons. Such a rise would have very serious consequences for mankind but, depending on the rate of change in sea level, it might be possible for saltmarsh to migrate landward in pace with the rise. However, at least on most temperate shorelines, such a migration of saltmarsh would be prevented by existing sea defences (and by others thrown up in response to the rising sea level). Before either scenario could unfold, it is likely that the area of saltmarshes will already have been further reduced by reclamation and development (see Chapter 7).

Biotic history

Saltmarshes are depositional environments and previous sections have outlined a generalised model of sedimentation and marsh development and discussed some approaches to studying the developmental history of particular sites. As inorganic sediment accumulates, biogenic material is also incorporated (rhizomes, seeds and pollen of plants, skeletal remains of fauna) providing a potential record of the ecological history of a site and offering the opportunity to test hypotheses about the relationships between sedimentary and biological processes. Such records will only cover the period of relative coastline stability, and while they might document transitions between marine and freshwater conditions, could not be used to address questions of the evolution of saltmarshes during the phase of rapid post-glacial sea-level rise. Nevertheless, detailed histories covering a few hundred (to, at most, a few thousand) years would be of great interest. To date, there have been few attempts to exploit this potential. A number of factors may have discouraged investigation:

> As material is transported into marshes by the tide, it may be difficult to determine whether subfossil remains are autochthonous or allogenic. If separation were possible, it would be difficult to estimate the size of the source area for allogenic material as it could have been carried into the estuary from the river catchment or from seaward.
>
> Reworking of material during erosive phases may result in a particular horizon containing fossils of different ages.
>
> Many saltmarsh plant communities are dominated by taxa from families whose individual members are difficult to differentiate from their pollen. [Clark & Patterson (1985) have demonstrated that, with careful observation, greater resolution of the pollen record than had been regarded as possible can be achieved.]

Clark (1986) and Clark & Patterson (1985) have shown in studies of saltmarshes on Long Island that close-interval sampling and pollen analysis permit detailed resolution of the historical record. In the sites investigated, there was no evidence for simple unidirectional change over the last few hundred years. Vegetation changes as interpreted from the fossil record were frequent and considerable and could be correlated with changes in the environment. Geomorphological changes (opening and closing of inlet mouths), cultural activity, clearing in the catchment, changes in the rate of sea-level change and the 18.6-year tidal cycles were all factors which could be related to changes in the pollen record. Storm events could also be invoked to explain certain features, such as the persistence of *Salicornia* in the record from high marsh sites (Clark 1986). The resolution of changes associated with periodic tidal cycles illustrates the precision with which fossil record can be studied and suggests that it would be of great interest to conduct studies using similar techniques on marshes with a range of vegetation types and in different geomorphological settings.

Zonation

One of the features of saltmarshes, and of mangroves, is that there is variation in species composition with elevation. Variation occurs in the microflora, the fauna, but most obviously in the vascular plants. Patterns can be seen at a number of scales but, when zonation is discussed, it most usually refers to the arrangement of species and communities in belts parallel to the shore (although pattern at this scale may be interrupted by other patterns associated with creeks or with local variation in microtopography).

Although the zoned nature of saltmarsh vegetation is frequently taken for granted, there are remarkably few data on the topic. The actual nature of the pattern perceived as zonation has only rarely been documented. Do species share common boundaries, so that zone boundaries are defined by community limits, or is species behaviour largely independent; is the downslope boundary of one species followed by the upslope boundary of another (contiguity – Dale 1984); is species behaviour consistent between sites? (See Fig. 1.24).

Pielou & Routledge (1976) investigated zonation of species in eastern Canada and found that the landward boundaries of species were more clustered than the seaward, and that the clustering of both landward and seaward boundaries increased with latitude. At higher latitudes, the zones of individual species become wider relative to the width of the marsh so their boundaries become more nearly coincident with each other and with the edges of the marsh. Pielou & Routledge (1976) suggested that this may reflect greater genotypic variability at higher latitude, although the smaller number of species at high latitudes may also be a factor. The clustering of upslope boundaries at all latitudes reflects the greater species richness of the upper marsh and the decline in species number in the more seaward part of a marsh (Pielou & Routledge 1976).

Many species, particularly those reaching dominance at low elevations, have broad ecological tolerance and may be found, at least on some sites, at any level in the marsh. Species characterising upper marsh communities are more likely to be restricted to higher elevations. Nevertheless, any attempt to define zones purely on species presence without taking relative abundance into account is likely to prove unsatisfactory.

Although there are still few studies which have recorded the elevation range of species, it is clear that, while some form of zonation is characteristic of saltmarshes, there is considerable variation between sites in the actual pattern of species distribution. Inversion of zonation patterns between fully marine and brackish reaches of estuaries is a frequent occurrence (Gilham 1957*a*; Beeftink 1966; Ranwell 1972). Changes in species behaviour between east and west coasts in Britain are reported by Adam (1981*a*), while Pielou & Routledge (1976) noted that the zone width for individual species increased with latitude in eastern Canada. Within single estuaries, there are often sites with apparently anomalous high or low elevation occurrences of particular species.

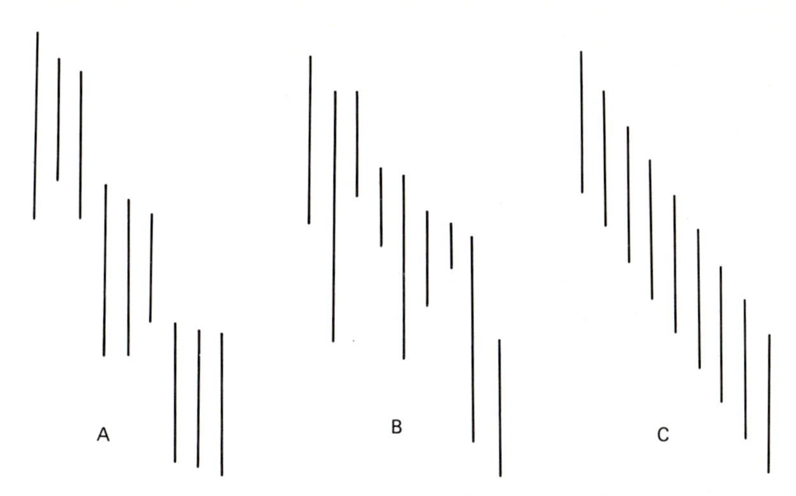

Fig. 1.24 Some possible arrangements of species on an elevation gradient. Each vertical bar represents the range of a species. In example A, zones could be defined by community boundaries and species' boundaries demonstrate approximate contiguity. (Redrawn from Pielou & Routledge 1976.)

In general, total species richness increases with elevation. In part, this reflects more species per community, but it is also related to a tendency for there to be a mosaic of communities at higher elevations rather than a single vegetation type. Although this mosaic of communities might blur the immediate impression of an overall zonation, if the distribution of species is examined there is, nevertheless, a tendency for particular species to be more abundant over a certain elevation range than others (Fig. 1.25).

The sparsity of detailed elevation data restricts the drawing of comparisons between zonation patterns. Even more generalised comparison can only be made with difficulty as there is no agreement on zone definition. Ecologists studying rocky shores have long acknowledged that zones cannot be defined by reference to tidal levels, but that biological boundaries can be used to delimit zones in such a way as to permit comparison between geographically disjunct areas (Lewis 1964; Stephenson & Stephenson 1972; Dring 1982). While there is evidence to indicate that zones defined directly by tidal levels would also be an inappropriate basis for comparison of saltmarshes, it has not yet proved possible to identify biological criteria which would serve to define universally comparable zones. In the absence of such a scheme, the criteria for defining zones vary between accounts, as does the number of zones recognised (although three – lower, mid and upper – is the most popular).

Despite these difficulties, variation in species composition with elevation

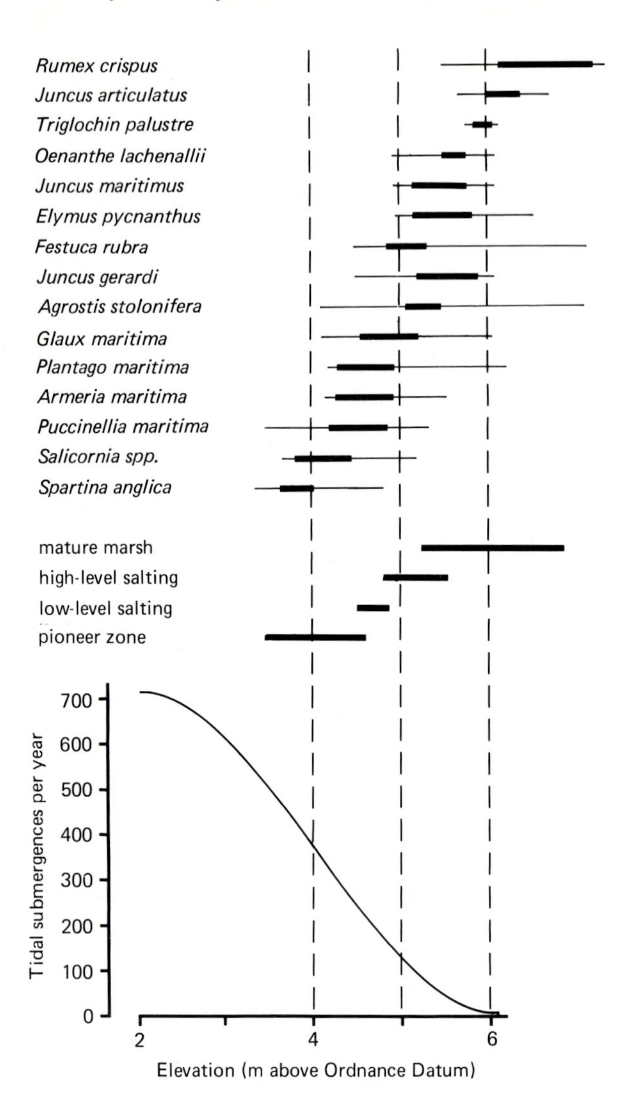

Fig. 1.25 Elevation ranges of some widespread saltmarsh species around Morecambe Bay. (The thicker parts of the lines represent the range where a species is most frequent.) (From Gray & Scott 1987.)

is a generalised property of saltmarsh vegetation which deserves explanation.

On rocky shores, there is agreement that, in very general terms, the upper limits of species are determined by absolute physiological tolerances to environmental conditions during emersion, while lower limits are set by competitive interactions (Pielou & Routledge 1976; Dring 1982), although exposure to wave action is a major modifying factor. For saltmarsh vascular plants, it has been suggested that the reverse applies – 'competition governs the locations of upslope boundaries, whereas each species' downslope boundary occurs at a level on the marsh set by its physiological tolerance for submersion in salt water' (Pielou & Routledge, 1976). Vince & Snow (1984) and Snow & Vince (1984) argued that their data from Alaska supported a similar conclusion.

The majority of saltmarsh species have very wide environmental tolerances when grown in cultivation and, in most cases, maximum growth under experimental regimes is unlikely to apply, except transiently, in the field situation. Thus, it is unlikely that the elevation gradient is partitioned so that each species' zone represents optimal conditions for growth. There is evidence to suggest that halophytes are restricted to saltmarsh habitats by poor competitive ability under non-saline conditions (see p. 274), while Gray & Scott (1977*b*) suggested that the relative abundance of the three grasses *Festuca rubra*, *Agrostis stolonifera* and *Pu. maritima* in saltmarshes on the west coast of England could be explained by their competitive abilities being modified by differing levels of waterlogging and salinity. A great deal of the patterning of saltmarsh species appears related to competitive interactions, but these interactions are influenced by environmental factors and the absolute limits of a species distribution are determined by physiological tolerances. What environmental factors correlated with elevation might influence zonation patterns?

Many factors will be consistently correlated with elevation – including physicochemical properties of sediments. However, for many of these factors, the range of variation within a site is likely to be less than that between sites (for example, on a 'sandy' marsh there will be considerable variation in particle size in the surface soil with elevation – however, the range of variation may not even overlap with that on a 'muddy' marsh). Factors of this type, although influencing the species composition, probably do not have much impact on zonation patterns.

The inversion of zonation shown at upper estuarine sites strongly suggests that soil salinity is an important factor in controlling zonation. It is often assumed that soil salinity is the controlling influence on zonation.

However, soil salinity varies both spatially and temporally in a complex fashion and, at many sites, the highest salinities are regularly experienced at mid and upper marsh levels. If tolerance of varying levels of soil salinity were the major factor determining the distribution of species, then lowest species richness would be expected to occur at high levels in the marsh and not in the low marsh. On arid coasts with prolonged hypersalinity, large bare saltflats in the upper intertidal suggest that soil salinity can, indeed, be the master factor. However, in general, even though upper marshes may regularly experience 'moderate' hypersalinity (Jefferies 1977*b*), the gradient in species and community diversity is from low to high marsh.

Although soil salinity is not predictably correlated with elevation, number of submergences and duration of submergence are. However, it is difficult to suggest the physiological mechanisms which would allow submergence *per se* to determine zonation. While the absolute seaward limit of marsh development might be reached when submergence is so long that photosynthesis is less than respiration, on many marshes even the lowest zones are submerged for only a comparatively small proportion of the time (see p. 000).

Dring (1982) summarised a discussion of algal zonation on rocky shores by saying:

> 'Recent laboratory experiments have reduced the wide range of possible explanations for these effects, but the *exact* physiological cause for a specific limit, or the *precise* environmental factor controlling it in the field, has not yet been established for any species at any site.'

Despite the recent advances in research on the ecophysiology of halophytes (see Chapter 4), this conclusion applies equally to plants on soft intertidal shores.

Snedaker (1982) reviewed studies of mangrove zonation and similarly concluded that explanations for zonations were as yet incomplete. In the case of mangroves Rabinowitz (1978) has suggested on the basis of studies in central America that there is a correlation between position in zonation and propagule size. This relationship was different from that in terrestrial forest communities as the most seaward ('pioneer') mangrove species had the largest propagules whereas pioneer forest trees generally have small seeds. There are few data on the relationship between seed size and position in the zonation of saltmarsh species. Saltmarsh species differ from mangroves in that many are capable of extensive vegetative propagation. Without knowledge of the relative importance of sexual and vegetative reproduction in species' establishment, assessment of the significance of seed size would be difficult.

Zonation and succession

The model of marsh development in which sediment is trapped and stabilised by vegetation would predict an increase in elevation with time. It is, thus, tempting to see in the spatial zonation of species an expression of successional changes through time. Such an interpretation provided the basis for Chapman's (e.g. 1941, 1974*a,b*) comparison of marshes. However, there are few studies which have documented long-term changes in saltmarsh vegetation or have reconstructed past vegetation from stratigraphic evidence so that most successional diagrams must be regarded as hypothetical.

In some circumstances, there is clear evidence that present zonation patterns do not reflect past successional processes. The best example of this is provided by the spread of *S. anglica* in northern Europe. In some cases, *Spartina* marsh has developed at sites previously lacking marsh, but very often it has extended to seaward of pre-existing communities. Although these communities now appear in a zonation at higher elevations than *Spartina*, they did not develop in a successional sequence through a *Spartina* stage. As the *Spartina* marsh develops and matures, it in turn may provide a habitat open for invasion by other species but, in most cases, it is not yet possible to determine whether subsequent development will repeat historic patterns. Although the spread of a 'new' species provides an exceptional test to any supposed general rule relating zonation to succession, the great diversity from site to site in species' behaviour also indicates that caution is required in interpreting zonations. This diversity (which also requires that considerable assumptions about 'normal' sequences have to be made in order to produce a generalised zonation diagram for a given region) is often not immediately explicable, but can sometimes be correlated with local peculiarities in environmental conditions. This suggests that the species' sequence in the evolution of a marsh will be influenced by many factors which may change over time. In many estuaries, changes in sediment loads and in water and sediment chemistry have occurred as a consequence of industrialisation. While possible affects on marshes are poorly documented, *a priori* these environmental changes may well influence succession. In addition, many marshes have undergone changes in management practices which will have influenced species composition and affected the possibilities for transitions between communities (compare the changes in grazed and ungrazed plots in Bridgwater Bay – see Fig. 1.26).

In a few instances, very rapid development of saltmarsh has been monitored (Ranwell 1964*a*; Packham & Liddle 1970), but in most cases community boundaries appear to be relatively stable for decades. Rapid

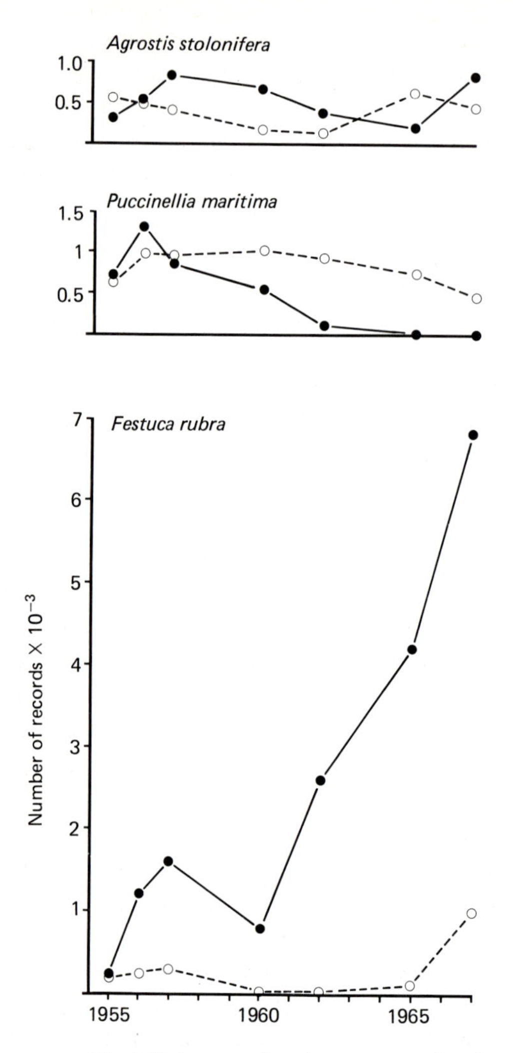

Fig. 1.26 A comparison between grazed and ungrazed plots at similar elevation at Bridgwater Bay, Somerset, UK. On the grazed marsh, ○ – – – ○, *Puccinellia* persists but on the ungrazed marsh, ● —— ●, it is replaced by *Festuca rubra*. In 1959 there was a severe summer drought which possibly explains the decline in records for *Festuca* on the ungrazed marsh. (From Ranwell 1968.)

vegetation changes have been recorded following changes in management (e.g., commencement or cessation of grazing) or some form of physical disturbance (storm damage or local death of species in hard winters – see Beeftink 1977*a*) but these changes, while they might determine the course of future succession, are not in themselves related to any overall successional scheme. A long-term study of permanent plots on the Wadden Sea Island of Terschelling suggested that, over a 30-y period, there were considerable changes in species composition within zones, but that interchanges between zones (which would classically be interpreted as succession) were few (Roozen & Westhoff 1985).

The physical development of a marsh is a constant process, even though the rate of accretion varies over time (with the incidence of storms, seasonally, and over the long-term, declining as elevation increases). The evidence would indicate that succession is not a matter of steady incremental change, rather there may be certain threshold conditions which prompt rapid change (Ranwell 1968; Packham & Liddle 1970) while, for most of the time, there is comparative stability. Randerson (1979), in modelling marsh development in the Wash, prescribed lower limits to each species and the resulting simulations indicated brief periods of rapid change in species composition and longer periods of stability. The time scale of the original simulations was, however, too long compared with the known age of the Wash marshes, but this was sensitive to changes in accretion rate. If succession really is a 'punctuated equilibrium', rather than a continuous process, the nature of the triggers which permit transitions between zones remain to be elucidated. For long periods of time, it may be more appropriate to recognise zonation as a steady-state phenomenon reflecting an environmental gradient (Lugo 1980; Wiegert 1979), with various temporal changes occurring at different elevations but with no overall directional change. Over a longer time interval, an overall directionality might be recognised, permitting zonation and succession to be related, but, at higher elevations, the greater number of possible species combinations makes it difficult to make generalised predictions about the course of succession – a situation summed up by Miller & Egler (1950): 'The present mosaic may be thought of as a momentary expression, different in the past and destined to be different in the future and yet as typical as would be a photograph of moving clouds'.

The description of saltmarsh vegetation

Vegetation, in its composition and structure, integrates the effects, both at the present and in the recent past, of environmental factors on the flora. Vegetation in its turn provides the environmental structure occupied by the fauna. Variation in the vegetation cover therefore provides a readily observable means of assessing the environmental complex of a particular marsh. Geographic patterns of variation in vegetation are discussed in more detail in Chapter 3. In this section the recognition of plant communities is discussed. An introduction to vegetation survey methods appropriate to saltmarshes is provided by Dalby (1987).

There are many descriptive accounts of saltmarsh vegetation; some involving single sites, others larger regions. The criteria used to delimit plant communities in these studies have been numerous, but two broad approaches can be recognised, those utilising structural/physiognomic attributes and those requiring floristic data.

The use of structure and physiognomy to delimit vegetation types underlies the recognition of vegetation formations and the classical subdivision of the earth's vegetation into a few broad phytogeographic zones. At this level of analysis, the initial similarity in appearance of saltmarshes precludes subdivison into more than a very small number of categories and saltmarsh is treated as an azonal (or transzonal) vegetation type. Despite the difficulty of defining units using structural criteria, there have been some attempts to accommodate saltmarsh vegetation in structural classifications (see for example Specht 1981*a*,*b*; Kirkpatrick & Glasby 1981).

The use of physiognomic data (referring to morphological and anatomical features of plants) may permit a basically structural classification to be considerably refined. This has proved the case in studies on rainforest (Webb, Tracey & Williams 1976). There has been no study of saltmarsh vegetation to see whether appropriate physiognomic attributes for use in structural/physiognomic classification can be defined.

It is possible to recognise three categories of saltmarsh plant communities on the basis of their dominant growth form:

Herb communities
Communities dominated by graminoid monocotyledonous plants (grasses, sedges and rushes).
Dwarf-shrub communities.

At most sites, all three types of community co-exist, although herb communities are generally limited in extent. However, there is geographical segregation in the importance of different community types. Arctic marshes are predominantly covered by monocotyledon (grass and sedge) communities, while dwarf shrub communities are most extensively developed along arid and semi-arid tropical and subtropical coasts.

There is some differentiation between communities and between geographical regions in representation within the flora of the various Raunkiaerian life forms (see p. 88). The geographical pattern in the distribution of life forms is similar to that revealed by consideration of floristics and vegetation, but the groupings are less obviously defined.

For the foreseeable future, it would seem that structural, physiognomic or growth form classifications will provide too coarse a subdivision of the macroscale variation in saltmarsh vegetation to be useful for most purposes. Floristic classifications, with potentially a much larger number of readily observed descriptive attributes, could provide a finer dissection of the macroscale variation.

A large number of methods of describing and classifying plant communities on the basis of floristic criteria have been developed. At one time and another, most have been applied to at least some examples of saltmarsh vegetation. Reviews of the methodology of plant community classification are provided by Shimwell (1971) and Whittaker (1973).

In the Anglo-American tradition, floristic classification has placed particular emphasis on community dominants, after which communities are also named (see, for example, Tansley 1939). Such an approach has a number of appealing features. If classification is on the basis of dominant species only, the non-specialist user of such a classification need learn to identify relatively few species. The number of community dominants being limited, a dominance-based community classification is likely to define relatively few communities. Again, this is of value to the non-specialist user; the classification is clear cut, easy to learn and an aid to communication.

On the other hand, the use of dominant species as the basis of vegetation classification has a number of disadvantages. Within vegetation units defined by dominants, there may be considerable variation in the less-abundant species (as in the case of *Juncus maritimus* communities – Adam, 1977*b*). In addition, particularly in surveys, the importance of dominants in classification frequently leads to under-recording of non-dominant species – from many accounts it is impossible to compile comprehensive species lists for the sites concerned.

The use of dominants as the basis for defining communities presupposes that it will be possible to recognise one or a few species as being the dominants of any particular community. In the case of saltmarshes, this is frequently the case, but it is not invariably so. In order to delimit communities when there is no clear dominant, recourse is frequently made to some non-floristic descriptive terms. The disadvantage is that these terms frequently lack precision and may be used as blanket descriptors. An example of such a name which has, over the years, been applied to very different assemblages of species is 'General Salt Marsh Community' (GSM). The term was originally introduced in England to accommodate communities in the mid-marsh without a clear dominant species (Tansley 1911). Nevertheless, it defined a heterogeneous assemblage of communities. Chapman (1934), in describing the plant communities of the North Norfolk marshes, used the term GSM for an homogeneous, distinctive herb-dominated community in which grasses are never a major component; a community which would have been included within the GSM of Tansley (1911), but which was only a subset of the earlier unit.

The definition of GSM, having been narrowed, has subsequently been broadened to cover communities differing considerably in floristic composition, but united only by the lack of a dominant species (Chapman 1960*b*). A GSM has been recognised in studies on saltmarshes outside Europe, communities differing totally in composition, even at the generic level, from the original GSM (Chapman 1960*b*; Chapman & Ronaldson 1958). There is an implication in some accounts that, despite the differences in species composition, communities included within a category are, in some (undefined) sense, similar. However, there is an almost total absence of data on tidal regime, productivity and regeneration of the various communities included under the umbrella of the GSM, so it is not possible to confirm that, in a functional sense, all these communities are vicariant.

It is appropriate for the basis of classification to be uniform across all communities; a classification based on the total floristic composition of vegetation would allow subdivision of more complex communities while at the same time permitting the recognition of monocultures. A full floristic classification would also have the advantage that its implementation requires more detailed study of the vegetation than a simple dominance classification.

A number of different approaches to the delimitation of plant communities defined by their full floristic composition has been developed (see Shimwell 1971; Westhoff & van der Maarel 1973; Mueller-Dombois & Ellenberg 1974). However, the aims and assumptions of the various methods currently used are similar.

Vegetation varies continuously in both space and time. However, the vegetation at any particular point is not a random assemblage of species, but reflects the selection, by the action of environmental and biotic factors, of those species from the available flora capable of co-existing under prevailing conditions. Certain combinations of species occur together more frequently than would be expected by chance. Classification resolves into:

The detection of these recurring groups of taxa
The bringing together of those stands containing particular recurrent groups of taxa (Mueller-Dombois & Ellenberg 1974)

Plant communities, defined in terms of recurrent groups of taxa, may encompass a range of variation, depending on the size of recurrent group of taxa compared with the total flora of the stands. Some of this variation may be 'noise' due to the essentially random distribution of rarer taxa but, in other instances, it may indicate patterns of environmental variation which, at present, cannot be interpreted.

A dichotomy has arisen between 'classical' phytosociological methods and numerical classificatory methods for the recognition of communities. Both approaches seek to delimit groups of stands with similar (but not necessarily identical) species composition. Numerical approaches (such as applied to saltmarsh vegetation by Gray & Bunce 1972), while effective in delimiting communities, tend to be used in such a way as to limit the possibility of direct comparisons between studies. The aim of the classical schools of phytosociology is to produce a framework applicable over a wider geographical area than the bounds of a single site. If the range of communities at any particular site can be accommodated within some agreed classification, then it becomes possible to compare sites on a uniform basis. However, there is no fundamental reason why numerical studies should not produce results which can be compatible with existing classificatory schemata. Regardless of the approach used to recognise communities, it should be possible to piece together various studies to produce a uniform classification.

While there are obvious advantages in having a system of saltmarsh plant communities defined by their floristic composition, this aim is not easily achieved. Although there are some species-rich saltmarsh communities, in general stands of saltmarsh vegetation are species-poor and individual species have a wide ecological amplitude.

At many sites, saltmarsh vegetation may present a mosaic appearance, the component parts of which would ideally be discriminated in any classification. However, while some of these visibly distinctive units may be easily defined in terms of presence and absence of particular species, others may be distinguished only by quantitative changes in the cover of certain species. Qualitative differences between communities are unambiguous; communities differing only quantitatively must be defined arbitrarily. In addition, patterns of quantitative change between communities are rarely recurrent between sites (Adam 1981*a*). For studies on restricted areas, it may be appropriate to describe communities of local occurrence. However, if each marsh has its own unique communities, the advantages of vegetation classification in simplifying the complexity of vegetation and permitting comparisons between areas are lost. For comparative purposes, it is desirable to define the basic vegetation units (associations) such that they can be delimited qualitatively.

Phytosociology of saltmarsh vegetation

A number of different schools of phytosociology developed in Europe in the early twentieth century (Shimwell 1971). The original distinctions between the various schools have become blurred over the years and, while there are still national differences in emphasis, the methodology outlined by Westhoff & van der Maarel (1973) and Mueller-Dombois & Ellenberg (1974) is the basis of most modern studies. Compatible phytosociological surveys are available from a number of European countries (Beeftink 1965; Gillner 1960; Tyler 1969*b*; Géhu 1976; Adam 1981*a*); in Japan (Miyawaki & Ohba 1965, 1969; Ishizuka 1974); and in Australia (Bridgewater 1982; Adam, Wilson & Huntley 1988). The phytosociological approach has not achieved wide acceptance in North America, but saltmarsh vegetation is described in the survey of Long Island by Conard (1935), while Thannheiser (1984) has provided a phytosociological treatment of coastal vegetation in eastern Canada.

The basic unit in the phytosociological system, the association, is an abstraction; a synthesis of data from a number of stands of vegetation which have a certain overall similarity. The concept of a particular association is expressed in the form of a table in which the species composition of a number of quadrats (relevés, Aufnahmen) is presented (Table 1.1). The association table illustrates not only the consistency of occurrence of the species defining the association but also the variation in representation of other taxa within the concept of the association. Associations are named after one or two species, normally from within the character combination.

In defining plant communities in qualitative terms, use is often made of character species. The latter are defined as those which are relatively restricted to stands of a particular community and, therefore, serve to characterise it and indicate its environment (see Westhoff & van der Maarel 1973). In saltmarsh vegetation, the wide ecological tolerance of many species means that the number of communities which can be defined by character species is limited. Saltmarsh communities are defined, not by the behaviour of individual species, but by the occurrence of a total assemblage of species – the 'character combination' (Kencombinatie – Westhoff & den

Held 1979 – 'differentiating species combination' of Beeftink 1962, 1965). The character combination is a group of taxa whose joint occurrence is exclusive to a particular vegetation type without any of the taxa necessarily being a character taxon.

Associations are conventionally grouped together into a hierarchy of higher units (Table 1.2). The basis for grouping units at particular levels in the hierarchy is floristic similarity. However, at the class level, the number of species in common to all subsidiary units is frequently small, and ecological criteria (habitat or community structure) may play a part in determining the limits to a particular class.

The adoption of a hierarchical framework has both advantages and disadvantages. The hierarchy gives a degree of flexibility in making comparisons and generalisations about vegetation. In comparing very different vegetation types, it may be appropriate for the comparison to be in terms of generalities at the class level. Within a particular habitat type, comparisons may be made at order or alliance level. Thus, in discussing saltmarshes in northern Europe, the mid and upper marshes with well developed Armerion communities can be contrasted with the lower marsh Puccinellion. In addition, the number of associations recognised is so large that some form of 'filing system' is required and a hierarchial system is ideal for this (see Table 1.2).

On the other hand, the pattern of variation between communities is complex. The relationships between communities would best be expressed in multidimensional classification reflecting a number of floristic (and, by implication, environmental) gradients (Webb 1954). This is impractical, but a hierarchial classification is unidimensional. While the axis of floristic variation expressed in the hierarchy frequently reflects a major environmental gradient, there is a danger that too ready acceptance of the classificatory framework may inhibit exploration of other patterns of variation.

The nature of saltmarsh vegetation poses particular problems in designing an appropriate classificatory framework. The wide ecological amplitude of certain saltmarsh species makes it possible to trace floristic links between most saltmarsh communities. Upper saltmarsh communities may consist largely of species which are also found in inland vegetation. In any phytosociological hierarchy, should these communities be related to those of the lower marsh, or should they be segregated from the saltmarsh and associated with various widespread inland communities? Whatever course is chosen, some upper saltmarsh communities will continue to fit uneasily in the classificatory framework.

Chapman (1953, 1959, 1960*b*, 1974*a*,*b*, 1977*a*) proposed a phytosocio-

Table 1.1. *An example of a phytosociological table*

	Relevé code no.																					
	132	175	184	325	153	323	147	262	51	264	52	50	185	70	187	54	266	12	265	267	281	53
Area (m²)	4	16	16	16	4	4	4	16	16	4	4	16	4	4	16	4	16	4	4	16	16	4
Height (cm)	5	25	25	10	6	4	15	9	3	25	40	3	20	15	20	5	4	20	5	40	30	40
Cover (%)	60	70	80	60	60	60	90	40	20	70	80	20	50	40	30	40	40	70	50	80	60	80
Number of species	13	11	13	11	5	13	11	6	7	8	7	8	12	7	8	6	5	9	8	10	7	5
Frankenia laevis												2	4	4	5	5	4	4	4	3	5	5
Suaeda vera	2	5	6	4		1	3	3	3	7	7	1	5	5	4			4	6	7	5	6
Limonium bellidifolium	3			1				3		4		2	3	1				6	4	5	4	4
Suaeda maritima	3		3	3						3							2	4	4	6	3	
Puccinellia maritima	5	3	7	5	4	5		3	3	3	4	4	4	4	3	5	5	5	5	5	6	4
Halimione portulacoides	3	6	5	2		1	2	3		5	5	1	5	1	2	1		2	4	4	5	7
Armeria maritima		4	4	3	4	3	7	3	3	3	2	4	3	4	2	5	4	2	2	2		
Limonium binervosum	2		4	5	7	7	5	6	3	2	3	4	4	3	2	4	5	2	3	3	1	
Plantago maritima	3	3	4	2	4	4	2					2	3									
Festuca rubra	5	5	5	6	5	6							3									
Artemisia maritima	4	1	4			2	3															
Cochlearia anglica										2	3		2									
Cochlearia officinalis													2									
Salicornia agg.	3										3									3		
Cochlearia danica		2							3					3	3							
Spergularia media			2																	1		
Triglochin maritima			2																			
Ammophila arenaria									2													

Agropyron junciforme									3						
Algal mat	9														
Limonium cf. *L. vulgare*	5	2	2												
Agrostis stolonifera						4									
Glaux maritima	3					3									
Spergularia marina		3		3		2				2					
Agropyron pungens		2	2			1	2								
Sagina maritima		2				3	3					2			
Plantago coronopus							3								3
Honkenya peploides							4								
Cladonia spp. inc. *C. rangiformis*							9								

Two associations are represented, the Suaedo–Limonietum binervosi and the Halimiono–Frankenietum laevis. Both associations form part of the alliance Frankenio–Armerion.
Data from Adam 1976*a*; further details provided in Adam 1981*a*.

Table 1.2. *A conspectus of noda forming part of the class Asteretea, to which much of the vegetation of north European saltmarshes belongs*

ASTERETEA TRIPOLII Westhoff & Beeftink 1962
  Glauco–Puccinellietalia Beeftink & Westhoff 1962
    Puccinellion maritimae (Christiansen 1927 p.p.) R.Tx. 1937
      Puccinellietum maritimae (Warming 1906) Christiansen 1927
      Halimionetum portulacoidis (Kuhnholtz-Lordat 1927) Des Abbayes & Corillion 1949
      Puccinellio-Halimionetum portulacoidis Chapman 1934
    Armerion maritimae Br.-Bl. & De Leeuw 1936
      Juncetum gerardii Warming 1906
      Juncus maritimus–Oenanthe lachenalii ass. R.Tx. 1937
      Artemisietum maritimae Hocquette 1927
    Eleocharion uniglumis Siira 1970
      Blysmetum rufi (G.E. & G. Du Rietz 1925) Gillner 1960
      Eleocharetum uniglumis Nordhagen 1923
    Puccinellio–Spergularion salinae Beeftink 1965
      Puccinellietum distantis Feekes (1934) 1945
    Frankenio–Armerion Géhu & Géhu-Franck 1975
      Suaedo-Limonietum binervosi Adam 1976
      Halimiono-Frankenietum laevis Adam 1976

In north temperate Europe, the major order within the class is the Glauco–Puccinellietalia; this can be divided into several orders. A major distinction is between the Puccinellion, communities of the low-marsh, and the Armerion, communities of mid- and upper-marshes.

logical classification which accommodated saltmarsh communities throughout the world. Unfortunately, most of the association names presented in Chapman's scheme have not been supported by association tables, nor is it possible to gain any detailed picture of the composition of the communities from the publications. Although a global phytosociological classification of saltmarsh vegetation is a desirable aim, any such scheme must be founded on associations established on the basis of appropriate floristic data.

General features of saltmarsh vegetation

Although there are considerable differences in species and community composition between different marshes, certain general features of saltmarsh vegetation can be recognised.

The most frequent generalisation made about saltmarsh vegetation is that it is species-poor. While the species paucity may have been overemphasised (as some upper marsh communities are, by grassland standards, relatively rich), much saltmarsh vegetation is species-poor and the impression of paucity is compounded by the obvious dominance of only a few species.

The lower marsh is universally species-poor in terms of vascular plants, with the lowest communities generally consisting of a very small number of species. Often the lowest community forms a clear zone dominated by a single species, but there may occasionally be a coarse mosaic of communities with each community forming large patches dominated by a single species.

With increasing elevation, species diversity tends to increase. There are some notable exceptions to this rule. In south east England, for example, the upper parts of many marshes are extensive pure stands of the shrub *Halimione portulacoides* with the only relief from the monotony provided by pure stands of the grass *Elymus pycnanthus* (*Agropyron pungens* auct.).

Not only does the species richness within communities tend to increase in the upper marsh, but so does the number of communities. Much upper-marsh vegetational variation can be related to recurrent microscale environmental patterns, such as the change from creek levee to inter-creek basins. Similar environmental patterns are also found in the lower marsh, but are frequently less obviously reflected in the vegetation. This may be a consequence of the species paucity of the lower marsh, but may also suggest that the frequency of tidal flooding in the lower marsh overrides any influence of point-to-point variation in the environment.

The extreme upper marsh, and the transition to other habitats, is a region of considerable environmental heterogeneity, strong environmental gradients and considerable temporal variability. This narrow zone supports a wide range of communities and a rich flora; and, in both California (Macdonald & Barbour 1974; Macdonald 1977*b*) and Australia (Bridgewater & Kaeshagen 1979; Adam 1981*b*), has proved to be particularly vulnerable to invasion by aliens. Patterns of diversity within the zone vary. Where the influence of drift litter predominates, dense stands of nitrophilous species may occur, but each species only dominates locally, and along the drift line of the whole marsh there may be a series of patches of vegetation each dominated by different species. On the other hand, where hypersaline conditions develop seasonally, there may be rich, if sparse, flora of small annuals with no dominants. Similarly, freshwater flushing over shallow soils tends to favour a diverse flora without clear dominants.

Some of the diversity of upper-marsh vegetation may be the product, not of extrinsic environmental heterogeneity, but of local environments created by the growth of plants themselves. Many saltmarsh plants have the ability to spread by vegetative means, leading to the formation of clones which may be of considerable extent. Tall-growing types such as *Juncus* spp. or some species of Cyperaceae may create a microclimate very different from that of the adjacent grassland, so allowing a distinctive subsidiary flora to flourish.

Species – area relationships

Saltmarsh takes the form of discrete 'islands' of habitat, sharply distinct from neighbouring terrestrial communities. It would be of interest therefore to investigate the relationship between plant species-richness and area of saltmarshes, but such an exercise does not seem to have been reported.

Rey (1984) tested various predictions of island biogeographic theory (MacArthur & Wilson 1967) using small islands of pure stands of *S. alterniflora* in Florida and studying the arthropod fauna before and after experimental defaunation. Prior to defaunation there was a highly significant species–area relationship among the islands and this was re-established 20 weeks after the start of the experiment. Whether a similar species–area relationship is shown by other fauna groups and in other marsh types remains to be tested.

Smith & Duke (1987) investigated correlates of mangrove species-richness in estuaries in northern Australia. Data on the intertidal area in each estuary were not available so as a proxy the length of the tidal estuary was used. In north western Australia this measure was not an important predictor of mangrove species-richness. In north east Australia, however, there was a significant correlation between estuary length and mangrove diversity. Smith & Duke (1987) suggested that the difference between regions was related to climatic factors. In the north east the climate is more stable and conducive to mangrove growth which permits greater niche differentiation. In the north west the effects of the harsher, more seasonal climate appear to override any species–area affect.

Data on species richness of British saltmarshes were collected by Adam (1976*a*; see also Adam 1978, 1981*a*). There is no indication in these data of a significant positive correlation between marsh area and species richness (either for total species richness or the halophytic element alone). For the halophytic element there is little variation in species number between sites while in the case of total species the richest marshes are Type C (see Adam 1978 and Chapter 3) which are mostly small in area. If the three marsh types recognised by Adam (1978) are studied separately there is still no consistent increase of species with increasing area. In Types A and B there is an increase in species with increased number of communities suggesting that habitat diversity rather than area is the significant factor controlling species richness. This relationship is not shown in Type C marshes, where in general there are relatively few communities, one of which, the Filipendulo–Iridetum may be very species-rich. Unfortunately precise data on the area of individual stands are not available so the species–area relationship within communities cannot be investigated. Rough estimates of areas from sketch maps and field notes would suggest that at least in low and mid marsh communities there is no relationship between stand area and species richness.

The area of saltmarshes can change very rapidly particularly during erosive phases. Although there has been no systematic recording the limited evidence from Morecambe Bay (p. 31) does not suggest that there is loss of vascular species following reduction in marsh area.

Mangroves

This book is primarily about saltmarsh. It is appropriate in this introduction, however, to make some reference to mangroves.

Mangroves occupy the tropical and subtropical habitat equivalents of the saltmarshes of more temperate regions. Mangrove stands are dominated by woody species which, under optimal conditions, may form tall closed forest with the canopy trees more than 30 m tall. Other mangrove stands are low shrub thickets, differing little in structure from communities dominated by the taller shrubby chenopods, which are regarded as saltmarsh.

The mangrove flora consists of two phytogeographically distinct groups. One is restricted to the Indian Ocean and West Pacific, while the other, and much smaller group of species, occurs in the Atlantic and on the west coast of South America (Chapman 1975, 1977*a*).

Taxonomically, the mangrove flora is diverse, with representatives from a number of different families (some 70 species in about 20 families). Some families are predominantly mangrove (e.g., Avicenniaceae); but, in other cases, only one or few species of a widespread family are mangroves (i.e. *Osbornea octodonta* in the Myrtaceae). Mangroves are recorded early in the angiosperm fossil record (Muller 1970) and the Rhizophoraceae are regarded as the dicot family showing the largest proportion of putatively 'primitive' character states (Sporne 1974). Nevertheless, most of the families with mangrove representatives today would not be regarded as closely related and, as with adaptations to the saltmarsh habitat, it would seem that the mangrove habitat has arisen independently a number of times.

The latitudinal limits of mangroves are generally regarded as being determined by the intolerance of most species to frost. Mangroves are best developed where the average temperature of the coldest month is above 20 °C and the seasonal temperature range is 5 °C or less. There is evidence that, towards the limits of distribution, infrequent severe frosts do cause death of mangroves (for discussion see Chapman 1975; Walter 1977), but detailed investigation of the climate of mangrove stands has not been carried out (Oliver 1982). Although mangroves can withstand frequent tidal flooding and tolerate soil salinities as high as that of seawater, in general they do not seem to tolerate hypersaline conditions. Except in the upper estuary or in sites influenced by percolating groundwater, soil salinities towards the landward edge of mangroves may reach high values. Mangroves in these zones may be stunted or totally absent, being replaced either by saltmarsh or salt flat (see, for example, Thom, Wright & Coleman 1975). The relative species paucity of tropical saltmarsh floras may largely be a reflection of their restriction to this extreme habitat above the mangrove zone.

The success of mangroves within the tropics provokes speculation as to why there are no temperate equivalents. The essential features of the mangrove habitat are a generally high, but variable, soil salinity and a poorly drained and frequently anaerobic substrate.

The taxonomic diversity of the mangrove flora indicates that a range of families possess the ability to evolve tolerance of these conditions. The families concerned in mangrove floras are essentially tropical or subtropical in distribution (at least so far as their arborescent members are concerned). Why should not families of more temperate distribution have evolved mangrove forms? In temperate regions, there is a restricted tree flora capable of growing in freshwater swamps. There have been few studies on the salt tolerance of temperate trees, but a range of woody species which occur on seacliffs and headlands can tolerate airborne salt spray. Recent studies on *Eucalyptus* have shown that seedlings of a small number of species can grow in saline solution culture (Blake 1981). (A number of species survived salinities in excess of 300 mmol^{-3} and so could be called halophytes.) Nevertheless, although eucalypts may occasionally be found around inland saline habitats, they are not a feature of saltmarshes. In southern Australia, brackish swamps may support woodlands of paperbarks (*Melaleuca* spp. – Myrtaceae) which might be regarded as temperate mangrove communities; but, although tolerant of periodic flooding, the salt tolerance of these communities seems limited.

Tomlinson (1986) suggests that physiological tolerance of saline conditions requires functional leaves to be present throughout the year and that mangroves are thus excluded from regions where climatic and other constraints favour deciduousness. Nevertheless, some mangroves are at least partially deciduous (for example *Xylocarpus* spp.) and Clough, Andrews & Cowan (1982) conclude that there is nothing currently known about the physiology of mangroves which would result in the growth form being confined to the tropics and subtropics.

2

The saltmarsh biota

Introduction

On casual inspection, saltmarshes might appear dull, monotonous places. After more detailed investigation, the range of organisms which will have been encountered is very diverse. The great diversity reflects the fact that saltmarsh straddles the boundary between land and sea and provides habitats for both terrestrial and marine organisms.

In this chapter, the biota of saltmarshes is introduced, with special reference to the flora. The relatively brief treatment of the fauna does not imply that animals are unimportant components of saltmarsh ecosystems. However, the biology of the saltmarsh fauna has been reviewed by Daiber (1982), a publication to which readers are referred for a more comprehensive treatment.

The genecology of some saltmarsh vascular plants is discussed. At a time when plant physiologists are increasingly looking to saltmarshes for experimental plants and systems and ecologists are seeking generalisations about species' behaviour to feed into ecosystem models it is appropriate to emphasise that species are inherently variable and that this variability is the basis for ongoing evolution.

The flora

The vascular flora

Saltmarsh floras are small when compared with those from most other habitats but it is nevertheless difficult to compile complete floristic lists for many parts of the world. In part, this reflects lack of scientific recording from many coasts but, more fundamentally, it is due to the lack of agreement regarding the definition of saltmarsh and the constitution of its flora.

The lower limit of saltmarsh can be defined unambiguously as the seaward margin of vascular plant communities, excluding those composed of seagrasses or other permanently submerged species. Intermingling of seagrass and saltmarsh communities is extremely rare. Determining the upper limit of saltmarsh is much more difficult unless the decision is pre-empted by an artificial structure such as a seawall or there is naturally a sharp break in slope between saltmarsh and the hinterland. The immediately obvious choice of a boundary would be the level of the highest tide, but this is not constant, varying from year to year and being strongly influenced by meteorological conditions. In consequence, there may be a broad zone which, if saltmarsh is defined as strictly intertidal, is only intermittently saltmarsh. Some of the species in this zone may be opportunistic inhabitants, temporarily disappearing after tidal inundation; others may be permanent residents and characteristic of the habitat. However, the literature on saltmarshes largely neglects consideration of the upper marsh transition and its flora is poorly documented.

Many of the plants which grow on saltmarshes are termed halophytes, as distinct from plants of non-saline habitats, which are known as glycophytes. [Jennings, 1976, has argued that glycophyte is an inappropriate, and potentially misleading, term which should be replaced by mesophyte. While the criticism is merited, the term glycophyte is so well established that to replace it carries the danger of engendering greater confusion than any caused by its retention – glycophyte is here used in its traditional sense.] Frey & Basan (1985) regard the presence of halophytes as a necessary part of the definition of saltmarsh.

Unfortunately, like many useful concepts in ecology, problems arise when attempts are made to define the term and there are a number of different definitions of halophyte. The most all-embracing of these is an ecological definition favoured by Flowers (1975), Flowers, Hajibagheri & Clipson (1986) and Jennings (1968, 1976) under which halophytes are plants which complete their life cycle in saline environments (i.e., the native flora of saline environments; Jennings 1976). Thus, the saltmarsh habitat is defined independently of the flora and all plants within that environment are halophytes. While this definition has the virtue of simplicity, some limiting value of salinity must be agreed, above which the habitat can be regarded as saline. In addition, salinity within a marsh is variable both spatially and temporally so that determining the status of any species would require an assessment of the range of salinity conditions it might encounter. Certain problems arise if categorisation is at species level – for example, *F. rubra* is a component of the natural flora of many northern hemisphere saltmarshes, but not all genotypes of the species would be capable of survival in saline habitats. There are also many physiological definitions of halophyte in which halophytes are not only distinguished from glycophytes, but in which frequently the halophyte category itself is subdivided. [The physiological definition of halophyte is discussed further in Chapter 4.]

It is clear from much of the literature that, despite claims of adherence to various physiological definitions, 'halophyte' is often applied in a much narrower, essentially ecological, sense. In these accounts halophytes are those species for which saltmarsh is a major and, in many cases, only habitat, while the glycophytic element in the saltmarsh flora comprises those species more generally thought of as occurring in inland, non-saline habitats (such a definition presupposes an external, non-biological definition of saltmarsh, contrary to the suggestion of Frey & Basan, 1985). In order to distinguish between this definition and ecophysiological definitions in this account, the terms halophytic and halophytic element will be applied to species more or less restricted to saltmarsh and other saline habitats; halophyte will be used in the wider sense of any species occupying saline habitats. [The disadvantages of this treatment are considerable, but it does fit in with pre-existing usage rather better than the all-embracing ecological definition proposed by Jennings and also has the virtue of not requiring the invention of novel, and potentially confusing, terms.]

The terms glycophyte and glycophytic element are used in their traditional sense of being species more widespread in non-saline habitats than on saltmarshes (glycophytes growing in saltmarshes are thus also halophytes). Nevertheless, it is hoped that the particular sense applicable to the terms in any instance will be obvious. Species of the glycophytic element may be as salt tolerant as many members of the halophytic element, and frequently the saltmarsh populations of the glycophytic species are genetically adapted to the saltmarsh habitat (pp. 124–127).

In many saltmarshes, but not all (see Chapter 3), the halophytic element predominates, at least in the lower and mid marsh. Characteristically, the halophytic element is made up of representatives of very few families and, additionally, the majority of individual species in the group are widespread, both ecologically within the zones of the marsh, and geographically. The families which supply the major part of the halophytic flora are the Gramineae, Chenopodiaceae (particularly the Salicornieae – Scott 1977), Juncaceae and Cyperaceae (see Beeftink 1984); members of the Plumbaginaceae and Frankeniaceae, although less widespread and often more minor components of the vegetation, are highly characteristic of the halophytic flora. There is variation in the representation of the halophytic families between the major types of saltmarsh (see Chapter 3) but within a particular type, even between geographically widely separated locations, there is a high degree of similarity (even to the extent that for each major marsh type it would be feasible to suggest 'assembly rules' at family level for the component communities).

In a survey of British saltmarshes Adam (1976*a* and unpublished data) recorded 325 species of vascular plants; when those species occurring only in the upper marsh at a single site are omitted, there are about 250 reasonably widespread species in the British saltmarsh flora. Of these, only 45 species constitute what would generally be thought of as the halophytic element (the exact size of the halophytic element depends on how species are defined – for example, whether *Salicornia europaea sensu lato* is split into its constituent taxa). These 45 species represent only twelve families with the best represented being Gramineae (10 spp.), Chenopodiaceae (7 spp.) and Cyperaceae (6 spp.). The total flora of 325 species includes members of 170 genera in 48 families, figures which probably coincidentally are similar to the 347 species from 177 genera in 75 families given by Duncan (1974) from the eastern United States. [Statistics for the halophytic flora of Europe are given by Beeftink (1984) – these data apply to a wider range of communities than saltmarsh alone.]

The halophytic element in Britain, although small compared with the total British flora is, nevertheless, comparable in size with that in other regions. On the west coast of North America, Macdonald (Macdonald 1977*a*; Macdonald & Barbour 1974) suggests that the maximum size of the halophytic element in any region is about 40 species, while Kirkpatrick & Glasby (1981) regard some 35 species as being characteristic of Tasmanian marshes. Altogether, the world halophytic flora of coastal saltmarshes probably numbers only a few hundred species.

The halophytic element is largest in temperate and mediterranean climates (Macdonald 1977*a*,*b*; Saenger *et al.* 1977), and is smaller both at high latitudes and in the tropics. In tropical Australia the halophytic element contains only seven species which constitute the bulk of total saltmarsh flora (Saenger *et al.* 1977). Tropical saltmarsh floras elsewhere are also small, but there has been little systematic recording. In the Arctic, in the upper tidal marsh, the glycophytic element is well represented.

While the glycophytic element in saltmarsh floras includes species of wide geographic distribution, it is often the case that they only occur in saltmarsh in part of their range (Duncan 1974). In Britain, this is strikingly demonstrated by a group of species characteristic of the upper saltmarshes in western Scotland. In this region, the halophytic element in the saltmarsh flora is much smaller than that in southern England (see Chapter 3). However, the upper marshes support species-rich fen-like communities

(Adam, Birks & Huntley 1977; Adam 1978, 1981*a*). A number of the species in these communities are approaching their northern limit in Scotland and, although occurring in inland communities in England, are strictly coastal in the north [such species would include, *inter alia*, *Samolus valerandi* (Ranwell 1972), *Lycopus europeus* and *Scutellaria galericulata*]. If a limited regional view were taken, then these species could be regarded as being part of the halophytic element which would then be comparable in size with that elsewhere. However, it has been usual to adopt a continental perspective when discussing these marshes and refer to the paucity of the halophytic element while virtually ignoring the rich upper marshes.

It is also difficult to characterise universally the ecological position of many members of the geographically widespread halophytic species. *Armeria maritima* is a species of the upper part of the marsh in mainland northern Europe and southern Britain, but is found at increasingly lower levels in northern and western Britain (Beeftink 1977*a,b*; Adam 1978, 1981*a*). To what extent this shift represents local population differentiation or a change in some major ecological factor remains unknown. At a wider scale, *Triglochin maritima*, one of the most widespread of all halophytic species, seems to be more prominent at lower levels in the marsh in the northern Pacific (in North America – Macdonald 1977*a,b*; McDonald & Barbour 1974; and in Japan – Ishizuka 1974) than in the north Atlantic where it is rarely a plant of the pioneer zones.

Rarity, endemism and aliens in saltmarsh floras

Rarity can have various meanings, depending on the spatial and temporal scales under consideration (see various authors in Synge 1981). In one sense, all plants restricted to saltmarshes could be regarded as rare in that they occur in a habitat which is, by definition, of limited extent. However, within this habitat many species are geographically widespread and, thus, from other points of view, would not be regarded as rarities.

Ranwell (1981*a*) points out that in Britain most rare species associated with saltmarshes are plants of the upper marsh and particularly emphasises the special nature of the flora in the transition between marsh and sand-dune (which he refers to as the arenohaline marsh). The mid and lower marshes in Britain are occupied by widespread species and, although a number of these reach geographical limits within Britain (Adam 1978), only one species, *S. maritima*, would be regarded as rare (Ranwell 1981*a*).

In northern Europe, many marshes have their upper limit defined by a seawall. Under appropriate management conditions, seawalls support a very rich flora, including a number of rare species (Beeftink 1975; Ranwell

1981*a*; Prince & Hare 1981). Many of these rarer species were presumably once more widespread in natural upper marsh – neutral grassland transitions prior to extensive marsh reclamation.

Even for these rare upper-marsh species, rarity is largely a function of the restricted nature of the habitat. Many of the species concerned are widespread geographically and narrowly restricted endemic species are very uncommon. However, the upper saltmarsh and habitats around the driftline are very prone to disturbance and many of the species characteristic of this zone have, undoubtedly, become rarer as a result of human activities. In some cases, species which may once have been more widespread are now limited to highly disjunct populations and are, thus, vulnerable to future disturbance.

The highest incidence of species of relatively restricted geographical distribution is found on saltmarshes experiencing mediterranean climates – particularly in the mid and upper zones subject to seasonally high salinities and at high latitudes. In the brackish marshes of the Baltic many endemic taxa below species rank have been recognised and these have limited geographical distributions (see Ericson & Wallentinus 1979).

Given the wide distribution of many saltmarsh species and the particular features of the environment, opportunities for the successful invasion and spread of alien species might appear limited. In most parts of the world this would be the case, with the exception of *S. anglica*. Ranwell (1981*a*) suggests that in Britain little threat from alien species exists and that deliberate introduction of additional species might be safely contemplated in management plans.

The position is similar elsewhere in Europe. Although very many alien species are recorded around docks, few are of more than casual occurrence and have not invaded natural communities such as saltmarshes.

In western North America (Macdonald 1977*a*,*b*; Macdonald & Barbour 1974) and Australia (Bridgewater & Kaeshagen 1979; Adam 1981*b*), introduced species are characteristic of a number of marsh communities, particularly in the upper marsh transition zone. The species concerned were probably spread in ships' ballast in the 19th century and, in some cases, are more successful in new continents than in the equivalent habitats of their original home. For example, *Parapholis incurva*, which is widespread and locally abundant in southern Australia in upper saltmarshes, is local in its occurrence in arenohaline marshes in Europe. In Australia, there is no evidence that the spread of introduced plants in the upper marsh has been at the expense of native species. To a limited extent, the aliens may be colonising microhabitats created as a result of changes in management

(Bridgewater & Kaeshagen 1979), but there is a case for arguing that there were 'vacant' niches available for which no native species was well adapted. A similar explanation has been put forward to explain the success of alien species in strand communities in California (Macdonald & Barbour 1974) and Australia (Heyligers 1984).

Apart from *Spartina* species, nearly all reported alien species on saltmarshes are restricted to the upper marsh. *Spartina* spp. are amongst the world's most aggressive aliens. *S. alterniflora* has become established, both naturally and as a consequence of planting (Bascand 1970), in various parts of the world. *S. maritima* may be an introduction in southern Africa (Pierce 1982), while *S. anglica*, discussed below, illustrates both the advantages and dangers of the release of new genetic combinations into natural environments.

Spartina anglica

The spread of *S. anglica* over the last century has so altered saltmarsh ecology in northern Europe and elsewhere that it is appropriate to discuss it at some length.

The common cord grass of northern European marshes was first collected in 1870 from Hythe in Southampton Water in southern England. Some twenty years later, the plant was observed to be spreading rapidly in the central Channel coast of England. Impressed with the ability of this plant, which had been given the name *Spartina townsendii* H. and J. Groves, to withstand frequent tidal submergence, and with the rapid rate of sediment accretion within stands of the species, several authorities promoted its planting in many estuaries. Reasons for planting included: the stabilisation of mudflats adjacent to shipping channels, the reduction of coastal erosion, and the promotion of accretion to aid reclamation of intertidal land. Extensive plantings occurred both in Europe and elsewhere (Ranwell 1967; Boston 1981; Chung 1982, 1985; Bascand 1970), and *Spartina* is now a major low-marsh species in northern Europe (Beeftink & Géhu 1973). From the initial plantings, spread by both vegetative means and seed, has occurred. The distribution and spread in the British Isles have been documented by Oliver (1925), Goodman, Braybrooks & Lambert (1959), Hubbard & Stebbings (1967) and Doody (1984*a*). (See Fig. 2.1.) The distribution at particular sites is discussed by Hubbard (1965), Marchant (1967), Taylor & Burrows (1968) and Smith (1982). The biology of *S. townsendii sensu lato* is reviewed by Goodman *et al.* (1969).

Prior to the discovery of the Hythe *Spartina*, two species of the genus had been recognised in Britain. *S. maritima* was a long-recognised native spe-

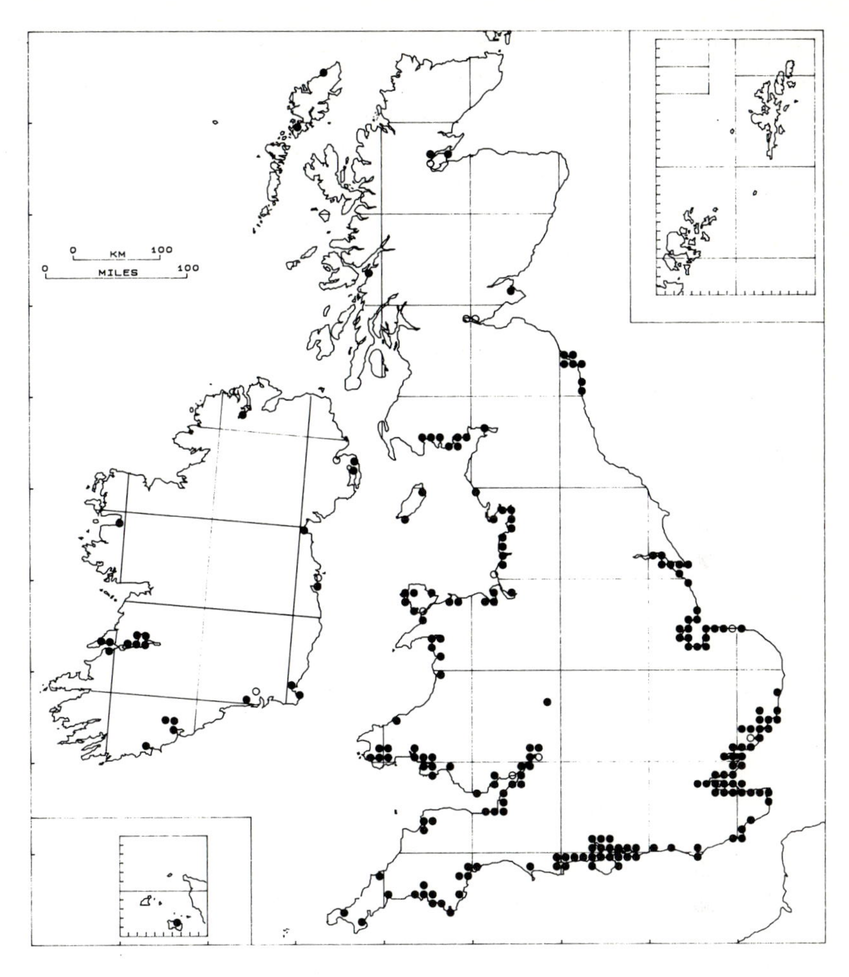

Fig. 2.1 Current distribution of *Spartina townsendii* and *Spartina anglica* in the British Isles [From BRC, Monks Wood.]

cies, restricted to south east England. It is difficult to judge from the early accounts how important a species it was, but it was probably rather local and not a major community dominant. The other species was the American *S. alterniflora*, believed to have been introduced in ships' ballast and first recorded from the River Itchen near Southampton in 1829. By the early years of the twentieth century, *S. alterniflora* was widespread and locally abundant on the Hampshire coast (Tansley 1911), but it is now virtually extinct in Britain (Marchant & Goodman 1969*b*).

The explosive spread of the newly recognised species of *Spartina* and its first discovery in an area where both *S. maritima* and *S. alterniflora* occurred led to the suggestion that natural hybridisation had taken place. This view, that the new species was of hybrid origin and that its successful spread was an example of 'hybrid vigour' soon became generally accepted. It was subsequently observed that the original specimens from Hythe, given the name *S. townsendii*, were sterile, while fertile forms were not collected until 1892 at about the time of the start of the natural rapid spread of the species. The sterile and fertile forms have subsequently been recognised taxonomically, the fertile form being given the name *S. anglica* C. E. Hubbard, while *S. townsendii* is restricted to sterile plants. The two species have often not been distinguished and the name *S. townsendii* applied to both forms. The existence of sterile and fertile forms was not recognised when material for planting was distributed. This may explain the relative failure of some plantings which might have consisted only of sterile individuals.

The most recent chromosome counts (Marchant 1968) suggest the following hypothesis for the origin of the *S. townsendii* aggregate:

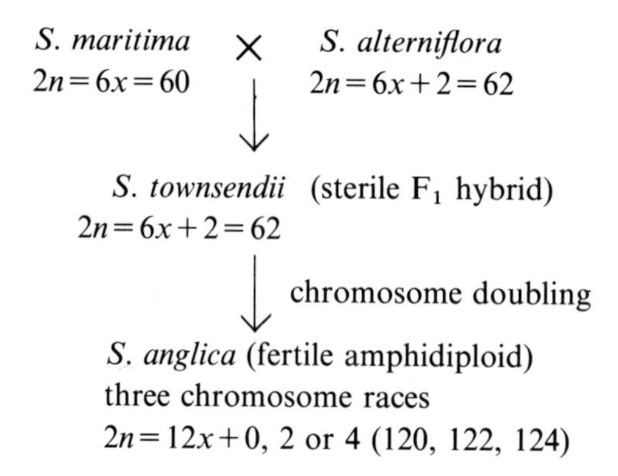

Marchant (1968) has also investigated plants with c76 and c90 chromosomes which, he suggests, may represent progeny of backcrosses as outlined below:

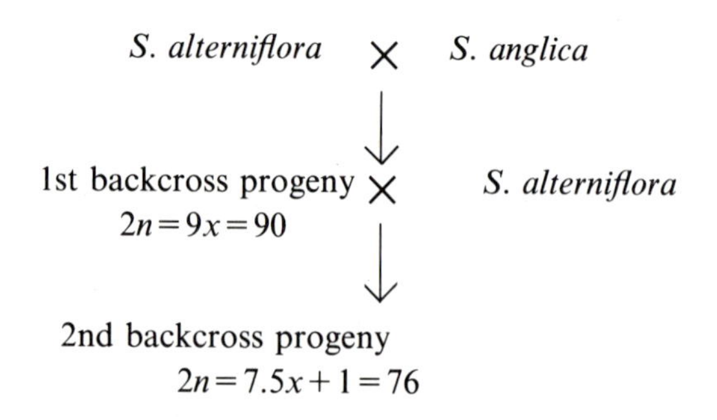

Plausible as this story is, it has not yet proved possible to recreate the putative hybrids artificially.

If *S. townsendii* is a sterile hybrid and *S. anglica* only recently arisen, then genetic differentiation within and between populations would not be expected. There is considerable variation between sites in plant height and stem density, but this may be of phenotypic origin. A very distinctive small form of *Spartina* in the Dovey estuary in Wales was reported by Chater & Jones (1951) and referred to as the 'brown dwarf' form (Chater 1965). Chater (1973) refers to this form as a mutant, but whether this is the true explanation of its origin is uncertain. Dwarf forms have also been recorded from New Zealand (Bascand 1970). Whether these are identical with those in Wales is unknown.

S. anglica has lived up to the claims of the proponents of its use in the reclamation and stabilisation of mudflats. However, the ecological consequences of the spread of *S. anglica* have been considerable and control and eradication, rather than further planting, have been the concern of many management authorities in more recent times.

S. anglica is more tolerant of tidal submergence than any other species on European saltmarshes. It has thus colonised mudflats which would otherwise have been incapable of supporting marsh. The area of intertidal mudflat has, therefore, been reduced as a consequence of the spread of *S. anglica*. Mudflats support a large fauna and the replacement of this community by a monoculture of *Spartina* with few associated animals has aroused concern. Evidence that the spread of *Spartina* may reduce the value of estuarine sites as habitats for wading birds has been provided by Davis & Moss (1984) and Millard & Evans (1984). The great tolerance of *S. anglica* to tidal submergence is further demonstrated in Westernport Bay, Victoria, where it has invaded mudflats to seaward of the *Av. marina* mangrove community (Bridgewater 1975; Boston 1981).

Locally around Britain, *S. anglica* has invaded and replaced communities of the seagrasses *Zostera noltii* and *Z. marina* (Ranwell 1981*a*; Bird & Ranwell 1964; Hubbard & Stebbings 1968; Haynes & Coulson 1982; Corkhill 1984).

Prior to the advent of *S. anglica*, the lowest marsh community at most sites in southern England was a monoculture of annual *Salicornia* sp. While *Spartina* can grow to seaward of *Salicornia*, it can also flourish at the same elevation. As a result, extensive *Salicornia* stands are now rare in estuaries with abundant *Spartina*.

The impact of *S. anglica* on its putative parents has, unfortunately, been little studied. *S. maritima* is now extremely rare in northern Europe. In Britain, its main occurrence is in pits and borrow pits in the upper marsh in south east England and it is only on Maplin Sands, Essex, that extensive development of stands in the low-marsh zone occurs (Adam 1978; Ranwell 1981*a*,*b*).

In northern Europe, *S. maritima* displays low vigour and partial seed sterility (Marchant 1967; Marchant & Goodman 1969*a*; Ranwell 1981*b*). It is apparently intolerant of high rates of accretion (Ranwell 1981*b*) and was severely affected in Holland by the exceptional frosts of the 1962–3 winter. *S. maritima* has a wide distribution between its current limits of 53°N in Britain and 35°S in southern Africa, although it may have been recently introduced into South Africa (Pierce 1982). Southern European and African plants are much more vigorous than those in northern Europe and, as Ranwell (1981*a*) suggests, at first sight might be thought to belong to a different species.

The evidence from local floras and herbarium specimens shows a considerable decline in the occurrence of *S. maritima* in southern England during the past century. There is evidence for the previous occurrence of a *Spartina* species either *S. maritima* or an unknown and now extinct form (Ranwell 1981*c*) in the Humber, further north than the current limit of *S. maritima*, between 2 and 3000 years ago (McGrail 1981). There is little evidence as to the former extent of *S. maritima* in British marshes. It is unlikely that it ever formed the extensive pure swards similar to those now formed by *S. anglica*. Rather, it seems probable that, in general, it occurred as scattered clones as it does today at Maplin.

The most widely held explanation of the recent decline in *S. maritima* is that it has been replaced by *S. anglica*. At many sites where *S. maritima* formerly occurred, the low marsh is now occupied by *S. anglica*, but this in itself does not prove competitive displacement.

The Channel and southern North Sea represent the northern limits of a species with optimal growth under much warmer climatic conditions. Marchant (1967) suggests that its distribution in Britain may have fluctuated in the past with changes in climate. A small change in climate may have tipped the competitive balance between *S. martima* and *S. anglica* in favour of the latter. Once displaced, it is unlikely that, even with climatic amelioration, *S. maritima* would be able to recolonise its former sites. At many sites, *S. maritima* appears to have declined before the spread of *S. anglica*, again suggesting that environmental change was the primary factor in the decline.

S. alterniflora spread rapidly in the Solent area in the late 19th century, but is now virtually extinct; a limited population being maintained only by active management. A considerable part of the decline, particularly in Southampton Water (Marchant 1967), may be due to reclamation but, in other localities, competition with *S. anglica* may have been important. *S. alterniflora* is still locally abundant in France (Beeftink & Géhu 1973). Early introductions in France probably occurred at about the same time as those in Britain, and there is a possibility of a polytopic origin for the *S. maritima–S. alterniflora* hybrid. The taxon known as *S. neyrautii* (see Marchant 1975), which was first recorded on the Biscay coast at about the same time as *S. townsendii* appeared in Britain, may represent a second episode of hybridisation.

S. anglica is a species capable of high rates of photosynthesis (Woolhouse 1981) and, as a result of its spread, the primary productivity of many marshes has probably increased considerably.

One consequence of this is that large amounts of *Spartina* litter accumulate at the driftline. Beeftink (1975) has suggested that this new uniformity in the composition of drift material has reduced the diversity of the strandline vegetation in Holland.

For the most part, *S. anglica* has formed extensive communities in the low marsh. In the mid and upper marsh, pans and creeks may be colonised, but there is little evidence for the widespread invasion of existing closed vegetation. Chater & Jones (1957) suggest a low rate of spread onto established communities in the Dovey estuary, but such invasion is relatively limited in occurrence (see also Beeftink 1977*a*,*b*). Whether the extensive development of *Spartina* marsh to seaward of existing communities has altered processes within the older communities remains little studied. The superior silt-trapping abilities of *S. anglica* when compared with other European low-marsh species has often led to the deposition of finer sediment than was previously the case in the lower marsh. This may result in the upper-marsh communities now being relatively starved of fine sediment. In addition, the poor drainage characteristics of the finer sediment may affect the movement of groundwater through the marsh with impacts on the upper-marsh vegetation. In Poole Harbour, Dorset, where extensive development of *Spartina* has resulted in the deposition of particularly fine sediment, it has been suggested that this marsh sediment causes a damming back of groundwater seeping out of the Bagshot Sands. This, in turn, has resulted in a rise of water table inland from the shore and possibly caused the death of trees in adjacent woodland (Ranwell *et al.* 1964).

Many *S. anglica* marshes present a picture of singular monotony as they are too young for succession to have allowed the development of other communities. The best-studied examples of successional development from *Spartina* are at Bridgwater Bay and Poole Harbour (Ranwell 1961, 1964*b*; Hubbard & Stebbings 1968) – at both localities succession to *Phragmites* reed bed being observed, although at Bridgwater there is a difference between the grazed and ungrazed portions of the marsh. In the grazed section, trampling and grazing has produced an opening of the *Spartina* sward at lower levels, allowing the invasion of *Pu. maritima* and the development of low grassy swards. Ranwell (1981*a*) has pointed out that the reed beds at Bridgwater Bay now provide a habitat for a diverse flora, including the nationally extremely rare umbellifer, *Sium latifolium*. It is too early to generalise as to successional paths on *Spartina* marshes. It is probable that reed bed development requires some reduction on soil salinities through groundwater influence. It is also possible that invasion by other species will be slow unless the tall dense stands of *Spartina* are opened up through grazing, the deposition of drift litter or frost action.

In the grazed marshes of the Dovey estuary in Wales, the development of sandy levees along creeks has allowed the invasion of *Spartina* marsh by *F. rubra*. Density of creeks is such that this ribbon development of *Festuca* may account for up to 43% of the total marsh area (Chater 1973). Initial replacement of *Spartina* in restricted microhabitats, rather than across broad zones, may prove to be the more widespread pattern of successional development (see Fig. 2.2).

In those cases where *Spartina* has developed to seaward of an existing marsh, successional development of the *Spartina* need not necessarily repeat the pattern implied by the zonation of the older communities in the mid and upper marsh.

It is ironic that, as increasing concern has been expressed about the continued spread of *S. anglica*, around the site of origin in the Solent die-back and subsequent erosion have occurred and large areas formerly *Spartina* sward have reverted to mudflat. The cause of die-back has not been completely explained, but it first became apparent in the 1930s (Goodman, Braybooks & Lambert 1959). There is no evidence of the primary involvement of pathogens (Goodman 1959; Sivanesan & Manners 1970). Goodman & Williams (1961) suggested that die-back was due to the toxicity of an aerobic root environment. Hubbard (1965) suggested that local impedance of drainage as a result of levee formation along creeks leads to die-back in Poole Harbour. However, Chater (1973) described the retention of surface water between levees in *Spartina* marsh in the Dovey estuary in Wales where no sign of die-back has appeared. Die-back appears to be restricted to the central Channel coast of England. In this region, the sediments which accumulated in the *Spartina* stands were very fine-grained and the physical nature of the sediment may be critical in the development of toxic conditions.

The reduction in the area of saltmarsh dominated by *Spartina anglica* in

Fig. 2.2 Replacement of *Spartina anglica* by *Festuca rubra* on creek side levées on a saltmarsh in the Dovey estuary, Wales.

south east England has been considerable (see Fig. 2.3) but, in many regions, poorly documented. The reduction in *Spartina* marsh occurs through three mechanisms: die-back in and around slight depressions in the sward; loss of vigour and erosion along the seaward edge of marshes; and, on the landward side of the *Spartina* zone, successional processes. On the Channel coast, all three mechanisms may be involved but, on the east coast, succession and erosion and fragmentation of the seaward margin are the major mechanisms. The apparent decline in vigour of *Spartina* at some localities is not readily explained (Haynes 1984). Growth of algal mats and recolonisation by *Zostera* on the mudflats created following the decline of *Spartina* may provide a habitat of greater value to wildlife than the previous *Spartina* saltmarsh (Haynes 1984; Haynes & Coulson 1982).

S. anglica is still actively spreading at many sites in the north and west of England, and colonisation of new sites continues. Further expansion northward is likely to be limited by intolerance of frost (Ranwell 1967). A number of successful plantings of *S. anglica* have been made outside Europe, and extensive stands are now found in New Zealand, Tasmania and, particularly, China (Ranwell 1967, 1981*b*; Chung 1982, 1985). However, plantings in tropical and subtropical regions have been unsuccessful, at least in part, because successful germination is favoured by cold pretreatment of seeds. It

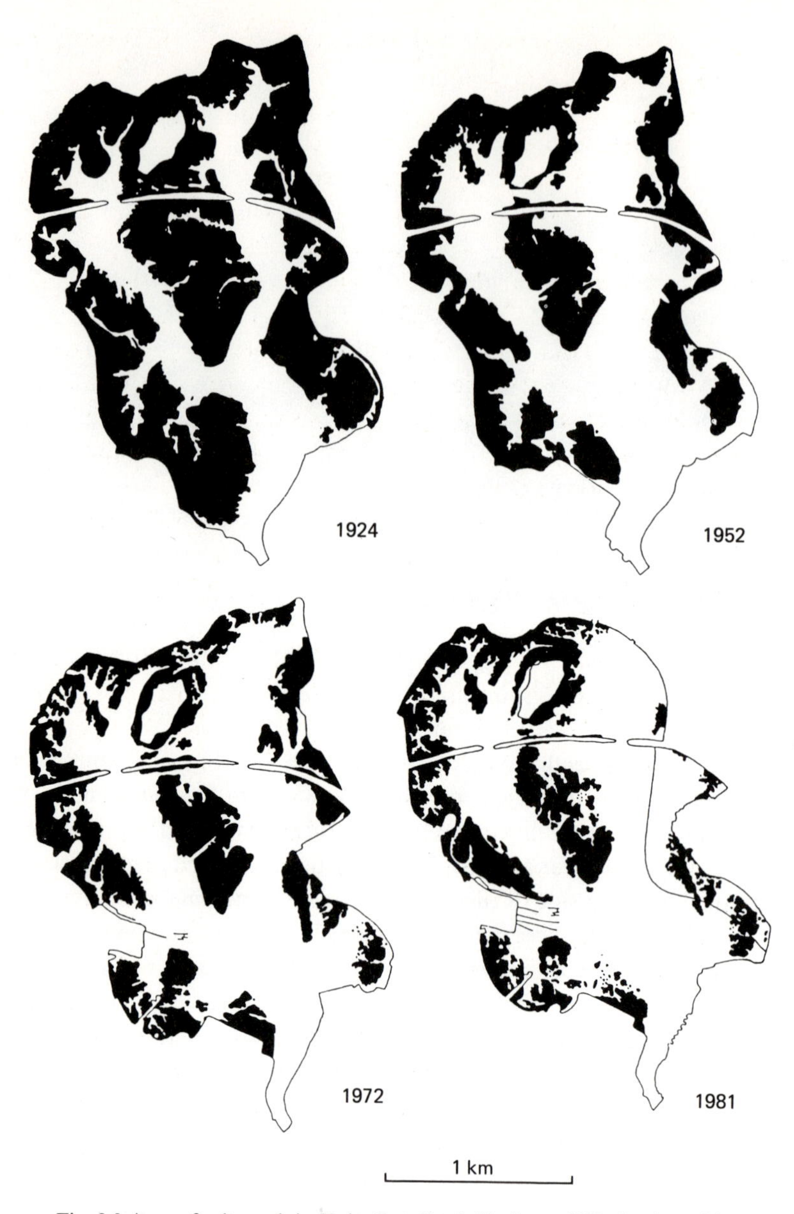

Fig. 2.3 Area of saltmarsh in Holes Bay, Poole Harbour, UK, dominated by *Spartina* at four dates. Some losses are due to reclamation but the major cause of loss is 'die-back'. (Redrawn from Gray & Pearson 1984.)

seems likely that *S. anglica* will continue to spread, albeit perhaps more slowly than in the past.

Life-form spectra of saltmarsh floras

In addition to comparing saltmarsh floras on a taxonomic basis, it is possible to do so on the basis of life form (Raunkiaer 1934). The life-form classification developed by Raunkiaer groups plants according to the position of the resting buds or persistent stem apices in relation to the soil surface, and provides a convenient method of indicating how a plant survives the unfavourable season. The spectrum of life forms in the flora of a particular region or habitat integrates the response of the flora to the environment, generally considered in terms of the prevailing climatic and ecological conditions, but also to some extent reflecting historical factors.

The life-form spectra have been calculated from a wide geographic range of sites (Chapman 1953, 1974*a*,*b*, 1976, 1977*a*; Waisel 1972; Beeftink 1985*b* – see Table 2.1). Spectra constructed from qualitative data will overemphasise the contribution of rare species. Ideally, spectra would be weighted by some measure of abundance of the taxa comprising the different life-form classes but appropriate data are rarely available. Similarly, comparisons of total marsh floras will be affected by differences in the range of vegetation types and habitat diversity between sites. Unfortunately, even if there were agreement as to methodology of community classification, it would be difficult to decide whether two communities on different continents were strictly comparable ecologically.

Some species may not be uniquely representative of a single life form. In the case of *Aster tripolium*, some upper-marsh populations may behave as annuals, whereas low-marsh representatives are invariably long-lived perennials (Gray 1971).

Bearing these limitations in mind, saltmarsh floras in general contain a predominance of hemicryptophytes (perennial herbs with buds at the soil surface). The chamaephyte–nanophanerophyte categories (dwarf shrubs) are also well represented, especially in regions with mediterranean climate. The representation of therophytes (annuals) is variable and is particularly low in New Zealand and New England. The proportion of therophytes is higher in regions of mediterranean and semi-arid climates which, to a large extent, reflects the spectrum of the total flora in these climatic regions. Many of the therophytes in mediterranean marshes occur in seasonally hypersaline sites and take advantage of the mild but wetter winters to complete their life cycles. In temperate marshes of Europe, vegetation dominated by therophytes is ecologically restricted within the marsh environment. The major occurrence of annuals in these marshes is in the low

Table 2.1. *Life form spectra of saltmarsh and halophytic floras (after Chapman 1960*b, *1974*a; *Waisel 1972)*

	Phanerophytes	Chamaephytes	Hemi-cryptophytes	Cryptophytes	Therophytes
Southern California	—	14.0	31.0	9.0	43.0
New England	3.5	3.5	60.5	11.0	14.5
Europe (ex Mediterranean)	—	10.0	40.0	10.0	30.0
New Zealand	8.0	4.0	56.0	8.0	16.0
Israel (halophytes)	17.0	29.1	34.5	—	19.4
S. of France (halophytes)	4.0	—	24.0	37.0	33.0
Normal spectrum	46.0	9.0	26.0	6.0	13.0

marsh and on creek banks, where communities dominated by *Salicornia* spp. and *Suaeda maritima* occur. The occupation and dominance of these zones by annuals may not reflect the optimal adaptation of the therophytic life form to prevailing environmental conditions but, rather, the historical absence of appropriately adapted hemicryptophytes. The successful spread of *S. anglica* (a hemicryptophyte) at the expense of *Salicornia* spp. may provide support for this hypothesis.

Annual species are also a major component of the driftline communities which mark the tidal limit on marshes. Many of the species in this habitat are nitrophiles but, during their growing season, may not be influenced by tides and so need not be halophytes in the physiological sense. Small annual plants also occur in openings in the sward in grassy upper marshes in northern Europe, where they constitute the plant communities assigned to the Saginetea maritimae (for details of the ecology of these communities see Beeftink 1975). For the most part, individual stands are small, but extensive areas may develop in response to human disturbance (for example, turf cuttings – see Gray 1972). Heavily disturbed (often poached) areas in the upper marsh which become hypersaline in summer provide a habitat in Europe for communities of the Puccinellio–Spergularion in which the annuals *Puccinellia distans* and *Spergularia marina* play an important part (Beeftink 1965, 1977*a*; Adam 1981*a*).

Among the more unusual life forms, there are a number of records of parasitic angiosperms in saltmarsh communities. On the Pacific coast of the USA, the dodder *Cuscuta salina* is a widespread constituent of saltmarsh vegetation (Macdonald & Barbour 1974; Macdonald 1977*a,b*; Purer 1942). In Europe, Beeftink (1977*a*) has discussed the local occurrence of *Cistanche lutea* (Orobanchaceae), a parasite on *H. portulacoides* and *Sarcocornia fruticosa* in Portugal. In the species-rich grass-dominated upper marshes of northern Britain and northern Europe, hemiparasitic members of the family Scrophulariaceae in the genera *Odontites* and *Euphrasia* are often a striking feature of the vegetation (Gillner 1960; Tyler 1969*a,b*; Adam 1976*a*).

In general, the data in Table 2.1 suggest that the spectra for saltmarshes co-vary with the spectra for regional floras with the obvious and important exception that saltmarshes do not provide a habitat for phanerophytes.

Origins of the vascular flora of saltmarshes

The total flora of coastal saltmarsh is not large; the halophytic element is even smaller, and is characterised by widespread species and the occurrence of closely related vicariant species in the different continents.

The glycophytic element in the saltmarsh flora includes representatives of many different families, suggesting that the development of some degree of salt tolerance has been a relatively easily achieved evolutionary step which has probably occurred independently a number of times. The glycophytes are rarely found in the low marsh, which is pre-eminently the habitat of halophytic species. The few families which dominate the halophytic element would not be regarded as closely related in any phylogenetic schemes. Evolution of traits appropriate for the habitat thus appears to have occurred independently in the different families, but has been a relatively rare event.

Chapman (1977*a*) has argued that the wide generic distribution of many halophytes can be explained in terms of continental drift, and suggests, for example, that the wide distribution of *Limonium*, *Suaeda*, *Puccinellia* and *Juncus* around the northern hemisphere indicates that the evolution of halophytic adaptations in these genera occurred prior to the opening of the North Atlantic. It is difficult, however, to account for the existence of all these genera on southern hemisphere marshes using this argument as Laurasia and Gondwanaland separated prior to the radiation of the angiosperms.

Unfortunately, the fossil record of the main halophytic families (Chenopodiaceae, Gramineae, Juncaceae, Cyperaceae) does not easily permit the differentiation of halophytic genera, so it is not possible to date the origin of the halophytic flora in relation to the known chronology of continental movement. In some cases of relatively restricted distributions, continental drift may provide the explanation for the observed distribution patterns. *Spartina* has its main diversity in the Americas, which may indicate the genus evolved after the opening of the Atlantic (Chapman 1977*a*). The occurrence of *S. maritima* in Europe and Africa could represent some rare long-distance dispersal event followed by evolutionary changes to the extent that the Old World plant can now be recognised as being a distinct species.

While continental drift provides one mechanism to explain the distribution of halophytic species, long-distance dispersal may also be proposed, particularly to explain occurrences in both northern and southern hemispheres.

Long-distance dispersal is frequently invoked to explain distribution patterns apparently inexplicable on other grounds. By its very nature, it is almost impossible to prove that it has been a factor in the development of distribution patterns but, in general, it is argued that the proposed mechanisms of long-distance dispersal are unlikely to be effective. However, special features of the saltmarsh environment make it reasonable to suggest that successful long-distance dispersal cannot be ruled out for halophytic species.

One of the classic proposed mechanisms for long-distance dispersal is by floating in oceanic currents. Undoubtedly, many seeds, fruits or other propagules can be carried across oceans in this way, as witnessed by the diversity of species carried across the Atlantic from the West Indies to Ireland (Nelson 1978). For most terrestrial species, the journey is in vain; the end point being an inhospitable shore environment which does not provide conditions suitable for germination and establishment. Saltmarsh plant propagules, which start their journey in an intertidal environment, might finish it in another, and so be in a position favourable for establishment. Prolonged immersion in seawater is unlikely to affect adversely the chances of subsequent germination for most saltmarsh plants. However, there are few data on the ability of seeds to float for prolonged periods. Such data as those provided by Praeger (1913) suggest that the small seeds of many saltmarsh species might not float long enough to cross oceans. However, dispersion by currents may be effective over shorter distances (in the order of several hundreds of kilometres), providing the possibility of effective long-distance dispersal by stages.

At the present day, an ideal mechanism for long-distance dispersal, including that between northern and southern hemispheres, would appear to be provided by migrating birds; species of waterfowl, waders and seabirds, all of which frequent saltmarsh habitats, perform some of the longest regular migrations.

Seeds might be transported both externally and internally. Seeds of halophytic species are not generally specialised for epizoochory, but may be carried in mud on the birds' feet. Olney (1963) recorded large numbers of seeds of *Su. maritima* on the feet of teal (*Anas crecca crecca*), while the very small seeds of *Juncus* would seem to be ideal candidates for dispersal in this way.

Long-distance dispersal internally has been discounted on the grounds that retention times are unlikely to be sufficient for transoceanic transport. Experimental and field studies by de Vlaming & Proctor (1968) and Proctor (1968) showed that seeds could, in theory, be transported internally over distances in excess of 1600 km by both ducks and waders. The retention time for seeds varied with both plant and bird species. Although some doubt is still attached to these findings, as there is a possibility of seeds being voided prior to migratory flights, the results are strongly suggestive that the long-distance dispersal by birds is likely to have occurred (whether dispersal would be followed by establishment is another matter). In terms of dispersal between hemispheres, waders seem to provide the most likely medium in view of their migration routes. However, waders are generally carnivorous, but Proctor (1968) showed that at certain times of year the gizzards of a number of wader species contain a variety of seeds.

Although birds might act as dispersal agents at present, is it realistic to invoke them to explain the development of the distribution patterns of the halophytic flora? The fossil record of birds extends back to the Jurassic period, but the earliest forms which may be related to existing waders appear in the upper Cretaceous, while the main lines of waders seem to extend back 50–70 million years (Hale 1980). The Anatidae (waterfowl) appear to have been distinct since the Eocene/Oligocene, but, possibly reflecting the bias of the fossil record, many of the earlier fossil records appear to be waterbirds of various types. Given that the main diversification into flowering plant families took place in the uppermost Cretaceous and early Tertiary periods, potential dispersal agents and plants were evolving contemporaneously. While it is not inconceivable, therefore, that birds could have spread the basic halophytic stocks around the world, it is impossible to test the hypothesis. One problem is that reconstructions of migration patterns would be largely speculation. Migration probably developed when climatic conditions in the nesting region became such that overwintering became impractical. It is clear in the northern hemisphere that major changes in migration patterns occurred during the Pleistocene glaciations and that for waders, at least, the formation of geographically distinct populations was promoted at this time (Hale 1980).

Notwithstanding the lack of hard evidence for dispersal by birds, it is frequently invoked both in general and as an explanation of particular distribution patterns. Clifford & Simon (1981) suggest that the distribution patterns of the grass genera *Distichlis* and *Puccinellia* could be explained by invoking transport between northern and southern hemispheres.

It is striking that freshwater aquatic habitats also have a relatively small flora containing a number of widespread elements, including several cosmopolitan species. Such habitats are similar to coastal marshes in being utilised by migratory waterfowl. This strengthens the argument, which must, however, remain circumstantial, for the importance of birds as dispersal agents between wetland habitats. The timing of any such dispersal events is likely to remain in the realms of speculation.

The relationship between coastal and inland saline floras has been studied in but little detail. The most extensive areas of non-tidal saline habitats are in arid or semi-arid regions, but other saline communities occur in wetter regions associated with saline springs or industry. In western Europe, inland saltmarshes have a flora similar to that of coastal marshes (Lee 1977). Given that these inland sites are frequented by waterfowl which may also visit the coast, it seems reasonable to suggest that they have been colonised by propagules transported by birds. A similar explanation was invoked by Adam (1977*a*) to account for the occurrence of a number of halophytic species at inland sites during the Devensian in England. It would be of interest to have genecological comparisons of populations of the same species in coastal saltmarshes and inland sites, to assess the degree of genetic divergence.

Arid saline habitats share a number of species and genera with coastal saltmarshes. Burbidge (1960) suggests that much of the Australian arid flora may have evolved from coastal ancestors, which were possibly stranded in the interior at the end of past marine incursions. This hypothesis is difficult to test in view of the paucity of the fossil record. Around southern and northern western Australia, the present arid zone extends to the coast so that direct interchange between coastal and inland floras is still possible. In addition, on those rare occasions when inland lakes in Australia contain water, they attract large numbers of nomadic waterfowl which, again, may act as distribution vectors.

Nevertheless, while there is considerable similarity at the family and generic level between the major saline arid areas of the world (Chapman 1974*a*), taxonomic studies do not provide strong evidence for recent interchange between the areas. For the genus *Atriplex*, Osmond, Björkman & Anderson (1980) showed that there are three major centres of diversity in North America, South America and Australia. While there is a high degree of similarity in the growth forms represented in the three continents, the proportion of endemic species in each region is high (70% or above), probably indicating separation of the basic stocks for a very long time. In addition, at least within Australia, there is little evidence of close taxonomic relationships between inland and coastal species.

The inland saline areas of south east Europe have a particularly rich flora, including a comparatively large number of taxa restricted to that region (Beeftink, 1984). The region is one where several major phytogeographic elements intermingle (Euro/Siberian – Mediterranean – Irano/Turanic – see Beeftink 1984), the high diversity and endemism may reflect the consquences of this intermingling.

The mechanisms of speciation amongst halophytes are poorly understood, but present understanding is reviewed by Beeftink (1984). The majority of European halophyte lineages appear to be old and most species appear to be of much earlier origin than the Pleistocene glaciation. A conspicuous exception is the recent origin through hybridisation of *S. anglica*. A number of halophyte genera form polyploid series and hybridisation in the past may have been the major mode of speciation in these groups (Beeftink 1984).

Vascular cryptogams

At the present day, vascular cryptogams are nowhere a major component of intertidal vegetation. Ferns of the genus *Acrostichum* occur widely as an understorey in the upper part of mangrove swamps; but in saltmarshes, vascular cryptogams are at best a minor component of the brackish upper marsh (from which a small number of species of fern and horsetails have been recorded).

This has not always been the case. Many fossil plants from prior to the rise of the angiosperms have been obtained from sediments which have apparently been deposited in intertidal environments. In many instances, the plants appear to have become fossilised in, or close to, their environment in life, and the vegetation they grew in has been commonly interpreted as being analogous to modern saltmarsh or mangrove (see Krasilov 1975). The extensive swamps which gave rise to coal deposits at various times were generally deltaic, the associated invertebrate fossils in some cases indicating marine, rather than brackish, conditions (for example, the occurrence of tubes similar to those of the marine worm *Spirobis* on the base of stems of Carboniferous *Lepidodendron* – Seward 1910). The stigmarian root axes associated with the Carboniferous swamp forests have been regarded as analogous to the modified root systems shown by many mangroves today.

Many of the fossil plant assemblages described from deltaic deposits appear to have been dominated by relatively large aborescent species which may thus be regarded as mangrove equivalents. Retallack (1975) in his reconstruction of the vegetation of the Triassic 'Gosford delta' (eastern Australia) envisaged a monodominant stand of the arborescent lycopod *Pleuromeia longicaulis* which is referred to as a mangrove. However, since *P. longicaulis* appears to have been a relatively small plant, perhaps 50 cm tall, it is probably more appropriate to regard the community as a saltmarsh, perhaps similar in structure to those dominated by various shrubby chenopods at the present day.

The rise to dominance by angiosperms in the terrestrial flora appears to have occurred relatively swiftly but, even so, vascular cryptogams are still widespread and a major component of the vegetation in many habitats. It is only in the intertidal zone that the triumph of the angiosperms has been so complete. In the absence of any physiological information about the earlier inhabitants of the saltmarsh and mangrove habitat, it is difficult to suggest reasons why angiosperms should be so much more successful.

Bryophytes

Saltmarsh is rarely thought of as a habitat for bryophytes. In most parts of the world, this would be true but there are now many reports of saltmarsh vegetation which contain reference to at least a few bryophyte species.

Both in terms of species richness and biomass, bryophytes play a more important role in marshes at higher latitudes. The occurrence of bryophytes in the vegetation of Arctic and Scandinavian saltmarshes has long been noted (see *inter alia*, Walton 1922; Nordhagen 1954; de Molenaar 1974; Gillner 1960; and Tyler 1969*b*).

The absence of bryophytes from marshes at lower latitudes is not easily explained. There are a small number of species which can tolerate extremely saline conditions in inland habitats in warmer climates. Saline flats in Australia, for example, provide a habitat for the mosses *Funaria salsicola* and *Pottia drummondii* and the liverworts *Riella halophila* and *Carrpos sphaerocarpus* (Catcheside 1980), but coastal saltmarshes in Australia are virtually devoid of bryophytes, and the few species that do occur (including *P. drummondii*) are either largely restricted to the extreme upper marsh and rarely extend far into the intertidal, or are epiphytes. Bridgewater (1975) records *Tortula papillosa* as an epiphyte on the shrubby chenopod *Sclerostegia arbuscula*.

Adam (1976*b*) discussed the bryophyte flora of British saltmarshes, where over fifty species of moss have been recorded. Liverworts were much rarer and only one species (a *Lophocolea* sp.) was widespread. The liverwort flora in Britain contains relatively few grassland species, and these have not evolved tolerance of the intertidal environment. Extremely salt-tolerant liverworts occur in Australia (Banwell 1951; Carr 1956); Engel & Schuster (1973) recorded a number of liverworts from spray-zone habitats in southern Chile.

Of the mosses recorded in Britain, only one, *Pottia heimii*, is largely restricted to saltmarsh, many of the other species being relatively widespread inland grassland species. Interestingly, *P. heimii* is the only species which fruits regularly and abundantly in the saltmarsh. Mosses were not generally recorded in the lower marsh, but a number of species were widespread at levels in the marsh flooded by about a hundred tides a year.

Mosses are rare on marshes in southern and eastern England and are totally absent from most sites. Those species which do occur are virtually restricted to the sides of footpaths in the upper marsh and along the foot of seawalls. All the species concerned are acrocarpous and there is considerable variation from year to year in their occurrence at a given site. The absence of mosses in closed communities may reflect the nature of the vascular vegetation. Much of the upper marsh in south east England is dominated by the shrub *H. portulacoides* or the tall grass *E. pycnanthus* (*A. pungens* auct.); the generally tall dense vegetation may not be conducive to bryophyte growth.

The marshes of western and northern Britain often have a diverse moss flora, although there is enormous variation between sites. A number of distinct ecological groups of species can be recognised.

Disturbed areas may support dense patches of acrocarpous mosses. Such habitats may have widely varying salinity, possibly becoming hypersaline in the summer. The most widespread, and often only, species present is *P. heimii*, but others include *Bryum* spp., *Barbula* spp., *Tortella flavovirens* and *Trichostomum brachydontium*. In the most-disturbed sites, for instance beside paths, this acrocarp flora is largely made up of species regarded as ruderal at inland sites (e.g., *Bryum bicolor*, *Archidium alternifolium*, *Ceratodon purpureus*), while in less-disturbed, more permanent microhabitats, the flora would be characterised as basicolous (*Barbula fallax*, *B. tophacea*, *Trichostomum brachydontium*).

The major habitat for saltmarsh bryophytes is in upper marsh grassland. The richest floras occur in short, heavily grazed turf, taller, ungrazed grassland having lesser amounts of fewer species. In England and Wales, the most widespread grassland bryophyte is *Eurynchium praelongum* (which has been recorded, as *E. stokesii*, from marshes in Washington and Oregon – Macdonald 1977*a*). Although acrocarpous species occur sparingly in grassland, the majority of species are pleurocarpous and are common in neutral or mildly basic grasslands inland.

The moss flora on Scottish marshes is different in character and, at most sites, richer in species. *E. praelongum* is less frequent, and the most abundant species is *Campylium polygamum*. The moss flora of Scottish marshes has many similarities to that of Scandinavian sites, although the occurrence as a turf component at some sites of *Grimmia* (*Schistidium*) *maritima* may be a uniquely Scottish feature.

Saltmarsh bryophytes offer many opportunities for research. Lacking roots, and so absorbing nutrients through the whole surface of the plant, bryophytes would seem to be faced with a number of problems in the saltmarsh environment. The rate and extent of salinity changes around saltmarsh moss are probably greater than those experienced by the roots of marsh angiosperms. Studies have been made on the ionic relationships of mosses from coastal rocks (Bates & Brown 1974, 1975; Bates 1976), but not on saltmarsh species. Very little work has been done on the genecology of bryophytes. Does the fact that most saltmarsh species are also widespread in a number of inland habitats reflect possession of a very robust basic physiology or the differentiation of a number of ecotypes? While absence is always less amenable to experimental explanation than presence, it would be of interest to explore reasons for the paucity of the moss flora on low latitude coastal marshes.

Algae

Although the immediate visual impression of saltmarsh is provided by the vascular plants, algae are often an abundant component of the flora. Algae are found attached to vascular plants, on and in the upper layers of the sediment, and as free-living forms.

The total number of saltmarsh algae species is very much larger than the number of vascular plants. There have been several studies which have described communities of saltmarsh algae (including Carter 1932, 1933*a,b*; Chapman 1937, 1939; Nienhuis 1970; Polderman 1975, 1978, 1979*a,b*, 1980*a,b*). These communities are rarely coterminous with vascular plant communities and are organised at different spatial scales (Fig. 2.4). In addition, the extent of temporal change in algal communities is considerably greater than within the vascular plant communities (Fig. 2.5). In the context of rapid general surveys, the vascular communities are probably a more conveniently studied means of characterising the environment than algae, although for particular purposes, algae may provide more precise estimates of environmental factors (and may, for example, be of value in pollution monitoring).

Although there have been far fewer studies of algae than of vascular plants, it would appear that the major genera share wide geographical distributions with the vascular plants, many algae assemblages being almost cosmopolitan. Notwithstanding the overall similarity of saltmarsh algal floras, there are regionally distinct assemblages. Polderman & Polderman-Hall (1980) showed that saltmarshes in western Scotland support a distinctive algal assemblage – vascular plant communities on this coast are also distinctive (Adam 1978).

Interaction between algal and vascular floras have been little studied, although shading of substrate and smothering by leaf litter may influence abundance and composition of the algal flora (Polderman 1979*a,b*). The

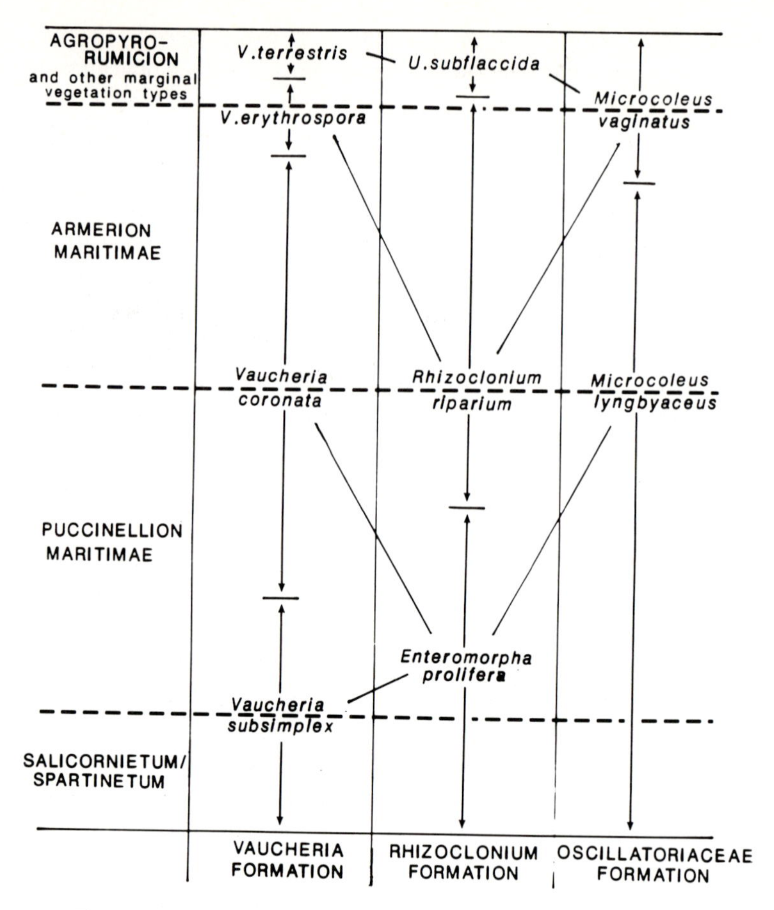

Fig. 2.4 Zonation of algal communities on Wadden Sea saltmarshes indicating the poor correlation between boundaries of communities defined by angiosperms and algae. (Redrawn from Polderman 1979*b*.)

epiphytic algal assemblages are often constant between sites, but to what extent this reflects a specific interaction between algae and host as distinct from a requirement for a particular physical environment which is coincident with the host's distribution has been little studied.

Occasionally, mats of algae may overgrow vascular plants and deleteriously affect them. It has been suggested that algal mats may provide favourable conditions for germination of vascular plants (Hill 1909; Poldermann 1979*b*). On the other hand, Jensen & Jefferies (1984) argue that algae were responsible for reducing *Salicornia europaea* populations on a Danish sand flat. Seeds germinating on the algal mat were likely to be carried away when the mat was dislodged, while seedlings establishing in the sediment were smothered and killed by algal mats.

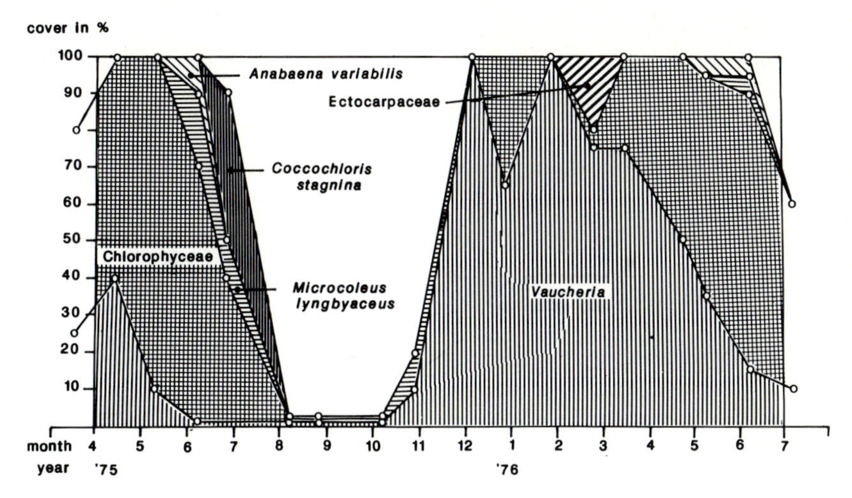

Fig. 2.5 Seasonal variation in the cover of algae within a permanent quadrat situated within Puccinellietum maritimae on the Wadden Sea island of Texel. (Redrawn from Polderman 1979*b*.)

In various estuaries, very dense growth of Chlorophyceae has been associated with increased nutrient loading, either from discharges directly into the estuary or by increased eutrophication of run-off from the catchment. It is suggested that there may be a relationship between dense algal growth and the decline of *Spartina* or the failure of saltmarsh to re-establish after *Spartina* die-back on the south coast of England (Ranwell 1981*b*).

Although biomass of algae is much lower than that of vascular plant communities, productivity of algae may, because of more rapid turnover, be a much larger fraction of the total. In addition, as algae lack the lignification of vascular plants they may break down and decay more rapidly than the higher plants and play an important role in nutrient cycling.

Algae are important as food sources, not only for fauna of marine origin but also for many of the invertebrates of terrestrial ancestry – for example, in north European marshes, the springtail *Hypogastrura viatica* and the beetle *Bledius spectabilis* (van Wingerden *et al.* 1981).

Algae may also play a crucial role in the establishment of saltmarshes. Coles (1979) has shown that stabilisation of mudflats is promoted by the mucilagenous nature of the dominant algae, particularly diatoms. Experimental reduction in grazing pressure on the mudflats, permitting greater algal growth, was followed by increased sedimentation. Algae can, therefore, be claimed to facilitate the subsequent colonisation by vascular plants. A role of algae in limiting erosion in the lower marsh during storms was suggested by Ranwell (1964*a*), while stabilisation of mudflat surfaces by blue-green algae was described by Kellerhals & Murray (1969).

Unattached macroalgae

Saltmarsh forms of 'seaweeds' are locally a highly visible component of saltmarsh vegetation. Despite being the subject of study over many years, very little is known about the biology of these forms. The biology of free-living seaweeds has been reviewed by Norton & Mathieson (1983) who recognise a number of distinct growth forms:

Entangled

These are usually much branched and (on saltmarshes) intertwined around vascular plants. Widespread examples in the northern hemisphere include the red alga *Bostrychia scorpioides* and a form of the fucoid *Ascophyllum nodosum* – ecad *scorpioides* (Chock & Mathieson 1976).

Loose lying

Norton & Mathieson (1983) describe these forms as being 'like living litter'. Included in this category are various fucoids sometimes found abundantly lying on the surface or aggregated into depressions in mid and lower saltmarshes.

These forms commonly exhibit a number of characteristics by which they differ from attached algae on rocky substrates (Baker & Bohling 1916; Norton & Mathieson 1983):

- absence of a holdfast;
- dwarfed thallus;
- profuse branching;
- curled and twisted thalli;
- the apical cell remains in the three-sided juvenile form rather than developing into the four-sided adult state;
- failure to reproduce sexually;
- vegetative fragmentation;
- often pale yellow in colour compared with the darker brown of attached forms.

Aegagropilous forms

Plants with this form have radially arranged branches and resemble spherical balls. This form is uncommon in saltmarshes, but the aegagropilous form of *A. nodosum* – ecad *mackaii* – which can form dense aggregates at the head of Scottish sea lochs (Gibb 1957) may be washed into saltmarsh creeks and pans.

Embedded forms

These seaweeds lack holdfasts, but have their bases embedded in sediment. The most extreme development of this form is exhibited by *Fucus muscoides* [*F. vesiculosus* var. (or ecad) *muscoides*] which is abundant on peaty saltmarshes in western Scotland and Ireland (Cotton 1912; Adam, Birks & Huntley 1977; Adam 1978, 1981*a*; Norton & Mathieson 1983). *F. muscoides* forms very dense moss-like carpets with the thallus about 10–15 mm tall (illustrated in Cotton 1912, and Fritsch 1935).

All free-living seaweeds were derived at some time from attached populations (Norton & Mathieson 1983) but the mode of origin and the reasons for their modified growth forms remain the subject of much speculation.

Some loose-lying fucoids in saltmarshes are known to be derived from fragments. Pieces of thallus stranded on mudflats in the marsh develop marginal proliferations which become independent plants after decay of the original fragment (Baker & Bohling 1916; Gibb 1957; Chock & Mathieson 1976; Oliveira & Fletcher 1980; Norton & Mathieson 1983). Whether such forms remain loose-lying or become embedded may depend upon the local environment in which the fragments find themselves.

Mathieson *et al.* (1982) have described how attached *A. nodosum* can be removed from rocky substrates in ice rafts, become fragmented and, after deposition in saltmarshes, can develop into the entangled form known as ecad *scorpioides*.

Explanation of the various features shown by saltmarsh fucoids remains obscure (Norton & Mathieson 1983), although the pale colour of some forms may indicate nutrient deficiency which could also be partly responsible for the failure of sexual reproduction (Norton & Mathieson 1983).

The taxonomic status of unattached seaweeds has long been debated. At different times, the same morphotype may have been given full species status or relegated to subspecific or varietal level. The most widely used taxonomic category currently is 'ecad' – defined by Davis & Heywood (1963) as 'the phenotypical product of environmental selection of genotypes which are able to grow in a range of habitats by consequence of their wide range of tolerance'. However, Oliveira & Fletcher (1980) concluded that saltmarsh forms of *Pelvetia canaliculata* were 'successive stages of depauperation of plants that were accidentally placed in an environment in which they are unable to survive permanently or to sexually reproduce' and that any formal taxonomic recognition was, therefore, inappropriate.

If saltmarsh fucoids are merely phenotypic modifications rather than genetically distinct ecotypes, it is surprising, as Norton & Mathieson (1983) point out, that some have 'defied attempts to experimentally reinstate the form of the attached plant from which they are presumed to have derived'. Norton & Mathieson (1983) suggest that

> 'it seems likely that many unattached seaweeds represent the permanent retention of a self-replicating juvenile state. Such neotony has the potential to give rise to new species as evidenced by the crucial part it is thought to have played in the evolution of vertebrates. Therefore, the uncertain taxonomic status of many unattached seaweeds may reflect the fact that we are witnessing new species in the making'.

Another major problem in our understanding of marsh fucoids is their distribution. Some forms are characteristic of a particular community and are repeatedly found over a wide geographic region, for example, *F. muscoides* in western Ireland and Scotland (Cotton 1912; Adam 1978, 1981*a*). In England and Wales, saltmarsh fucoids are absent from many sites, although extremely abundant at others. If saltmarsh populations can be established from fragments of previously attached plants washed in by the tide then, given the widespread occurrence of *Fucus* spp., *A. nodosum* and *P. canaliculata* on rocky shores, it is difficult to explain why fucoids have become established on so few marshes. This phenomenon is not restricted to Europe – in south east Australia the fucoid *Hormosira banksii* is widespread on rocky shores, but only a few free-living populations in mangrove stands are known (King 1981*a,b*).

Despite our ignorance of so many aspects of the biology of saltmarsh fucoids, there is no doubt that when present in abundance they may make an appreciable contribution to total stand productivity (King 1981*a,b*).

Many of the other macroalgal components of saltmarsh vegetation are also, technically, unattached. The various skins and mats of algae on sediment surfaces (Polderman 1979*a*) are more widespread than free-living fucoids. Pans in the saltmarsh can become covered with free-floating masses of macroalgae, particularly *Enteromorpha* spp.

Saltmarsh fungi

Fungal floras of saltmarshes have been subjected to study at few sites, and most of the information available relates to North American and British marshes, so that any broad generalisations may be unjustified.

Knowledge of marine fungi in general has been reviewed by Kohlmeyer & Kohlmeyer (1979) who suggest that a distinction should be made between:

> obligate marine fungi – 'those which grow and sporulate exclusively in a marine or estuarine habitat'

and

> facultative marine fungi – 'those from freshwater or terrestrial milieus able to grow (and possibly also to sporulate) in the marine environment'.

Kohlmeyer & Kohlmeyer (1979) argue that fungi isolated from marine habitats using various culture methods should not be considered to be marine species unless their growth in the marine environment is also demonstrated. Propagules of many terrestrial and freshwater species may be able to survive, but not develop, for varying periods in marine conditions. These species may, however, be revealed by some culturing methods. As is well known from studies of terrestrial fungi, culture methods are liable to give a selective and biased picture of the abundance of species in a particular habitat. However, such methods are often the only practical means of obtaining identifiable material.

In most cases, fungi identified in saltmarsh habitats have not been grown experimentally under a range of salinity conditions, so they cannot be categorised into classes such as obligate marine, facultative marine, terrestrial, etc. Kohlmeyer & Kohlmeyer (1979) suggest that the vast majority of the fungi isolated from saltmarshes are widespread terrestrial species, and there is no evidence that they are active in saltmarsh habitats. The number of marine or facultative marine species cultured from saltmarsh substrates is small. While much experimental work needs to be done on the growth of fungi from saltmarshes, 'terrestrial' species need not necessarily be inactive in saline habitats. Just as many widespread vascular plant species of the glycophytic element of the saltmarsh flora have salt-tolerant ecotypes, it is possible that salt-tolerant strains of terrestrial species may occur. Pugh & Lindsey (1975) obtained evidence pointing to probable differentiation of physiological races in the phylloplane fungus *Sporobolomyces roseus*, with coastal isolates growing and respiring at higher rates at high salt concentrations than inland isolates.

In studies of soil fungi from British saltmarshes, Pugh (1960, 1962) divided the flora into two categories:

> 'saltmarsh inhabitants' which were isolated with increasing frequency in an upshore direction;

and

> 'saltmarsh transients' which were regularly carried into the marsh by the tide (see also Hendrarto & Dickinson 1984) and showed a downshore increase in frequency of isolation.

The fungal flora increase in abundance and species richness upshore (Pugh 1960, 1962, 1974). The relative paucity of the rhizosphere flora in the pioneer zone is regarded by Pugh as reflecting the waterlogged anaerobic state of the soil and the low content of organic matter.

The range of soil fungi isolated from various British and European marshes is similar (Elliott 1930; Pugh 1960, 1962, 1974, 1979; Pugh & Beeftink 1980). Many of the species isolated from under continuous saltmarsh vegetation are widespread terrestrial species, so that the provisos raised above apply. However, one widely recorded species, *Dendryphiella salina*, would be regarded as an obligate marine species in Kohlmeyer & Kohlmeyer's (1979) terminology. Pugh (1974) has shown that *D. salina* is replaced under less saline conditions by *Gliocladium roseum*. As salinity increased, the growth of *Gliocladium* rapidly declined, and it was competitively replaced by *D. salina*. Pugh & Beeftink (1980) grew *D. salina* and *G. roseum* at constant osmotic potential in a medium in which the main components of the potential were sodium chloride and mannitol, and varied the salt/mannitol ratio. They showed that the decline in growth of *G. roseum* was related to changes in salt concentration and not to osmotic changes *per se*. Pugh & Beeftink (1980) also demonstrated that cellulytic activity of *G. roseum* declined with increasing salt concentration, while activity in *D. salina* was low at low and high salinities, but maximal with 2.3% NaCl in the culture medium.

A considerable number of cases of parasitic relationships between saltmarsh angiosperms and fungi have been reported (Ellis 1960*a*; Beeftink 1977; Kohlmeyer & Kohlmeyer 1979; Pugh 1979). Pugh (1979) states that ergots, mildew and rusts are the most easily noticeable diseases of saltmarsh plants. In some cases, attacks by rust fungi may reach what appear to be epidemic proportions. However, the impact of fungi upon their hosts has not been studied. In Britain, some populations of *Limonium* spp. are regularly attacked by the rust *Uromyces limonii*, to the extent that in late summer the general hue of the vegetation is rusty orange. Nevertheless, there is no evidence for any decline in *Limonium* abundance at these sites. In one of the few cases of a major die-back of a saltmarsh species which has been investigated, that of *S. anglica* in the Solent, no evidence of the primary involvement of fungi has been found (Goodman 1959; Sivanesan & Manners 1979). Pegg & Foresberg (1981) showed that a *Phytophthora* sp. could cause extensive die-back of the mangrove *Av. marina* in an area subject to severe disturbance from industrial development, although, in undisturbed conditions, *Avicennia* shows a high level of resistance to attack. The possibility that saltmarsh plants in disturbed sites may be more vulnerable to fungal attack should be borne in mind when developments in estuarine habitats are considered.

Many fungi have been isolated from the rhizosphere of saltmarsh angiosperms. The extent of root parasitism has not been studied and, in any case, the boundary between saprophytism and parasitism is often difficult to define. Elliott (1930) pointed out that the tap root of *Arm. maritima* was more susceptible to decay than the underground organ of any other species she investigated; possibly the distribution of *Armeria* in marshes is partially controlled by the distribution of pathogens.

There have been only a few studies which have looked for possible mycorrhizal associations on saltmarshes. Mason (1928) recorded vesicular–arbuscular (VA) mycorrhizas in a number of species on Welsh saltmarshes, while Fries (1944) showed VAs to be common in many plants on the Swedish coast. Klecka & Vukolov (1937) found VAs in a range of species from inland saline sites on the Neusiedlersee and other sites in Bohemia. Among their records were findings of VAs in *S. europaea*, *J. gerardi* and *Triglochin maritima*, three species in which Mason (1928) had failed to find mycorrhizas.

The grass, *Distichlis spicata*, widely distributed in both inland salines and on coastal saltmarshes in North America, frequently has VA mycorrhizae (Allen & Cunningham 1983). In a short-term glasshouse study, growth of both inland and coastal populations of *D. spicata* over a range of salinities was little influenced by the presence or absence of mycorrhizae (Allen & Cunningham, 1983). However, long-term field experiments are required before it can be determined whether or not *Distichlis* gains any advantage from being involved in a mycorrhizal association.

Rozema *et al.* (1986) recorded VA mycorrhizas from a number of species from Dutch saltmarshes. The incidence of infection varied from species to species but there was no obvious relationship between elevation in the marsh and degree of infection. Rozema *et al.* (1986) also carried out an experimental study on the effects of mycorrhizal infection on the growth of *A. tripolium*. Experimental raising of soil salinity did not cause any significant decline in fungal infection. Growth of *Aster* at higher salinities was significantly improved in mycorrhizal plants. The sodium content of shoots was lower in mycorrhizal plants at higher salinities but potassium and phosphorus levels were not affected by infection. Rozema *et al.* (1986) suggested that the improved growth was not related to any effect of mycorrhizas on mineral nutrition (the argument advanced to explain the advantage gained by mycorrhizal plants in most terrestrial situations) but to the improved water relations of the host. Mycorrhizal *Aster* at high salinities showed a faster rate of leaf extension and smaller changes in leaf thickness than uninfected plants, indicating increased water uptake.

Greater uptake could be accompanied by a reduction in stomatal resistance and hence greater photosynthesis. Rozema *et al.* (1986) discuss a number of possible mechanisms by which mycorrhizas could influence water uptake but more research is needed on this topic.

The taller saltmarsh plants may never be completely immersed, even by the highest tides. The uppermost leaves of such plants may thus provide a habitat for truly terrestrial fungal species. Nevertheless, parasitic fungi attacking saltmarsh plant foliage would encounter conditions of low water potential. The smaller plants and the lower parts of the taller plants will be immersed for varying lengths of time at various periods. While some terrestrial species may be able to exploit the plant surface in the period between flooding tides, the more permanent inhabitants of the niche must be capable of enduring immersion in saline water.

A wide range of fungi has been recorded from the aerial parts of saltmarsh plants (Kohlmeyer & Kohlmeyer 1979). Apinis & Chesters (1964) recorded the ascomycetes on the senescent leaves of a number of British saltmarsh grasses, while the mycoflora of leaves of the shrubby chenopod *H. portulacoides* was investigated by Dickinson (1965). Gessner (1977) studied the distribution of fungi on *S. alterniflora* and showed that the fungal flora varied with position on the plant with the lower parts of the plants, regularly inundated by the tide, being occupied by marine ascomycetes and deuteromycetes, while the upper parts of the plants provided a habitat for terrestrial species. In addition, there were seasonal changes in the species recorded, although to a large extent this reflected seasonal patterns in the appearance of reproductive structures rather than temporal replacement of species (Gessner 1977).

Spartina spp. are associated with many fungi; a survey by Gessner & Kohlmeyer (1976) showed that 101 species have been recorded. A number of species are found almost wherever the host occurs but others, although widespread on other species, have been recorded only rarely on *Spartina.*

Some of the fungi isolated from leaf surfaces may be utilising leaf exudates, while others may commence decomposition of senescent leaves before leaf fall.

Decomposition and mineralisation processes in saltmarshes have been the subject of very considerable research effort. It seems to have been generally assumed that bacteria are the main agents of decay in the saltmarsh environment. The role of fungi in these processes has probably been underestimated. The various studies on *S. alterniflora* indicate that fungi predominate in the early stages of decomposition (Gessner & Goos 1973; Gessner, Goos & Sieburth 1979).

Lichenised fungi (lichens) are a negligible component of marsh floras. A small number of species have been recorded as epiphytes on upper-marsh shrubs (for example, on *Su. fruticosa* (*Su. vera*) – Ellis 1960*a*). In contrast, mangroves support a very rich epiphytic lichen flora.

Bacteria

Bacteria are an important component of the saltmarsh microflora. Studies on mineral cycling in estuaries have emphasised the vital role of bacteria. As far as ecosystem functioning is concerned, it could be claimed justifiably that it is bacteria that are responsible for those features of nutrient cycles which are unique to saltmarsh. Although there have been many studies in bacterial activity, there have been relatively few taxonomic studies but the major part of the saltmarsh bacterial flora appears to be cosmopolitan.

Genecology of saltmarsh plants

Environmental heterogeneity, both in space and time, is the rule rather than the exception in all habitats. This environmental variation may be presumed to impose a variety of selection pressures upon plant populations. There are now many examples in the literature of patterns of genetic differentiation in plant populations which are closely correlated with variable environmental factors and it is frequently assumed (although less frequently confirmed) that this correlation reflects the response of the population to the imposed selection pressures.

Some of the best known and most convincing demonstrations of intraspecific genetic differentiation in relation to habitat come from studies of plants (particularly grasses of the genus *Agrostis*) growing on soils contaminated with heavy metal ions (see summary by Bradshaw & McNeilly 1981). Work on heavy-metal-tolerant grasses has demonstrated that genetic differences between individuals of the same species may be maintained over very short distances (of the order of centimetres) despite the apparent lack of barriers to gene flow between individuals; genetic differences may evolve very rapidly and may be expressed both in morphological and biochemical/physiological characteristics. The patterns of variation may be abrupt, with sharp boundaries between different ecotypes, or clinal.

Saltmarshes are heterogeneous habitats *par excellence*. The saltmarsh flora is limited, with most species displaying a wide ecological amplitude. It is thus reasonable to anticipate that many saltmarsh species will consist of a large number of genetically distinct races with restricted geographical or ecological distribution. Historically, recognition of the differences between inland and coastal populations of a number of species in Sweden played an important part in Turesson's development of the concept of the ecotype (see Turesson 1922), while studies on variation in *Plantago maritima* were the basis of Gregor's (1938, 1939, 1944, 1946, 1956) work on patterns of ecotypic variation and the recognition of clines. More recently, there have been genecological studies of a number of saltmarsh species (see Gray 1974).

Widespread species do not necessarily consist of numerous, genetically different, restricted populations. Another means of accommodating environmental variability is for a species to be developmentally plastic, a response allowed for by the indeterminate growth form of most plants.

Levins (1962, 1963) has argued that genetic differentiation is favoured by environments which are spatially heterogenous but stable (or at least predictable), whereas unpredictability favours phenotypic plasticity. The major patterns of spatial variation in the saltmarsh environment are predictable, while the point-to-point mosaic of variability imposed over the predictable, repeated, gradients is far less so. Similarly, the major temporal changes in the environment are imposed by tidal rhythms and the climate. The tides provide predictable cyclic variation at a number of time scales, daily, biomonthly, seasonally. Some of the longer-term cycles are clearly of biological significance, for example, in north temperate marshes phenology of establishment and of seed dispersal is related to the occurrence of the equinoctial spring tides. Jefferies (1977*b*) and Jefferies, Davy & Rudmik (1979, 1981) suggest that the reduction in hypersaline conditions in the upper marsh by flooding spring tides in the late summer is sufficiently predictable for the growth patterns of a number of species to be genetically controlled such that the major period of growth occurs towards the end of summer. On the other hand, the climate imposes both predictable cycles (the march of seasons) and unpredictability. The coincidence of the extremes of the tidal cycle with extreme climatic conditions is likely to have major biological repercussions, but will be extremely unpredictable (Bleakney 1972).

The expression of phenotypic plasticity is itself under genetic control and within a single species adaptation to the environment may take the form of both specialised genotypes and phenotypic plasticity. Three examples of such behaviour can be given. *H. portulacoides* is a variable species and, on the basis of leaf morphology, three varieties are formally recognised in some taxonomic treatments: vars. *parvifolia*, *angustifolia* and *latifolia*. Sharrock (1967) showed that varieties *parvifolia* and *angustifolia* remained morphologically distinct in cultivation. Variety *latifolia* proved to be very plastic: in muddy sediments it was distinct from the other two varieties, but when grown in sand it resembled var. *parvifolia* and on shingle, var. *angustifolia*. Thus, wild populations phenotypically var. *angustifolia* may represent genetically determined var. *angustifolia* or be an extreme in the range of variation of var. *latifolia*. The second example is provided by *Succisa pratensis*, a widespread glycophyte found in upper saltmarsh grasslands in southern Sweden. Saltmarsh specimens tend to be dwarf compared with those from inland population and had been given taxonomic recognition as form *nana*. Turesson (1922) showed that in cultivation all examples of these dwarf saltmarsh forms increased in height but, whereas some populations reached the stature of plants from inland sites, others remained relatively small and were presumably hereditary dwarfs.

Pu. maritima is a widespread grass species on north European saltmarshes. Phenotypically, the species is variable and a large proportion of this variability reflects underlying genotypic differentiation (Gray, Parsell & Scott 1979; Gray & Scott 1980; Gray 1985*a*). In northwest England, phenotypic variation within and between populations on upper marshes for a number of vegetative and reproduction attributes has a genetic basis; in north Norfolk, upper marsh *Pu. maritima* exhibits two distinct phenotypes, a tall form growing amongst *H. portulacoides* and a small form in low-herb dominated vegetation. In this case, the two distinct forms are the result of direct modification of the phenotype rather than genetic differentiation (Gray 1985*a*).

Phenotypic plasticity is clearly of value in allowing a plant to accommodate minor fluctuations of the environment. Given that most plants will possess some ability to respond to a variable environment by alteration of phenotype, why should not the majority of adaptation to extreme environments be at the phenotypic level? The evidence suggests that genetic differentiation is more common than phenotypic plasticity (Heslop-Harrison 1964) so there must be some disadvantage in plasticity. Turesson (1922) argued that, in terms of physiological economy, genetically adapted forms were more efficient than phenotypic plasticity – the cost of plasticity was the permanent possession of physiological machinery which might never be required.

Patterns of variation in the marsh environment

The most obvious gradient of variation in the marsh environment is that of frequency of tidal submergence which is directly correlated with elevation of the marsh surface. Also correlated with marsh elevation, but not linearly, are many other factors including soil salinity, soil aeration and drainage characteristics, and what may be called soil maturity (a complex variable including such factors as organic matter, calcium content, and nitrate and phosphorus status – Gray & Bunce 1972). Although complex, there is likely to be an overall gradient from low to high marsh and, in essence, this gradient will be repeated from marsh to marsh. If factors associated with this gradient impose strong selection pressure on plant populations, then it is possible that gradients of variation (clines) in a given species will be found correlated with marsh elevation and that the patterns will be repeatable between sites.

Superimposed upon the overall elevation gradient there will be a considerable amount of point-to-point variation in many factors (Gray 1972; Gray & Bunce 1972), such that any marsh provides a complex mosaic of microhabitats. Some of this variation may be inherently produced by the processes of marsh development, for example, the variation in soil conditions between a creek levee and an intercreek low (Beeftink 1966), but other components of the variation may be site-specific. Even prior to marsh establishment, a mudflat may present a habitat with considerable local diversity in sediment properties consequent on the activities of burrowing animals; this initial heterogeneity may be enhanced or obscured following the development of plant cover. The total microhabitat diversity on a marsh may include repeated short gradients of variation (as away from creeks) and sharp environmental discontinuities.

In spite of the similarities in patterns of environmental variation between marshes, there are many site-specific factors which are active for sufficient time for them to mould the genetic constitution of local populations. In addition, saltmarsh is not continuous along the coast; rather, contiguous areas of marsh are separated by stretches of other coastal habitat (cliff, dune or shingle beach), which does not necessarily provide suitable growth conditions for saltmarsh species. Saltmarshes can, thus, be viewed as a chain of habitat islands. Saltmarsh species occupy a habitat more conducive to long-distance dispersal than most terrestrial communities but, nevertheless, successful transport of seeds from an area of marsh in one estuary to a developing marsh in another area is still unlikely to be of frequent occurrence. Thus, the genetic diversity of pioneer populations could be low, leading through the processes of genetic drift to permanent site-specific variation between sites.

In northern Europe, a factor which, although not site-specific, is disjunct in its operation is grazing pressure. Heavy grazing pressure has been imposed on some marshes for sufficient length of time (at least several hundred years) for it to be likely that selection for grazing-tolerant genotypes to have occurred.

Many saltmarsh plants have a wide latitudinal range. While the influence of the sea is likely to mean that the climatic variation over this range is less than would be experienced between the same latitudes at inland sites (Macdonald & Barbour 1974), the climatic tolerances of some species do seem to be large. A number of latitudinal clines of variation have been described for widespread inland species and it would not be unlikely for similar clines to occur in saltmarsh species.

As well as being geographically widespread, many halophytes are also found over a wide range of conditions in the saltmarsh with a number of species occurring throughout the marsh from the pioneer low marsh to the upper marsh. For such species, it can be asked whether the marsh is occupied by a number of subpopulations which are adapted to specific microhabitats and are effectively isolated from each other or whether the full range of genetic variation is found in the low-marsh populations from which only some genotypes are selected by the changing conditions during marsh accretion.

Not all species found on saltmarshes are restricted to that habitat. Arbitrarily, we can recognise two different patterns of distribution: (a) species which are widespread on saltmarshes but with restricted, disjunct distributions in other, inland habitats; and (b) species which are widespread inland but which occur in the upper saltmarsh. This latter category can be further subdivided into ruderal species found on the driftline and in disturbed nitrogen-rich sites inland (a distribution pattern shown by a number of the more serious weeds of temperate agriculture), and meadow species which are also found in upper-marsh communities.

In this account, a number of studies of different species is considered. These studies are not comparable in that the characters investigated and the sampling strategies differ; it is not yet possible to decide whether there are identifiable syndromes of adaptative characters and patterns of variation among saltmarsh plants. In considering the examples given, it must be borne in mind that sampling strategy determines the types of variation that may be revealed; clines are difficult to detect from random samples over a wide geographical range of sites, while no conclusions about geographical variation can be drawn from intensive studies of one or two sites. Equally, it must be remembered that character states may vary independently and the pattern of variation in one character may be very different from that in others. This was well illustrated in the pioneer studies of Gregor on variation in *Pl. maritima*. In this species there is a cline from low marsh to high marsh in growth habit and scape length (Gregor 1938, 1939). However, if the degree of spotting on the leaf is studied, in Britain there is no consistent relationship with marsh elevation; rather, the degree of spotting seems to be site-specific, possibly reflecting the disjunct distribution of marshes and the possibility that certain character states may be determined by the few initial colonists. Interestingly, at a larger geographical scale, some trends in the degree of spotting can be detected (Gregor 1939).

Aster tripolium

A. tripolium is a member of the Compositae of wide ecological and geographical amplitude in coastal marshes in Europe. In addition, populations also occur at inland saline sites and, more rarely, on seacliffs. Variation in this species has been the subject of investigation by Gray (1971, 1974), who reported the existence of both geographical and topographical clines in various characters. The general pattern of variation shown in Britain has also been demonstrated in the Netherlands by Huiskes, van Soelen & Markusse (1985).

Fruits from populations covering the latitudinal range of the species were germinated by Gray (1971) and the seedlings grown to maturity under standardised experimental conditions. Various parameters associated with the reproductive cycle – time to first flowering, date of flowering season, fruit weight, germination requirements – were scored.

Time to first flowering displayed a cline from north to south in Europe. Populations from southern Europe (Austria, Portugal and Rumania) contained no plants flowering within two years of germination. Populations from England showed a mixture of first and subsequent-year flowerers. Time of flowering was correlated with a number of other features. First-year flowering plants tended to lose the basal rosette of leaves early in development and be generally less woody than plants which overwintered before first flowering and in which the rosette of basal leaves was persistent.

The obvious ecological trends paralleling this gradient are climatic, although which particular climatic factors are likely to be most important are unknown. It is also true that other ecological conditions (e.g., soil types, tidal regimes) are different at the two ends of the cline and the competing floras are very different.

Within the British Isles, there was a similar cline between low and high marshes with no low-marsh plants flowering in the first year while up to 60% of high-marsh plants were first-year flowerers. First-year flowering plants flower significantly later (in late September–October) than low-marsh plants which flower in August–early September. However, flowering in upper-marsh populations is influenced by a number of other factors.

Under experimental conditions variation in simulated tidal regimes or growth in poor soil can delay flowering until the second year after germination. Mid- and low-marsh populations do not show this plasticity in the onset of flowering.

In a number of populations there is a parallel cline in fruit size, upper-marsh plants producing significantly smaller fruit than those from the lower marsh, although the range of fruit weights is large at all sites. Studies in cultivation suggest that fruit weight has a high within population stability so that this cline is likely to reflect genetic differences between high- and low-marsh populations.

Studies on the germination of fruits from high- and low-marsh sites show that the majority of fruit from low-marsh sites have no innate dormancy, whereas fruit from upper-marsh plants require prolonged prechilling (at 2–5°C) to break dormancy.

In summary, the repeated cline in English marshes is from low-marsh populations of plants which first flower two or more years after germination, flower in August–September, and produce heavy fruit with no innate dormancy to upper-marsh populations with a high proportion of first-year flowering plants which flower in later September–October and produce light fruits – a high proportion of which require a chilling treatment to break dormancy.

Although *A. tripolium* is both self- and cross-fertile, it is probably largely outcrossing in wild populations (Gray 1971). The topocline is apparently maintained in the face of the possibility of gene flow (both as pollen and seed) between low- and high-marsh populations and between adjacent marshes, although the difference in flowering time between upper and lower marsh may be responsible for a degree of reproductive isolation between populations. If this is so, there must be strong selection pressure in favour of the character states found at various positions along the cline.

For the characters so far considered, there is evidence for clinal variation; in other characters, there may be disjunctions in distribution. An example of disjunction may be shown in the expression of ray florets in the inflorescence. *A. tripolium* generally has inflorescences with yellow disc flowers, but plants exist in which the ray flowers are absent (distinguished in some floras as var. *discoideus* Rchb.) Gray (1971) showed that both the fully rayed and the rayless states are homozygous, but that crosses between them are variable in the expression of the ray character. In south east England the rayless forms of *Aster* can form pure stands which are found at much lower levels on the marsh than rayed forms. The populations of rayed and rayless *Aster* are to a large extent spatially segregated and although intermediate heterozygotes can be found, on most marshes they are relatively uncommon. Gray (1971) suggests that the rayless form of *Aster* has spread considerably in recent years. Why this should be so remains obscure.

On the north European mainland coast, the distribution of rayless *Aster* has been studied by Sterk & Wijnands (1970), Duvigneaud & Jacobs (1971), Huiskes, van Soelen & Markusse (1985). Sterk & Wijnands (1970) showed that although rayless *Aster* was first recorded in Belgium in 1581 it was apparently local in distribution until the 20th century; in recent years, as in Britain, it has spread considerably. In Dutch populations, the frequency of heterozygous partially rayed forms is high and Sterk & Wijnands (1970) question the utility of affording the rayless form taxonomic status. However, as in Britain, the ecological optimum of the rayless form appears to be in the lower marsh.

Puccinellia maritima

Puccinellia is a large genus of grasses, most of whose members are halophytes. Geographically, the genus is widespread with the majority of species in the northern hemisphere, but with a number of representatives in southern temperate regions. Many species are found in inland saline habitats.

Pu. maritima is a widespread perennial species on European coasts (Gray & Scott 1977*a*) where it is frequently a dominant species in low- and mid-marsh communities and, as a minor component, may also be found in the upper marsh at many sites.

Phenotypically, *Pu. maritima* is very variable and studies of numerous collections from the British Isles by Gray & Scott (1980) suggest that much of this variation is genetically determined. A range of chromosome numbers had been reported for the species, so it was possible that some of the range of variation was due to the occurrence of different cytological races. It had also been suggested that the species was probably apomictic, which would be essential breeding behaviour for some populations if the reported chromosome counts (for example, $2n = 63$) were correct but could also tend to produce considerable differences between sites. However, the series of counts by Scott & Gray (1976) indicates that the chromosome number in Britain is a constant $2n = 56$ (octoploid on the basis that $x = 7$ for the genus) while breeding studies by Gray & Scott (1977*a*) showed a high degree of outbreeding and little ability for self-fertilisation.

Gray & Scott (1980) studied the variation between 56 plants of *Pu. maritima* from sites around the British Isles. These plants were a subsample from a larger collection of over 300 plants. The plants had originally been collected at random within each marsh so that material from a range of different habitats was included in the study. The plants had been in cultivation for some time before the start of the study and seven single tillers were taken from each plant and established to provide replicates. By the nature of the experimental design, this study would reveal little about clinal variation between low- and high-marsh populations; however, broad geographical trends, differences between grazed and ungrazed marshes, variation correlated with soil nutrient status, and broad differences between high and low marshes could be investigated economically.

Some 20 parameters of vegetative and reproductive growth were measured and the correlation matrix between characters was subject to principal-components analysis. The first three components extracted by this analysis accounted for 66% of the total variance, suggesting that variation in the growth of *Puccinellia* could be expressed in terms of the contrast between few trends (Fig. 2.6). The first axis accounted for 37.9% of the variance and reflected what Gray & Scott (1980) termed the 'vegetative biotype' of the plant. At the negative end of the axis were plants of lax habit, forming tall, loose clumps of few long-leaved tillers; these plants were contrasted with prostrate, short-leaved, many-tillered forms. The second axis, accounting for 17.2% of the variance, measured the yield of plants with high dry matter production at the positive end of the axis. The third component provided a contrast in reproductive behaviour between plants producing a large number of flowering tillers and those which reproduced mainly vegetatively.

Information about the site conditions of the original collecting localities can be superimposed on the ordination diagram to ascertain whether particular habitat factors are associated with particular biotypes of *Puccinellia*.

On the first two axes, plants from grazed sites tended to be grouped together, most being prostrate, with numerous tillers and, at least in cultivation, a low potential yield. Not all of the prostrate plants came from grazed marshes, but many of the marshes ungrazed at sampling may have been grazed in the past. The prostrate habit is a combination of a number of the characteristics measured and it is difficult to identify which are responding directly to grazing pressure. However, assumption of a prostrate growth form is a feature of many species which tolerate intense grazing.

On the third axis, which measures reproductive strategy, there is no obvious effect of grazing. At first sight, this is unexpected as flower heads are eaten by grazing stock so that a large proportion of the reproductive potential of a plant producing many flowering tillers would, apparently, be wasted on a grazed marsh. The differences between vegetative and sexual

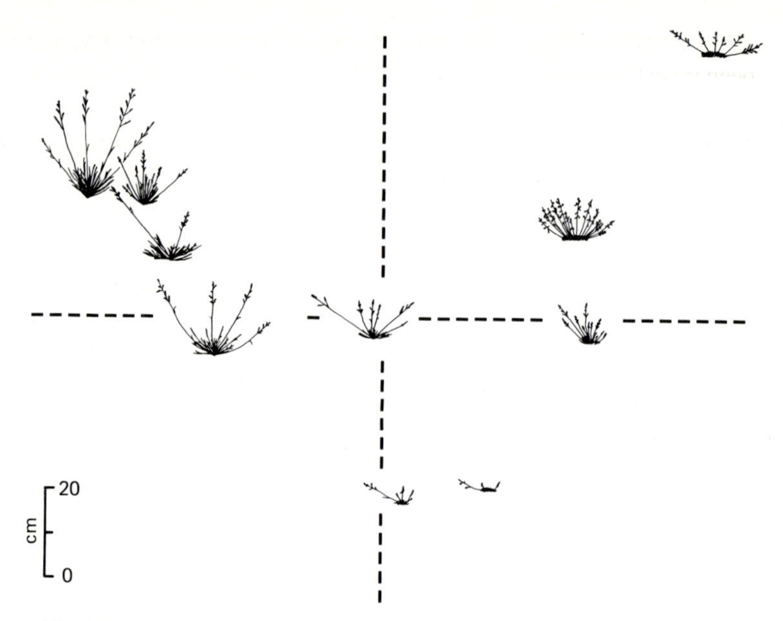

Fig. 2.6 Variation in *Puccinellia maritima*, showing the appearance, in cultivation, of some genotypes, positioned relative to the first two axes of a principal components analysis of 20 morphological and phenological characters in 55 genotypes. (Based on Gray & Scott 1980.)

reproductive strategies were detected in cultivation; behaviour in the field may be different. As Gray & Scott (1980) point out, if these differences are not expressed in the field, then opportunities for selection against 'free flowering' biotypes on grazed marshes may be much reduced. Gray & Scott (1977*a*) showed that flowering may be suppressed by high salinities. Field observations suggest that the incidence of flowering of many *Puccinellia* populations is variable from year to year but that occasional seasons of very prolific flowering do occur.

Variation on the yield axis could be related to the nutrient status of the soil at the collecting sites. Plants from mature upper-marsh soils or from organic silts or muds outyielded those from sandy low-marsh sites. There is no obvious relationship between soil type or position on marsh and the vegetative biotype or reproductive strategy axes.

Over and above the variation which can be related to local habitat factors there is some indication of regional differentiation, particularly on the reproductive strategy axis. It is interesting that there is a significant positive correlation between latitude and ear emergence, suggesting the possibility of climatic or day-length control of the onset of flowering.

The variation in *Pu. maritima* between upper- and lower-marsh sites has been investigated in greater detail by Gray, Parsell & Scott (1979). Populations in Morecambe Bay in northwest England and at Wells in north Norfolk were studied. In both regions, *Pu. maritima* is found in open, pioneer sites and as a component of mature upper-marsh communities. Plants from both mature and pioneer marsh populations from the various sites were grown in cultivation; after a period of growth under standardised conditions, leaf length and plant height were measured. These two measurements serve to define the position of the plant on the 'vegetative biotype' axis of variation discussed above. With the exception of one site, the variability between the mature marsh plants was less than that shown in the lower marsh, and was within the wider range of variation in the lower marsh (see also Gray 1985*a*). Particularly interesting was the comparison between two sites, Meathop and Grange, in Morecambe Bay. These two marshes are of the same age separated by the narrow channel of the River Winster. The Meathop marsh is heavily grazed by sheep, but the Grange marsh had not, at the time of sampling, been grazed in its history (Gray 1972; Gray & Adam 1974). The pioneer populations at both sites contained a comparable range of biotypes, but the mature, grazed Meathop marsh sward contained mainly small prostrate plants and the upper marsh at Grange was mainly composed of large, erect, long-leaved plants. As *Pu. maritima* is a perennial, which is potentially long-lived (Gray & Scott 1977*a*), these findings prompt the question as to whether the mature marsh populations are composed of genotypes selected from the pioneer pool during marsh development.

Gray *et al.* (1979) suggested that the range of biotypes on mature marshes could have arisen in one of two ways. During marsh development, some of the individuals present in the open pioneer community may be unable to withstand the competition of a closed community (or defoliation on a grazed marsh) and be eliminated. Thus, a mature marsh should contain fewer individuals per unit area than the pioneer marsh. An alternative model would be that the individuals present in the lower marsh are replaced by new genotypes as the marsh develops so that there would be no predictable relationship between marsh maturity and number of genotypes per unit area.

Tillers were sampled within 250-cm-square quadrats from pioneer and mature communities at Grange and Meathop. These quadrats were subdivided by a 25-cm grid and the plant at each grid intersect collected. The plants were grown in culture solution and the root tips extracted for electrophoretic analysis of variation in non-specific esterases, peroxidases and acid phosphatases. Variation in these enzymes is regarded as an index of genetic heterogeneity rather than of intrinsic significance. By this method, it is possible to be certain that two individuals are genetically different. However, so few loci are studied that it would be impossible to be certain that two individuals electrophoretically identical were the same. Bearing this limitation in mind, it is interesting that the grazed Meathop marsh had only 12 and 17 'types' in upper and lower sites respectively, while the Grange marsh had 26 and 31 'types' in the equivalent quadrats. When the spatial distributions of these 'types' are mapped, the structures of the populations at the two sites appear different (Fig. 2.7).

In the Meathop quadrats, most of the space is occupied by 3 or 4 'types' with the spaces between them filled by a small number of 'types', each covering a medium-sized area. At Grange, there were also 3–4 principal 'types', but the gaps between are occupied by a large number of 'types', each covering a very restricted area. Gray *et al.* (1979) argue that this difference could be due to the space between the colonists on the grazed marsh being filled by vegetative spread, while on the ungrazed marsh there is seedling establishment into the gaps. Unfortunately, the limitations of electrophoresis give us no way of being certain that the same genotypes are present in high- and low-marsh quadrats at either site.

The scale of patterning of genotypes shown in this study is certainly comparable with the scale of patterns shown in a number of soil factors. However, it would be inappropriate to assume that the genetic heterogeneity at this scale could be interpreted in terms of fitness to the microenvironment. Interpretation of fine-scale patterns of electrophoretic variants is difficult (Lewinton 1977).

The conclusion from this study is that there is a characteristic difference between the range of morphological biotypes in lower- and upper-marsh populations of *Pu. maritima* and these differences are presumed to reflect selection during marsh development. Although there is a reduction in morphological diversity as the marsh develops, the results from Grange show that total genetic diversity is not necessarily reduced to the same extent.

Sporobolus virginicus

Sp. virginicus is a saltmarsh grass of wide geographical distribution in subtropical and warm temperate regions. Morphologically, there is considerable variation in the phenotype of the species. Genecological studies of *S. virginicus* have been undertaken in Australia by Smith-White (1979, 1981, 1984).

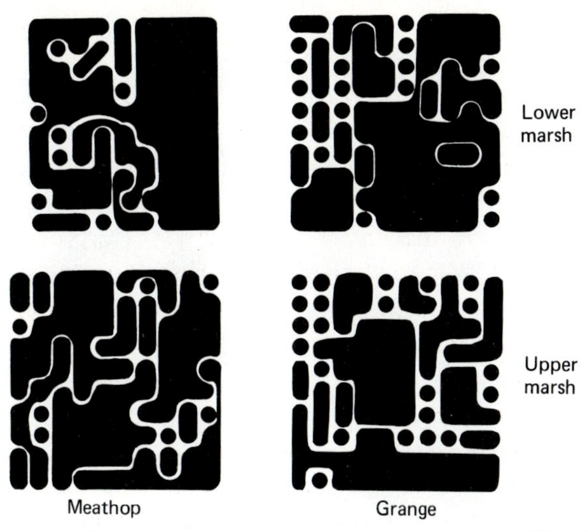

Fig. 2.7 A stylised representation of the clone structure of *Puccinellia maritima* populations within 2.5 × 2.5 m grids on a grazed (Meathop) and ungrazed (Grange) saltmarsh. Each continuous black shape represents the distribution of an identical enzyme phenotype (detected by electrophoresis), which is assumed to represent a single individual. (Redrawn from Gray, Parsell & Scott 1979.)

In contrast to *Pu. maritima*, which in Britain exhibited great variability within a single ploidy level, *Sp. virginicus* in Australia is a polyploid complex, with all ploidy levels between diploid and hexaploid being recorded (Smith-White 1984). To some extent, morphological variation and ploid level are correlated and there is considerable geographical segregation of ploids (Smith-White 1979, 1984 – Fig. 2.8).

In addition to geographical separation of the ploidy levels, Smith-White (1979, 1984) suggests there is also ecological differentiation with diploids favouring sandy, well-drained, sites while tetraploids are found on poorly drained, often anaerobic, muds. Experimental studies on growth under different imposed salinity regimes revealed considerable genetically determined differences in shoot production between populations of tetraploid *S. virginicus* (Smith-White 1981). Populations with the most vigorous growth under conditions of low salinity showed considerable reduction in growth as salinity was increased. On the other hand, those populations with low shoot production at low salinity showed little reduction in their growth at higher salinities. These findings suggest genotypic differentiation of populations to optimise growth under conditions prevailing in particular microhabitats.

Unfortunately, there are few chromosome counts of *S. virginicus* outside Australia, so whether the broadly latitudinal correlation in the distribution

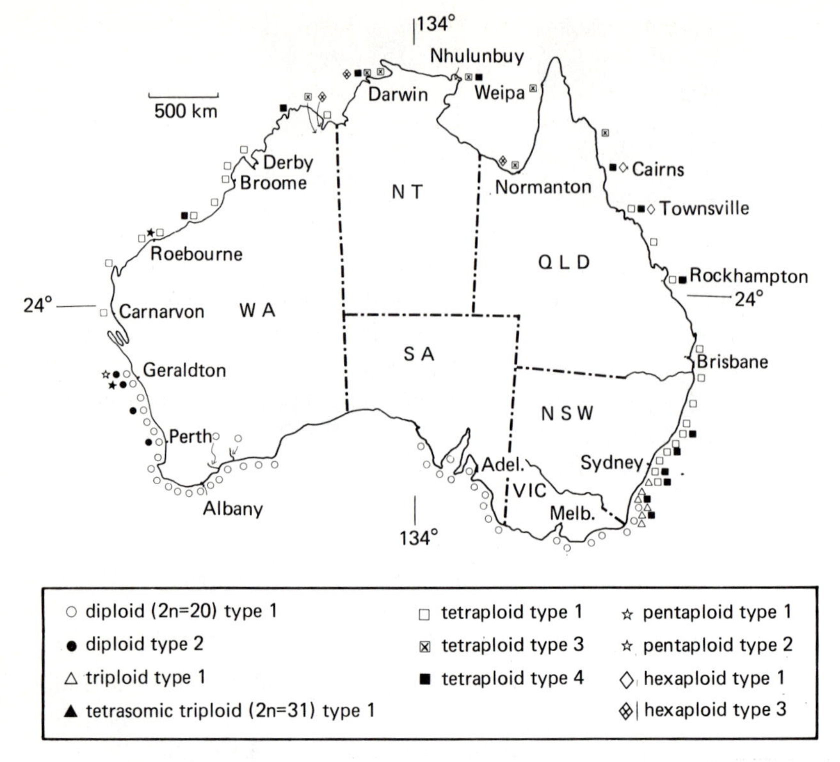

Fig. 2.8 Distribution of races of *Sporobolus virginicus* around coastal mainland Australia. Races are defined by chromosome number and phenotype (Type 1, short to tall generally erect plants with short, narrow leaves; Type 2, similar to Type 1, but with broader leaves; Type 3, tall plants with long narrow leaves; Type 4, very robust plants with long broad leaves). (Redrawn from Smith-White 1988 with additional unpublished information from A.R. Smith-White.)

of ploidy levels (or even the existence of the various ploidy levels) is repeated in other continents remains unknown.

Spartina alterniflora

S. alterniflora is by far the most important single species on saltmarshes of the North American Atlantic coastline. Throughout its range, the species exists in two distinct forms, a tall form up to 3 m tall found in the low marsh and along creek levees, and a dwarf form normally 10–40 cm tall found on the upper levees and in the depression between creek levees. In some accounts, an intermediate form is also recognised. Shoot density in stands of the tall form is generally lower than that of the smaller form. These two forms have long been recognised, but there has been disagreement as to the relationship between them, some authors implying that they are genetically distinct, others that they are an expression of phenotypic plasticity.

On the basis of short-term reciprocal transplants, Stalter & Batson (1969) suggested that the two forms represented different ecotypes. Mooring, Cooper & Seneca (1971) studied the germination and seedling growth of the two forms and proposed that there was no fundamental difference between them. Shea, Warren & Niering (1975) also carried out reciprocal transplants but, unlike Stalter & Batson (1969), were able to continue their observations over three years. Over this time period, the transplants did not remain distinct but, rather, came to resemble the vegetation into which they had been transplanted. These results, together with an electrophoretic comparison for forms, led Shea *et al.* to suggest that the forms were not genetically distinct, but merely reflected a phenotypic response to environmental differences between the habitats in which the two forms occurred.

Valiela, Teal & Deuser (1978) observed in the course of fertilisation experiments that under a high fertiliser enrichment regime stands of dwarf *S. alterniflora* came to resemble the tall form. After four years of treatment, the original dwarf stands had almost the same height and stem density as tall stands, and Valiela *et al.* (1978) agreed with Shea *et al.* (1975) in attributing most, if not all, of the differences between the forms to phenotypic plasticity. The environment of the tall and dwarf stands differ in many respects, but Valiela *et al.* (1978) suggested that, in terms of growth response, the most important difference is in nitrogen availability, the habitat of the tall form supplying more nitrogen to plants than that of the dwarf form.

The general consensus from the most recent studies is that tall and dwarf forms of *S. alterniflora* are not genetically distinct. This does not rule out population differentiation in respect to other characters along other environmental gradients; indeed, such a widespread species would be exceptional if it did not consist of a range of ecotypes. It is unfortunate that other aspects of variation in the species have not been studied.

The growth patterns of *S. alterniflora* in response to varying nitrogen supply and other environmental factors are discussed in detail by Smart (1982) and Morris (1982).

Salicornia spp.

The annual species of *Salicornia* form a notorious taxonomic complex. Identification of species within the complex is difficult, partly because there are few gross morphological differences between species and partly because growth patterns can be altered drastically by damage to the stem apex. In many ecological studies no attempt to recognise segregates has been made and a single taxon of wide ecological amplitude is recorded.

Even when a number of species are recognised, differences in taxonomic treatments and nomenclatural confusion make it difficult to compare studies. Nevertheless, there are differences between *Salicornia* populations in different habitats, suggesting that ecologists should persevere in attempting to recognise a number of species.

It is generally accepted in taxonomic treatments that the recognised species can be divided into two series, one of which is diploid ($2n = 18$) and the other tetraploid. Populations of *Salicornia* in the River Dee estuary in northwest England were studied by Ball & Brown (1970). Two species were recognised, *S. europaea* L. *sensu stricto*, a diploid, and *S. dolichostachya* Moss, a tetraploid. Apart from chromosome number, the two species could not be distinguished by any single character, but a combination of morphological characters gave reliable segregation. The two species were widespread on the marsh. Both species germinated in April or May and grew vegetatively until early autumn when flowering occurred. *Salicornia* spp. are most frequent in open sites – the pioneer zone, disturbed areas in the upper marsh and on creeksides – where they may be community dominants but are also found scattered through closed perennial communities of the mid and upper marsh.

Ball & Brown (1970) showed that *S. dolichostachya* was more frequent than *S. europaea* in open habitats but that the order of frequency was reversed in areas with a high cover of perennial grasses. *S. dolichostachya* was shown to have a more rapidly elongating radicle than *S. europaea*, and it is suggested that this is advantageous in establishment, enabling seedlings more quickly to withstand the uprooting forces of tidal currents. The greater frequency of *S. europaea* in closed communities was ascribed to its ability better to withstand competition from perennial species. What aspects of the biology of the species confer such advantage remain unknown.

Ecological differentiation between diploid species of *Salicornia* has been studied by Jefferies (1977*b*), Jefferies, Davy & Rudmik (1979, 1981) and Davy & Smith (1985), working mainly in North Norfolk in southeast England. Two populations were investigated, a low-marsh population conforming to *S. europaea* and one from the upper marsh, corresponding to *S. ramosissima* J. Woods.

Growth of the lower-marsh *S. europaea* was continuous throughout the growing season, and addition of nitrogen produced a growth response at any time. Growth of *S. ramosissima*, on the other hand, was very slow until late July when there was a sharp increase in the growth rate. Additions of nitrogen did not affect growth during the very slow early growth phase, but did lead to dramatically increased yield at the end of the growing season.

Reciprocal transplants of the two species between the upper and lower marsh failed to change this pattern of growth, suggesting that there is genetic control of the phenology of the two species (Jefferies *et al.* 1981).

The lower-marsh site in this study was flooded by all but the lowest neap tides and, in consequence, soil salinities remained relatively constant and hypersalinity did not develop. The upper-marsh site was exposed for a long period in the early summer during which hypersaline conditions developed (Jefferies, 1977*b*), and Jefferies *et al.* (1981) argue that this pattern of variation in soil salinity is highly predictable and also suggest that the late summer onset of rapid growth in *S. ramosissima* follows this lowering of soil salinity. Growth appears to be programmed to coincide with favourable environmental conditions.

The response of the two species to additional nitrogen is of interest. Jefferies (1977*b*) showed that, although there was temporal variation in the available nitrogen in the soil solution, the total available nitrogen in the two sites was very similar. Nevertheless, the upper-marsh population responded more to additional nitrogen, although this increase did not take effect until after the onset of the rapid growth phase. Jefferies *et al.* (1981) suggest as an hypothesis to explain this response that in the upper marsh nitrogen is utilised during the early part of the growing season, when the soil solution is hypersaline, for osmoregulation at the expense of growth. They argue that there has been selection for individuals which do not show a growth response to addition of nitrogen in the early part of the season. When growth becomes possible in late summer, there is a severe nitrogen limitation because much of the available nitrogen will have been utilised by other perennial species in the community. The response to added nitrogen is, thus, considerable. In the low marsh where growth is continuous throughout the summer, more available nitrogen can be diverted to growth rather than osmoregulation and the growth response to added nitrogen is much smaller.

Jefferies (1977*b*) showed similar differences between upper low-marsh populations of a number of widespread perennial species (*Pl. maritima*, *A. tripolium* and *Triglochin maritima*); the upper-marsh populations showed lower growth rates and less response to added nitrogen than populations from the lower marsh. In addition, species restricted to the upper marsh (*Arm. maritima* and *Limonium vulgare*) had lower growth rates than low-marsh species. These results, and those of Smith-White (1981) for *Sporobolus*, suggest that, as a general rule, tolerance of high salinity is achieved at the expense of growth.

Populations of *S. ramosissima* and *S. europaea* are strongly cleistogamous; the anthers are rarely exserted and self pollination occurs in unopened flowers, limiting the opportunity for gene flow between individuals. There is also limited seed dispersal, most seed being deposited close to the parent plant, or germinating *in situ* without being shed, although some longer-distance dispersal does take place, particularly with the low-marsh *S. europaea* (Jefferies *et al.* 1981; Jensen & Jefferies 1984; Watkinson & Davy 1985). The capacity for inbreeding permits new populations to be established by single plants.

Effectively, the two species are reproductively isolated one from another. Electrophoretic studies demonstrate that inbreeding is associated with extreme genetic homogeneity (Jefferies & Gottlieb 1982). Examination of 800 individual plants from marshes around Britain revealed only two electromorphs, one forming *S. europaea*, the other *S. ramosissima*. Although electrophoresis, being limited to an examination of selected gene loci, does not permit the conclusion that the total genotype is constant within the two species, it does suggest there is likely to be relatively little genetic differentiation within the species. This is in marked contrast to the considerable variation between individuals in the outbreeding *Pu. maritima*, and to the variation within the largely vegetatively reproducing *Pu. phryganodes* (Jefferies & Gottlieb 1983).

Ingrouille & Pearson (1987) showed that the conventional taxonomic recognition of *S. europaea* and *S. ramosissima* could not be supported by morphological measurements and suggested that the two taxa should be combined under *S. europaea*.

Festuca rubra and *Agrostis stolonifera*

F. rubra and *A. stolonifera* are two grasses with similar ecological and geographical distributions on European marshes. The two species are particularly important on grazed marshes in north west Europe; on heavily grazed marshes much of the mid and upper marsh consists of swards dominated by one or both of the species (Adam 1978, 1981*a*; Gray & Scott 1977*b*). Both species are also very widespread in inland, non-saline, habitats where they are found under a wide range of conditions, including both acid and alkaline soils.

The two species differ in that *A. stolonifera* will grow even if partially submerged for long periods but *F. rubra* is intolerant of prolonged waterlogging, although Davies & Singh (1983) demonstrated that saltmarsh populations of *F. rubra* were somewhat more tolerant of poorly drained soils than inland forms. *F. rubra* often extends further seaward on marshes than *A. stolonifera* (Hannon & Bradshaw 1968), but in the upper estuary *Agrostis* is sometimes a pioneer in marsh development.

Hannon & Bradshaw (1968) compared upland and saltmarsh populations of the two species from North Wales. Plants were grown in a range of concentrations of seawater in both sand and solution culture with relative salt tolerance being assessed by total yield and root/shoot ratio in sand and root length in solution. The results for the two species were similar: upper-marsh populations were the most tolerant. However, as we have seen with other species, growth of these plants under non-saline conditions was less vigorous than that of plants from non-saline sites; salt tolerance thus seems to be achieved at the cost of intrinsically slow growth. Hannon & Bradshaw (1968) point out that although the upland populations were very much less salt tolerant than plants from saltmarshes their salt tolerance compared favourably with that of many glycophytes. Thus, the inland populations seem to possess some potential for salt tolerance which has been selected for in the saltmarsh habitat. It would be interesting to know whether this also applies to the many other glycophytes found in the upper-marsh *Festuca/Agrostis* grasslands – does the range of species include only those whose inland genotypes possess some, even if limited, salt tolerance, or do most species have the ability, given time, to evolve saltmarsh ecotypes?

Differentiation with respect to salt tolerance in *F. rubra* has been further investigated by Rozema *et al.* (1978), who compared saltmarsh, sand-dune and inland populations in the Netherlands. *F. rubra* is morphologically variable and this variation has been recognised in a number of taxonomic treatments at either varietal or subspecific level. Rozema *et al.* (1978) regarded their populations as representing three subspecies; ssp. *litoralis* on the saltmarsh, ssp. *arenaria* in the dunes, and ssp. *rubra* for the inland race.

Seedlings and transplanted adult plants were grown in culture solution augmented with various levels of sodium chloride. The three subspecies showed a graded response with ssp. *litoralis* the most tolerant and ssp. *rubra* the least. In fact, 60 mM sodium chloride in the external solution appeared to represent the survival limit for the inland population. Growth of all three subspecies was reduced with increasing salt, but ssp. *litoralis* still survived fairly well at 300 mM sodium chloride. Analysis of the mineral content of the shoots of the plants showed that ssp. *litoralis* accumulated much less sodium chloride than ssp. *rubra*, with ssp. *arenaria* again intermediate. This suggests that in *F. rubra* salt tolerance is a function of salt exclusion in the roots.

Measurement of shoot proline levels showed an increase with increasing external salinity in all three subspecies, with the increase being greatest in ssp. *arenaria*. The dune habitat is often dry and growth of ssp. *arenaria* was much less affected by experimental drought conditions than ssp. *litoralis*. Rozema *et al.* (1978) suggest on this basis that proline accumulation in *F.*

rubra is primarily a response to osmotic stress rather than a specific response to salinity.

Further studies on ecotypic differentiation on *A. stolonifera* have been made by Tiku & Snaydon (1971), Ahmad & Wainwright (1976, 1977), Ahmad, Wainwright & Stewart (1981), Hodson *et al.* (1985) and Robertson & Wainwright (1987).

Ahmad & Wainwright (1976) studied variation in the leaf surface between inland, saltmarsh and seacliff populations of *Agrostis*. Seacliff plants are exposed to salt in the form of spray, while saltmarsh plants are periodically immersed in seawater. Seacliff and saltmarsh plants in South Wales retained less salt after immersion in saltwater than those from an inland population, whereas, after spraying with salt water, the seacliff plants retained least salt and the inland plants the most. These differences reflect variation in wettability of the leaf surface, a parameter which can be assessed as the contact angle of droplets on the leaf, the higher the contact angle the lower the wettability. Seacliff and saltmarsh plants had higher contact angles than those from inland. Interestingly, there was no statistical difference between leaf surfaces in saltmarsh plants, which will both be covered by tidal flooding; on seacliff plants the contact angle for the upper leaf surface, which will be more exposed to sea spray was higher than that for the lower surface.

These differences in wettability are correlated with the form of the epicuticular wax crystals. In the saltmarsh plants, and on the upper surface of seacliff leaves, there is a dense covering of upstanding flakes of wax, whereas on the more wettable surfaces there is a random deposit of rodlets. Similar cuticular features have been reported from seacliff populations of the grasses *Dactylis glomerata* and *Holcus lanatus* (McNeilly, Ashraf & Veltkamp 1987).

Ahmad & Wainwright (1977) used measurements of root growth in different culture solutions as indices of salt tolerance and confirmed Hannon & Bradshaw's (1968) view that this was an appropriate technique, although Tiku & Snaydon (1971) had suggested the contrary. The saltmarsh population was the most salt-tolerant and also grew better under partly anaerobic root conditions than the other two populations.

The same populations as were studied by Ahmad & Wainwright (1976, 1977) were subsequently investigated by Ahmad *et al.* (1981) who obtained results in many ways similar to those of Rozema *et al.* (1978) for *F. rubra*. When grown hydroponically with different external sodium chloride concentrations, the saltmarsh populations accumulated less internal sodium chloride than the other populations and showed a greater increase in proline, asparagine, glutamine, serine and glycinebetaine. When subject to osmotic stress imposed by mannitol or polyethylene glycol, the saltmarsh ecotype showed greater ability to resist losses in shoot water content than inland and seacliff plants. Ahmad *et al.* (1981) suggest that salt exclusion is an important part of salt tolerance in *Agrostis*, while excess ion accumulation, possibly causing internal osmotic inbalance, is responsible for the sensitivity of non-tolerant ecotypes.

The investigations of *Agrostis* and *Festuca* have concentrated on differences between saltmarsh and other populations. Given that both species are widespread on saltmarshes, it is likely that there will be variation between populations in different parts of a marsh, as has been demonstrated for *Festuca* by Rhebergen & Nelissen (1985).

Armeria maritima and *Plantago maritima*

F. rubra and *A. stolonifera* are species which while common on saltmarshes, are even more widespread inland and the saltmarsh populations can be regarded as occupying the extreme of tolerated habitats for the species. *Arm. maritima* is a European species which is widespread and frequently abundant on saltmarshes but which also occurs inland in rather specialised habitats. *Armeria* is found on seacliffs, on high mountains, on nutrient-poor sands and on outcrops of rocks rich in heavy metals (such as serpentine). Morphologically, the species is very variable and a number of subspecies are recognised (Pinto da Silva 1972).

The present distribution of *Armeria*, which is very similar to that of *Pl. maritima*, is but a fraction of its extent in early post-glacial times (Godwin 1975). The subsequent contraction in range has been to sites likely to have remained treeless throughout the time of general expansion in forest cover. *Armeria* is essentially a plant of open habitats intolerant of shade. It is interesting that at many of its present inland sites in Britain it occurs with a suite of other species now rare or local, but widespread in late and early post-glacial periods (Pigott & Walters 1954).

Armeria is likely to have undergone several expansions and contractions of range during the Pleistocene glaciations. During these events, have coastal and inland genotypes remained distinct or have populations become so mixed that there is a basic genetic uniformity? Unfortunately, there have been very few comparative studies on the physiology of different populations. Stewart & Lee (1974) have studied the accumulation of proline in an inland and a coastal population from North Wales. When grown at low salinities (less than 100 mol m^{-3} sodium chloride), both populations accumulated proline to the same extent, but at higher salinities the coastal population accumulated far more. At 200 mol m^{-3} sodium chloride, the inland population died after three weeks' treatment while the saltmarsh plants, although showing a reduction in growth compared with that in freshwater, survived. These results suggest a major disjunction in salt tolerance between inland and coastal populations, but more populations over a wider geographical range need to be studied before firm conclusions can be drawn.

In the case of *Pl. maritima*, Sheehy Skeffington & Jeffrey (1985) present evidence that inland populations in Ireland, although not exposed under natural conditions to high salinity, nevertheless contain individuals tolerant of relatively saline conditions – behaviour suggesting a contrasting pattern of adaptation to that shown by *Armeria* in North Wales.

Rumex crispus

A. maritima was discussed as an example of a species, widespread on saltmarshes, with a restricted inland distribution. Another group of species occurs which is characterised by occupying a limited habitat on saltmarshes, but which is widespread inland. These are plants of the driftline, many of which are common inland weeds. Again, these are plants of open habitats but, unlike *Armeria*, would have been unlikely to have occupied refugia on poor soils or at high altitude. Current inland habitats are essentially man-made and, during the forest maximum, the species are likely to have been restricted to nitrophilous coastal sites.

There have been few comparative studies of coastal and inland populations of these driftline species. The most detailed is that of Cavers & Harper (1967) on *R. crispus*. The coastal populations involved were from the driftline on shingle beaches, but morphologically similar forms (recognised taxonomically as var. *littoreus* Hardy) are found on saltmarsh driftlines. Driftline and inland populations of *R. crispus* are normally clearly isolated spatially. Cavers & Harper (1967) showed that 'seed' of maritime *R. crispus*, which has large corky tubercles, floated in seawater for longer than inland seed and gave more rapid germination when transferred from sea- to freshwater. The coastal populations rooted better under a mulch of seaweed, but when grown mixed with the inland form on garden soil were at a disadvantage. The very dense panicles of the coastal form suffered less gale and salt-spray damage than those of inland plants. Inland plants frequently flowered in the first year of growth (clearly of advantage to a weed in a potentially ephemeral habitat), but no driftline plants flowered before the second year.

It would seem that driftline and inland populations differ in many ways at all stages in the life cycle.

Discussion

The examples discussed above by no means exhaust the cases of genecological studies of saltmarsh plants. However, they serve to demonstrate a number of different patterns of variation from a taxonomically diverse range of plants.

Many saltmarsh plants have been shown to consist of a range of ecotypes, each adapted to a particular niche within the wider saltmarsh environment. If it were possible to recognise such genotypes easily, the precision with which plants could be used as environmental indicators would be much improved. Unfortunately, much of the variation is in terms of physiological characteristics, and phenotypic plasticity may disguise the genotype, so that, for practical purposes, formal taxonomic recognition of subspecific categories is not possible in most cases. Sometimes physiological races and morphological taxa coincide as in the case of *F. rubra* (Rozema *et al.* 1978), but even here there is a strong possibility of physiological variation within the various forms of the species. In other cases, morphologically defined taxa have been recognised which describe only part of the variation within a widespread species (for example, the recognition of *A. tripolium* var. *discoideus*, while the differences between upper and lower marsh populations of rayed *Aster* are not amenable to formal taxonomic treatment).

Notwithstanding the arbitrariness and unevenness of taxonomic treatments, interpretation and appreciation of intraspecific variation are liable to be influenced, even if subconsciously, by current taxonomic practice. If the annual *Salicornia* is regarded as a single variable widespread species, then the pattern of variation, both within and between ploidy levels is not greatly dissimilar to that shown by other species. If a number of species are recognised, then there is a tendency for our perspective to change; the species are restricted in their distribution and, at least in the cases of *S. europaea* and *S. ramosissima*, apparently with little internal variation. The species now appear unusual when compared with other saltmarsh species.

Our understanding of the significance of many of the patterns of variation shown by saltmarsh plants is limited. While it can easily be appreciated that physiological salt tolerance will confer advantages in a saltmarsh to genotypes possessing it compared with non-tolerant inland races, it is not immediately obvious what advantages possession of certain character states may confer. Why, for example, should raylessness in *A. tripolium* apparently be favoured in low-marsh habitats? In studies of electrophoretic variation, the choice of enzymes to be investigated is rarely related to any perceived advantage of particular isoenzymes in a given environment.

Rather, the variation revealed is regarded as a generalised indicator of the differences between populations. Most genecological studies of saltmarsh plants have been involved with identifying components of variation that are genetically determined and correlating them with environmental conditions. Experimental manipulation of populations to assess the fitness of particular genetic combinations is only rarely practised (but see Davy & Smith 1985). We are also largely ignorant of mechanisms which allow the maintenance of patterns of variation; the study of gene flow in saltmarsh plants has been largely neglected (see Chapter 5).

The various examples quoted do not allow for many generalisations – for species also occurring in non-saline habitats then the saltmarsh forms are likely to be constitutively more salt-tolerant (but slower-growing in non-saline habitats); species with a wide habitat tolerance on saltmarshes are likely to contain numerous genotypes. Selection for greater tolerance of anaerobic soil conditions is also likely in saltmarsh populations of widespread species (Davies & Singh 1983; Burdick & Mendelssohn 1987). Although we might predict that widespread species will show genetic variation, we could not *a priori* predict how this variation might be expressed; recognition of syndromes of adaptation to particular micro-environments, even if they occur, is not yet possible.

Gray (1985*a*) suggests that there is some evidence for consistant change in the life cycle characterisation (allocation of resources to sexual reproduction and longevity of individuals) as species approach their habitat limits. Thus, *Pu. maritima* and *A. tripolium* both allocate a larger proportion of their resources to flowering and fruiting in the upper marsh. Gray (1985*a*) also hypothesises that widespread inland species occurring on upper marshes (components of the glycophytic element) may show adaptations less highly evolved to fit the environment and involving fewer loci than those in more halophytic species. He argues that salt tolerance in *F. rubra* and *A. stolonifera* is 'physiologically less complex than that of the "true" saltmarsh perennials'. As discussed on pp. 275, the physiological responses of these grasses to increasing salinity is different from that of more halophytic species; whether they can be regarded as less evolved is a difficult question and it is difficult to estimate the number of loci involved in increased salt tolerances, not least because there is still imperfect understanding of the range of features which might be involved in conferring salt tolerance.

The widespread occurrence of genetic variation in saltmarsh plants is potentially of practical significance. Various engineering works in estuaries may cause extensive disturbance to saltmarshes or even, on occasion, lead to the development of new marshes. Rehabilitation and colonisation of such sites is generally uncontrolled but, in future, there may be instances where it is desired to speed up the process by planting appropriate species. For such an exercise to be cost efficient it would clearly be sensible to utilise appropriate genotypes rather than to rely on whatever material happened to be available. In certain circumstances, it may even be necessary to carry out a breeding program to tailor a genotype to a particular environmental situation.

Also of potential use is the occurrence, particularly in northwest Europe, of a number of glycophyte species with salt-tolerant upper-marsh races. Such races may be of immediate value in amenity planting on, for example, salt-damaged road verges (Humphreys 1982), but are more likely to be of value as sources of genetic material in breeding programmes for salt tolerance in agriculturally important species. Few attempts seem to have been made to utilise this resource but, in addition to pasture grasses, European upper saltmarshes also include potentially valuable legumes such as *Trifolium* spp. and *Lotus corniculatus*.

The variation within saltmarsh species is such as to limit the generalisations which can be made about many aspects of a species' biology. Thus, although we are justified in arguing that a saltmarsh species is more salt-tolerant than a glycophytic species, quantitative generalisations about tolerance limits at the species level would be unjustifiable unless the species concerned had been subjected to detailed genecological investigation.

Although the discussion on genetic variation has concentrated exclusively on vascular plants, there is no reason to doubt that similar variation would also be found in other members of the saltmarsh biota.

Saltmarsh fauna

Saltmarsh as a habitat for animals

The saltmarsh fauna can be subdivided in various ways: taxonomically, by ecological affinity (the saltmarsh fauna includes groups with marine, freshwater and terrestrial affinities and evolutionary origins), trophically, by subhabitat occupied, or by residence status. Some animals are permanently resident in saltmarshes; others are visitors, some seasonally (as are many migratory birds), some only at high tide, others only at low tide, some at particular stages in their life cycle, many only casually. For some visiting species access to saltmarsh is essential, for others it may be important (or even essential) for particular populations but is not essential for the species as a whole; for yet other species exploitation of saltmarsh resources is opportunistic.

The biology of the saltmarsh fauna (with particular emphasis on North America) is reviewed by Daiber (1982). The north European fauna is discussed by Smit *et al.* (1981) and Dijkema (1984*b*) and a general introduction is provided by Long & Mason (1983). Geographically, available data on fauna provide a less-complete coverage than those on flora. Although the most comprehensive data are from North America and northern Europe, even in these regions relatively few sites have been studied in detail and geographic patterns in the distribution of species are, for most taxonomic groups, imperfectly understood. In addition to the poor geographic coverage, very few studies are comprehensive in the range of subhabitats or taxonomic groups surveyed. Sampling methods required to census different groups of organisms differ and even for studies of particular ecological subgroups recording is often selective – thus while there are many instances of sampling the benthos of marine origin most records are of the macrobenthos and there are far fewer studies which include micro- and meio-fauna.

The open pioneer zone of saltmarshes and the floor of creeks provide an extension of the estuarine mudflat habitat with a basically similar fauna (Jackson, Mason & Long 1985). The biology of the estuarine fauna has been the subject of several reviews (including *inter alia* Green 1965; McLusky 1981; Barnes 1984). The vegetated saltmarsh surface provides a range of microhabitats for fauna (see Fig. 7.1). Many of these microhabitats, particularly those associated with different plant organs, are not unique to saltmarsh, although the physical conditions prevailing within each microhabitat will be strongly influenced by their saltmarsh surrounds. Saltmarsh pans, however, provide a specialist environment (Nicol 1935). Pans vary considerably one from another in their environmental conditions and also may experience major temporal fluctuations. Even in temperate latitudes pans in the upper marsh may become hypersaline for considerable periods. Pans may provide breeding grounds for mosquitoes and other biting insects, the larvae of which may tolerate very high salinities (Foster & Treherne 1976). Attempts to control such insects have led to considerable modification of the saltmarsh environment and the expenditure of large sums of money (see Daiber 1982, 1986; Kay, Sinclair & Marks 1981).

Driftline accumulations of plant material, some from the saltmarsh but much composed of inwashed algae and sea grasses, support a large fauna (Rammert 1981), although one probably not differing significantly from that of drift litter on sandy or shingle beaches. These drift-litter beds may be important sites in the nutrient cycling processes of saltmarshes but quantification is lacking. However, the fauna clearly play a major role in the processes which ultimately result in mineralisation of the litter.

Unlike the flora many members of the fauna are mobile and thus can avoid experiencing adverse environmental conditions by movement. However, sedentary species, or those whose powers of movement are limited, face a number of particular problems in surviving in saltmarsh environments. [Strategies of terrestrial invertebrates in saltmarshes are analysed by Heydemann 1981; Schaeffer 1981; and van Wingerden, Littel & Boomsma 1981.]

For organisms of marine affinities there is a risk of desiccation when the marsh surface is exposed to air. These organisms are most abundant in the lower marsh or in microhabitats which remain moist for long periods of time. Gastropod molluscs may endure dry periods by sealing off the shell with their opercula while many other animals retreat into burrows in the sediment. The density of burrows, of, for example, crabs, may be high in some areas of marsh and the presence of burrows appreciably increases aeration of the soil (Clarke & Hannon 1967; Montague 1982; Bertness 1985).

Invertebrates of terrestrial origin, which are air-breathing, are at risk when marshes are inundated by the tide. Many saltmarsh arthropods trap bubbles of air under epidermal hairs, permitting respiration to continue during submergence. In some small spiders (spiders are well represented on saltmarshes and are probably the major invertebrate predators of terrestrial origin) the epidermis over the lung books acts as a tracheal gill, again permitting respiration when submerged (Heydemann 1979).

The waterlogged anaerobic nature of many saltmarsh soils restricts utilisation of the habitat by air-breathing invertebrates, although the improved aeration of the immediate rhizosphere by oxygen diffusion from roots may provide a microhabitat for some species. However, in general, soil-dwelling, air-breathing invertebrates are most numerous in the upper marsh or in comparatively well drained creek edges (Treherne & Foster 1979). The aphid *Pemphigus trehernei*, for example, lives on the roots of *A. tripolium*, but is restricted to plants growing on the edges of creeks (Foster & Treherne 1975).

Phytophagous species, if restricted to the aerial parts of taller-growing plants, may avoid the direct effects of tidal inundation altogether. Nevertheless, the low water potential and high salt levels of much plant tissue remain problems for any saltmarsh herbivores.

The flooding and ebbing tidal currents may dislodge many animals from their habitat and carry them away from the marsh. Collapse of creek banks, which may occur on ebb tides, can destroy populations of soil-dwelling invertebrates (Treherne & Foster 1979).

Many herbivores avoid contact with tidal currents by spending much of their life cycle within plant tissue (Heydemann 1979; Long & Mason 1983). Many terrestrial invertebrate species show behavioural patterns which reduce the possibility of the adverse effects of tidal inundation (discussed in various contributions to Smit *et al.* 1981). The springtail *Anurida maritima* shows a true circatidal rhythm of 12.4 h (Foster & Moreton 1981) emerging to forage shortly after the tide ebbs and returning underground at least an hour before the tide returns. Many parts of a saltmarsh will be free from tidal flooding for considerable periods but the activity rhythm of *Anurida* persists during periods of non-submergence so that it remains in synchrony with the tides when flooding next occurs. A rather different activity pattern has been demonstrated in the carabid beetle *Dicheirotrichus gustavi*, which like most carabids is active at night and returns to a soil burrow before dawn. However, nocturnal activity is suppressed during periods of submerging tides (Treherne & Foster 1977), similar behaviour patterns being shown both on the North Norfolk and Essex coasts even though light tide in these regions is about six hours out of phase (Foster 1983).

For animals inhabiting saltmarsh pans fluctuations in salinity are a major feature of their environment. The relatively restricted fauna of this habitat demonstrate a number of physiological adaptations to changing salinities (Foster & Treherne 1976; Lockwood & Inman 1979). Additional stresses in the pan environment are provided by fluctuations in the oxygen concentration of the water and in temperature.

Invertebrate fauna

The fauna of saltmarshes in the Wadden Sea is described in Smit *et al.* (1981) and Dijkema (1984*b*); while the details of their accounts (summarised below) are specific to north west Europe, the general pattern applies more widely.

In the pioneer zone of the Wadden Sea marshes (frequently open and with *S. anglica*, *Pu. maritima* and *Salicornia* spp.) the meio- and micro-zoobenthos comprises about 300 species with an average abundance of 200–500 individuals cm^{-2}. The macro-zoobenthos comprises about 100 species but only a few of these contribute significantly to the total biomass – the polychaete *Nereis diversicolor*, the bivalve mollusc *Macoma baltica*, the gastropod *Hydrobia ulvae*, and the amphipod *Corophium* sp. The density of these few species may be very considerable – 120 000 individuals m^{-2} for *Hydrobia*, 400 000 m^{-2} for *Corophium* sp.

The majority of species are either detritivores or consumers of the microflora (algae, cyanobacteria and bacteria); few species are directly dependent on higher plants with only three phytophagous species each being recorded from *S. anglica* and *Salicornia* spp. (Heydemann in Dijkema 1984*b*).

The continuously vegetated lower marsh offers much greater habitat diversity than the pioneer zone and supports a richer fauna. Within the low-marsh zone there is a shift in the composition of the fauna with increasing elevation – below mean high water 95% of the arthropod species are crustaceans, above this level 80–90% are insects (Heydemann in Dijkema, 1984*b*). About 130 of the insect species are phytophagous (see Table 2.2), some generalist but the majority specialised in their feeding habits, not only to particular host species but to restricted parts of those hosts.

Although crabs occur in the low-marsh zone in Northern Europe, they are a less prominent component of the fauna than in other parts of the world – for example, eastern North America (Montague *et al.* 1981) or eastern Australia (Robinson *et al.* 1983) (despite the abundance of crabs in eastern USA they are far less common on the west coast – Seliskar & Gallagher 1983).

The fauna of the midmarsh is considerably richer in species than that of the lower marsh, and, except in creeks and pans, is almost entirely of terrestrial origin. The total number of phytophagous insects recorded is about three times that in the lower marsh and comparable increases in species diversity are recorded for other trophic groups (Heydemann in Dijkema 1984*b*). However, the number of species dependent on saltmarsh is fewer than in the low marsh. Heydemann (in Dijkema 1984*b*) suggests that some 75–80% of the lower-marsh invertebrate fauna is marsh-dependent, compared with 25–50% in the mid marsh – the figure drops further to only 5–10% in the extreme upper marsh. Amongst the phytophagous insects many are generalists but some are host-specific (Table 2.3).

The major invertebrate carnivores in the mid-marsh are spiders of which there is a considerable diversity of species (Heydemann 1979; Schaefer 1981); the importance of spiders in the trophic structure of saltmarshes appears to be a general feature and has been recorded in New Zealand (Paviour-Smith 1956), Australia (Grimshaw 1982) and the east coast of North America (Barnes 1953; Davis & Gray 1966).

The composition of the upper marsh strongly depends on the nature of

Table 2.2. *Numbers of phytophagous insect species occuring on halophytes of lower saltmarshes (from Dijkema 1984b)*

	Puccinellia maritima	*Halimione portulacoides*	*Suaeda maritima*	*Aster tripolium*	*Spergularia media + salina*	*Triglochin maritima*	*Limonium vulgare*	*Plantago maritima*	*Atriplex hastata*
Aphidoidea	4		1	4		1	1		1
Coleoptera–Chrysomelidae					1			4	
Coleoptera–Curculionidae	2						1	6	
Lepidoptera	3	6	7	6		4	2	6	8
Cecidomyiidae	2							1	
Agromyzidae	2			6		2		1	1
Chloropidae	7			1		2			
Drosophilidae				1	1				1
Cicadina	1	1	1	1					1
Other groups	1			6	1				1
Total number of insect species	22	7	9	25	3	9	4	18	13

Table 2.3. *Numbers of phytophagous insect species occurring on halophytes of middle saltmarshes (from Dijkema 1984b)*

	Festuca rubra	*Glaux maritima*	*Artemisia maritima*	*Elytrigia pungens*	*Juncus gerardii*	*Armeria maritima*	*Plantago coronopus*
Aphidoidea	14	1	2	1	2		
Coleoptera–Chrysomelidae	1				1	1	1
Coleoptera–Curculionidae	2	1		2	1		3
Lepidoptera	6	1	2	2	2	3	
Cecidomyiidae			2		1		
Agromyzidae	2		2	3	1		1
Chloropidae	7			3	3		
Drosophilidae			1				
Cicadina	2		1	1	1		
Other groups	2		1				
Total number of insect species	36	3	11	12	12	4	5

the vegetation – for example, the fauna of brackish upper-marsh reedswamps is very different in character from that in a dry grassland transition between saltmarsh and sand dune (Heydemann in Dijkema 1984*b*).

In dry sandy upper-marsh burrowing staphylinid and carabid beetles may be prominent and a number of species of ant may be found, sometimes forming large mounds (Woodell 1974; Kay & Woodell 1976; Nielsen 1981). The dominant phytophagous insects are leafhoppers (Heydemann in Dijkema 1984*b*).

In brackish reedswamp communities a number of specialist pool-living species occur, while *A. stolonifera*, *Scirpus maritimus* and *Phragmites australis* are all important hosts for a range of phytophagous insects.

Vertebrates

Reptiles and amphibia are not normally characteristics of saltmarshes although they may be more numerous in some brackish situations. The other major vertebrate groups are important and characteristic components of the saltmarsh fauna.

Fish

At low tide, fish are restricted to pans and permanently flowing creeks but at high tide they are able to move over the main marsh surface. At high water the fish fauna is augmented by individuals moving into the marsh from adjacent waters.

The fish community of saltmarshes contains a number of species which are permanent residents and juveniles of other species which, as they mature, move offshore. Saltmarshes provide an important nursery habitat for juveniles of many species which are of economic importance in coastal fisheries (for example, see Morton, Pollock & Beumer 1987).

There is no direct grazing on higher plants by fish, which, however, may be major consumers of algae, either on the sediment surface (Polderman 1979*b*) or epiphytes.

Birds

The frequency of tidal flooding limits the value of saltmarsh as habitat for nesting birds. Nevertheless, upper and midmarsh may be free from tidal flooding for periods of weeks, or even months, sufficient to allow a number of species to breed (Smit 1981). Fuller (1982) divides the breeding birds of British saltmarshes into six groups – similar but not identical subdivision would be possible elsewhere in the world (for example, eastern

USA, – Burger, Shisler & Lesser 1982; western USA, – Seliskar & Galagher 1983). The six groups are:

> Colonial species nesting in limited areas of sites and feeding mainly outside saltmarsh. The major example on British saltmarshes is the black-headed gull, *Larus ridibundus*.
>
> Non-colonial species nesting throughout the saltmarsh but with a preference for the slightly higher creek levees. The redshank, *Tringa totanus*, is an example.
>
> Middle- and high-marsh species, often ground nesting, which nest at levels normally free from submergence during the breeding season, for example, skylark, *Alauda arvensis*, and meadow pipit, *Anthus pratensis*.
>
> Species confined to grazed upper-marsh grasslands, notably lapwing, *Vanellus vanellus*, and yellow wagtail, *Motacilla flava*.
>
> Species associated with the transition between saltmarsh and other habitats. A number of species may utilise brackish reedswamp.
>
> Species which nest outside saltmarsh but which utilise saltmarsh during the rearing of the young. In northern Europe, an example of this behaviour pattern is provided by the shelduck, *Tadorna tadorna*.

Saltmarshes are probably best known, at least by the public at large, not for their breeding birds but for the large flocks of other species which may utilise saltmarsh at various times of the year. The value of saltmarsh to these species has been a major factor in the conservation of some major saltmarsh nature reserves in northern Europe and elsewhere.

Although small numbers of waders breed on European saltmarshes, a much larger number of individuals and species utilise estuaries during the winter. However, few waders feed on saltmarshes, although a number may utilise creek beds; feeding occurs mainly on intertidal mudflats. Saltmarshes provide high-tide roosts and the provision of secure, undisturbed high-tide roosts close to low-tide feeding grounds is important to the long-term conservation of waders.

Wildfowl, both ducks and geese, congregate on saltmarshes in large numbers, for both feeding and roosting. The impact of grazing and trampling by wildfowl may be considerable (Jefferies, Jensen & Abraham 1979; Smith & Odum 1981; Smith 1983).

Winter use of north European saltmarshes is not restricted to waders and wildfowl, which in fact contribute only 40% of the total species on British saltmarshes in winter (Fuller 1982). A number of passerines are characteristic of saltmarshes in winter and may be present in large numbers. Seed-eating finches are prominent, particularly linnets (*Acanthis cannabina*) and greenfinches (*Carduelis chloris*), while saltmarshes in eastern England, particularly around the Wash, are the major overwintering habitat for twite (*Acanthis flaverostris*) which feed on seeds of *Salicornia* spp. and *A. tripolium*.

Mammals

The impact of grazing by domestic livestock on saltmarsh vegetation in various parts of the world is considerable. The extent of grazing by large mammals prior to grazing by man's livestock can only be speculated upon but may have been locally important. At the present day, for example, red deer (*Cervas elephas*) graze at times on saltmarshes in western Scotland, while in southern Australia various macropods, kangaroos and wallabies, are occasional visitors to upper saltmarshes.

On European marshes grazing by rabbits, *Oryctolagus cuniculus*, may be intense locally in upper-marsh zones, particularly where sand dunes are adjacent to saltmarsh (Rowan 1913) (see Fig. 6.1).

Hares may graze at lower zones on the marsh than rabbits and on occasion may be seen on mudflats. In the upper marsh, particularly in areas dominated by rank growth of *F. rubra*, voles (*Microtus* spp.) may be common.

In North America, muskrat, *Ondatra zibethicus*, favours brackish upper saltmarsh habitats, where it may be present in large numbers. Frequent burning of these areas, to facilitate trapping and with the intention of promoting vegetation development favourable to muskrat may have been an important factor determining present-day vegetation composition (Hackney & de la Cruz 1981).

Zonation in the fauna

While visiting elements in the fauna may range widely across the marsh and the resident macrofauna may also travel extensively, much of the fauna is relatively restricted in its occurrence within a single marsh. While there is ample evidence that the fauna associated with different vegetation zones is often very different (Paviour-Smith 1956; Davis & Gray 1966; Heydemann in Dijkema 1984*b*), in most instances, there are inadequate data to determine whether there are sufficient coincident species' boundaries to define faunal zones independently from vegetation zones or, if such faunal zones can be recognized, whether they share the same limits as those of the floral zones.

The same questions as could be asked of the zonation patterns in the flora (p. 55) are applicable to the fauna. While explanations can be given to account for the distributions of some species, such information is lacking for the majority of the saltmarsh fauna and any overall synthesis of the phenomenon of zonation is a long way off.

Faunal communities

Plant communities, however defined, are units which can be related to areas occupied. The concept of an animal community, while it may have spatial dimensions, is more concerned with functional interrelationships between species. In faunal studies the term community is often used even more loosely than it is in vegetation studies and covers grouping of related organisms (i.e. the bird community of a saltmarsh), groupings at the same trophic level, as well as the whole range of organisms within a particular area.

Few studies of saltmarsh fauna have adopted a comprehensive approach. One of the earliest attempts to unravel the relationships amongst the whole fauna was Paviour-Smith's (1956) study of the trophic structure of a New Zealand saltmarsh. If trophic diagrams are to be more than intelligent speculation, then very considerable painstaking research is required. While techniques such as the measurement of stable isotope ratios (p. 352) may facilitate the tracing of food webs they cannot indicate the trophic level at which utilisation occurs. It is clear that some organisms cannot easily be assigned to any single trophic level, either because feeding habits change during development or because, even by adults, food from a whole range of resources is consumed (Paviour-Smith 1956; Montague *et al.* 1981).

Flora–fauna–environment interactions

The maintenance of saltmarsh ecosystems is based on interactions amongst and between the biota and the environment. A variety of these interactions are discussed as appropriate in other chapters. However, in this section examples of the role of the flora in creating specific faunal habitats, and of the fauna in producing environmental conditions influencing the flora, are discussed.

Fauna may play an important role in the sedimentary processes leading to marsh growth (p. 32, Frey & Basan 1985; Bertness 1984). Elements of the fauna may, through their activities, modify the soil environment so as to influence plant growth. Particular attention has been given to the importance of crabs as habitat modifiers, both in saltmarsh (Montague 1982; Bertness 1985) and mangroves (Warren & Underwood 1986). Burrowing crabs may be present in large numbers (Bertness & Miller 1984), significantly influencing microtopography, sediment chemistry and drainage.

Bertness (1985) has investigated in some detail the interactions between the fiddler crab *Uca pugnax* and the growth of *S. alterniflora*; *U. pugnax* is absent from soft, uncohesive sediment at lower levels than the marsh as burrows cannot be maintained in such sediment. It is also largely absent from stands of the short form of *S. alterniflora* where the dense root mat prevents burrowing. *Uca* is therefore most abundant in stands of the tall form of *Spartina*, where the substrate is sufficiently cohesive for burrows to be maintained but where root density is lower than in stands of the short-form plants.

Bertness (1985) was able to set up pens in the different habitats in a marsh from which crabs could be excluded. Even in one growing season reduction in crab activity in stands of the tall form of *Spartina* resulted in substantial reduction in above ground-production and an increase in below-ground biomass. Assessment of sediment properties showed that crab burrows increased soil drainage, soil oxidation and the decomposition of below-ground debris. Bertness (1985) suggested that *Uca* and *Spartina* enjoyed a facultative mutualism. In soft sediment where burrows could not be otherwise maintained, the invasion of *Spartina* and the presence of roots and rhizomes increases sediment firmness and permits crabs burrowing. At higher elevations, burrowing activity creates soil conditions favouring productivity and prevents the establishment of a root mat. In high-elevation but poorly drained sites, the dense root mat in short-form *Spartina* stands precludes crab burrowing. Bertness (1985) concludes

> 'while initially dependent on *S. alterniflora* in soft sediments, *U. pugnax* burrowing activity appears to maintain a habitat suitable for continued burrowing, and as a by-product, increases cordgrass production and maintains tall-form *S. alterniflora* stands'.

Crabs are not equally abundant in all saltmarshes, but, nevertheless, it is probable that other burrowing animals may have a similar, but somewhat lesser, effect on sediment properties and plant growth.

While animals can create local heterogeneity in sediment which can advantage plants, plants may modify the rhizosphere so as to create microhabitats favourable for some animals, but the currently available data on the relationship between faunal distribution and roots do not permit generalisations. Teal & Kanwisher (1966) and Teal & Wieser (1966) suggested that oxygenation of the rhizosphere of *S. alterniflora* roots created conditions favourable to nematodes, although Bell, Watzin & Coull (1978) showed non-significant or negative correlations between meiofaunal (chiefly nematodes and harpacticoid copepods) abundance and root biomass. However, Bell *et al.* (1978) did not differentiate between live and dead roots. Osenga & Coull (1983) working at the same marsh as Bell *et al.* (1978) found that nematodes were positively correlated with live-root density, while in contrast harpacticoid copepods were more abundant in areas with larger numbers of dead roots. The abundance of nematodes near living roots could be related to several factors – the presence of an oxygenated rhizosphere or because exudates from the roots favour microorganisms which provide a food supply for nematodes.

At first glance therefore, the different results reported by Osenga & Coull (1983) and Bell *et al.* (1978) appear to be explained by Bell *et al.* not differentiating between living and dead roots. However, Teal & Kanwisher (1966), Teal & Wieser (1966), and Osenga & Coull (1983) all worked in stands of the tall form of *Spartina* with relatively low root density, while Bell *et al.* (1978) studied short *Spartina* with a very high root biomass. It is possible that the differences in underground biomass also effect faunal distribution.

Two further examples of possible interactions between fauna, flora and the environment, involving aphids and ants, can be discussed.

In temperate saltmarshes, a relatively large number of aphid species have been recorded. These include species feeding on roots as well as those feeding on aerial parts (Heydemann 1981).

Aphids are amongst the most serious of agricultural pests, and severe infestations can greatly reduce crop yields (Dixon 1973). The importance of aphids in natural communities has not been studied as extensively. However, the low floristic diversity and, consequently, large dense populations of host plants may mean that saltmarshes are not dissimilar from agricultural systems and that aphids might be expected to reduce host-plant growth. Owen & Wiegert (1976) suggest, on the contrary, that aphids could benefit their hosts. Honeydew secreted by aphids might act as a carbohydrate source for soil-living, nitrogen-fixing bacteria, so increasing the rate of nitrogen fixation and, thus eventually, the nitrogen available to vascular plants.

This hypothesis has appeal as an explanation of how saltmash plants can sustain the very high aphid densities often recorded. Many saltmarsh species use nitrogenous compounds as cytoplasmic osmotica (see Chapter

4) and, yet, available nitrogen is often present in the soil in only small quantities. Jones (1974) showed that addition of glucose to saltmarsh soils could increase bacterial nitrogen fixation.

Foster (1984) has investigated the effect of the aphid *Staticobium staticis* on its hosts *Limonium vulgare* and *L. humile*. Very high densities of this aphid can occur and the potential maximum honeydew production is considerable. Foster (1984) estimates that up to 20 g dry weight of honeydew per m^2 could fall on to the soil every day at the peak of the aphid season. Such an amount of additional carbohydrate should drastically increase soil nitrogen fixation, although no measurements are available. Jefferies & Perkins (1977) showed that added nitrogen could increase flowering and seed production in *L. vulgare*.

It might be expected, therefore, that infestation with *Staticobium* would increase seed output by *L. vulgare* and *L. humile*. However, Foster (1984) did not record such an effect; dense infestations by aphids could totally prevent successful reproduction, while plants kept free of aphids by use of insecticides produced seed.

It is possible that although aphids prevent reproduction in the years of high aphid densities, the plants benefit from higher soil-nitrogen levels in intervening years, although Foster (1984) considered this unlikely.

Foster (1984) suggested that the *Limonium*/*Staticobium* relationship did not provide support for the Owen & Wiegert (1976) hypothesis. While synthesis of osmotica in *Limonium* requires nitrogen, it also requires carbon sources; the magnitude of the diversion of carbohydrate through aphids at maximum densities may more than outweigh any advantages gained from greater nitrogen availability. It remains possible that at lower aphid densities the honeydew can still stimulate nitrogen fixation and that vegetative growth of *Limonium* may be enhanced, so giving it an advantage in those years when reproduction is favoured.

While the Owen & Wiegert (1976) hypothesis is not supported by the *Limonium*/*Staticobium* example, other grazers may stimulate primary productivity by increasing the availability of nitrogen (Bazely & Jefferies 1985, see p. 361).

Ant nests are of frequent occurrence in the upper saltmarsh, at least in northern Europe (Woodell 1974; Kay & Woodell 1976; Heydemann 1981, Nielsen 1981; van Windergen, Littel & Boosma 1981). The most frequent species recorded is *Lasius flavus*, but other *Lasius* and *Myrmica* species also occur.

Ants interact with the flora in two major ways – by providing physical habitat, and by having associations with aphids.

Terrestrial anthills provide a particular set of habitat conditions and may support distinctive plant communities. Woodell (1974) showed the anthills in the upper marshes on the north Norfolk coast had more *Pu. maritima* and *F. rubra* and less *J. gerardi*, *L. vulgare* and *Pl. maritima* than the immediately surrounding saltmarsh vegetation. In addition, the anthills provided a habitat for the dwarf shrub *Frankenia laevis*, which was restricted to south-facing aspects on the hills. *F. laevis* is at its northern limit on the Norfolk coast, and Woodell (1974) suggested that the southern side of the anthills was microclimatically more favourable than other aspects for growth. Woodell (1974) showed that the anthill soils were lighter textured and better drained than those in the surrounding marsh.

The scarcity of *L. vulgare* and *Pl. maritima* on anthills was attributed by Woodell (1974) to their exclusion by repeated burial, while the abundance of *F. rubra* and *Pu. maritima* may have been a consequence of rabbit grazing, there being some indication that grazing was concentrated on anthills.

Anthills on saltmarshes in Wales also supported a distinctive floristic assemblage, different from that in the adjacent marsh (Kay & Woodell 1976). Rosette, prostrate or creeping perennials were excluded from the anthills, while annual species were significantly more abundant than in the marsh.

Several species of root-feeding aphids are closely associated with saltmarsh ants (Heydemann 1981; Nielsen 1981; van Wingerden *et al.* 1981). The main food source for *L. flavus* on a Danish saltmarsh was aphid honeydew (Nielsen 1981). During the summer, the ants keep aphids on roots in the nest and adjacent galleries; in the winter aphids and their eggs are aggregated in the nest (Nielsen 1981; van Wingerden *et al.* 1981).

Ants may interact with the flora in at least two other ways; as pollinators or as distributors of seeds. Pollination by ants is rare as ant secretions may have an antibiotic effect on pollen (Beattie 1985). *Glaux maritima* is one of the few species for which ant pollination has been suggested (Faegri & van der Pijl 1979); this claim requires experimental confirmation. Myrmecochory, the dispersal of seeds by ants, is an important process in many communities (Beattie 1985) but its occurrence in saltmarshes has not been investigated.

Although anthills can be considerably higher than the surrounding marsh, they may be completely submerged by flooding spring tides (Kay & Woodell 1976). Survival of ants may be assisted by the trapping of air within the nest mound, but the species found on marshes have considerable capacity to endure prolonged submergence to brackish water (van Wingerden *et al.* 1981).

3

Variation in saltmarsh vegetation

Introduction

Saltmarshes are found on many of the world's coasts and experience a wide range of environmental conditions.

There is a tendency, however, to regard all saltmarshes as being very similar and to base broad generalisations on the results of studies of particular sites. It is important to be aware of the variability of saltmarshes if generalisations about their ecology are to be sustained.

In this chapter variation in saltmarsh vegetation at various spatial scales is discussed. After a broad-brush account of global variation, the saltmarshes of Britain and temperate Europe are discussed in detail and some factors which might be responsible for variation between sites are explored.

The extent to which other attributes, particularly those related to ecosystem functions, co-vary with the flora and vegetation remains to be documented.

Patterns of variation

Although in certain features there are resemblences between all saltmarshes, there are also major differences. Frey & Basan (1985) suggest that differences between saltmarshes are related to eight factors:

(1) Nature of the local marsh flora (and fauna);
(2) Effects of climate, hydrology and soil factors on the flora;
(3) Availability, composition, mode of deposition and compaction of sediments;
(4) Organism–substrate interrelationships;
(5) Topography and areal extent of the depositional surface;
(6) Tidal range;
(7) Wave and current energy; and
(8) Tectonic and eustatic stability of the coastal area.

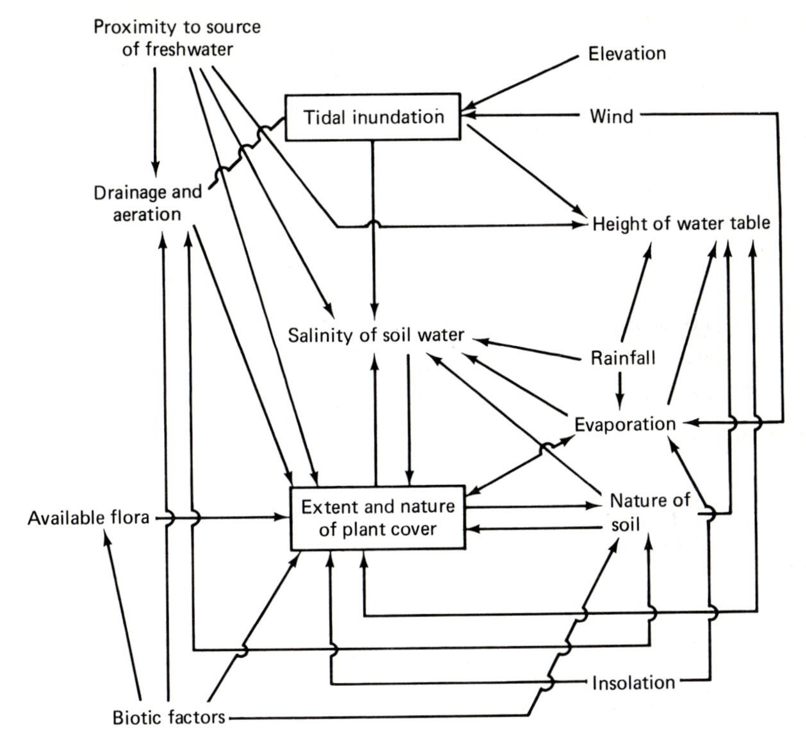

Fig. 3.1 Environmental factors influencing saltmarsh vegetation (the holocoenotic complex of Clarke & Hannon 1969.) (Modified from Clarke & Hannon 1969.)

Clarke & Hannon (1969) identify a number of these factors as components of the holocoenotic complex of factors determining the nature of saltmarsh vegetation at any particular site (Fig. 3.1).

The literature contains two main approaches to saltmarsh typology – one giving priority to biological (usually flora and vegetation) criteria, and the other stressing the physical environment (hydrological and geomorphological features). Ideally, any recognition of marsh types should involve consideration of all potential differences between sites. However, many of these factors have been little studied and most classifications of marsh types have reflected bias in choice of attributes.

There are likely to be complex interrelationships between the various factors determining variation between saltmarshes, so that the biota (part, at least, of which can be observed comparatively easily) may not only provide the means for recognising marsh types, but also serve as indicator of various physical environmental factors. Because of different evolutionary histories, it is possible that the biota of separate geographical regions will differ despite an otherwise similar physical environment; thus it is unlikely that a classification based solely on physical features will always prove to be a predictor of the biota.

Of the various biotic attributes, the vascular flora and vegetation are perhaps most amenable to rapid survey and, because of the relatively small size of the flora, surveys can frequently be conducted by non-specialists. The degree to which biogeographic patterns in the fauna are congruent with those in the flora remains unknown, although in the case of phytophagous fauna with a limited host range, distribution will be a function of that of the host plant.

Patterns of variation in saltmarsh vegetation may be perceived at a range of spatial scales (Beeftink 1965; Westhoff 1985).

Microscale

Patchiness in the distribution of individuals of particular species at a scale of several centimetres to (in the case of shrubby plants) several metres.
Pattern associated with recurrent microtopographic variation, for example, between creek levees and intercreek basins, at a scale of metres to tens of metres (Beeftink 1966).

Mesoscale

Zonation, the occurrence of distinctive communities, or assemblages of communities, arranged in belts parallel to the shoreline, at a scale of tens to hundreds of metres.
Estuarine gradients, changes in community and species distribution from the seaward end of an estuary to the tidal limit (see Ranwell 1968), operating over a scale of kilometres.

Macroscale

Variation at a scale of hundreds of kilometres. There is spectrum of variation, from essentially regional variation (such as that described by Adam 1978), to much broader scale, continental patterns, normally correlated with latitude.

At this broadest scale of variation, although some species may be in common between several areas, the pattern we perceive is a reflection of differences in the pool of species available for the colonisation of saltmarsh habitats. Within each of the various geographical areas there may be considerable variation in both flora and vegetation between sites in response to smaller-scale patterns of environmental variability.

The broadest-scale patterns in saltmarsh vegetation presumably reflect the currently acting effects of climate and the legacy of past events expressed as the pool of available species. Microscale variation, although modulated by competitive interactions at both intra- and inter-specific levels, is influenced by flooding and salinity regimes and by sediment type – factors strongly correlated with tidal regime. Mesoscale variation is, again, in large part, a reflection of variation in sediment type, salinity, and flooding regimes and can be correlated with tidal regime, position along the estuarine gradient and source and nature of sediment supply. However, the expression of these factors in the vegetation may be strongly modified by other, particularly anthropogenic, factors.

Variation between sites includes components of both macro- and mesoscale patterns. For many purposes, it would be useful to segregate out both purely local and macroscale variation from mesoscale variation, much of which will be in response to repeated environmental patterns – that is, to separate out the singular from the recurrent.

This could be achieved either by identifying recurrent mesoscale patterns of variation and assigning marshes to a series of classes on the basis of these patterns regardless of any broader-scale patterns of variation or by establishing a descriptive framework to accommodate macroscale variation and recognising mesoscale patterns within the geographically extensive groups. Both approaches have been attempted, although, for many ecological purposes, an approach giving priority to mesoscale patterns is potentially more appropriate. However, the currently available data are such that, whichever approach is taken, only an approximation to a classification can be attained.

The units of any classification of marsh types can be treated as independent, in a non-hierarchial scheme, or arranged into a hierarchial framework. Even if the units themselves are determined by botanical criteria, it is possible that the framework in which the units are disposed may be better based on other criteria (for example, geomorphological or hydrological).

Geomorphological classifications of coastal types are many and varied (see Bird 1984; Davies 1980; King 1982) with considerable differences in emphasis on particular attributes between schemes. There is also a number of classifications specifically of estuary types (Kennish 1986). Many such classifications produce insufficient categories of saltmarsh to be useful in an ecological context. Nevertheless, a simple subdivision of saltmarshes ac-

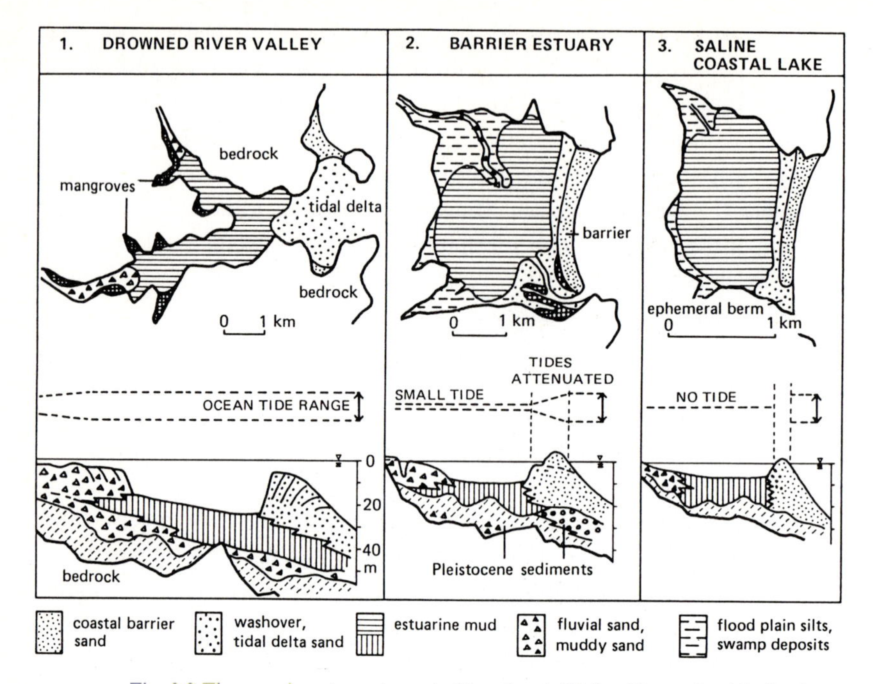

Fig. 3.2 Three main estuary types in New South Wales, illustrating idealised sediment distributions in plan and in section. Tidal ranges within estuaries are shown relative to that in the open ocean. (From Roy 1984.)

cording to their geomorphological setting (bayhead, estuarine, barrier island, etc.) has historically provided a framework for discussion of saltmarsh vegetation (Chapman 1960*b*).

For some purposes, it has been thought desirable to incorporate saltmarshes within a broader classification of wetlands (e.g., Cowardin *et al.* 1979; Cowardin 1982). Such classifications frequently embody a basically geomorphological framework, with further input from vegetation and soil and water chemistry. These classifications may become complex and, to date, have not achieved great popularity outside the United States. For most purposes, discussion of variation in saltmarshes is likely to be independent of any consideration of other wetland types and a classification which concentrates upon saltmarsh to the exclusion of extraneous information is likely to be of more immediate value.

Roy (1984) has proposed a classification of estuaries in New South Wales in terms of three basic types and a number of stages of geomorphological evolution (see Figs. 3.2, 3.3). He has suggested that a number of physical and biological attributes of estuaries will vary with position in this classifi-

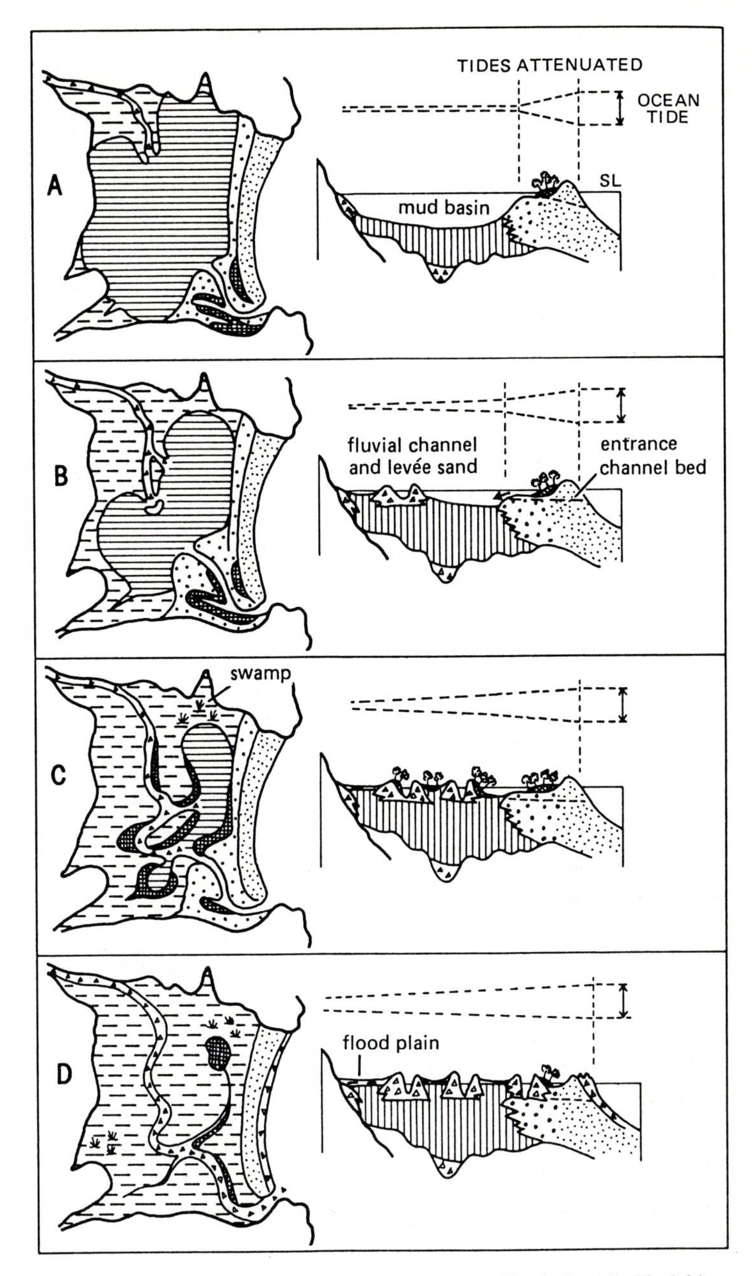

Fig. 3.3 Stages in the evolution of a barrier estuary. (Symbols as in Fig 3.2.) (From Roy 1984.)

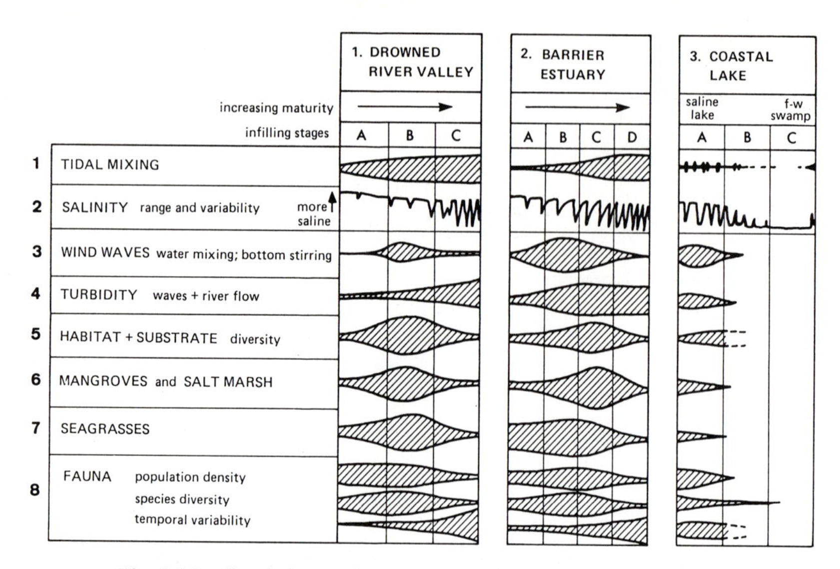

Fig. 3.4 Predicted changes in various properties of estuaries during geomorphological evolution. No absolute correlations are implied between estuaries of different types. (From Roy 1984.)

cation (Fig. 3.4). These hypotheses are amenable to verification, although this has yet to be done. However, preliminary observations are not at variance with Roy's proposals.

New South Wales estuaries have limited development of saltmarsh and fall into a restricted range of geomorphological types. Nevertheless, Roy's (1984) approach would appear to have considerable potential, both in describing current biological patterns and in predicting future changes (particularly in response to accelerated geomorphological change induced by human activity). It would be valuable to extend his approach to a much wider range of estuarine types and to explore correlations between the geomorphological classification and biological features in far greater detail.

Dijkema (1984*a*) has adopted a geomorphological framework for the classification of European saltmarshes. The primary division in this classification is on the nature of the sedimentary material forming the substrate for marsh development. The classification is intended primarily as a means of subdividing the variation in European saltmarshes and does not carry the predictive element of future marsh development which is a feature of Roy's (1984) scheme. Nevertheless, a number of biological attributes are correlated with the geomorphological types recognised by Dijkema (1984*a*).

At a mesoscale, the pattern of salinity within an estuary is normally predictable, although the absolute position of particular isohalines will vary during the tidal cycle and, over the longer term, seasonally and between

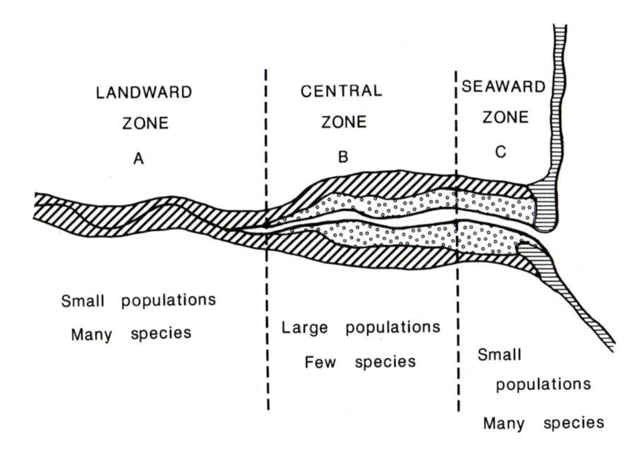

Fig. 3.5 A model of biotic diversity within estuaries. (Redrawn from Ranwell 1968.)

years. Despite this variability, it is frequently possible to divide estuaries into sections on the basis of the average salinity of the water. Although the relationship between salinity of interstitial water within the saltmarsh sediment and that of the flooding water is complex, and the interaction between vegetation, microtopography and climate can produce high salinities within upper estuarine marshes, some plant communities are best developed within upper estuarine reaches (Beeftink 1965, 1977*a*). Ranwell (1968) has proposed a tripartite division of estuaries in which diversity of the saltmarsh biota varies from the mouth to the head of the estuary (Fig. 3.5). In this model, diversity varies not only with the average salinity of the flooding water, but is also influenced by habitat diversity – the most seaward portion of the estuary is likely to present greater habitat diversity (and, in particular, to provide transition zones between saltmarsh and either sand dunes or shingle beach) which promotes greatest biotic diversity. However, in many estuaries, it is the marshes in the upper estuary which have been most heavily modified by human activity so that the pattern illustrated by Ranwell (1968) is often not readily apparent.

There are in the literature studies on particular estuaries which propose mesoscale classifications based on biological criteria. Such classifications are frequently unique to the area concerned and are not immediately capable of generalisation to other sites (see, for example, the study by Gray & Bunce 1972 in Morecambe Bay, northwest England). At the macroscale of variation, the most frequently used classification of marsh types is that of Chapman (1953, 1960*a*, 1974*a*,*b*, 1977*a*) which is based on the distribution of flora and plant communities. It could be argued that, at the global scale, it is not necessary to define geographical groupings for saltmarshes. Rather, boundaries could be imposed on the basis of the total terrestrial biota and saltmarshes compared between conventionally defined biogeographic realms. To some extent, the grouping of saltmarshes agrees with standard biogeography, but the homogeneity at family and generic level of the saltmarsh flora permits recognition of interhemisphere affinities which would be swamped by the discontinuities in the total flora. Elsol (1985) has pointed out that Chapman's biogeographic classification of saltmarsh fails to emphasise intercontinental affinities in saltmarsh floras. However, the existing data base for comparison of saltmarsh floras is variable in quantity and does not yet permit a formal analysis of the type favoured by Elsol.

A global view of saltmarsh vegetation

Any synoptic biogeographic classification of saltmarshes must embody many generalisations developed from less than complete data. Although information on flora and vegetation is more readily available for many saltmarshes than for the fauna, there are still extensive areas for which only the barest sketch of the vegetation exists.

Saltmarshes could be compared in terms of floristic elements (groups of species with approximately the same geographic distribution) and floristic regions (geographic areas with similar floras). However, even in countries where the marsh flora is well documented such an exercise has rarely been performed (but see Barbour & Macdonald 1974). Nevertheless, the distribution of certain species and plant communities provides the basis for the few attempts to categorise saltmarsh types at the global scale.

Chapman (1953, 1960*b*, 1974*a*,*b*, 1977*a*) proposed a classification of saltmarsh types based on floristic and vegetational criteria which recognised nine major geographical groupings of coastal sites:

(1) Arctic Group
(2) North European Group
  - Scandinavian subgroup
  - North Sea subgroup
  - Baltic subgroup
  - English Channel subgroup
  - South-west Ireland subgroup

(3) Mediterranean Group
  - Western Mediterranean subgroup
  - Eastern Mediterranean subgroup
  - Caspian subgroup

(4) Western Atlantic Group
  - Bay of Fundy subgroup

New England subgroup
Coastal Plain subgroup
(5) Pacific American Group
(6) Sino-Japanese Group
(7) Australasian Group
Australian subgroup
New Zealand subgroup
(8) South American Group
(9) Tropical Group

Although there have been alternative proposals for particular continents (for example, Frey & Basan 1985; Glooschenko 1980*b*), Chapman's classification has provided the basis for most general discussions of world saltmarshes. The characteristics of the various groups are discussed most comprehensively in Chapman (1974*a*).

The classification was based on an informal analysis of both community and species distributions. In general, geographically disjunct regions remain separate in the classification. For example, in the temperate Northern Hemisphere, four major marsh groups are recognised (North European, Western Atlantic, Pacific American, and Sino-Japanese). However, to achieve this segregation, emphasis is placed on differences at the species level, although there are many genera in common between the regions.

It is possible to suggest other groupings which attempt to highlight parallels between geographically separated groups. One such is presented here: the proposal can be regarded as an hypothesis, but testing the hypothesis will require more rigorous comparative studies between regions than have been performed to date.

The revised classification discussed below recognises the following types of saltmarsh:

Arctic
Boreal
Temperate
European
Western North American
Japan
Australasian
South African
West Atlantic
Dry coast
Tropical

As with Chapman's earlier scheme, this classification is based on a mixture of floristic and vegetation data. In order to cut across conventional geographical boundaries, it has been necessary to postulate that certain communities and taxa are, in the most general sense, vicariant between regions. Data to validate such assumptions are almostly completely lacking. Indeed, it is not even clear that it is realistic to assume that there will be vicariant taxa and communities between regions – given different biotic assemblages and evolutionary histories, is it likely that the saltmarsh habitat will be partitioned so as to produce the same pattern of available niches? In order to justify a claim of vicariance for two species (or communities) in different regions, at the very least, it should be demonstrated that they occupy habitats with similar tidal immersion characteristics and patterns of variation in soil salinity. In the absence of data, however, vicariance is claimed on the basis of subjectively determined similarity of position in marsh zonation, taxonomic affinities, similarity of physiognomy and similarity in associated species. On such bases, for example, it seems reasonable to equate the European *J. maritimus* with the North American *J. roemerianus* and the Southern Hemisphere *J. kraussii* but not, as has been done on a number of occasions, to claim vicariance between either *J. kraussii* or *J. roemerianus* and *J. gerardi*.

Many of the accounts which provide the basic data for a biogeographic analysis of saltmarshes refer to single sites. There is, thus, a problem of interpretation in that the flora and vegetation of any site are not only a function of geographical regional marsh characteristics, but are also a reflection of the site's position on mesoscale gradients of variation. In some instances, these mesoscale gradients may be so strongly expressed as to obscure macroscale relationships. Macroscale patterning is more readily apparent in the more saline marshes – pattern in brackish marshes is less apparent and may transgress boundaries based on overall marsh floras. This may reflect the fact that many important brackish marsh species have even wider geographical distributions than many fully halophytic species.

While a broad regionalisation of marsh types is possible, the boundaries between the major groupings are far from sharp, but are diffuse and spread over broad latitudinal zones. In addition, mesoscale and local variation in environmental conditions may permit different marsh types to occur in close proximity. The problem of defining the limits of the major marsh types is particularly difficult in western North America where, in addition to a latitudinal pattern of variation, there are strong local distinctions between more brackish and saline marshes (Macdonald & Barbour 1974; Glooschenko 1980*b*).

At a practical level, comparison between sites is made more difficult by taxonomic uncertainties. Although saltmarsh floras are relatively small, they contain a number of species groups which are taxonomically critical. In addition, taxonomic revision has resulted in name changes for many species. As nomenclature in general ecological accounts frequently does not strictly follow any particular taxonomic treatment, the problems of determining synonymies is further confounded.

Notwithstanding the 'fuzziness' of the classification, the suggestion of relationships between geographically disjunct regions may provide a basis for hypotheses on the evolutionary history of saltmarsh floras.

The accounts of the various types are not comprehensive. For further details, reference should be made to Chapman (1974*a*). The aim is to illustrate both the diversity and similarity of saltmarsh vegetation and to provide a frame of reference within which more detailed studies can be placed. In the absence of comprehensive studies on saltmarsh fauna and environmental factors, vegetation may provide a proxy measure of total biotic and environmental diversity. In view of the comparative ease by which vegetation surveys can be performed and the increasing need to make rapid judgements on the relative merits of sites, it is highly desirable that the value of vegetation as a broader indicator of site quality be assessed for the range of saltmarsh types.

Chapman's classification (1960*b*, 1974*a*) also included a number of groupings of inland saline areas. In this account, inland marshes are not considered. There are many species in common between coastal and inland saline areas, but there are also differences. For example, the inland saltmarshes of central Europe support a rich and diverse flora, many species of which do not reach the western European coast (Géhu 1984*b*; Polunin & Walters 1985). While a detailed understanding of the origin of the halophyte flora would require study of both coastal and inland habitats, such a study is beyond the scope of this account.

Arctic type

Arctic saltmarshes are the most floristically uniform of all major marsh types. They are species-poor and support few communities.

A general account of Arctic marshes is given by Chapman (1974*a*), while the phytosociology of the plant communities is reviewed by de Molenaar (1974). The majority of communities which have been described are widespread (many being circumpolar) in distribution. While there are relatively few ecological studies of Arctic saltmarshes, the distribution of the Arctic flora is well documented and this information allows extrapolation of community distribution away from the well-studied areas.

Saltmarshes on Spitzbergen have been subject to a number of investigations (Walton 1922; Dobbs 1939; Hadač 1946; Hofmann 1969), while the marsh vegetation on Jan Mayen is briefly discussed by Russell & Wellington (1940). Saltmarsh vegetation in Greenland has been subject to a number of studies with detailed recent accounts being provided by de Molenaar (1974) and Vestergaard (1978). Marshes in the Canadian Arctic are discussed by Polunin (1948) and Jefferies (1977*a*). The Alaskan Arctic marshes are reviewed by Macdonald & Barbour (1974) and Macdonald (1977*a*).

Much of the Arctic coast is exposed to ice action, and along such coasts well-developed saltmarsh is rare, being limited to patches only a few tens of metres across. In bays and inlets relatively sheltered from ice abrasion (although still subject to cryoturbation), extensive saltmarshes with continuous vegetation cover are developed (Jefferies 1977*a*; Macdonald 1977*a*). Estuaries are an important physiographic feature of Arctic coasts, and the rivers flowing into them transport considerable amounts of gravel, sand and silt, much of which is deposited in the estuaries and along adjacent coasts. On the more gravelly saltmarshes, the vegetation that develops has strong affinities to shingle strand communities (Jefferies 1977*a*).

The physical instability of open-shore marshes precludes any long-term successional development of marsh communities. Although mosaics of communities can be found, their distribution reflects the time since last disturbance (and possibly also substrate conditions) rather than giving any indication of long-term successional trends.

In more sheltered areas, a regular zonation of communities may be observed. In discussing the marshes of the Truelove Lowland, Devon Island, Jefferies (1977*a*) suggested that the zonation reflected a transition from brackish to freshwater conditions and not the successional development of communities within a brackish environment. However, in eastern Greenland, de Molenaar (1974) suggested that the observed zonations could be interpreted as reflecting a succession and that this view was at least partly supported by the stratigraphy of plant remains in the soil profile.

The characteristic community of low marshes throughout the Arctic is one dominated by the grass *Puccinellia phryganodes*. This species forms mats by vigorous vegetative growth, but has never been observed to set seed. Plants from eastern North America have been demonstrated to be triploid ($2n=21$) (Bowden 1961), but some populations in Europe may be tetraploid (Sørenson 1953). The apparent absence of seed suggests that colonisation of new sites must be by vegetative means; ice rafting of stolons may provide an effective long-distance transport mechanism (Sørenson 1953). The triploid chromosome number and absence of seed set suggests that North American populations would be geneticallv uniform. Electrophoretic studies by Jefferies & Gottlieb (1983) disprove this hypothesis, demonstrating a high level of variability, both within and between populations. Jefferies & Gottlieb (1983) suggest that this genetic diversity arose as a result of rare successful events of sexual reproduction, although the possibility of such an event occurring remains to be confirmed.

Stands dominated by *Pu. phryganodes* (the Puccinellietum phryganoidis Hadač 1946; Hofmann 1969 – see de Molenaar 1974) are species-poor; the only other species approaching a constant occurrence is the dicot *Stellaria humifusa*. The bryophyte flora of the community is small, but individual species may be locally abundant (Hofmann 1969; Walton 1922; de Molenaar 1974). The community is found on a range of substrates and may be capable of withstanding prolonged periods of submersion.

In situations where freshwater flushing through the sediment occurs, a widespread low-marsh community is the Caricetum subspathaceae Hadač 1946 em de Molenaar 1974, dominated by the sedge *Carex subspathacea*. This community tends to occur on finer sediments (clay-silty sand rather than gravel), and even at low tide the substrate is generally waterlogged (de Molenaar 1974). *S. humifusa* is a widespread, although generally minor, component of this community. The bryophyte flora tends to be richer than that in the Puccinellietum, and pleurocarpous species are more abundant. At higher latitudes, a floristic variant of the community, characterised by the presence of *Carex ursina* and *Dupontia fisheri* occurs (de Molenaar 1974). While *C. subspathacea* tends to replace *Pu. phryganodes* under more brackish conditions on finer sediments in the low marsh, its elevation range is greater and it is frequently recorded towards the upper limit of the marsh.

The upper marsh community in the Greenland marshes studied by Vestergaard (1978) and de Molenaar (1974) is one dominated by the sedge *Carex glareosa* (Caricetum glareosae – de Molenaar 1974). This community is more species-rich than those of the lower marsh, although *S. humifusa* is still very frequent. Other characteristic species include: *Pl. maritima* (a number of Arctic taxa occur within this species); *Potentilla egedii*; and, at Angmagssilik, *F. rubra* (de Molenaar 1974). At Disko in western Greenland, *F. rubra* occurs rarely in the upper marsh, but is approaching its geographic limits in this region (Vestergaard 1978). As with the Caricetum subspathaceae, there may be a rich bryophyte understorey. The Caricetum glareosae is widespread in the Arctic, but is absent from much of northern Canada (de Molenaar 1974; Jefferies 1977*a*).

A number of other *Carex* species have been recorded, either as community dominants or mixed with other species in various Arctic marshes.

In sheltered, shallow, brackish waters, a number of emergent species have been recorded, of which the most widespread is *Hippuris tetraphylla*.

Arctic marshes may be characterised, not only by species' presence, but also by absences. Jefferies (1977*a*) pointed out that annuals such as *S. europaea* agg. and *Suaeda* spp. are absent from the Arctic – the short growing season and low temperatures precluding annual species generally at these latitudes (Savile 1972). A number of widespread Northern Hemisphere perennial species are also absent, most notably *Triglochin maritima*, but others are found, at least locally, in the Arctic (including *F. rubra*, *Pl. maritima* s.l. and *Cochlearia officinalis*).

The salinity of coastal waters in the Arctic is frequently low (Jefferies 1977*a*) and, in addition, many marshes are subject to freshwater seepage and run-off from the hinterland. Nevertheless, moderately high soil salinities may develop in depressions in the upper marsh.

As well as the environmental factors operating in all saltmarshes, plants in the Arctic must contend with a short, cool growing season during which sufficient reserves must be laid down to permit perennation through the extreme winter conditions.

In the Southern Hemisphere, absence of vegetated lands at appropriate latitudes precludes there being an Antarctic analogue of the Arctic marshes.

Boreal type

Boreal type marshes are found over a very wide latitudinal range. Within this range, climatic conditions will vary considerably with latitude, but there is a moderating influence of the maritime conditions so that the climatic gradient is less than that over the same latitudes in fully terrestrial habitats.

Given the wide geographical range of the Boreal type, it is not surprising that there is considerable variation in the vegetation and flora. The type could be subdivided to emphasise both regional and broadscale latitudinal patterns of variation. At the higher latitudes, there is considerable circumpolar similarity, but at lower latitudes there is greater regional differentiation in both flora and vegetation. Nevertheless, despite considerable variation within Boreal saltmarshes, there is a certain integrity to both flora and vegetation and points of difference can be found with both Arctic and more temperate marsh types. Boundaries between marsh types are, however, arbitrary and the whole Boreal type could be seen as a broad ecotone between temperate lower latitudes and the high Arctic.

Boreal type marshes are found over a wide range of environmental conditions. There is variation in type of tidal range regime (semi-diurnal in the Atlantic, mixed in the north Pacific) and prevailing salinities. Many of the more extensive marshes of the Boreal type experience either permanently, or seasonally, brackish salinity regimes. For example, the very extensive marshes on the west side of Hudson Bay, the southern Alaskan and British Columbian marshes, and the marshes of the northern Baltic all occupy coasts with prevailing low salinities and a strong seasonal influence of meltwater.

The boundary between Arctic and Boreal marshes varies in latitude and is not well defined. The biogeographic boundary between the high Arctic and low (or sub-) Arctic is difficult to establish in a whole range of habitats and saltmarsh is no exception. However, it occurs further south in eastern Canada than it does in northern Eurasia.

The Boreal type may be differentiated from the Arctic type by the occurrence of *T. maritima* and *S. europaea* s.l., although at their northern limit both species are rare with disjunct local populations. From marshes at more temperate latitudes, the Boreal type is characterised by the importance in more brackish conditions of various Cyperaceae (particularly species of the *Carex paleacea* complex) and by the absence in more saline sites of *Limonium* spp. Widely distributed Northern Hemisphere species well represented in Boreal marshes include: *G. maritima*, *Su. maritima* (or related species such as *Su. depressa*), *J. gerardi*, *Pl. maritima*, *F. rubra* and *A. stolonifera.* All occur in more temperate marshes and, in the case of *Plantago* and *Festuca,* also extend into the high Arctic.

Arctic marshes were characterised by relatively few species. The majority of these species have distributions which extend south into the Boreal marshes. The species concerned do not share a common southern limit, rather there is considerable variation between species and also between continents. In general, Arctic species occur further south in America than in Eurasia. *Stellaria humifusa*, for example, although of sporadic occurrence on the Pacific coast of America, extends as far south as Washington (Burg, Tripp & Rosenberg 1980), while in the east it reaches the southern end of James Bay (the southern extension of Hudson Bay) (Ringius 1980) and Cape Breton (Nichols 1918). In northern Europe, *S. humifusa* is restricted to northern Fennoscandia, with one isolated occurrence at the head of the Gulf of Bothnia (Tyler 1969*b*; Ericson & Wallentinus 1979). *Pu. phryganodes* and *C. subspathacea* are restricted to northern Fennoscandia in Europe and to Alaska on the Pacific coast of America (Macdonald & Barbour 1974), but extend to southern James Bay in the east of America (Ringius 1980; Riley & McKay 1980; Glooschenko & Harper 1982).

De Molenaar (1974) in his revision of the phytosociology of Arctic and Subarctic marshes suggests that a series of community pairs exist with one of each pair in the Arctic and the other further south. Thus, the Arctic Puccinellietum phryganodis can be differentiated from the more southern Triglochlino – Puccinellietum phryganodis by the presence in the latter of *T. maritima.* The low-marsh sedge community under more brackish conditions in the Arctic, the Caricetum subspathaceae, is replaced further south by the Triglochino–Caricetum subspathaceae, which has *T. maritima* and *A. stolonifera* as differentiating species. The high-marsh Arctic Caricetum glareosae is matched by the Agrosto–Caricetum glareosae with *A. stolonifera* as the differentiating species.

Marshes with strong relationships to the Arctic type have been described from N.W. Alaska (Hanson 1951, 1953), northern Hudson Bay (Jefferies, Jensen & Abraham 1979), northern Fennoscandia (Nordhagen 1954) [although Kristiansen (1977) emphasises that the major communities in northern Norway are to be regarded as subarctic rather than arctic in character], and the Kola peninsula (Kalela 1939) with a rather fragmentary occurrence at the extreme northern end of the Gulf of Bothnia. Somewhat similar communities have been described in northern Iceland (Steindorsson 1954). The marsh vegetation of eastern Hokkaido in which *Puccinellia kurilensis*, *C. subspathacea* and *C. ramenskii* are important species is also similar in character (Ishizuka 1974; Miyawaki & Ohba 1969).

In sheltered brackish conditions, emergent aquatic communities may develop in which, at high latitudes, *Hippuris tetraphylla* is characteristic. Various sedges and other Cyperaceae are also important and, in the particular conditions of the Baltic, *Ruppia maritima*,*Potamogeton pectinatus*, *Zannichellia palustris*, *Naias marina* and various charophytes may form dense submerged communities.

Some of the most extensive Boreal marshes are found on the western side of Hudson Bay and in James Bay (Kershaw 1976; Ewing & Kershaw 1986; Glooschenko 1978, 1980*a*,*b*; Glooschenko & Martini 1978, 1981; Glooschenko & Harper 1982; Ringius 1980; Riley & McKay 1980; Jefferies, Jensen & Abraham 1979; Jefferies *et al.* 1983; Cargill & Jefferies 1984*a*,*b*). This region is characterised by generally low prevailing salinities in the offshore waters (although, through evaporation, high salinities may be attained in depressions), a very gentle gradient from the tidal limit seawards, and active isostatic uplift (about 1 m per century). Kershaw (1976) noted that these marshes are very little dissected by creek systems, although locally pan development occurs [Riley & McKay (1980) illustrate the very high density of pans which may occur]. Jefferies, Jensen & Abraham (1979) showed that Lesser Snowgeese are important agents in the enlargement and creation of pans in these marshes although Riley & McKay (1980) suggest that ice scour is the major pan-forming agent. Conditions in the upper marsh are often essentially those of a freshwater mire during the growing season when this zone is irrigated by run-off of meltwater from above the permafrost (Kershaw 1976). Both Kershaw (1976) and Jefferies *et al.* (1979) indicate that the proportion of the tidal range occupied by these marshes is

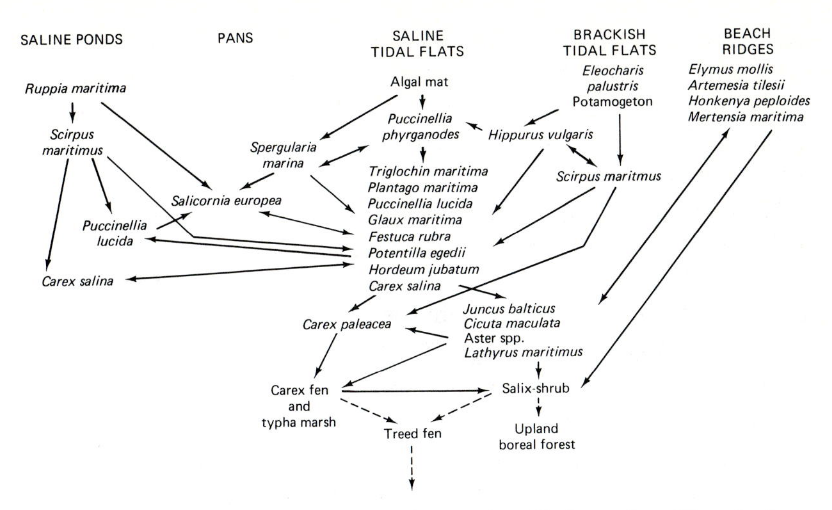

Fig. 3.6 Variation within Hudson Bay marshes. (Redrawn from Glooschenko 1980*a*.)

small and that they may remain above the reach of tides for much of the summer; extensive flooding is most often associated with storms.

Depressions in the upper marsh, which become hypersaline during the summer, provide the habitat for *S. europaea* (Jefferies *et al.* 1983). These populations are at the limits of the species' range and are restricted to the south-facing slopes of the depressions. These slopes are significantly warmer than north-facing ones and the improved microclimate permits greater seed set (Jefferies *et al.*, 1983).

The upper marsh flush zone of the Hudson Bay marshes is frequently dominated by the grass *Calamagrostis neglecta*, the lower limit of the zone often coinciding with the storm strandline (Kershaw 1976). *C. neglecta*, or closely related species, is characteristic of the freshwater/brackish marsh transition at the higher latitudes within the Boreal zone (Southern Alaska – Crow 1971, 1977; Hokkaido – Ishizuka 1974; and northern Fennoscandia – Nordhagen 1954).

Many of the Boreal marshes, particularly in more brackish conditions, consist of communities dominated by various Cyperaceae and grasslands, often in mosaics rather than clear-cut zones. The micro- and mesoscale variation within Hudson Bay marshes is illustrated in Fig. 3.6. Similar distribution patterns have been described from southern Alaska by Vince & Snow (1984).

There are strong similarities between America and Europe in both the sedge and grassland communities. Ringius (1980) notes the resemblance between the James Bay marshes and those in southern Norway and south west Sweden (Dahl & Hadač 1941; Gillner 1960, 1965). Similarities include the occurrence of *Eleocharis* spp. as emergents in sheltered shallow water to seaward of the main marsh and the importance of *Carex paleacea* (or related species) and of rich grassland with *Deschampsia cespitosa* in the uppermarsh, while *S. europaea* is restricted to saline depressions. Related grasslands have also been described from the Pacific American coast. The Hedysarum alpinum–Deschampsia beringensis community described from southern Alaska by Crow (1971, 1977) has clear affinities to these uppermarsh grasslands, although it has a number of geographically restricted taxa. The brackish and freshwater tidal marshes of the Fraser River delta on the Canadian/US border support a range of communities which show affinities to brackish communities in the Baltic (c.f. Bradfield & Porter 1982; Hutchinson 1982; Tyler 1969*a*; Ericson & Wallentinus 1979).

The similarity of Boreal marshes is best expressed in brackish sites and in the upper marsh zone. In more saline sites, particularly in the lower marsh, there is more variation. A range of *Puccinellia* species has been recorded in different regions. These species have a reputation for taxonomic difficulty, and whether all the names employed by ecologists have been validly applied is questionable. In Europe, both *Pu. maritima* and, at higher elevations, *J. gerardi* are found at high latitudes and provide links to the Temperate marshes further south.

Establishing the southern boundary of Boreal marshes is difficult as there is a gradual gradation to marshes of the Temperate type. In Europe, Boreal marshes extend to southern Norway and S.W. Sweden. The occurrence in N.E. Scotland of *Carex recta* in fresh and brackish marshes at the tidal limit in the Wick and Thurso rivers suggests a connection with the Boreal type. Interestingly, *C. recta* does not occur in Norway, where it is replaced by other species in the *C. paleacea* group, but does occur in North America (for example, in James Bay – Glooschenko & Harper 1982). The distinctive marshes of western Scotland (Adam 1978) have many features in common with Boreal marshes and may represent a further geographic extension of the type. The coastal marshes in Iceland (Steindorsson 1954, 1974, 1976) contain some communities which are clearly of the Boreal type, but others which show affinities to grazed temperate marshes. The occurrence of Boreal and temperate saltmarsh communities in Iceland is discussed by Thannheiser (1987).

In eastern Canada, Boreal marshes occur as far south as James Bay. On the Atlantic coast there is a complex situation with communities characteristic of Boreal marshes occurring southwards to New Brunswick and Nova Scotia but mixed with West Atlantic and Temperate elements (Thannheiser

1984). On the west coast of North America, the main southern limit of the Boreal types appears to be in the Fraser River Delta area just north of the Canadian-USA border, although a number of boreal elements are locally important much further south. For example, *Carex lyngbei* extends as far south as northern California (Macdonald & Barbour 1974; Macdonald 1977*a,b*) and the more brackish marshes in Oregon and Washington have many affinities with more northern marshes.

In Japan, Boreal marshes seem to be restricted to eastern Hokkaido. The distribution and composition of marshes on the Russian Pacific and northern Chinese coasts are insufficiently described for recognition of marsh types.

Chapman (1953, 1959, 1974*a,b*, 1977*a*) argued that the marshes of the American Pacific coast were very different from any other marshes and formed a distinct group. The classification advanced here does not support this view. The general character and patterning of the plant communities and their composition at generic (and, in some cases, species) level clearly indicate affinities with both eastern Canada and northern Europe. While these affinities are most clear in the brackish marshes, links can also be detected in the more saline communities.

In the Southern Hemisphere, there are few areas of saltmarsh at comparable latitudes to those of the Boreal type. However, in the south of the South Island, New Zealand, and on the islands of Foveux Strait, Cockayne (1921, 1967) described a community, which he called coastal-moor, which may represent an Austral equivalent of the Boreal marsh type. This community occurs on peaty soil and is exposed to drenching by salt spray rather than tidal flooding. In addition to widespread Australasian saltmarsh species such as *Sarcocornia quinqueflora*, *Samolus repens* and *Selliera radicans*, the flora contains a number of species which Cockayne refers to as subantarctic in character, including *Rumex neglectus*, *Montia fontana*, *Myosotis pygmaea*, *Euphrasia repens* and *Gentiana saxosa*. During the summer, the flowering of the various herbs makes this a colourful community. Although, in terms of latitude, the coastal-moor is found on coasts equivalent to the central Bay of Biscay, it may not be fanciful to see parallels in the plant community composition between the coastal-moor and the *Gentiana dentosa* saltmarsh community described from northern Norway by Gillner (1955).

Temperate type

Regional floristic differentiation is greater within Temperate marshes than in the Boreal type. The various regional types can be recognised as subgroups, but there are sufficient similarities in floristics (particularly at generic level) and physiognomy to justify the amalgamation into a single major group.

The Temperate marshes differ from the Boreal type by an increase in the representation of halophytic species and a decrease in the occurrence of bryophytes. The increased representation of halophytic species may be no more than a reflection of the subjectivity of the definition of species groups. The Temperate marshes of northern Europe have been studied for longer than those elsewhere and act as the benchmark against which other marshes are compared. The suite of species which are widespread in northern Europe are regarded as 'typical' for marshes in general. The most widespread halophytic species in Temperate marshes are also found in the Boreal type. These include: *S. europaea*, *Su. maritima*, *G. maritima*, *T. maritima* and *Pl. maritima* (s.l.). However, Temperate marshes may be distinguished from Boreal marshes by the occurrence of *Limonium* spp. – including, *L. vulgare* and *L. humile* in northern Europe, *L. carolianum* in eastern North America, *L. californicum* in western North America, and *L. tetragonum* in Japan.

The vegetation of Temperate marshes is dominated by graminoids and herbs (and the relatively high diversity of halophytic herbs is one of the most characteristic features of the type), but a number of sub-shrubs occur and the proportion of these increases at lower latitudes. The dicotyledonous herbs of Temperate marshes are frequently gregarious in habit and synchronised in their flowering, producing striking displays such as those of *L. vulgare* in northern Europe. [In other major groups, visually distinctive patterning of saltmarshes is largely a function of foliage colours rather than of flowers.]

The total flora of Temperate marshes is larger than that of any other type. Not only is the halophytic component larger than in Arctic or Boreal types, but there is also a substantial glycophytic element. The greater diversity of flora and communities permits the recognition of subdivision within the type. The great diversity of European marshes is in major part a reflection of the long history of marsh utilisation or modifications, but the overall greater richness of Temperate saltmarshes noted by Davies (1980) reflects more favourable climatic conditions than those at higher or lower latitudes.

European marshes

There have been more studies on the floristics and phytosociology of saltmarshes in Europe than elsewhere.

Temperate European saltmarshes reach their northern limit in southern Sweden, although Temperate elements are found intermingled with Boreal elements further north. Further east in the Baltic, prevailing low salinities make delimitation of Boreal and Temperate marshes more difficult. Two species whose main northern limit serves to characterise the limit of Temperate marshes are *Pu. maritima* and *J. maritimus*. Thus, the occurrence of *Pu. maritima* in the relatively simple marshes in southern Iceland (Hadač 1970; Steindorsson 1974, 1976) suggests that these marshes might be best treated as depauperate Temperate marshes with Boreal elements rather than being of the Boreal type (see also Thannheiser 1987).

From southern Denmark southwards, the dwarf chenopod shrub *H. portulacoides* plays an increasing role in saltmarsh vegetation. The extensive stands of *Halimione* are the major feature of European Temperate marshes distinguishing them from those of other temperate regions.

Western North America

On the Pacific coast of North America, Temperate marshes are best defined by the occurrence of the grass *Distichlis spicata* (Glooschenko 1980*b*; Hutchinson 1982; Dawe & White 1986). However, prevailing brackish conditions in major estuaries allow *C. lyngbei* to remain a community dominant much further south (Seliskar & Gallagher 1983) extending into northern California (Macdonald & Barbour 1974; Macdonald 1977*a,b*). Thus, over a long stretch of coastline, Boreal and Temperate marsh types are intermingled, with salinity conditions being the major discriminating factor between the two types. Where higher salinity conditions prevail, species such as *S. virginica*, *T. maritima* and *Spergularia* spp. (Seliskar & Gallagher 1983) are important in the lower-marsh zone. While several of these species are also found further north, their combined occurrence in the lower marsh may be regarded as a Temperate feature.

From northern California southwards, *Spartina foliosa* becomes the major low-marsh species and, thus at generic level provides a link between the east and west coasts of America. However, *S. foliosa* does not impose the same extreme dominance over such a large proportion of the intertidal as does the east coast *S. alterniflora*. [*S. alterniflora* has been introduced to the west coast (Macdonald 1977*a*) and, should it spread, may cause profound change to western marshes.] *Limonium californicum* also becomes a feature of marsh vegetation at about the same latitude as *S. foliosa* (Macdonald & Barbour 1974; Macdonald 1977*a,b*) and provides a link with Temperate marshes in other continents.

Japan

The Japanese archipelago covers a considerable range of latitude – in the north, the saltmarshes in eastern Hokkaido can be assigned to the

Boreal type; in the far south, there are limited occurrences of mangroves. In southern and central Japan, there are now few large marshes, reclamation for rice cultivation and for industrial development having considerably reduced the once more-extensive area (Ishizuka 1974). The vegetation of these marshes is described by Ishizuka (1974) and by Miyawaki & Ohba (1969). The detailed phytosociological account by Miyawaki & Ohba (1969) suggests that the vegetation lacks the diversity of both species and communities shown in Europe. Nevertheless, the presence in Japan of a number of widespread Northern Hemisphere species (such as *Su. maritima*, *T. maritima* and *A. tripolium* encouraged Miyawaki & Ohba (1969) to assign the communities to positions within the phytosociological hierarchy developed in northern Europe. The rather open low-marsh communities were assigned to the alliance Thero–Suaedion, the mid- and upper-marshes to the class Asteretea tripolii and the brackish reed and sedge communities to the class Phragmitea. However, the dominant grass in the upper-marsh Asteretea communities is *Zoysia sinica* var. *nipponica* which does not have relatives in Europe. In order to accommodate these communities, a separate order, the Zoysietalia, was recognised.

Chapman (1953, 1960*b*, 1974*a,b*, 1977*a*) recognised a distinct Sino-Japanese group of marshes. Little information on the flora and vegetation of marshes in China and Korea is available, but it does appear that they have considerable similarities with those in Japan. Many Chinese saltmarshes have been reclaimed or otherwise modified (Daquan 1983) and, in recent decades, there has been an extensive programme of marsh creation involving planting of *S. anglica* and *S. alterniflora* (Ranwell 1981*b*; Chung 1982, 1985). In the Nakdong estuary in Korea prevailing conditions are fresh to brackish during the monsoon and *Scirpus triqueter* forms extensive stands, with only limited areas of more halophytic vegetation (Doornbos, Groendijk & Jo 1986).

Australasia

Chapman (1953, 1960*b*, 1974*a,b*, 1977*a*) recognised an Australasian group of marshes which, by implication, encompassed all coastal saltmarshes in Australia and New Zealand. However, much of the saltmarsh vegetation in Australia can be assigned to the Dry Coast and Tropical types and is clearly distinctly different from that in New Zealand. There are, however, considerable similarities in both flora and vegetation between south-eastern Australia and New Zealand, and these allow the continued recognition of an Australasian Temperate group of marshes.

Bridgewater (1982) included the marshes of Victoria and Tasmania within a broad grouping of all southern Australian saltmarshes. While there are many species occurring throughout southern Australia, it is possible to recognise a phytogeographic boundary in South Australia separating regions with a markedly seasonal Mediterranean climatic region to the west from the less strongly seasonal temperate regions to the east.

The saltmarshes in Victoria and Tasmania may be characterised by the dominance in the lower marsh of *S. quinqueflora*, the local occurrence of other succulent chenopod shrub communities and upper-marsh communities, including *Stipa stipoides* tussock grassland, *Gahnia filum* tussock sedge and *J. kraussii* and *Leptocarpus brownii* rush lands (Kirkpatrick & Glasby 1981; Kirkpatrick 1981; Bridgewater 1982; Bridgewater, Rosser & de Corona 1981). [*J. kraussii* is very similar in appearance to the European *J. maritimus* and in the older literature was referred to as *J. maritimus* var. *australiensis*.] Grasses play a comparatively minor role in the main marsh, although locally communities dominated by *Distichlis distichophylla* and *P. stricta* occur. In recent decades, the introduced *S. anglica* has spread vigorously at a number of localities (Boston 1981).

Marshes in New Zealand have many species in common with those in south-eastern Australia. *S. quinqueflora* is the major low-marsh dominant, while in the upper marsh *J. kraussii* and, at least locally, *S. stipoides* occur. The herbs *Samolus repens*, *Selliera radicans*, *Cotula coronopifolia* and *T. striata* are frequently abundant (Cockayne 1921, 1967; Chapman & Ronaldson 1958). Grasses, apart from introduced species, are relatively uncommon. However, there has been extensive growth of *S. anglica*, particularly in estuaries on the South Island of New Zealand (Bascand 1970).

Saltmarshes in New South Wales have much in common with those in Victoria, but also show some considerable differences. A number of species widespread in Victoria are absent or rare in New South Wales, and communities dominated by the grass *Sp. virginicus* become increasingly extensive at lower latitudes. The phytogeographic affinities of these marshes remain to be fully explored.

While the existence of an Australasian Temperate group of marshes can be sustained, what grounds are there for regarding this grouping as a subset of a broader global Temperate marsh type? There are some floristic elements in Australasian marshes for which it is difficult to suggest analogues elsewhere (*Samolus repens*, *Selliera radicans* and, in Australia only, *Wilsonia* spp.). However, the occurrence of *T. striata*, *L. australe* and *Su. australis* suggests, at the generic level, a relationship to the Northern Hemisphere Temperate saltmarshes. *J. kraussii*, both in appearance and in its ecology, appears to be the equivalent of the European *J. maritimus*.

The flora of brackish habitats is more uniform geographically than that in more saline habitats. Nevertheless, it is striking that a number of species in brackish marshes in south-east Australia are apparently conspecific with species in similar habitats in the Northern Hemisphere. Many of these species are members of the Cyperaceae (including *Schoenoplectus validus*, *S. litoralis*, *S. pungens*, *Bulboschoenus fluviatilis*, *Isolepis cernua*), but a number of herbs also fall into this category (for example, *Bacopa monnieri*).

A particular affinity can be traced at the generic level between Australasia and North America. *D. spicata*, which is one of the characteristic species of Temperate marshes in Pacific North America and also occurs on the eastern coast of the continent, has a vicariant in the Australian *D. distichophylla*. Although a relatively minor component in saltmarsh vegetation, similar relationships are seen in the unusual umbelliferous genus *Lilaeopsis* (Hill 1927) (*L. brownii* in Tasmania, *L. occidentalis* on the Pacific coast of North America, *L. chinensis* in north-east North America) – historical interpretation of such a pattern is difficult.

A link between south-east Australia and Temperate marshes in Japan may be seen in the occurrence of *Zoysia* spp. in upper-marsh grasslands in both regions.

Over the last century, there has been a considerable invasion of exotic species into the upper zones of Australian saltmarshes (Bridgewater & Kaeshagen 1979; Adam 1981*b*). The geographical origins of these species are various and little idea about the overall geographical affinities of south-east Australian marshes can be gained from a study of the alien flora. However, the success of *Cotula coronopifolia* as an alien in Pacific North America and Europe may provide some confirmation of the temperate affinities of marshes in south-east Australia.

South Africa

The saltmarshes of southern Africa have been relatively poorly studied compared with those in other continents, but are better documented than those elsewhere in Africa. The saltmarshes have a number of species in common with south-east Australia – *C. coronopifolia*, *J. kraussii*, *T. striata* and *Sp. virginicus* (Day 1981). The prevalance of *Sarcocornia* spp. in the lower marsh also indicates similarity with Australia, while the presence of *L. linifolium* further strengthens the Temperate links of these marshes.

The marshes clearly have some affinities with more tropical areas – as seen by their relationship to mangroves and by the extent of *Sporobolus* grasslands. The closest analogues thus appear to be with the New South Wales marshes.

A feature of particular interest is the presence of *S. maritima*. This not only provides a direct relationship with Europe, but also suggests a close similarity, in terms of structure and distribution of communities, with California where *S. foliosa* occurs.

Pierce (1982), however, questions the previous assumption (Marchant 1967) that *S. maritima* is a native to South Africa. The comparatively late first record of *Spartina* in South Africa, and evidence of recent spread in a number of estuaries, suggest that the species may have been introduced by early white settlers (Pierce 1982).

West Atlantic type

This marsh type corresponds to Chapman's West Atlantic group and is defined by the dominance in the lower marsh of *S. alterniflora*.

Despite considerable attrition at the hands of man (Gosselink & Baumann 1980), the total area of the marshes of this type is still large (in excess of 2 million hectares – Frey & Basan 1985), and constitutes a large proportion of the global coastal saltmarsh resource. The type is found over a large latitudinal range and experiences a wide range of climatic conditions from the subtropical regime of the Gulf of Mexico to the harsh winter conditions of southern Canada (more than 80% of the Gulf and Atlantic coasts of the USA is topographically appropriate for saltmarsh development). The variation in tidal range is the largest possible from the microtidal conditions of the Gulf to the macrotides of the Bay of Fundy. Throughout this range of conditions *S. alterniflora* is the dominant species over much of the intertidal. *S. alterniflora* can grow up to 2.5–3 m tall and physiognomically a tall stand is similar to a reed bed. This is a different structural form from most other low-marsh communities experiencing full seawater salinities (although reedbeds frequently occur at low elevations under brackish conditions). Other *Spartina* species found in the low-marsh zone in Northern Hemisphere marshes (*S. anglica*, *S. maritima*, *S. foliosa*) rarely exceed 1 m in height; *S. maritima* and *S. foliosa* do not dominate over as large a proportion of the intertidal – *S. anglica* currently dominates over the entire vegetated intertidal at some sites, but this is unlikely to prevail in the longer term. Thus, stands of *S. alterniflora* form an environment unique in the context of the world's saltmarshes. [Claims are occasionally made for other species being 'ecological equivalents' of *S. alterniflora* – for example, Eilers (1979) argues this for *Carex lyngbyei*, a characteristic species of Boreal marshes on the Pacific coast. While this species is often the most luxuriant member of the flora, it occupies a much more restricted portion of the intertidal range (Reimold 1977; Vince & Snow 1984), and has a lower maximum biomass (Burg, Trip & Rosenberg 1980; Gallagher & Kibby 1981). Although, in certain contexts, there is some justification for equating *S. alterniflora* and *C. lyngbei*, the differences probably outweigh the similarities, and recognition of a separate West Atlantic marsh type serves to highlight the ecological distinctiveness of *S. alterniflora*]

Three subtypes have been recognised within the saltmarshes of eastern America – Bay of Fundy, New England and Coastal Plain (Chapman 1953, 1960*b*, 1974*a*,*b*, 1977*a*; Reimold 1977); the distinction between the groups being based partly on floristics and partly on differences in physiography. On floristic grounds alone, two major subtypes can be recognised – the Coastal Plain, southwards from North Carolina, and a Northern subtype. There is considerable variation within the Northern subtype and further subdivision may be advisable when appropriate data are available.

In both Coastal Plain and Northern marshes *S. alterniflora* occupies the low marsh and creek banks but in the Northern region it is succeeded in the high marsh by *S. patens*. In the Coastal Plain marshes *S. alterniflora* occupies a wider proportion of the intertidal and *S. patens*, although geographically widespread, does not dominate such an extensive zone. The upper marsh of Coastal Plain marshes is frequently dominated by *J. roemerianus* (Eleuterius 1976). There is also a change in physiography between the two groups; *S. patens* dominated communities are often interrupted by numerous pans while, in the Coastal Plain, pans are far less frequent, although creek systems are more elaborately developed (Teal & Teal 1969). The boundary between the two groups corresponds fairly closely to the limit of Pleistocene glaciation in eastern America and, at the present, winter ice action continues to be an important ecological factor in the Northern marshes.

In the north-eastern United States and southern Canada, there is an intermingling of floristic elements. *S. alterniflora* occurs as far north as Labrador where the main marsh communities clearly have Boreal affinities (Roberts & Robertson 1986). The Boreal element occurs southwards into Nova Scotia (Reed & Moisan 1971; Ganong 1903; Glooschenko 1980*b*; Grondin & Melançon 1980; Thannheiser 1984). However, while *S. alterniflora* dominates the lower marsh, the extent of Boreal communities at higher elevations reflects the prevailing brackish conditions, which are particularly evident in much of the Gulf of St Lawrence (Deschênes & Sérodes 1985).

In addition to the intermingling of West Atlantic and Boreal communities, there is also a recognisable Temperate element in the flora. Despite the continuing presence of *S. patens* northwards into southern Canada (Roberts & Robertson 1986; Thannheiser, 1984), the upper zones of saltmarshes in this region are floristically diverse. Many widespread Temperate species occur (including *T. maritima*, *Pl. maritima*, *G. maritima*, *Su. maritima*, and *L. carolinianum*). Affinities with marshes on the Pacific coast are shown by the occurrence of *D. spicata*, while similarities with Europe are evidenced by the importance of *J. gerardi* in the upper marsh (although the Juncetum gerardii recognized by Thannheiser (1984) is species-poor compared with many European examples). This temperate element achieves greatest diversity in northern eastern USA, but extends northwards into Canada; it becomes depauperate and sporadic in occurrence with increasing latitude. *L. carolinianum* is recorded from as far north as Newfoundland (Roberts & Robertson 1986). Further details of the distribution of this Temperate element are provided by Ganong (1903), Johnson & York (1915), Conard (1935), Miller & Egler (1950), Chapman (1960, 1974) and Niering & Warren (1980) and Thannheiser (1984).

S. europaea, an important constituent in the lowest zones of European marshes, is prevented from occurring in that position by the dominance of *S. alterniflora*. It is, nevertheless, widespread in the region but, as a ruderal halophyte (Conard 1935), restricted to openings within vegetation rather than as a component of the pioneer zone.

If the dominance by *S. alterniflora* of much of the intertidal were downweighted and all elements in the flora were given equal weight then it would be possible to recognise a Temperate marsh type in north-east America similar to Temperate marshes in northern Europe and on the western coast of America but, as with these other regions, with complex local distributions of Temperate and Boreal elements.

From North Carolina southwards, the two major communities on the saltmarsh are those dominated by *S. alterniflora* and *J. roemerianus*, both species forming tall dense stands of vegetation with few associates, although under more brackish conditions floristic diversity increases. *S. patens* occurs at or above the level of *J. roemerianus* (Reimold 1977). The total flora recorded from these marshes is large (Duncan 1974) but the majority of species are restricted to the upper-marsh fringe or to brackish areas; the main part of the marsh area, which includes some of the most extensive marshes in the world, is dominated by only one or two species. A particular feature of the upper-marsh fringe is the abundance and local dominance of a number of shrubby composites, particularly *Iva* spp. and *Baccharis* spp.

Accounts of the vegetation of the Atlantic coast marshes are provided by Teal & Teal (1969), Ursin (1982) and Reimold (1977), while much ecological information is provided in the numerous studies of productivity of the *S.*

alterniflora marshes. In Florida, saltmarsh is largely replaced by mangrove but, as in most regions of mangrove, there is an intermingling of saltmarsh and mangrove depending on local site conditions. On the Gulf coast of the southern states there are again large areas of saltmarshes essentially identical in composition with those of the Atlantic coast. The most extensive of these marshes are in the Mississippi delta region where the saltmarshes form but one part of a complex of salt, brackish and freshwater marshes and swamp forests. Despite the long history of alternation and modification of this area by man, the diversity of communities along the gradient from saline to brackish conditions is still well displayed. Accounts of these southern marshes are given by Penfound & Hathaway (1938), Penfound (1952), Eleuterius (1972) and West (1977).

Dry Coast type

This marsh type is found in subtropical and tropical latitudes along coasts which have either a seasonally or a permanently dry climate. The soil salinities are frequently high for long periods of time. The vegetation is characteristically fairly open and dominated by low shrubs, most frequently succulent stemmed shrubby chenopods (from genera such as *Sarcocornia*, *Arthrocnemon*, *Sclerostegia*), but also commonly *Frankenia* spp. and more-geographically restricted taxa from the Plumbaginaceae and Compositae. This vegetation is referred to in the south of France as 'sansouire' and the similarity in structure and physiognomy of saltmarshes along arid and semi-arid coasts to sansouire has been pointed out by a number of authors (see Davies 1980).

At species and genus level there are considerable differences between continents in the composition of the vegetation. The degree of differentiation is at least as great as that between the subtypes of the Temperate type. However, subtypes are not formally recognised because, while some dry coasts are well known floristically (particularly those around the Mediterranean), others are virtually unexplored botanically, and it would be inappropriate to single out the few well known regions for subdivision. In addition, there is considerable similarity between these dry coast marshes and the vegetation of saline semi-desert habitats at the same latitudes – indeed, at some localities, intertidal marsh may merge directly into supratidal salt desert. It would seem premature to devise a phytogeographic classification of the coastal saltmarshes without comparison with inland areas.

The flora of the saline marshes is composed overwhelmingly of halophytic species, the glycophytic element is rare, except in the case of predictably Mediterranean climatic regions where the gaps between the perennial shrubs may be occupied by short-lived annuals during the wet season. Nevertheless, there will be localised habitats within the saline marshes where brackish, or even freshwater, conditions prevail. The vegetation of such habitats is markedly different from that of the saline areas and is rarely biogeographically distinctive – dense stands of *Phragmites australis* or various colonial Cyperaceae are virtually cosmopolitan in distribution. [It is highly probable that there is considerable ecotypic differentiation within these species, but this remains to be investigated.] In more species-rich brackish communities, there is greater geographic differentiation but, even so, many of the component species may have very wide distributions.

The most extensively studied of the brackish communities occur around the Mediterranean (particularly in the Camargue) (see Braun-Blanquet, Roussine & Nègre 1952; Molinier & Tallon 1970, 1974; Bassett 1978, 1980; Britton & Podlejski 1981; Corré 1979; Gradstein & Smittenberg 1977). Most of these brackish fen-type communities can be assigned to the vegetation class Juncetea maritimi which is essentially Mediterranean in its distribution (Beeftink 1965, 1972). While a large component of the vegetation consists of species of southern European distribution, there are also many widespread wetland species. It is striking, for example, that amongst the site types recognised by Britton & Podlejski (1981) in the Camargue is one (Group A) which is characterised *inter alia* by *Lycopus europaeus*, *Oenanthe lachenalii*, *Lythrum salicaria*, *Samolus valerandi* and *Rumex crispus* – species which are also characteristic of the brackish upper-marsh communities of north-western Scotland (Adam 1978, 1981*a*).

The boundary between Temperate and Dry Coast marsh types is not sharp, rather there is an extensive transition, much modified by local conditions, over which a reduction in the importance of grass-dominated communities and an increase is chenopod shrubs occurs. Delimitation of boundaries is further complicated in many areas by the coincidence of this transition zone with the limit of mangrove distribution producing local mosaics of saltmarsh and mangrove.

A particular marsh type which may deserve separate recognition is that found in southern Africa (Day 1981), California (Macdonald 1977*a*,*b*) and the southwest Iberian peninsula (Géhu 1984*a*,*b*) where there is a low-marsh *Spartina*-dominated community (*S. maritima* or *S. foliosa*) and a range of low shrubs (principally chenopods) in the mid and upper marsh. Although these marshes could be regarded as one extreme of the variation within the temperate group, delimitation of a distinct type occurring in areas with a semi-Mediterranean climate has value in emphasising the intercontinental affinities of much saltmarsh vegetation.

Within the Dry Coast type, diversity is greatest in seasonally wet regions and decreases considerably on permanently arid coasts. This decrease in species diversity is clearly documented for the Pacific North American coast (Macdonald & Barbour 1974; Macdonald 1977*a,b*), and Macdonald (1977*a,b*) recognises the southern Californian (Mediterranean) marsh flora as forming a different biogeographic grouping from that on the arid coasts of Baja California. Such a division could also be made elsewhere, but absence of detailed floristic data prevents the position of the relevant boundaries from being established. It is interesting that the diversity of the southern Californian marshes is increased by introduced species (Macdonald & Barbour 1974; Macdonald 1977*a,b*). A similar increase in diversity (involving many of the same alien species) can also be observed in the upper marsh zone at many sites in the Mediterranean climate zone of southern Australia (Bridgewater & Kaeshagen 1979).

In Mediterranean climatic regions, it is clear that the short-lived annual and ephemeral component of the flora 'takes advantage' of the cooler and less saline conditions of autumn and winter to complete its life cycle. In the case of *Lasthenia glabrata*, there is evidence that increasing salinity (which in the field would be a predictable event of spring) acts as trigger for flowering and seed set (Kingsbury *et al.* 1976). There is much less information as to how the phenology of perennials is related to climatic conditions. In southern California, Purer (1942) showed that a number of perennials had autumn/winter periods of maximum growth, but that others (for example, *Frankenia*) had maximum growth during the summer. Clearly, there would be advantages, in terms of resource capture, to species which could avoid competition with both annuals and other perennials in the winter; on the other hand, hypersalinity during the summer is likely to be a severe constraint on plant growth. It would be of interest to investigate the phenology of Dry Coast perennials in relation to their physiological mechanism for coping with salt – it may be significant that *Frankenia* spp. have well developed salt-excreting glands, whereas many of the other perennial species are succulents.

The majority of Dry Coast marshes are at relatively low latitudes and experience climates ranging from warm to hot. However, in Patagonia and in the Straits of Magellan, semi-arid conditions prevail at high latitudes and the saltmarshes are subject to cool winters and temperate summers (Dollenz 1977; West 1977). In the Straits of Magellan, the lower, more regularly innundated marshes have a flora characteristic of the Dry Coast type with the major species being *Frankenia chubutensis*, *Suaeda argentinensis*, *Sarcocornia perennis* (*Salicornia ambigua*) and *Atriplex reichei*. The rarely flooded upper-marsh fringe supports a number of low shrubs of geographically restricted distribution; the composites *Lepidophyllum cupressiforme* and *Baccharis magellanica*, and the cushion plant umbellifer *Azorella caespitosa* being particularly prominent (Dollenz 1977). Of considerable phytogeographic interest is the occasional incidence of *Pl. maritima* and *A. maritima* (spp. *andina*) – two species from the small bipolar biogeographic element found at high latitudes in both hemispheres.

At lower latitudes, saltmarshes of the Dry Coast type are found around the Mediterranean itself [where Chapman (1953, 1960*b*, 1974*a,b*, 1977*a*) recognised three distinct subgroups in the western and eastern Mediterranean and around the Caspian Sea)]. The dominant dwarf shrubs include *Sarcocornia* spp., *Suaeda* spp., *Limoniastrum* spp., and *Frankenia* spp.; grasses are comparatively rare, but the sansouire may be fringed by dense stands of *Elymus pycnanthus* (*Agropyron pungens* auct.). In European phytosociological schemata, the sansouire vegetation is generally assigned to the class Salicornietea fruticosae (see Braun-Blanquet *et al.* 1952; Beeftink 1965; Molinier & Talon 1970, 1974; Géhu & Rivas-Martinez 1984; Géhu 1984*a,b*).

Dry Coast marshes are widespread in north Africa (Chapman 1977*b*), the Red Sea (Kassas & Zahran 1962, 1965; Zahran 1977; Evenari, Gutterman & Gavish 1985), the Arabian Gulf (Basson *et al.* 1977) and on regions of the coast of the Indian subcontinent too arid to support mangroves (Blasco 1977). The vascular plant communities of these coasts have an open structure and the majority of species are low shrubs. Even following periods of rain, ephemeral herbaceous species are absent (Kassas & Zahran 1965).

In northern Sri Lanka, with a monsoonal climate and prolonged dry season, the marsh vegetation has strong affinities to the Dry Coast type, although grasses are more prominent than is normally the case (Pemadasa *et al.* 1979). The marshes of southern Australia (west from the Coorong in south-east South Australia) are also clearly of the Dry Coast type (Bridgewater 1982).

Saltmarsh vegetation of the Dry Coast type has been described from southern California and Baja California (Purer 1942; Macdonald & Barbour 1974; Macdonald 1977*a,b*; Mahall & Park 1976*a,b,c*; Neuenschwander, Thorsted & Vogl 1979). These marshes differ somewhat from other Dry Coast areas in that the most frequently inundated areas of the low marsh are dominated by dense stands of the grass *Spartina foliosa*. On the dry coast of south-west Texas and north-east Mexico, *Spartina* spp. are generally absent from coastal marshes, except for the local occurrence of *Spartina spartinae* where hypersalinity is ameliorated (West 1977). Dry

Coast saltmarshes presumably also occur in Chile, but little detailed information is available on the vegetation and dry marshes are likely to be small because of the topography and tectonic activity of the coastline (West 1977, 1981). Dry Coast marsh vegetation is recorded from the Atlantic coast of South America from Uruguay southwards (West 1977). With increasing latitude, in Patagonia the coastal marsh vegetation becomes more similar to that already discussed from the Straits of Magellan (West 1977; Hauman 1926).

Very little information is available on saltmarsh vegetation around much of the African coast, but it is probable that saltmarshes of the Dry Coast type are found in southern Africa. A number of the sites discussed in Day (1981) appear similar to many in southern Australia in that the combination of communities present suggests a classification transitional between Dry Coast and Temperate marsh types.

Tropical type

In the tropics, most environment suitable for the development of saltmarsh have been usurped by mangroves and the extent of saltmarsh is, thus, limited. However, saltmarshes are far more widespread in the tropics than maps such as those in Chapman (1977*a*) or Long & Mason (1983) would suggest. Whether such saltmarshes occurring adjacent to mangrove stands should be treated as a separate saltmarsh type or whether they can be considered as part of the mangrove complex is a moot point; for present purposes, they will be regarded as saltmarshes in their own right.

Where hypersaline conditions develop in the upper intertidal zone, extensive salt flats may occur above the level of mangroves. These saltflats may support patchy development of extremely species-poor saltmarsh vegetation of the Dry Coast type – such vegetation has been described from many regions (see Thom, Wright & Coleman 1975; Saenger *et al.* 1977; Blasco 1977).

Continuously vegetated saltmarsh also occurs in the tropics and, at least in South America, is far more extensive than is acknowledged in most of the literature (West 1977). West (1977) recognises three distinct situations in which continuous saltmarsh cover may occur:

As a pioneer community seaward of mangroves;
As a grassland on the inner edge of, or within, the mangrove forest;
As a secondary community on disturbed, degraded mangrove areas.

The occurrence of *Spartina* marsh to seaward of mangroves has been described from many sites from Florida to southern Brazil (West 1977;

Thom 1967). Although few of these sites have been investigated in detail, the evidence appears to suggest a dynamic relationship between mangrove and saltmarsh; as the saltmarsh advances seawards, so the upper part of the marsh is invaded and replaced by mangroves. The only *Spartina* species concerned appears to be *S. alterniflora*; many accounts referred to *S. brasiliensis*, but this is regarded by Mobberley (1956) as a synonym for *S. alterniflora*.

It is interesting that at the southernmost limit of mangroves, in Westernport Bay, Victoria (a temperate rather than tropical site), the introduced *S. anglica* has formed a sward to seaward of *A. marina* (Bridgewater 1975, 1982). Whether a similar successional relationship as occurs in South America will develop is presently uncertain.

West (1977) provided a number of examples of grassland developing above or within mangroves. Salinities in these communities are apparently sufficient, at least seasonally, to retard mangroves, but not sufficient to promote Dry Coast type communities. Species represented include: *Spartina* spp., *D. spicata*, *Sp. virginicus* and *Paspalum vaginatum*. *Sporobolus* and *Paspalum* have been recorded in similar habitats in Africa (Chapman 1977*b*), while extensive *Sporobolus* grasslands (in which *P. vaginatum* is a local constituent) occur above mangroves at several places on the Queensland (Anning 1980; Dowling & Macdonald 1982) and northwest Australian coasts. It would seem possible to define a basic saltmarsh type characterised by *Sp. virginicus* and *P. vaginatum*. *Batis maritima* and *Sesuvium portulacastrum* may also be characteristic of this type. Both *Sp. virginicus* and *P. vaginatum* occur widely in Temperate marshes (although very much at the warmer, subtropical end of the range of variation in Temperate marshes), but extensive, almost pure, swards of the two species appear to be a Tropical feature.

West (1977) records a number of instances in South America where cutover mangroves have been replaced by *S. alterniflora*, but indicates that the usual invader of such sites is the fern *Acrostichum aureum*. In many parts of the tropics, mangroves are an intensively utilised resource and, in south-east Asia, may have been subject to heavy logging over a considerable period with no signs of replacement by saltmarsh even if the composition of the mangrove itself may be changed (Blasco 1977).

In general, Tropical saltmarshes are species-poor and virtually all the species would be regarded as halophytic.

Other marsh types

Information on the saltmarsh vegetation of many regions is still scanty and certain assumptions have been made in assigning long stretches of poorly studied coast to particular types. Data are particularly scarce from Africa and much of Asia. It is also difficult to compile a synthesis of South American saltmarsh vegetation despite the valuable accounts by West (1977, 1981). Marshes of the Dry Coast and Tropical types clearly occur on that continent. However, between southern Brazil and northern Patagonia there are extensive saltmarshes on estuaries with large discharges and prevailing brackish conditions. Inversion of zonation patterns occurs on a grand scale in these estuaries (West 1977). The lower, regularly inundated, zones are dominated by sedges and grasses, while the upper zones are occupied by more halophytic species. In wetter regions, the halophytic communities are dominated by *Spartina* spp. and in drier areas by various low shrubs. While these marshes have affinities with Temperate marshes elsewhere, the scale of the zonation inversions may make it advisable to recognise these marshes as comprising a separate type.

The recent spread (both by human agency and by natural means) of *S. anglica* may make it necessary to recognise a new subtype of the Temperate type, with a distribution transgressing the boundaries of previously distinct subtypes. At this juncture, the course of marsh development in different regions with newly established *S. anglica* marshes is uncertain and it would be premature to assume they would remain similar. However, if it does prove appropriate to unite *S. anglica* marshes (from such regions as northern Europe, China, New Zealand and south-east Australia), then the distinctiveness of *S. alterniflora* marshes might be questioned and the West Atlantic type incorporated as a subtype within the Temperate type.

Discussion

The distribution of saltmarsh types indicates a relatively simple pattern with the major axis of variation being correlated with latitude. In the tropics, there is a widespread, species-poor, saltmarsh flora with little regional differentiation in the vegetation. At higher latitudes in temperate climates, the total halophytic flora is much larger and there is considerable regional differentiation in both flora and vegetation. Nevertheless, at family and generic level, and in terms of vegetation structure and physiognomy, there is a strong overall similarity of Temperate saltmarshes. At higher latitudes still, Boreal saltmarshes are characterised by a relatively small number of widely distributed halophytes and a larger, more regionally differentiated, glycophytic element. Arctic marshes show considerable circumpolar similarity and are characterised by a small flora with an even smaller halophytic element. The distribution pattern strongly suggests that the major determining factors are climatic.

The dominance of mangroves in the tropics restricts saltmarsh to less-favourable sites where persistent high soil salinities limit the diversity of the flora. At high latitudes, cool summer and cold winter climatic conditions limit the diversity of the saltmarsh flora and vegetation. Climatic conditions are clearly also the determining factor in the distribution of the Dry Coast marsh type. In Temperate latitudes, where climatic conditions are comparatively equable and the development of hypersaline conditions is limited in space and time and where, for whatever underlying reason, saltmarsh is not in competition with mangroves, the most diverse halophytic flora has developed.

Unfortunately, the fossil record is largely silent on the history of the saltmarsh flora. The dominance of angiosperms in vegetation is more pronounced in saline habitats then almost anywhere else. When and where this predominance was first expressed remains unknown. The similarity between disjunct saltmarsh floras at family and generic level has been explained in terms of a very early origin of an ancestral halophytic assemblage. However, this similarity extends between northern and southern hemispheres, while Laurasia and Gondwanaland were well separated at the time of the origin and early radiation of the angiosperms. Again, although the bulk of the halophytic flora is included within a few families, these families are not taxonomically closely related. The taxonomic distribution of halophytes, and the diversity of physiological adaptations to saline conditions, is best explained by the independent evolution of halophily on a number of occasions. However, the ability to evolve halophytic adaptations is limited to comparatively few families and the widespread distribution of the few halophytic lineages is probably a reflection of successful long-distance dispersal.

An exception to the rule of widespread distribution (and then only a partial exception) is provided by *Spartina* and the distinctive West Atlantic marsh type. The advent of *Spartina* was a 'local' evolutionary event and the present relatively restricted distribution may reflect either an inherently low dispersal potential or that the origin was, in geological time, far more recent than of other major halophytic lineages.

Geographical variation at the continental scale – temperate Europe

Variation in Britain

Within the major marsh types there is considerable variation in both flora and vegetation between individual sites. Some of this variation may reflect conditions unique to individual sites, other components may indicate the position of the sites along recurrent mesoscale gradients (such as between the mouth and head of estuaries). However, there may also be variation on what can still be regarded as a macroscale geographically but which is, nevertheless, on a smaller spatial scale and involves lesser floristic changes than that reflected by the distribution of the major marsh types.

When discussing general features of marshes in a particular region, or in making comparisons between sites, it would be useful to separate these various components of intersite variation. In this section, variation between British coastal marshes is discussed and hypotheses advanced to explain the observed distribution patterns. Variation in Britain is then related to that in Europe.

This existence of considerable variation between British saltmarshes has long been recognised. Chapman (1941) delimited three types of saltmarsh on mainland Britain with a fourth type restricted to Ireland. Division between the types was based on differences in supposed successional patterns (determined largely on the basis of zonation) and the major types were geographically separated (there being an east coast, west coast and south coast sere).

The dangers of interpreting zonation as direct spatial representations of temporal change were pointed out by a number of authors (see Ranwell 1972; Jefferies 1972; Davies 1980), but the basic framework proposed by Chapman was generally accepted. However, as more sites were studied in detail, the apparent simplicity of the pattern described by Chapman began to be questioned and it became appropriate to re-investigate variations between sites at the national level.

Using data on the occurrence of plant communities, Adam (1978) examined differences between 133 saltmarsh sites (details of the communities are provided in Adam 1981*a*). There are certain limitations to this study. The data used were purely qualitative (presence or absence of communities); thus, the presence of communities restricted to specialised sub-habitats is effectively given extra weight compared with those occupying extensive areas. The geographical coverage of sites did not encompass all stretches of the coast with extensive saltmarshes. Nevertheless, it was hoped that the data were sufficient to reveal the major trends of variation in the British Isles. Sampling concentrated on sites at the more seaward end of estuaries and on open (but low wave energy) coasts, thus any confounding effects of mesoscale variation along estuarine gradients were avoided. However, the extent of geographical variation between upper estuarine sites remains unknown (extant sites in the upper reaches of estuaries are relatively few, this section of the estuary being the most subject to extensive modification).

Analysis of the site v. community data matrix by cluster analysis and ordination techniques (Adam 1978) indicated the existence of three major site groupings (A,B,C). Each major group could be further subdivided, but differences within each major group were much less than those between them.

Type A marshes are characterised by a relatively low community diversity and a flora comprised largely of halophytic species. The low marsh is frequently Spartinetum townsendii, but the *A. tripolium* var. *discoideus* nodum is also widespread (community nomenclature follows Adam 1981*a*). Extensive stands of Halimionetum portulacoidis occur in the mid-marsh, while the upper-marsh fringe is frequently occupied by the Elymetum pycnanthi (Agropyretum pungentis). The majority of sites for Type A marshes occur in south-east England (Fig. 3.7).

In Type B marshes, the major low marsh community is the Puccinellietum maritimae (although Spartinetum townsendii stands are also widespread). Mid and upper marsh zones are predominantly grassland, referable to the Juncetum gerardii (*sensu lato*) in which glycophytes are a major component. The extensive saltmarshes in north-west England and Wales are of this type, which also occurs at widely scattered sites elsewhere around the coast.

Type C marshes have relatively few communities but, as they are characterised by the occurrence of fen-like communities in the upper marsh, they are often very species-rich (Gillham 1957*b*; Adam, Birks & Huntley 1977; Adam 1981*a*). The halophytic element in the total flora is small. The majority of sites with this marsh type are in western Scotland.

Types A and B marshes in Adam (1978) correspond fairly closely to Chapman's (1941) east- and west-coast seres. However, there is no equivalent in Adam's scheme to Chapman's south-coast sere. The south-coast sere was distinguished largely by the extent of the Spartinetum townsendii – not only has *S. anglica* spread considerably since the 1930s, but the qualitative nature of the data employed by Adam (1978) precludes distinction of marsh groupings based on areal extent of communities.

Type C marshes are not readily related to any of Chapman's groupings. By implication, Chapman (1941) assigned the marshes of western Scotland to the west-coast sere, but the analysis presented by Adam (1978) points to a major change between western Scotland and the marshes of the Irish Sea coast. Type C marshes have some features of Chapman's Irish sere (based on the account by Rees 1935), but this grouping was never described in detail in Chapman's accounts.

Despite the differences between the classifications advanced by Chapman (1941) and Adam (1978), the longheld view of major differences between the saltmarshes of south-east England and those of the west coast is supported in both accounts.

There are a number of differences in environmental features between the

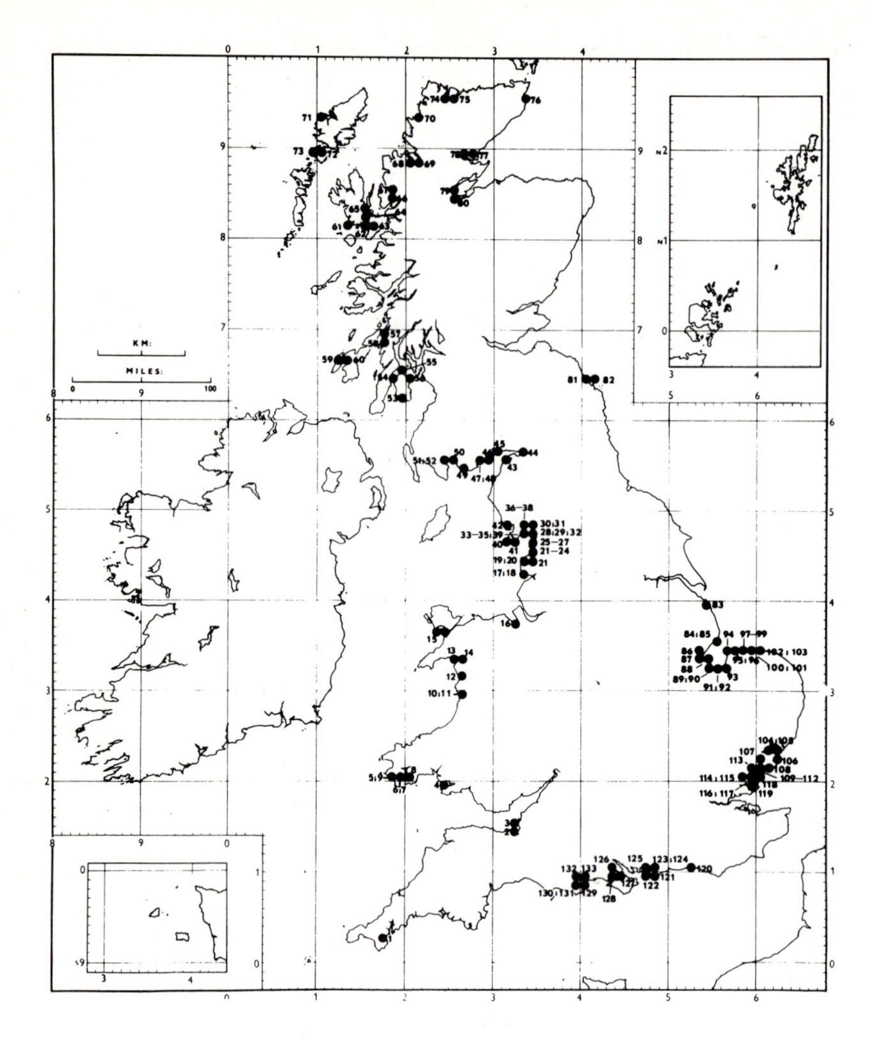

Fig. 3.7a–d Distribution of major marsh types around the British coast. (Fig. 3.7a shows the distribution of sites studied.) (From Adam 1978.)

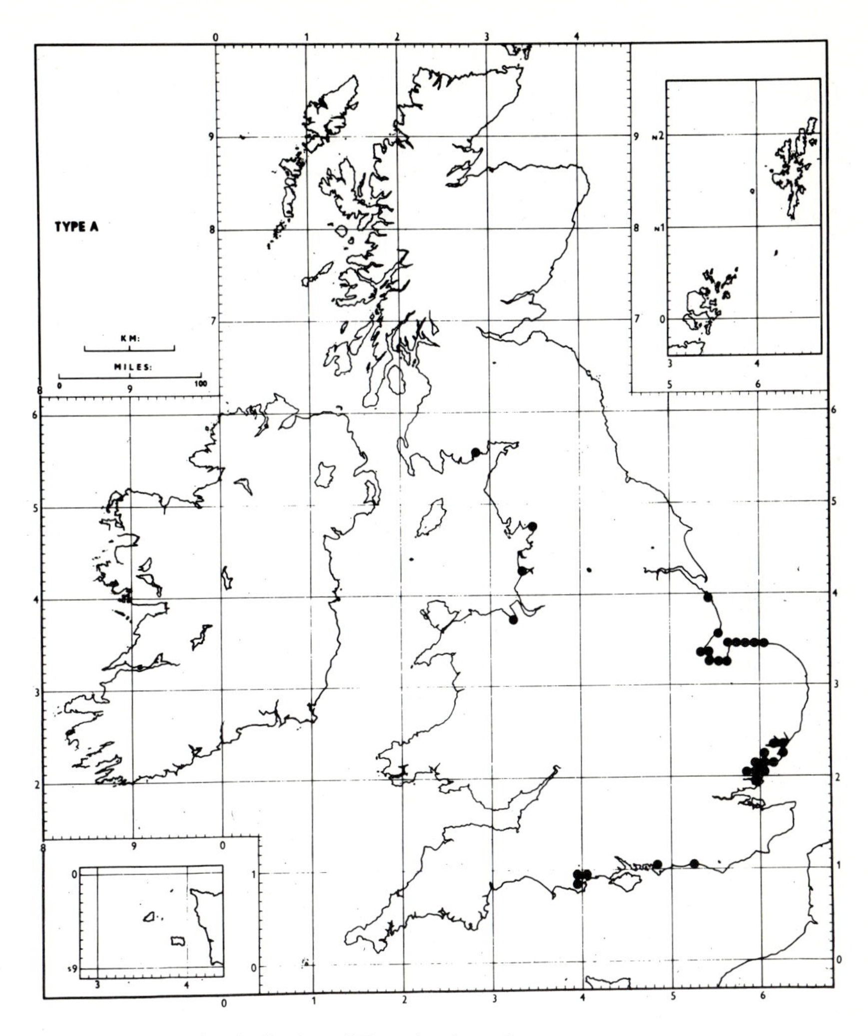

Fig. 3.7b Distribution of Type A saltmarshes.

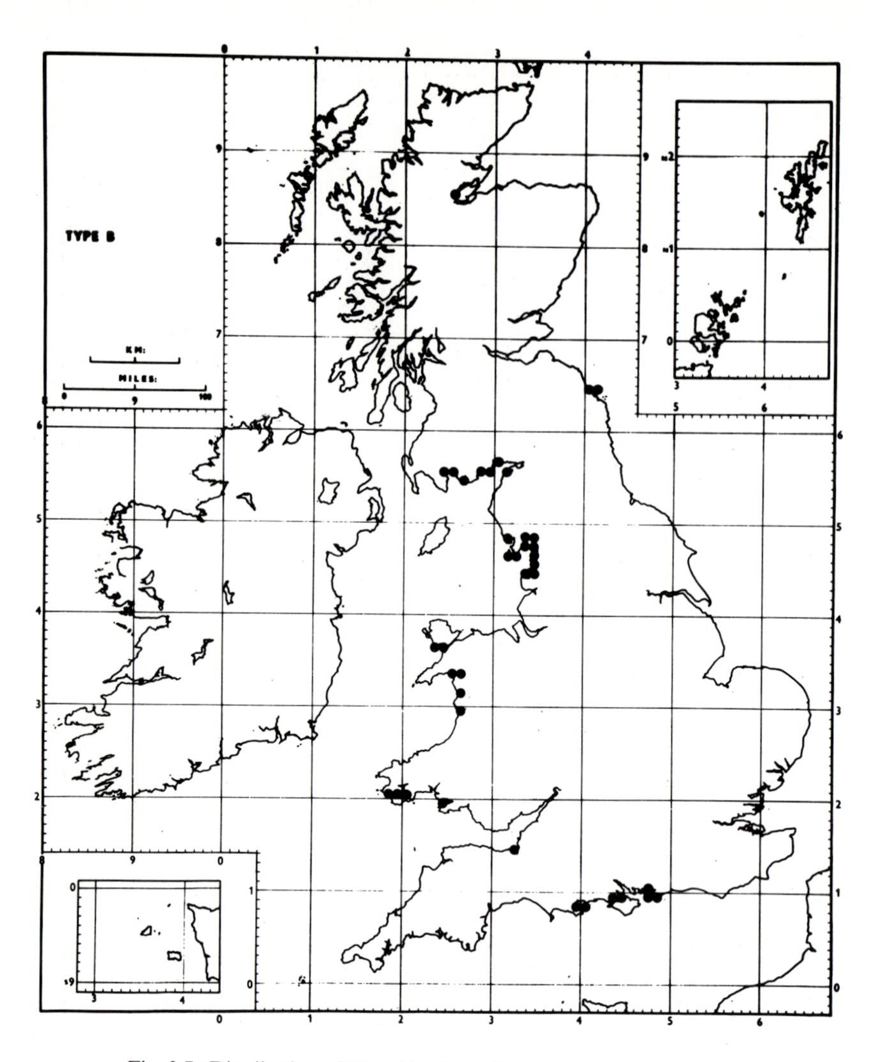

Fig. 3.7c Distribution of Type B saltmarshes.

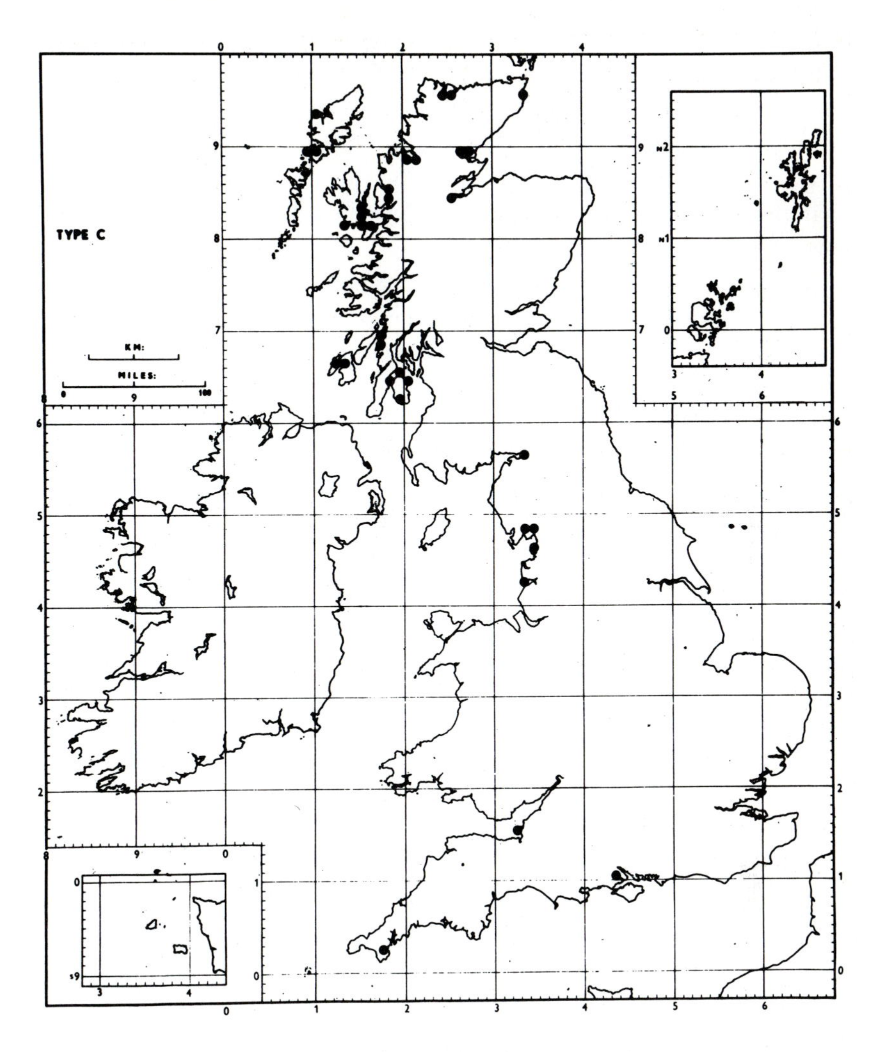

Fig. 3.7d Distribution of Type C saltmarshes.

two regions. Marshes on the west coast occur predominantly on sandy substrates, those on the south coast on finer silts and clays. Creek systems on the west coast are comparatively simple; those in the south-east are frequently tortuous. There are also climatic differences; marshes on the west coast receive higher rainfall and experience less variation in temperature than those in the south-east which may be exposed to higher summer and lower winter temperatures.

Two principal hypotheses have been advanced to explain the differences between Type A and Type B saltmarshes – that substrate is the major determinant of the range of communities present or that grazing is the most important factor.

Chapman (1941) regarded substrate as the most important factor, the dominance of *Pu. maritima* in the low marsh of Type B marshes was seen as a consequence of the sandy substrate and that as a result 'these marshes are particularly suitable for grazing' rather than *vice versa*.

A relationship between abundance of *Pu. maritima* and sandiness of substrate is demonstrated in the Wash, with *Puccinellia* being more abundant in the north-west sector of the embayment where the substrate is sandier than at muddier localities where *S. europaea*, *A. tripolium* var. *discoideus* and *S. anglica* are the major low-marsh species (see also Randerson 1979).

The importance of sediment type in moulding the vegetation was also emphasised by Tansley (1939), while variation in vegetation with sediment type was demonstrated in detailed studies of single sites by Marsh (1915) and Chapman (1934).

The importance of grazing as an influence on saltmarsh vegetation has been described by a number of authors (see Ranwell 1961, 1968; Gray 1972; Gray & Scott 1977*b*), and grazing was regarded as the most important cause of the differences between marshes on the west coast and in south-east England by Jefferies (1972) and Pigott (1969). The striking difference in representation of the Juncetum gerardii between the west coast and the south east has been attributed to grazing. The role of grazing in promoting the development and maintenance of Juncetum gerardii swards has been stressed by many authors (see Dahlbeck 1945; Mikkelsen 1949; Gillner 1960; Beeftink 1962, 1965; Tyler 1969*a*; Mäkirinta 1970; and Westhoff 1969, 1971). In general, the halophytic element in these grazed communities is small, but the total species richness often far exceeds that of ungrazed marshes (Westhoff 1971; Adam 1981*a*). Within the Juncetum gerardii variation in grazing pressure may be responsible for controlling the balance between different species (Gray & Scott 1977*b*). Variation within the community may also depend on whether grazing is by sheep, cattle or horses, while both Ranwell (1967) and Gray (1972) have suggested that different breeds of sheep may have differing grazing preferences on saltmarshes.

The distribution of grazed marshes in the early 1970s showed a high degree of concordance with that of Type B marshes (see Fig. 3.8), a correlation lending support to the view that grazing is the major factor controlling the composition of saltmarsh vegetation.

Both hypotheses can, therefore, be supported – can other arguments be advanced to favour one hypothesis over the other?

Although the majority of Type A marshes were found in south-east England, four were on the west coast. At the time of sampling, all were ungrazed and one, Holme Island in Morecambe Bay, was known never to have been grazed (Gray 1972). All were on sandy substrates but supported very different vegetation from that on adjacent grazed marshes. Extensive stands dominated by *H. portulacoides* occur in the mid marsh and *Limonium* spp. and *Arm. maritima* occur. At other sites on the west coast, populations of these species are generally small and restricted to situations where access by grazing animals would be difficult.

In south-east England, the few marshes currently grazed do not resemble grazed west coast marshes. The most extensive grazing pressure on Type A marsh is around the Wash, where all marshes are depauperate in both species and communities (Gray 1976) (this depauperate character possibly reflecting the frequency of reclamation and the comparatively young age of most extant marshes). There are no consistent differences between grazed and ungrazed marshes, but as many of the ungrazed marshes have been grazed in the recent past, it is not possible to claim that grazing has had no effect. Nevertheless, *H. portulacoides* is abundant on both grazed and ungrazed marshes, despite its almost complete absence from grazed west coast marshes. This apparent difference in the effect of grazing may be related to substrate differences, to variation in grazing pressure (the Wash marshes being cattle grazed – most of those in western Britain being sheep grazed), or genetic differences in the *Halimione* populations.

Although there is currently a sharp geographical separation of grazed and ungrazed sites, this is a modern phenomenon. Most British saltmarshes have been grazed at one time or another, even if only on a casual basis. Many of the Type A marshes have been grazed comparatively recently (including those on the north Norfolk coast – Jefferies 1976), mainly by sheep and cattle, but locally by large flocks of domestic geese (Chadwick 1982). Unfortunately, there are few accounts of these marshes when they

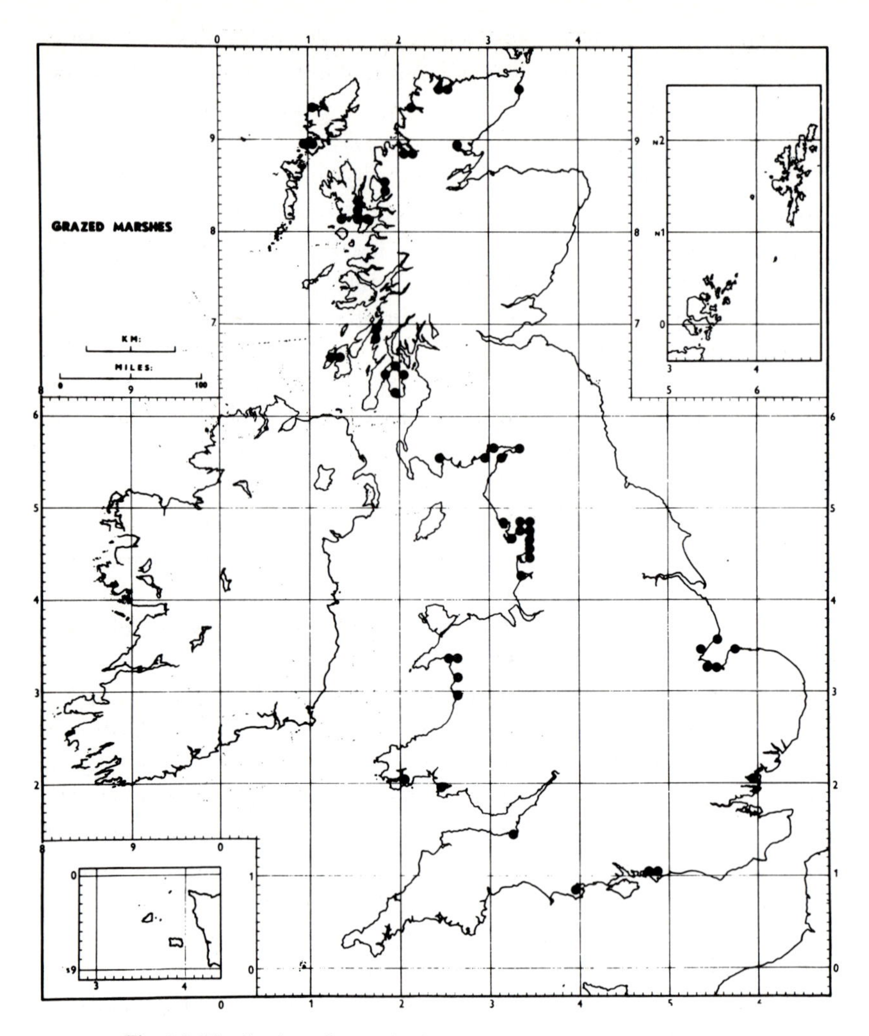

Fig. 3.8 Distribution of grazed saltmarshes in Britain in the early 1970s. (From Adam 1978.)

were grazed, but a number of anecdotal accounts suggest that, even under grazing regime, large populations of *Limonium* spp. survived, species which in western Britain seem to be particularly sensitive to grazing.

The survival of species such as *Limonium* under grazing in southern England seems more likely when the changes that have been documented following the cessation of grazing are considered. Evidence presented by Ranwell (1968), Cadwalladr *et al.* (1972) and Cadwalladr & Morely (1973, 1974) indicates that reduction in sheep grazing in western Britain is followed by a spread of *F. rubra* and the development of *F. rubra* dominated 'mattresses'. Similar evidence was advanced by Westhoff (1969) and the development of species-poor *F. rubra* dominated communities, when grazing pressure is eased, has also been reported from British chalk grasslands by Smith (1980). These dense *F. rubra* stands provide few opportunities for the invasion of other species. While diversity may be restored by resumption of grazing (Bakker, Dijkstra & Russchen 1985), it is difficult to envisage that the present range of communities found in south-east England have developed from a *Festuca* mattress post-grazing phase.

It seems likely that, in the absence of grazing, variation in the low marsh would have been correlated with substrate conditions, *Puccinellia* being more prominent on sand. However, mid-marsh vegetation would have been fairly uniform on both east and west coasts. The impact, both of grazing and the cessation of grazing, may have been very different on the two coasts; grazing promoting Juncetum gerardii grasslands and cessation of grazing leading to species-poor *F. rubra* dominated grasslands on the west coast while, in the south east, halophytic species appear to have persisted under grazing with consequently fewer opportunities for the development of *F. rubra* stands after grazing ceased. Whether the impact of grazing is mediated more by substrate or by climate is uncertain.

Saltmarshes in Britain provide important winter feeding grounds for ducks and geese, and several sites are of international significance for this reason. Wildfowl favour short grassland communities and avoid tall, rank *Festuca* (Cadwalladr *et al.* 1972; Cadwalladr & Morley 1973, 1974). On the west coast there is, thus, a synergism between wildfowl and stock grazing. It is interesting to speculate that wildfowl numbers may have increased following human utilisation of saltmarshes; maintenance of current flocks requires a continuance of grazing intensities sufficient to prevent development of *Festuca* mattresses.

There remain some differences between south-east and west coasts which do not appear to be related to either grazing or substrate conditions. *Arm. maritima* and *G. maritima*, which are characteristic of the upper marsh in the south east, are found much lower in the zonation in the west (Beeftink 1977*a*,*b*; Adam 1981*a*). It has been suggested that this reflects a response to the generally wetter, more oceanic climate of the west, but confirmation of this is lacking.

Most of the Type C marshes in western Scotland are grazed, but it is possible that grazing, although altering the relative abundance of species, has had a lesser impact than on marshes further south. Many of the halophytic species which are apparently so grazing-sensitive further south on the west coast are absent, presumably for climatic reasons, in western Scotland. Even in the absence of grazing, it is not possible to suggest other potential low-marsh species. The high regional rainfall and the influence of seepage from terrestrial communities would ensure provision of habitat for a diverse, largely glycophytic mid- and upper-marsh flora regardless of the grazing intensity.

‘Natural’ saltmarsh vegetation in Britain

The present distribution and composition of saltmarsh communities in Britain are in large measure a reflection of historical and continuing human activities. Is it possible to find traces in the present vegetation of distribution patterns which seem to reflect influences other than man and so provide a basis for speculation on the ‘natural’ vegetation?

Within the British saltmarsh flora, a number of species have restricted distributions. By studying the distribution patterns of these species, it is possible to recognise certain regions of the coast where a number of these species share the same limits to their range (Ranwell 1968). From a knowledge of the role of these species in saltmarsh vegetation, it may be possible to define regions where a major vegetational change would have occurred in the absence of man’s influence.

The distribution of the British flora has been mapped in detail by Perring & Walters (1962). Using their data, it is possible to recognise two groups of saltmarsh species with approximately coincident northern limits. There are few saltmarsh species reaching a southern limit in Britain and no obvious shared boundary for the few species concerned (the most widespread northern elements in the British saltmarsh flora are *Blysmus rufus* and *Centaurium littorale*). Interestingly, Macdonald (1977*a*) also suggests that on the Pacific North American coast, species reaching southern limits do so independently of each other, while there is greater coincidence of northern boundaries of distribution.

The most clearly defined of these northern limits runs from just north of the Solway Firth in the west to somewhere between the Firth of Forth and the Scottish border in the east (the boundary being sharper in the west). Species reaching their northern limit here include: *H. portulacoides*, *Art. maritima*, *L. vulgare*, *L. humile*, *Elymus pycnanthus*, *Parapholis strigosa* and probably *S. anglica* (see Fig. 3.9). In addition, *Centaurium pulchellum*, *Apium graveolens* and *Trifolium fragiferum* which become restricted to upper-marsh habitats towards their northern limit also fail to extend north of this Solway Line. The 'Solway Line' has long been recognised (Chapman 1941, 1950; Gimingham 1964; Boorman 1967, 1971; Ranwell 1968, 1972) and, as well as marking a major change in the potential flora of saltmarshes, seems also to coincide with major changes in the flora of seacliffs.

The halophytic species reaching their limit at the Solway are sporadic in their occurrence on the west coast and their distribution seems to reflect a sensitivity to grazing. The Solway Line does not, however, coincide, either at present or historically, with a major change in land use practice on saltmarshes. So it is perhaps reasonable to postulate that the Solway Line reflects a natural distributional limit.

Climatic factors which might be responsible for the expression of the Solway line have not been subject to experimental investigation. Chapman (1950) pointed out that the line was roughly coincident with the July 60 °F (15 °C) isotherm, but whether it is summer warmth, winter frost or some other factor which is the critical determinant of distribution is unknown.

The second group of species is found south of a line roughly from the Wash to the Bristol Channel (Fig. 3.10). Ecologically, the majority of these species fall into two subgroups, those of drier (and, at least seasonally, more saline) habitats, and those of wetter, more permanently, brackish conditions. Included in the first subgroup would be *Frankenia laevis*, *Suaeda vera* (*Su. fruticosa* auct.), *Limonium bellidifolium*, *L. binervosum* (which extends further north on seacliffs and saltmarshes in the west), *Bupleurum tenuissimum*, *Puccinellia fasciculata*, *Pu. rupestris*, *Hordeum marinum* and *Halimione pedunculata* (thought to be extinct in Britain but recently rediscovered in Essex – Leach 1988). The second subgroup would include *Chenopodium botryodes*, *Althaea officinalis*, *Carex divisa*, *Alopecurus bulbosus* and *Polypogon monspeliensis*. Also part of the south-eastern marsh flora are three lower-marsh species (*S. maritima*, *Sarcocornia perennis* and *A. tripolium* var. *discoideus*), while *Inula crithmoides* occurs over a range of communities at different levels on the shore.

Few of the species in this south-east group are of more than local importance in present vegetation. Within the region of their distribution, most of the species show disjunctions with many absences from apparently suitable localities. The upper saltmarsh is that region of the marsh most

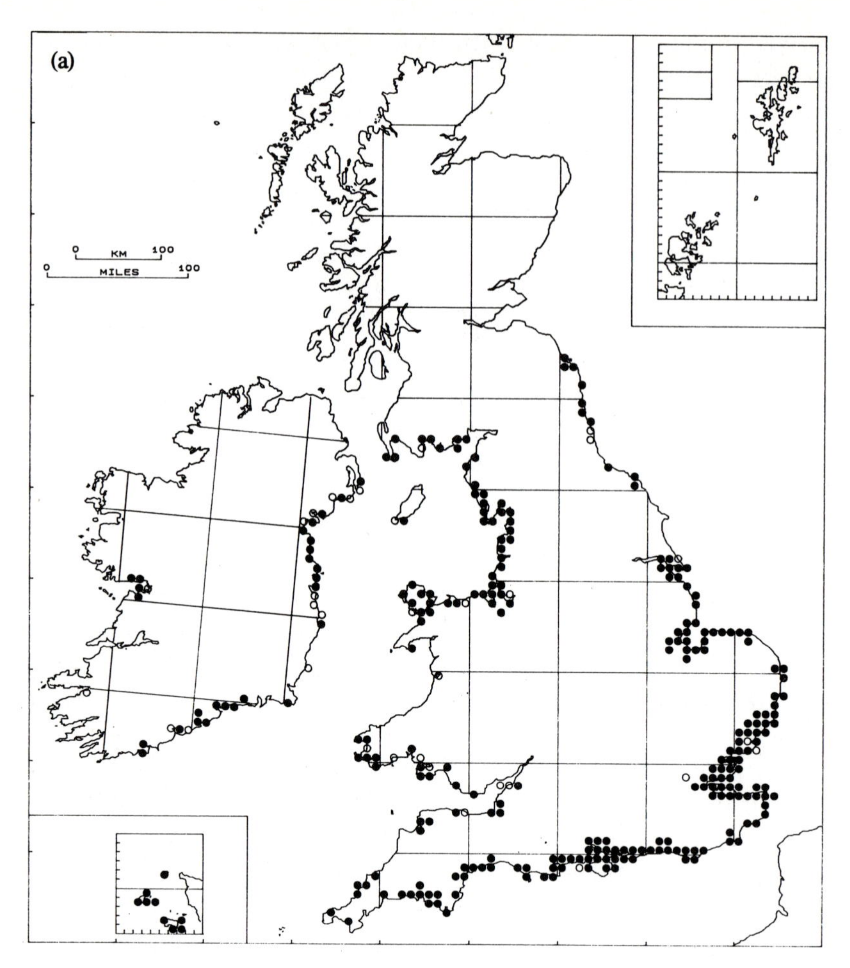

Fig. 3.9 Distribution of some species reaching a northern limit at the 'Solway Line'. [Maps from the BRC: (a) *Halimione portulacoides*; (b) *Artemisia maritima*; (c) *Limonium vulgare*; (d) *Centaurium pulchellum*.]

frequently subject to human interference and few marshes in south-east England show natural transitions from tidal to non-tidal conditions. The present distribution of species in the upper marsh could represent the remnants of a wider occurrence, fragmented by disturbance. A number of species of the drier sub-group (including *H. marinum*, *B. tenuissimum*, *Pu. fasciculata*, *Pu. rupestris*) are now more common on sea walls, or on tracks behind sea walls, than in true inter-tidal marsh.

Within the dry subgroup of species there is an assemblage which is characteristic of the north Norfolk coast; *F. laevis*, *Su. vera*, *L. bellidifolium*

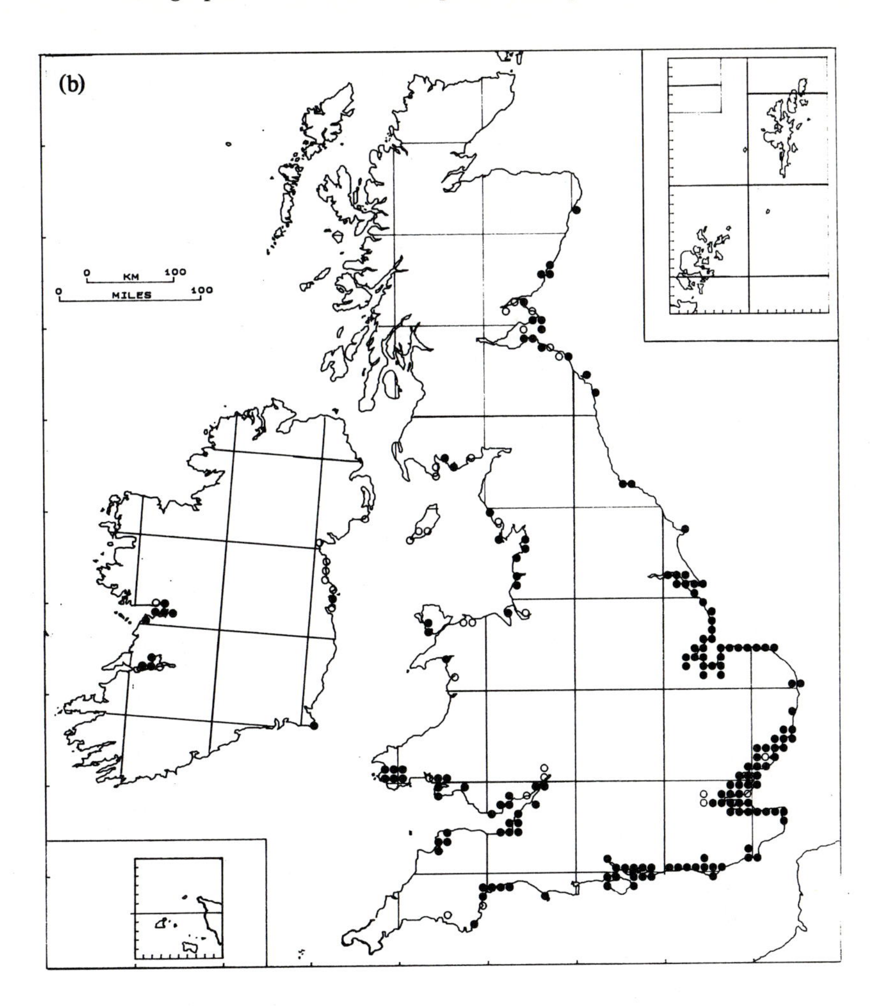

and *L. binervosum*. These species are characteristic of the communities of the Frankenio–Armerion and represent a Mediterranean element in the British flora (Matthews 1955). The species are found in the transition between marsh and either dunes or shingle. The whole assemblage is found only along the north Norfolk coast and there is no evidence to suggest a wider distribution in the recent past. However, a related, more depauperate community, with *F. laevis*, does occur in similar habitats on the Sussex coast. Whether the assemblage is limited climatically in Britain or is restricted by a requirement for a specialised transitional habitat is open to debate. The discovery of thriving *Frankenia* populations in Wales (Roberts 1975; Waldren 1982) suggests that the current distribution of the assem-

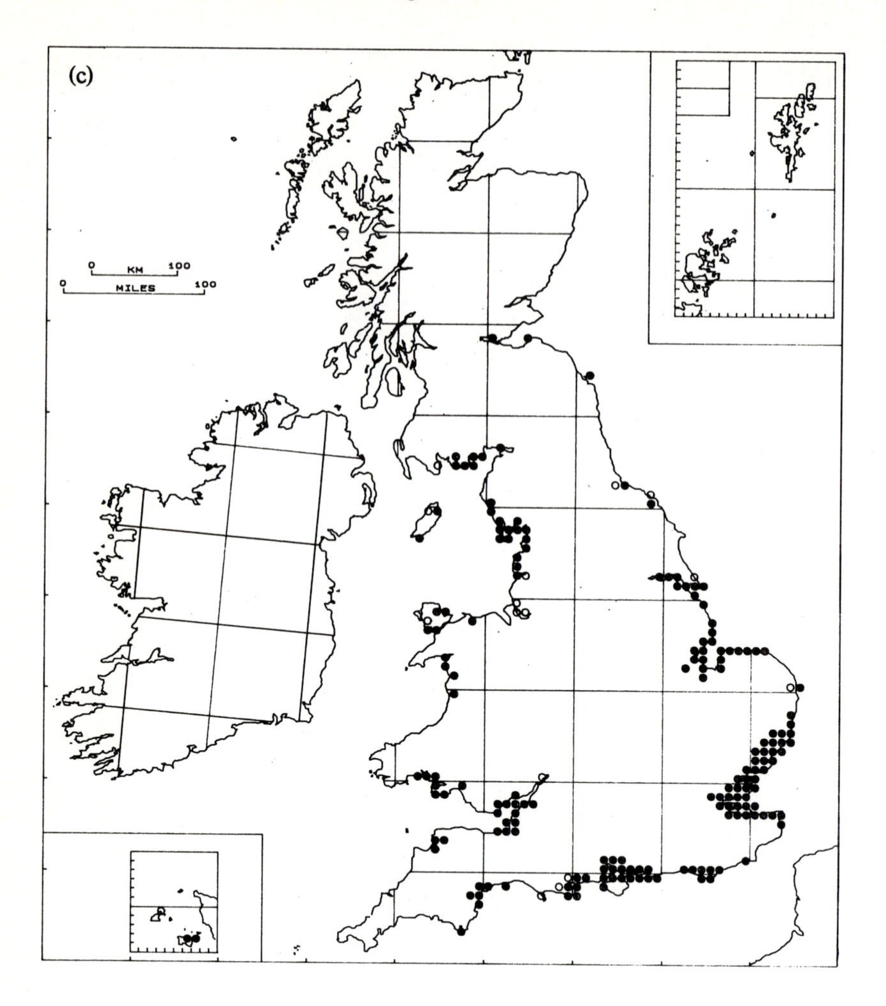

blage may not be limited by contemporary climatic conditions. Stands dominated by *L. binervosum* occur in similar open habitats transitional between saltmarsh and sand dunes at a number of localities in south Wales and also at Malahide Island in eastern Ireland (Ní Lamhna 1982). These represent a depauperate geographical extension of Frankenio–Armerion communities.

The marshes of western Scotland were probably similar in character to their present condition even in the absence of grazing.

The physiography of the east coast of Scotland differs from that of the west. Much of the coastline is not conducive to extensive development of saltmarsh, although some large areas of marsh occur in the larger firths. Along rocky and shingle coasts, there are numerous small patches of

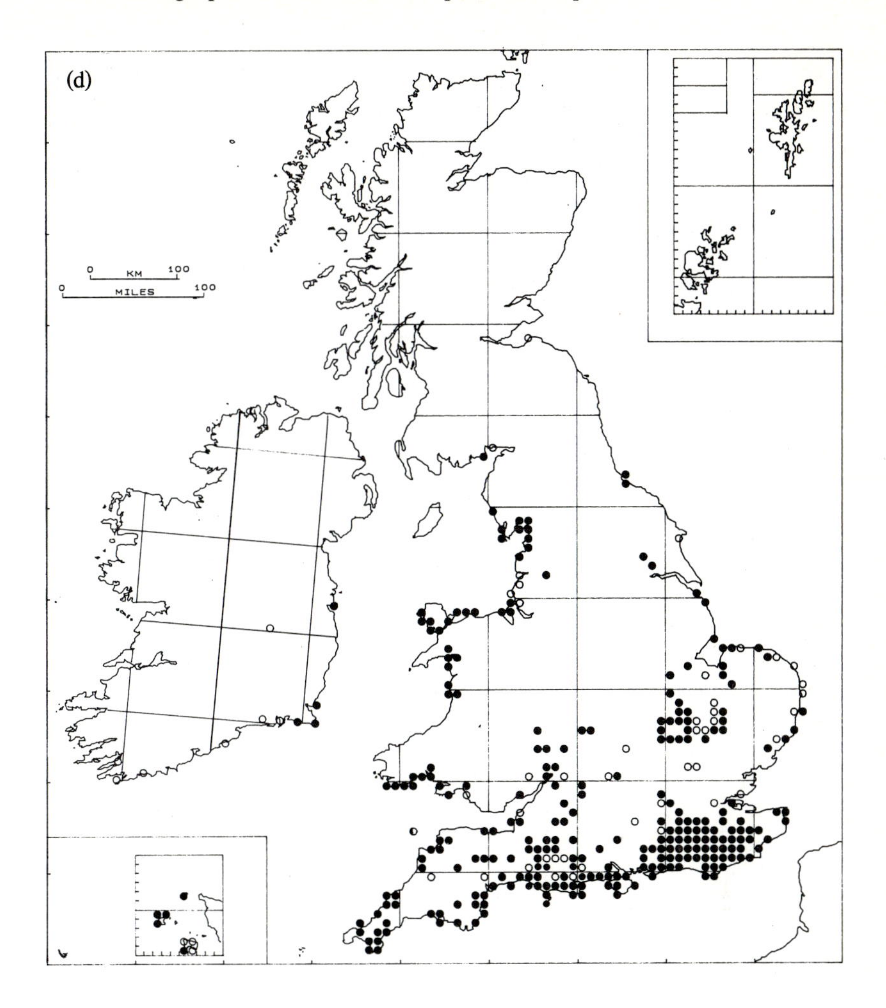

saltmarsh vegetation, referred to by Doody, Langslow & Stubbs (1984) as 'beach head' marshes. In terms of environmental features, flora and vegetation, these areas are very similar to the loch head marshes of western Scotland (Doody *et al.* 1984; Leach & Phillipson 1985). As with loch head marshes in the west, these areas are likely always to have supported a rich and diverse flora including a mixture of widespread saltmarsh species and brackish fen elements. However, in the Firth of Forth there are some southern elements (such as *Art. maritima*) found locally in these assemblages, suggesting that this region is of biogeographic interest as marking the northern limit of southern vegetation types. There are too few records to determine whether the distribution of species is stable.

There is relatively little information about larger marshes north of the

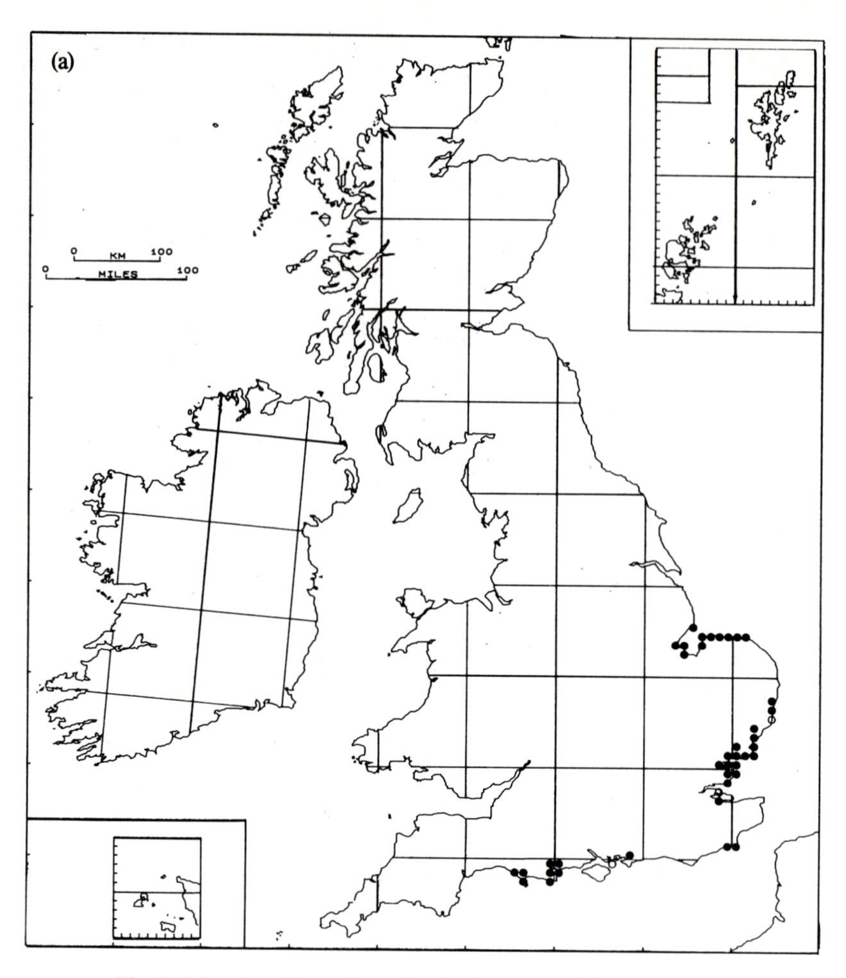

Fig. 3.10 Species with southern distributions on British saltmarshes. [Maps from the BRC: (a) *Suaeda vera*; (b) *Alopecurus bulbosus*; (c) *Althaea officinalis*; (d) *Sarcocornia perennis*.]

Firth of Forth. However, the available low-marsh flora is limited so it is probable that the lower marsh would have been dominated by *Pu. maritima*. In the absence of *H. portulacoides* and *Limonium* spp., it is likely that variants of Juncetum gerardii would have been the major mid- and upper-marsh communities. Thus, these marshes may have been naturally similar to the present grazed Type B marshes of the Irish Sea coast. However, they may have been relatively species-poor, as are present marshes in the inner Firth of Forth (Proctor, Fraser & Thompson 1983).

The boundaries of the distribution of the 'natural' saltmarsh types

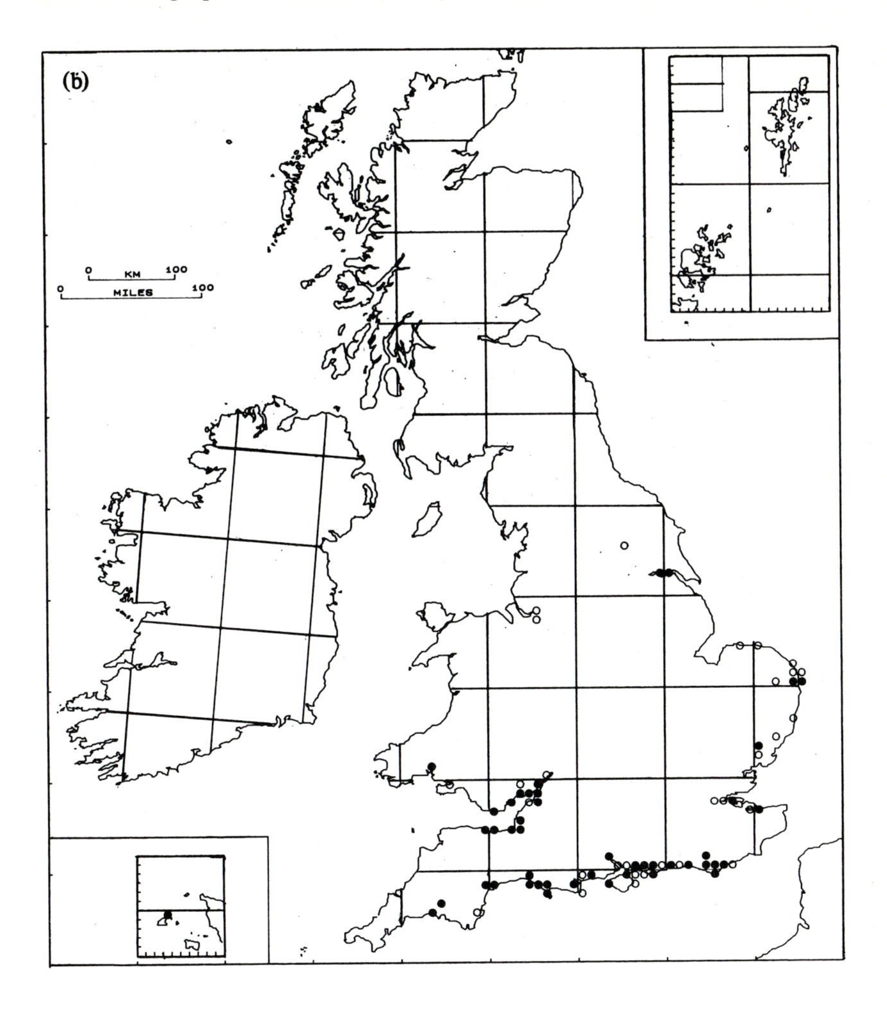

presented by Adam (1978, see Fig. 3.11) are unlikely to have been sharp; rather, there would have been broad transitional zones where the marsh type depended on local site conditions.

North of the Solway Line, marshes would have been generally similar to those of the present, with Type C marshes in the west and Type C marshes on beach head situations and Type B marshes in more extensive estuarine sites in the east.

South of the Solway, marshes on the west coast would have been rather different from the present grazed grassy swards. *H. portulacoides* and *Limonium* spp. would have been important in the low and mid marsh. At the lowest limits of marsh, *Pu. maritima* would have been the major species,

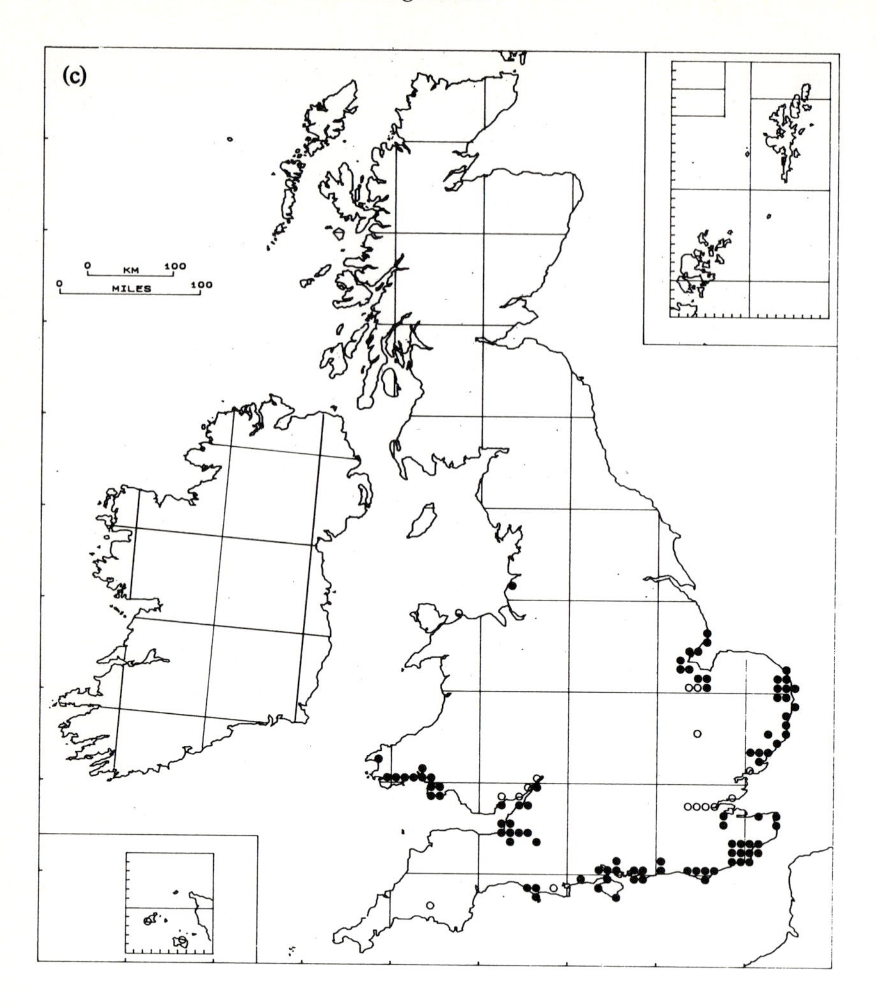

partly as a reflection of the sandy sediment, but also because other important low-marsh species, such as *S. maritima* and *A. tripolium* var. *discoideus* may be limited climatically to the south east. In the upper marsh, the more generally moist climate may have permitted a diversity of glycophytic species to become established. In the south east, the general marsh vegetation would have been similar to that of present Type A marshes but with a greater diversity in the upper marsh and distinction between the flora and vegetation on sites with a 'dry' transition to terrestrial vegetation and those with brackish conditions.

Thus, the major natural determinants of marsh type would appear to be climatic with a major division at the Solway Line possibly reflecting the

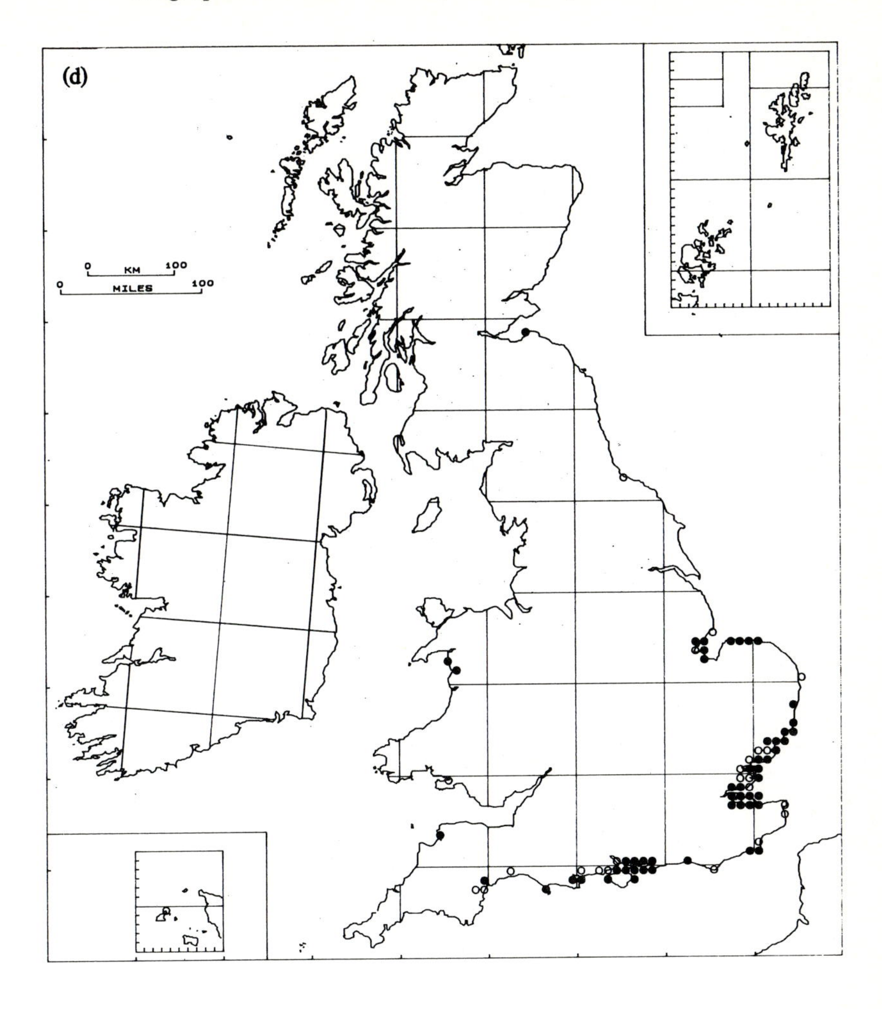

response of species to various temperature parameters and a difference between east and west coast affected by the higher rainfall in the west. In addition, sediment differences between east and west coast would also have influenced the vegetation. The influence of climate and sediment can still be seen in the present vegetation, but on the west coast the grazing regime has been the major determinant of the extensive occurrence of Type B marshes. In the south east, human activity has reduced the diversity of the upper marsh, while on both coasts the spread, encouraged by man, of *S. anglica* has introduced a new community, the development of which could alter the present distinction between marsh types.

The Solway Line marks the northern limit of a number of widespread

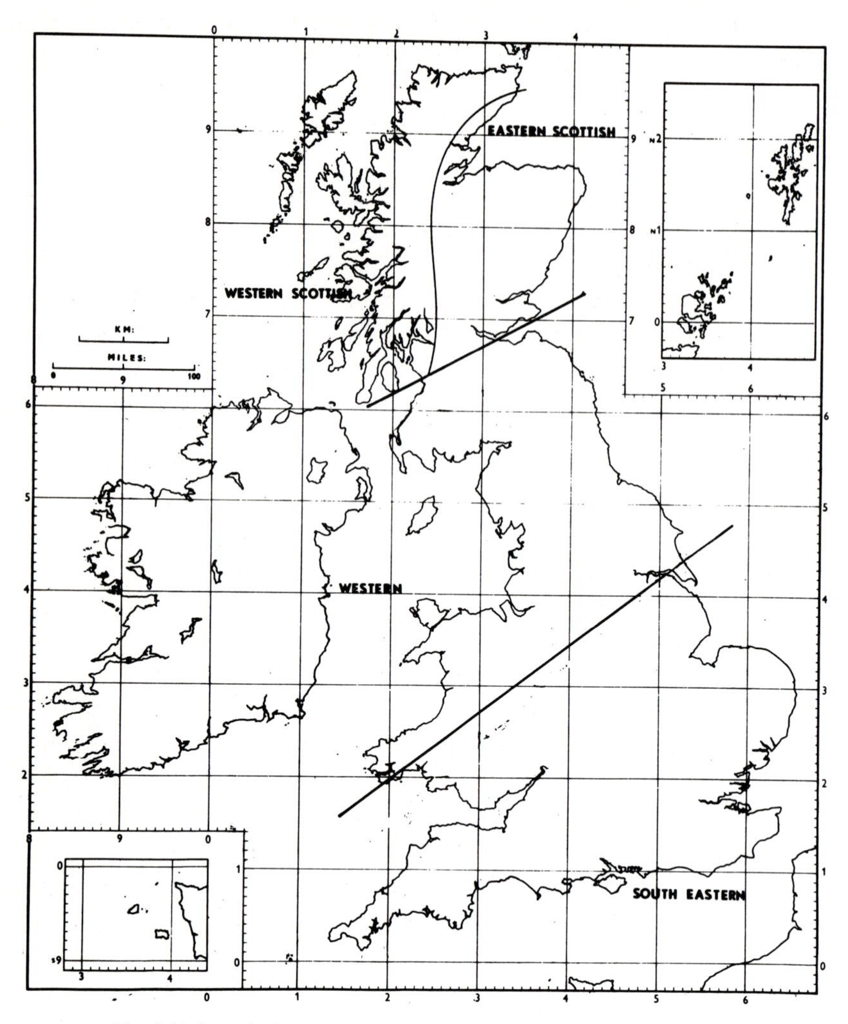

Fig. 3.11 Boundaries of hypothetical 'natural' marsh types in Britain. (Modified from Adam 1978.)

temperate halophytes. It could be argued that the Line also marks the limit of Temperate type marshes – marshes to the north of the line, particularly those in western Scotland, have a number of characteristics of the Boreal type (relatively small halophyte flora, diverse glycophyte flora and relatively large bryophyte flora).

Variation in European mainland marshes

The biogeographic subdivision of the European saltmarsh flora is discussed by Beeftink (1984) and Géhu & Rivas-Martinez (1984). These two accounts differ slightly in the positioning of boundaries, but in general terms neither is in serious conflict with the outline presented here.

Most of the plant communities on British saltmarshes can also be found on marshes elsewhere in north-west Europe (Adam 1976*a*, 1981*a*; Beeftink 1972, 1977*a*). Although a formal analysis of community distribution on the mainland has not been performed, marsh types broadly similar to the British Type A and Type B occur. However, the coincidence of species boundaries, which allowed the recognition of distinctive assemblages in the British saltmarsh flora, is less apparent on the mainland coast. In the case of the Frankenio–Amerion species, their occurrence in north Norfolk represent a considerable northerly extension from their main area of distribution.

Type B marshes, essentially grass-dominated throughout, are found in south-west Sweden, Denmark, north-west Germany, beach plain marshes in the Netherlands, and on sandy marshes in north-western France and the Bay of Biscay (Gillner 1960, 1965; Iversen 1936; Dijkema 1984*b*; Chapman 1941, 1960, 1974*a,b*, 1977*a*; Beeftink 1977*a*; Oliver 1907; Wagret 1968; Verger 1968; Géhu 1976). In all these areas there is a long history of grazing and other uses of the marshes and, although the marshes are on sandy substrates, the roles of sediment and grazing in determining the nature of the vegetation are inextricably confounded.

In Britain, Type B marshes are perhaps natural only in parts of eastern Scotland. On the mainland, Type B marshes may be similarly limited in natural occurrence to Denmark and south-west Sweden. In the absence of grazing, it is probable that what are now Type B marshes in western Britain would have extensive stands of the shrub *H. portulacoides* in the mid-marsh. The northern limit of *Halimione* on the mainland is in southern Denmark. There is good evidence (Iversen 1936; Beeftink 1959) for a considerable increase of *Halimione* during this century in Denmark.

H. portulacoides has a distribution extending southwards to the Mediterranean. Over this range, it exhibits a shift in its position in the marsh zonation (Beeftink 1985*b*), occupying lower positions further south. In Britain, however, there is considerable variation between sites in the position of *Halimione* (Adam 1976*a*). Interestingly, there are localities on the north Norfolk coast where *H. portulacoides* and *Sarcocornia perennis* form an open community at the lowest marsh level – which, along with the Frankenio–Armerion communities, provides a link with Mediterranean vegetation types.

While the marshes in south-west Sweden (Gillner 1960, 1965; Dahlbeck 1945) have extensive stands of the Puccinellietum maritimae and can, on this basis, be related to other areas in northern Europe, they have a considerable southern element in their flora, particularly in Scania (see Dahlbeck 1945; Tyler 1969*b*). *J. maritimus* and *Iris spuria* here reach their northern limit as members of the upper marsh flora, *Art. maritima* is close to its northern limit, as is *Lotus tenuis*, locally a striking component of mid-marsh Juncetum gerardii grasslands. *Halimione pedunculata*, a species whose distribution is centred on the salt steppes of south central Europe and western Asia, reaches its north-western limit in the region. It has been rediscovered in Britain – Leach 1988 – after being presumed extinct; it is decreasing elsewhere in north-west Europe – Perring & Farrell 1983. The occurrence of this southern assemblage reflects the warm dry summer climate, but the complex intermingling of species with northern and southern phytogeographic affinities illustrates the difficulties in delimiting the boundaries of saltmarsh types.

Type A marshes are characteristic of the muddier estuarine marshes on North Sea coasts and are extensive along the French Channel coast (Beeftink 1962, 1965, 1968, 1975, 1977*a*; Chapman 1960*b*; Géhu 1976).

The species-rich fen grasslands of Type C marshes appear in their detailed composition to be unique, but general similarities can be seen with communities in southern Norway (Dahl & Hadač 1941) and overall these marshes have strong Boreal affinities, despite providing the northernmost sites for a number of southern species. [Géhu & Rivas-Martinez (1984) assign marshes in the extreme north of Scotland to a Boreo–Atlantic grouping; it seems more appropriate to extend the distribution of this grouping much further south, at least as far as Argyll.] However, species are also shared with brackish fens in the Mediterranean, and it is possible to delimit a distinct group of fen grassland to rush communities occurring along the western seaboard of Europe from the Mediterranean to southern Scandinavia (Bellot 1966; Braun-Blanquet & de Ramm 1957; Vanden Berghen 1965*a*,*b*; Adam, Birks & Huntley 1977).

The clearcut distribution patterns of marsh types in Britain, both those of the present and those of the earlier hypothetical natural types suggested by Adam (1978) are not paralleled on the mainland. For the most part, Type A and B marshes are found in close proximity, with segregation being in terms of sediment type, physiography, and grazing pressure. Taking the European coast as a whole, there are no major coincident floristic boundaries, rather a steady dilution of southern elements and vegetation types with increasing latitude.

An historical perspective

In geological terms, saltmarshes occupy the most ephemeral of habitats. Shorelines have rarely been constant in position, but have changed in consequence of global eustatic changes in sea level, local isostatic changes and the processes of geomorphological coastal evolution. The consequent changes in saltmarsh ecology have received little attention.

During the Quaternary, there have been several periods of glaciation, each accompanied by major reductions in sea level and major latitudinal shifts of bioclimatic zones. At the lowest sea levels, any saltmarshes would have been far from present shorelines. Were the saltmarsh types at these times similar to those of the present, or was there a major re-ordering of species assemblages? Unfortunately, much of the available evidence must now be inaccessible under continental shelf seas.

For Arctic and Boreal marsh species, survival during glacial maxima must have involved either major migration or refuge on isolated ice-free areas. Apart from speculation by Nordhagen (1954) on the possible history of some of the distinctive species of Fennoscandian marshes, little attention has been given to the problem.

A number of what are now predominantly saltmarsh species were more widespread at inland sites during interstadial; whether these occurrences reflect development of saline conditions or the opportunism of halophytes in the absence of competition is unclear (Adam 1977*a*; Beeftink 1985*b*).

Most coastlines do not provide continuous opportunity for saltmarsh development; instead estuaries are separated by cliff or sand-dune coasts. Saltmarshes can be regarded as habitat islands, and this disjunct distribution will have applied throughout changes in sea level. Recolonisation of sites may have required considerable long-distance dispersal, and differences in flora between estuaries may be a reflection of the stochastic element in re-establishment.

The total area of saltmarsh will have changed considerably as coastlines have developed. Woodroffe, Thom & Chappell (1985) have documented the brief occurrence of vast mangrove stands in northern Australia at about the end of the post-glacial rise in sea level; similar changes in saltmarsh area may have taken place in other regions at various times.

Climatic effects on species' distribution

While it is possible to define geographic limits of species, there are few studies on those factors which determine the position of these limits. While a species' distribution limits may be correlated with various climatic parameters, how the plant responds to climate is largely a matter of speculation.

One of few cases in which the mechanism by which climate imposes a limit on a saltmarsh species has been determined is *S. europaea*. This species does not extend into the high Arctic, and its presence is one of the distinguishing features between Arctic and Boreal marshes. Towards its northern limit, the distribution becomes patchy and some populations may fail to set seed, presumably being sustained by colonisation from nearby areas (Vince & Snow 1984). Jefferies, Jensen & Bazely (1983) showed that in Hudson Bay *S. europaea* was restricted microtopographically to south-facing slopes, which were significantly warmer during the brief growing season than north-facing slopes. Germination on north-facing slopes was poor and late, while transplants to north-facing slopes showed low seed production compared with that on south-facing slopes. At more northerly sites with colder soils, germination would be delayed further and the growing season would become too short for successful seed set.

In the case of perennial species, climatic conditions necessary for seed set and establishment need occur only infrequently for populations to be maintained, but climatic effects on the vegetative phase may determine distributions (by effects such as the alteration of competitive interactions between species or by killing plants at a time when propagules are not available for recolonisation). Unfortunately, climatic effects on the growth of perennials have been little studied. There is evidence that frost adversely effects *S. anglica* and *H. portulacoides* (Ranwell 1967, 1972; Beeftink 1977; Beeftink *et al.* 1978), but most northern Temperate species seem to be capable of tolerating occasional hard frosts.

It is interesting that at a time when there is no convincing evidence for a consistent trend in climatic change over continental northern Europe, a number of species appear to be extending their range northwards (Beeftink 1985) – these species include: *Su. maritima*, *Atriplex littoralis*, *Spergularia media*, *Carex extensa*, *H. portulacoides* and *E. pycnanthus*. The 'Greenhouse Effect', although threatening the continued existence of many marshes through its effects on sea level, may as a consequence of global warming be accompanied by considerable changes in species distribution.

4

Coping with the environment

Introduction

The saltmarsh environment is far removed from the optimum for growth of most plant species. Environmental characteristics of saltmarshes which would be inimical to plants include:

High (but variable) salinity in the soil solution;
Essential nutrient ions present as a low proportion of the total ionic composition of the soil solution;
Anaerobic soil conditions.

Tidal immersion may have a number of effects including:

Temperature shock;
Changes in photoperiod;
Mechanical effects of tidal currents;
Deposition of sediment on leaf surfaces.

Interest in the physiological ecology of saltmarsh plants has a long history but most research has been on the effects of salinity and other aspects of the environment have received little attention.

It is the salinity of saltmarshes which distinguishes them from freshwater marshes and fens but the concentration of interest on salinity has possibly led to a narrow view of the important factors operating to differentiate between saltmarsh communities. While it may be appropriate to see the selection of the saltmarsh flora from the broader terrestrial flora as being on the basis of salt tolerance, selection for particular microhabitats within the saltmarsh may have favoured other traits. Within the framework of a salt-tolerant flora selection for occurrence in different communities may have been in terms of tolerance to varying degrees of soil anaerobiosis, ability to thrive under particular soil nitrogen levels, or ability to withstand currents of particular velocity. In view of the fluctuations in soil salinities in many communities it seems unlikely that differential salinity tolerance *per se* is a major discriminating factor between species unless the discrimination is achieved during the short period of germination and seedling establishment (although certain communities of brackish habitats may be restricted in their distribution by the inability of the component species to tolerate high salinities).

Although there have been many studies on the physiology of halophytes much of the work on the effects of salinity on plants has been on the markedly adverse impact of salinity on non-tolerant species. Such plants are, except ephemerally, absent from saltmarshes. Nevertheless, it is useful to compare halophytes with intolerant species in order to see which features of halophytes may reasonably be assumed to be related to their salt tolerance. Comparative studies also allow the 'costs' of achieving salt tolerance to be assessed: does being a halophyte impose constraints upon the ability of a species to respond to other factors?

In this chapter, the responses of plants to both salinity and waterlogging are discussed.

Unfortunately, there are few studies which have investigated the physiological adaptations of plants to the whole complex of environmental factors operating in field conditions. Any synthesis of the ecological adaptations of saltmarsh plants will require greater attention to the interactive effects of various factors.

The chapter concentrates almost exclusively on flowering plants. The effects of fluctuating salinity and water table would have considerable impact on other elements in the flora although much less is known about the physiological responses of the other groups in the saltmarsh flora. The environmental challenges to the fauna are equally great and the range of adaptations and responses exhibited by animals is probably wider than that displayed by plants.

Plants and salinity

The effects of salinity on plants have long attracted the interest of both ecologists and physiologists. Over the last twenty years, considerable research has been carried out on biochemical and physiological aspects of plants' response to high salinity and it is now possible to present an overview of the range of adaptations exhibited by halophytes. Recent reviews of the effects of salinity include Epstein (1980), Flowers (1975, 1985), Flowers, Troke & Yeo (1977), Flowers & Yeo (1986), Flowers, Hajibagheri & Clipson (1986), Goodin (1977), Gorham, Wyn Jones & Mcdonnell (1985), Greenway (1973), Greenway & Munns (1980), Munns, Greenway & Kirst (1983), Jeffrey (1987), Jefferies & Rudmik (1984),

Jennings (1976), Queen (1975), Rains (1979), Stewart & Ahmad (1983), Stewart & Popp (1987), Tal (1985), Wainwright (1980, 1984), Waisel (1972), Winter, Osmond & Pate (1981), Wyn Jones *et al.* (1977) and Yeo (1983).

In order to discuss responses to salinity, it is first necessary to document the sorts of conditions to which halophyte roots are exposed. Given that these conditions are very different from those prevailing in most terrestrial communities, it is appropriate to ask whether halophytes differ fundamentally from other angiosperms in their nutrient requirements; has evolution resulted in such specialisation that halophytes would be unable to grow outside saline environments?

Historically, study of the effects of salinity was limited to the end result, the total growth of plants. The growth responses of halophytes to raised salinity are discussed and the difficulties of interpreting much of the available data are introduced. In order to understand the basis of the reported growth responses, it is necessary to look in more detail at the mechanisms by which metabolism is protected from the potentially adverse effects of high salinity and most of this chapter is devoted to a detailed consideration of this topic.

Salinity

The composition of seawater is given in Table 4.1.

For many purposes the waters flooding saltmarshes may be regarded as more, or less, dilute seawater. The relative concentration of major ions in the interstitial soil water reflects that of the flooding water, although at very low salinities exchange processes between the soil solution and cation-exchange sites may affect the divalent/univalent ion ratio (Kuraishi *et al.* 1985). Nitrogen and phosphorus levels in the soil solution may be considerably different from those in the adjacent water body. Levels of micronutrients may differ between estuarine and coastal waters and may also vary between estuaries but any consequences for saltmarsh plant nutrition remain to be investigated. In most major estuaries, direct discharges from industry, and run-off from urban areas, has resulted in an increase in various heavy metals in tidal waters. Some of these metals may accumulate in saltmarsh sediments and the foliage of saltmarsh plants – reports of directly toxic effects on plants are lacking but any such accumulation in the vegetation may be further concentrated at higher levels in the food chain (Beeftink *et al.* 1982; Ragsdale & Thorhaug 1980; Gallagher & Kibby 1980).

The upper reaches of an estuary are the site of numerous and complex chemical changes (Morris *et al.* 1978). Whether the particular chemical

Table 4.1. *The major constituents in solution of oceanic seawater having a salinity of 35‰ (after Harvey 1957)*

	$g\,kg^{-1}$	$mol\,m^{-3}$
Sodium	10.77	483.0
Magnesium	1.30	54.5
Calcium	0.41	11.5
Potassium	0.39	10.0
Strontium	0.01	0.1
Chloride	19.37	558.0
Sulphate	2.71	29.0
Bromide	0.07	0.4
Borate	0.03	0.5

condition of upper estuarine waters is reflected in the fringing vegetation is not clear but in general the nature of the vegetation in low-salinity brackish conditions seems to be influenced more by the total ionic concentration rather than by any specific ions.

The ionic composition of the rooting environment for saltmarsh plants is dominated by sodium and chloride. While chloride is an essential ion for plant growth, neither ion would be quantitatively important in 'normal' inland soils. The second most common cation and anion are magnesium and sulphate which again would not be, proportionately, as important in 'normal' soil. For many plants the magnesium concentrations found in saltmarsh soils would be toxic. There has been little work on the effect of high levels of magnesium on halophytes although Hodson *et al.* (1982) demonstrated that a saltmarsh population of the grass *A. stolonifera* was more tolerant of high levels of magnesium than an inland clone of the same species. Nevertheless, of the major nutrient ions, potassium (the cation normally present in glycophytes in largest amount) and calcium are present in seawater at absolute concentrations which would be regarded as high in most inland soils.

In non-tidal saline communities there is considerable variation in ionic composition, particularly in the anions. To what extent the differences in the flora between inland saline areas reflects the varying tolerance to different ions remains uncertain despite much research over many years.

Bearing in mind the uniformity of ionic ratios in coastal marshes and their constancy over geological time the relevance, in an ecological context, of many physiological studies must be questioned. Many investigations have involved challenging plants with varying levels of NaCl against a background of a standard nutrient solution. This exposes plants to a range of conditions very different from those of any coastal marsh (see also Shennan, Hunt & MacRobbie 1987*a*). In the saltmarsh, the ionic ratios remain nearly constant over a wide range of salinities. The experimental imposition of salinity using NaCl, while possibly providing information upon the robustness of a species' physiology, may provide results which are difficult to extrapolate to the field situation. In particular, there is evidence for an ameliorating effect of calcium upon the tolerance of vascular plants to increased salinity (LaHaye & Epstein 1969, 1971; Leopold & Willing 1984; Munns, Greenway & Kirst 1983; Gorham *et al.* 1986); calcium has also been shown to have a major influence on the control of cation levels in bryophytes (Bates & Brown 1974; Bates 1976). Under natural conditions the Ca/Na ratio will remain constant with fluctuating salinity (although differential uptake of ions may alter ratios within the immediate rhizosphere) and some of the results of experiments challenging plants with NaCl may reflect the alteration in the Ca/Na ratio as much, or more, than they reflect plant response to absolute Na levels. Greenway & Munns (1980) and Munns, Greenway & Kirst (1983) suggest that some halophytes may be insensitive to Ca/Na ratio over a wide range of values while for non-halophytes low Ca/Na ratios may be one cause of salt toxicity. However, it is clear that there is considerable variation between species in response to calcium and salinity. In rice, a species very susceptible to increased salinity, there is no amelioration of the effects of salinity by calcium over a very wide range of Na/Ca ratios (Yeo & Flowers 1985).

Lynch & Läuchli (1985) showed that in barley, increased soil salinity resulted in decreased calcium content of shoots. This effect was apparent at relatively low concentrations of sodium chloride which inhibited calcium transport from root to shoots by interfering with the release of calcium into the root xylem. If this proves to be a widespread phenomenon amongst glycophytes exposed to high salinity the question then arises as to how calcium status in the shoots of halophytes is maintained.

Nutrient requirements of halophytes

All plants require certain mineral elements for growth; some of these elements are required in comparatively large amounts (macronutrients – N, P, K, S, Ca, Mg) while the others (micronutrients or trace elements – Cu, Zn, B, Cl, Mo, Mn and Fe) are required in very much smaller amounts (and in some cases are toxic at relatively low concentrations).

Even in the case of the macronutrients most plants can obtain sufficient from very low external concentrations (see Barbour 1970). There is little evidence to suggest that under non-saline conditions halophytes have a greater requirement than other plants for either macro- or micronutrients. However, it would appear that some halophytes have a constitutive ability to absorb large quantities of ions even under non-saline conditions (Schimper 1903; Collander 1941; Flowers, Troke & Yeo 1977) and to varying degrees this may be a general characteristic of halophytes serving to distinguish them from glycophytes.

The absorption of large quantities of ions at low external salinities could be of adaptive significance even though, at low salinities, it might be viewed as a disadvantage. The uptake of such large quantities of ions will generate far lower leaf osmotic potentials than 'required' to maintain turgor – assuming the ions are largely sequestered in vacuoles (see p. 249) and osmotic balance is maintained by organic solutes in the cytoplasm then there will be a diversion of carbon (and probably also nitrogen) from growth into the synthesis of cytoplasmic osmotica. Nevertheless, although conditions of low salinity can be maintained experimentally, in the field they are likely to be a transient phenomenon. If low-salt tissue developed during these brief low salinity phases, the plant would be faced with a need for rapid osmotic adjustment when high salinity conditions returned (Yeo & Flowers 1986*b*).

Amongst the essential micronutrients is chlorine (Broyer *et al.* 1954) which is the major anion in the saltmarsh environment. Several reports have suggested that some halophytes may have particularly high requirements for chloride in some aspects of their photosynthetic metabolism. However, high chloride concentrations reported in chloroplasts of several species now appear to be artefacts (Ball 1986). The apparent requirement for high chloride levels in chloroplasts for optimum photosynthesis in the mangrove *Av. marina* (Critchley 1982) results from damage during isolation of the thylakoids (Andersson *et al.* 1984; Ball & Anderson 1986; Ball 1986). In general, growth of plants is not precluded by the chloride levels available in 'normal' non-saline soils. Epstein (1980) has suggested that, while chlorine requirements are greater than those for other micronutrients, no case of chlorine deficiency has been demonstrated in field-grown plants. Nevertheless, there is some evidence that some species may have chloride requirements considerably higher than those of other species. Particularly high requirements have been reported for kiwifruit (*Actinidia deliciosa*), even though this species is adversely affected by low sodium concentrations in the rooting zone (Smith, Clarke & Holland 1987). The physiological basis for this chlorine requirement is obscure.

Sodium is not believed to be generally an essential micronutrient. However, owing to the difficulties of excluding minute amounts of sodium, experimental demonstration that growth of plants is possible under sodium-free conditions is difficult. Development of experimental systems allowing the rigorous investigation of sodium nutrition has shown conclusively that sodium is an essential micronutrient for some plants (Brownell & Wood 1957; Brownell 1979). Since the original demonstration of a sodium requirement in the saltbush, *Atriplex vesicaria*, sodium has been shown to be essential for a range of species. Taxonomically, these species are a diverse assemblage but they are united by carrying out photosynthesis with the first fixation products being a four-carbon dicarboxylic acid (in most cases the C_4 pathway but in some cases CAM) (Brownell 1979; Jennings 1976). The physiological role of sodium in the photosynthetic mechanisms of these species remains obscure, but appears to be related to the organisation of the light harvesting and reaction centres in the mesophyll chloroplasts (Grof, Johnston & Brownell 1986). The actual levels of sodium required to overcome deficiency symptoms are very low (less than 1 mol m^{-3}). As such, the sodium requirement may be a physiological curiosity rather than a determinant of field distribution.

Types of halophyte

Various studies over many years have suggested that there is variation in the salt tolerance of species. The reaction of both ecologists and physiologists to this diversity of plant behaviour has been to impose classifications defining degrees of halophilism. The history of such schemata is discussed by Waisel (1972) and Chapman (1960*b*, 1974*a*). The criteria on which classifications were based varied but included the range of salinity conditions over which a species occurred in the field (Iversen 1936; Chapman 1942), the response of a species to different levels of external salinity (van Eijk 1939; Tsopa 1939), or the internal salt concentration of plant tissues (Steiner 1939). (The topic of classification of halophytes into 'physiotypes' is discussed further on p. 272.)

A problem with such classifications is the definition of terms such as high salinity. The chosen levels vary between classifications and in some schemes (e.g. Tsopa 1939) the levels are left undefined. The choice of a dividing salinity between glycophytes and halophytes remains arbitrary.

Flowers, Hajibagheri & Clipson (1986) defined halophytes as species found growing where the salinity exceeds 100 mol m^{-3}. In the face of a continuum of plant responses to increasing salinity any such limiting condition may be questioned but its general acceptance may facilitate discussion. A range of other limiting values have been suggested in the literature (see Iversen 1936; Stocker 1928; Chapman 1942; Munns, Greenway & Kirst 1983 adopt as a limit a water potential of -0.33 MPa, the equivalent of 70 mol m^{-3} monovalent salts).

In discussion of types of halophyte, it is frequently implicit (and sometimes explicit – Waisel 1972) that at least some species (obligate halophytes) actually require high levels of salt for growth. Barbour (1970) suggested that an obligate halophyte would be a species with optimal growth at moderate or high salinity, and incapable of growth at low salinity (low salinity in this case being less than 2% salt). Although Chapman (1960*b*, 1974*a*) and Waisel (1972) suggest that certain species (most particularly succulent chenopods) are obligate halophytes, Barbour argued that there was no evidence for any saltmarsh or mangrove species conforming to his definition of an obligate halophyte; indeed there were many demonstrations that both saltmarsh and mangrove species could be grown successfully under garden cultivation. In terms of ecological requirements it would be necessary to demonstrate that these species could complete their life cycle successfully under non-saline conditions. Unfortunately, this has not been done in all cases but there is evidence in most instances for good seed set under non-saline conditions.

The pioneers of ecophysiology (for example, Schimper 1903; Warming 1909) avoided the implication that halophytes have an obligate requirement for salt; Schimper (1903) specifically acknowledged that most halophytes could be grown successfully under non-saline conditions. On the basis of our current understanding of the physiology of halophytes it seems unlikely that an obligate requirement for high levels of salt would occur in angiosperms (although in some extremely halophilic bacteria obligate requirements for high salt levels are well attested).

One species which is arguably an obligate halophyte in ecological terms is *Suaeda aegyptica*. Eshel (1985) showed that growth stimulation in this species was a specific response to sodium and not potassium and, importantly, plants only reached the flowering stage under a sodium chloride treatment. If flowering is indeed only possible under saline conditions then populations could only be maintained at sites experiencing high salinities. Confirmation that other conditions cannot promote flowering would be desirable.

Salinity and growth

Many of the studies of the effect of salinity on plants have been at the whole-plant level of investigation and have concentrated on the effects on growth.

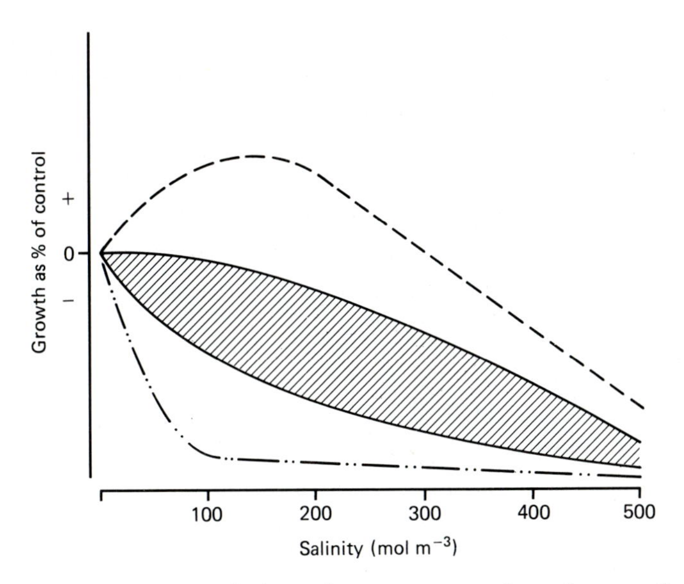

Fig. 4.1 Generalised growth response curves of several groups of species subject to various salinity treatments. (Based on Greenway 1973 and other sources.) – – – species showing salt stimulation with maximum growth at *c.* 200 mol m^{-3} external NaCl; //// envelope of curves for halophytes not exhibiting salt stimulation; – · · – salt-sensitive glycophytes.

Greenway (Greenway 1973; Greenway & Munns 1980) suggested that on the basis of growth responses, it is possible to recognise four groups of plants (see Fig. 4.1); very salt-sensitive non-halophytes with a steep decline in growth with increasing salinity; halophytes and non-halophytes with a decline in growth with increasing salinity and little growth above 2–300 mol m^{-3} external chloride; halophytes with no growth stimulation but a small decline in growth with increasing salinity up to 3–500 mol m^{-3} external chloride when a steep decline in growth occurs; and halophytes showing growth stimulation with increasing salinity, up to about 200 mol m^{-3} followed by declining growth at higher salinities.

Such a classification, while useful, is arbitrary as there is a continuum of plant responses. Different genotypes of a single species may show different responses, while responses to increased salinity may be mediated by other conditions (for example, nitrogen availability – Smart & Barko 1980; Haines & Dunn 1976; humidity – Gale, Naaman & Poljakoff-Mayber 1970; light intensity – Longstreth & Nobel 1979; Kemp & Cunningham 1981; and temperature – Kemp & Cunningham 1981). Unfortunately, the possibility of interactions between environmental factors is rarely allowed for in experimental design and experimental conditions are not always adequately defined. Caution is therefore needed when comparing different studies.

The various halophyte categories as delimited by Greenway are not sharply differentiated in the field and communities may contain species from several of the categories. Species exhibiting growth stimulation are not necessarily the most 'successful' in the saltmarsh and the salinity at which optimal growth is attained experimentally is often low compared with that which might prevail in the field for much of the growing season.

Growth is a complex phenomenon and the analysis of various facets of growth may provide more insight into the environmental control of resource allocation than a single (and often vaguely defined) measure of growth (Shennan, Hunt & MacRobbie 1987*a*,*b*). Most studies of the effects of salinity on growth have involved single harvests after long exposure to experimental treatment (for example Glenn 1987) – Shennan *et al.* (1987*a*,*b*) have shown that sequential harvests at short time intervals can provide much useful information not obtainable from the more frequently employed single-harvest experimental designs.

Growth stimulation

Although it is unlikely that there are obligately halophytic saltmarsh plants (Barbour 1970), some species appear to show growth stimulation at moderate salinities.

While there are a number of apparent demonstrations of stimulation, it is not always certain on examination of the data that the claims can be attested. The first problem resides in the measurement of growth. Fresh-weight changes are clearly inappropriate as changes in water content (increase or decrease in succulence) may mask any changes in organic matter content. In most circumstances dry matter changes are an easy and appropriate measure of growth but under saline conditions most plants contain appreciable quantities of salts. Changes in plant mineral content may be of such magnitude that changes in organic matter could be obscured. To overcome this problem ash-free dry weights should be measured but this is not always the case in examples reported in the literature. Flowers (1985) suggested that the interpretation of ash-free dry weight measurements in the assessment of growth of halophytes depends on knowing the energy requirements involved in salt transport and distribution. Data on these energy costs are limited. Hodson, Opik & Wainwright (1985) stressed the importance of considering several different bases for measuring growth when assessing the effect of salinity on plants.

The second problem lies in the concentration of interest in most studies in the above-ground portions of plants only. Changes in ash-free dry weight of above-ground parts may not indicate changes in growth of the whole plant but rather reflect changed patterns of allocation between root and shoot [over time, increase in the root/shoot ratio may indeed cause a reduction in growth rate (Pearcy & Ustin 1984) even if assimilation per unit area of active tissue was unaffected].

Fischer & Turner (1978) showed that species in mesic and arid environments exhibit an increased root/shoot ratio as soil water potential decreases. Such a response is intuitively seen as adaptive in that it simultaneously leads to a reduction in the transpiring surface and an increase in the water absorbing area, although on the debit side a large root system imposes extra demands on the shoot for assimilation.

Munns & Termaat (1986) suggested that when non-halophytes are exposed to salinity, there is a rapid decline in shoot growth. Root growth is less affected so the root/shoot ratio increases. Data presented by Termaat & Munns (1986) indicate, however, that there may be a taxonomic component in species response. Wheat and barley showed an increase in root/shoot ratio but in neither *Trifolium repens* nor *T. alexandrinum* did the ratio change significantly with increasing salinity over a short time period (over a longer period shoot growth of clovers may be reduced – Winter & Läuchli 1982).

Osmond (1980) suggested that in halophytes there may be a decreased root/shoot ratio with increased salinity. This has been reported in a number of instances. Montfort & Brandrup (1927, 1928) showed that in *A. tripolium* the root/shoot ratio declined with increased salinity and that root growth was maximal under freshwater conditions, whereas there was evidence for a slight stimulation of shoot growth up to the equivalent of about half-strength seawater. This response pattern for *Aster* was confirmed by Bickenback (1932) and was demonstrated for the grasses *Sp. virginicus* by Gallagher (1979) and *S. alterniflora* by Haines & Dunn (1976), and for the succulent chenopod *S. virginica* by Pearcy & Ustin (1984).

However, although growth patterns of some species apparently support Osmond's (1980) contention, other data deny it. Clarke & Hannon (1970) reported considerable increases in root/shoot ratios with increasing salinity for a number of Australian saltmarsh species while Parrondo, Gosselink & Hopkinson (1978) suggested for *D. spicata*, *Spartina cynosuroides* and *S. alternifolia* that in general increased salinity caused a greater reduction in shoot growth leading to an increase in root/shoot ratios at higher salinities. [The generally high proportion of total biomass in the soil in saltmarshes may indicate that this is a widespread phenomenon.] Cavalieri (1983) and

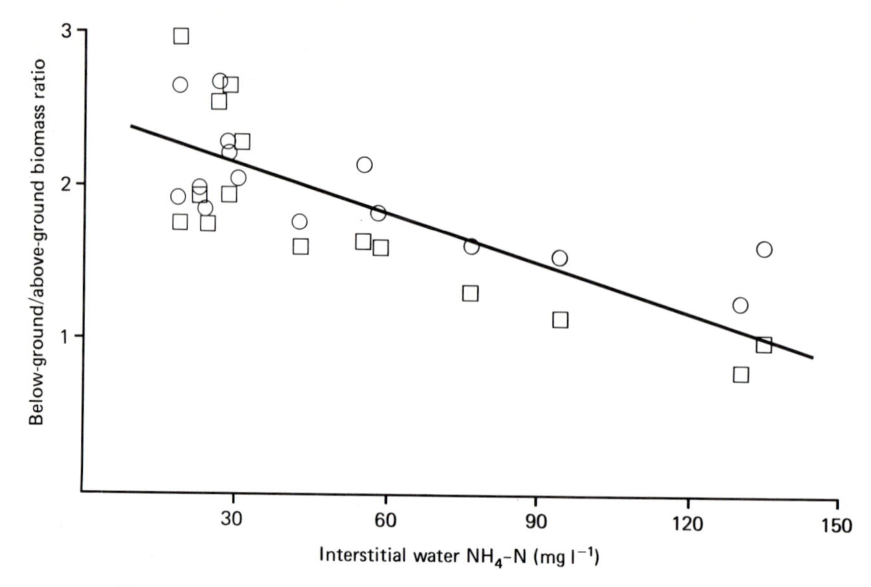

Fig. 4.2 Relationship between nitrogen supply and above ground to below ground biomass ratio for *Distichlis spicata* (□) and *Spartina alterniflora* (○) ($r = -0.78$, significant at the 99% confidence level). (Redrawn from Smart & Barko 1980.)

Smart & Barko (1980) demonstrated for *S. alterniflora* and *D. spicata* that root/shoot ratios were strongly influenced by the availability of nitrogen and phosphorus, greater resources being devoted to root growth under lower nutrient conditions (Fig. 4.2). Okusanya & Ungar (1984) showed that the root mass of *Spergularia* spp. declined with increasing salinity but that this could be reversed by the addition of nutrients. Mineral nutrient composition of the shoots was related to the size of the root system; the larger the root system the better the balance between ions was maintained. In the absence of information on nitrogen and phosphorus supply, interpretation of data on salinity effects on root growth becomes difficult.

However, the possibility of effects on the allocation of resources between above- and below-ground parts of plants emphasises the dangers of studying only changes in above-ground biomass when the effects of salinity on plant growth are under consideration.

While total root mass (as distinct from root/shoot ratios) in *S. alterniflora* showed no clear relationship to salinity, root length decreased with increasing salinity (Cavalieri 1983). For some species, however, there is evidence for stimulation of growth in individual roots (Waisel 1972). Studies on a number of grass species growing at an inland saline site in

Britain by Venables & Wilkins (1978) suggest that there is a variety of root growth responses (measured as extension) to increased salinity, comparable with those shown by shoots (Fig. 4.3). Hajibagheri, Yeo & Flowers (1985) showed that the growth rate of roots of *Su. maritima* growing in culture solution was lower (6.57 mm d^{-1}) than that of roots exposed to 340 mol m^{-3} sodium chloride (7.42 mm d^{-1}).

Salinity may also affect allocation of carbohydrate within the aerial portions of a plant. Jefferies (1972) suggests that in sites experiencing high salinities, a number of European perennial halophytes may have markedly reduced flowering (see also Jefferies, Davy & Rudmik 1979). Studies on the winter annual *Lasthenia glabrata* in California (climatically a highly predictable environment) indicate that an increase in soil salinity may be followed by the onset of flowering (Kingsbury *et al.* 1976). More work on the morphogenetic effects of changing salinity is needed.

Notwithstanding the points discussed above, in some instances a salinity-induced growth stimulation is incontestable, particularly in various members of the Chenopodiaceae (see review by Jennings 1976) although even here the effect may be mediated by other factors. Gale, Naaman & Poljakoff-Mayber (1970) demonstrated that in *Atriplex halimus* growth stimulation occurred at low relative humidities but not under more humid conditions.

Photosynthetic mechanisms of saltmarsh plants

Growth represents the balance between photosynthesis and respiration. While the ratio of photosynthetic to non-photosynthetic tissue and the rate of respiration will have a major influence on growth, the ultimate determinant of maximum possible growth is the maximum photosynthetic capability of a plant. Three photosynthetic systems are known from terrestrial plants, the C_4, C_3 and CAM pathways (with some species showing characteristics intermediate between C_4 and C_3, and some species alternating between C_3 and CAM depending on environmental conditions). Potentially C_4 plants are the most productive while CAM plants have much lower intrinsic growth rates. The different photosynthetic systems are associated with different water-use efficiencies and transpiration rates. Low transpiration rates would reduce salt uptake and it might be expected that plants in the saltmarsh would exhibit low transpiration rates (see p. 245). Species with the C_4 and CAM pathways characteristically show higher water-use efficiency than C_3 species and it might be expected that these pathways would be well represented in the saltmarsh flora.

The CAM pathway is strongly associated with succulence and many

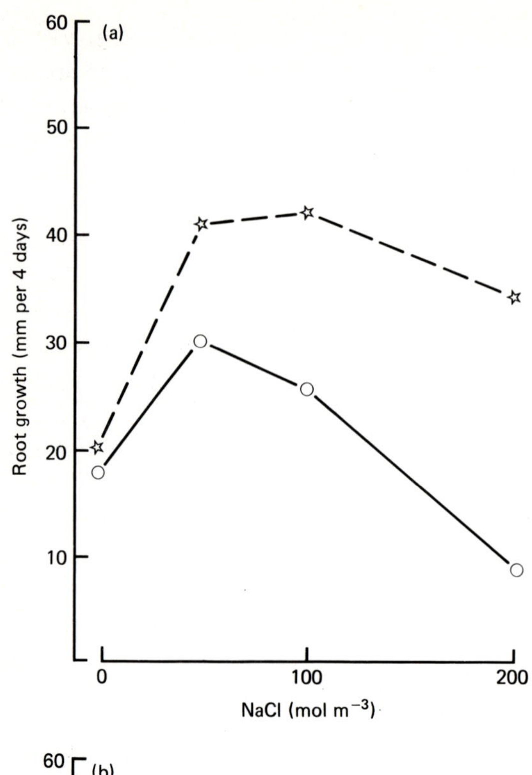
(a)
Root growth (mm per 4 days)
60
50
40
30
20
10
0
100
200
NaCl (mol m⁻³)

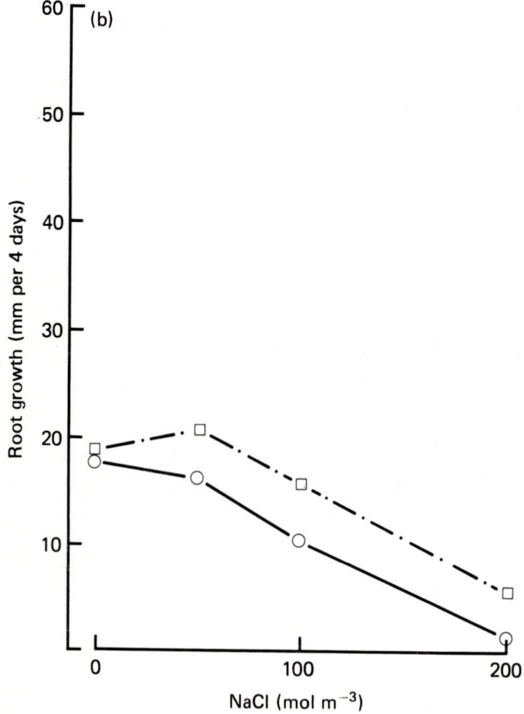
(b)
Root growth (mm per 4 days)
60
50
40
30
20
10
0
100
200
NaCl (mol m⁻³)

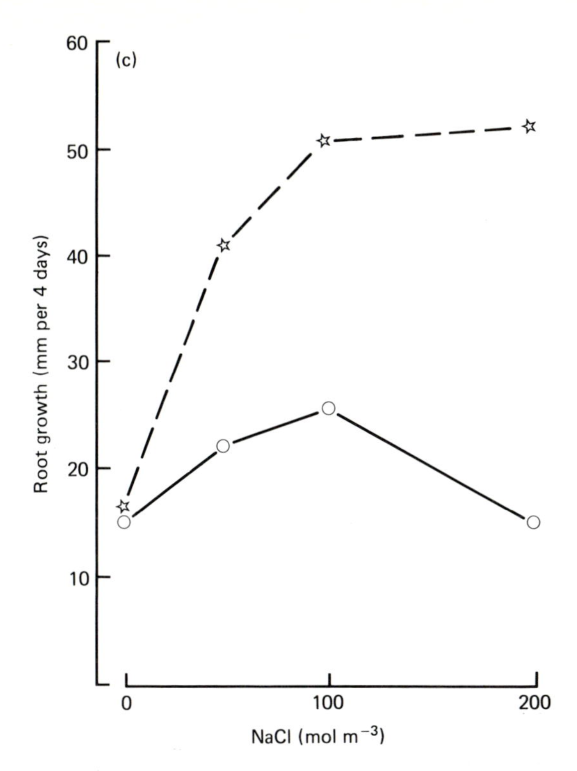

Fig. 4.3 Variation in rooth growth of a number of grass species grown hydroponically in saline solutions. Each point is the mean of values for eight individual roots. For each species a number of populations were studied, originally collected from soils of dil.erent salinities: (a) *Agrostis stolonifera*; (b) *Lolium perenne*; and (c) *Puccinellia distans*. ○ – ○ population collected from non-saline site; ☆ – ☆ population from highly saline soil; □ – · □ population from soil of moderate salinity. (*L. perenne* was absent from the most saline sites.) (Redrawn from Venables & Wilkins 1978.)

coastal plants are markedly succulent. In the case of *Mesembryanthemum crystallinum*, a species of cliff tops and sandy sites near the sea, it has been demonstrated that under favourable conditions photosynthesis proceeds by way of the C_3 pathway, but that salinisation leads to a switch to CAM photosynthesis (Winter 1979), although this response appears to be primarily a response to lowered water potential rather than a specific response to salt (Lüttge & Smith 1984). For saltmarsh species, the theoretical advantages of CAM seem considerable, not only the reduction in salt uptake but also the possibility of photosynthesis continuing when the plants are submerged by flooding tides and direct gas exchange to the atmosphere is temporarily curtailed.

Nevertheless, it would appear that all intertidal succulent halophytes carry out photosynthesis only by the C_3 pathway (Antlfinger & Dunn 1979; Kuramoto & Brest 1979) and transpiration rates are relatively high. Ganzmann & von Willert (1972) reported that *A. tripolium*, which has markedly succulent leaves, showed features of CAM plants, notably an increase in the malate content of the leaves during the night, but the pattern of carbon dioxide exchange was that of a C_3 species, not a CAM plant.

The majority of grasses of tropical and subtropical saltmarshes are C_4 species. However, even in temperate latitudes, C_4 species may be important, and even dominant, components of the vegetation. On the east coast of North America, *S. alterniflora* is found as the major species over a wide range of climatic conditions from subtropical to regions in Labrador, where winter conditions are extremely cold, and stands may be exposed to ice action (Roberts & Robertson 1986). *D. spicata* extends into high temperate latitudes on both eastern and western seaboards of North America. In the southern hemisphere, *D. distichophylla* is found in temperate latitudes in Tasmania and Victoria, being replaced at lower latitudes by *Sp. virginicus* (another C_4 species). Northern Europe, until recently, was exceptional in that C_4 species were not prominent in saltmarsh vegetation. *S. maritima*, although occurring as far north as southern England and the Netherlands, does not appear to have played a major role in marsh vegetation. However, over the last century there has been the dramatic spread, aided by human intervention, of *S. anglica*, a species with aggressive colonising ability (Gilbertson, Kent & Pyatt 1985) and now dominant over large areas.

Plants with C_4 photosynthetic pathway have, at least theoretically, a number of competitive advantages over C_3 species:

> Higher potential productivity;
> Higher water-use efficiency (which has been demonstrated in comparisons of *S. alterniflora* with C_3 species – DeJong, Drake & Pearcy (1982);
> More efficient use of available nitrogen. [In saltmarshes where some of the available nitrogen may be diverted from growth to cytoplasmic osmotica, this could be a trait of particular adaptive value.]

However, as Long (1983) emphasises, the superiority of C_4 species in all three respects declines with lower temperatures.

In general, C_4 species are most abundant in tropical latitudes, and saltmarshes are exceptional in providing habitats for, and being dominated by, C_4 species at high latitudes. Long (1983) points out that of the very few

C_4 species at high temperate latitudes, none occurs in a mesic terrestrial habitat (apart from saltmarsh, other C_4 species occur on strandlines and in freshwater marshes). Amongst C_4 species, *S. anglica* is of particular interest in that it not only evolved at temperate latitudes but also, despite various plantings, has not become established in the tropics (Ranwell 1967).

C_4 species of tropical origin grow badly, if at all, at low temperatures, but this is not an inevitable consequence of possession of the C_4 pathway, and *S. anglica* shows photosynthetic rates exceeding those of C_3 species at both 5° C and 10° C (Long 1983). However, leaves developed at low temperatures do not have high photosynthetic capacity (Dunn *et al.* 1987) and, in any case, active leaf growth requires temperatures of about 9–10° C (Long 1983). In consequence, at higher latitudes the development of the canopy of C_4 species is delayed compared with that of C_3 species; *Puccinellia* commences growth in March, several months before the growth phase of *Spartina* (Fig. 4.4). The net primary productivity of the two species over the whole season is similar, but concentrated in a much shorter period in *Spartina*.

No C_4 species is found where the mean minimum temperature of the warmest month is below 8° C (Long 1983). Along east Atlantic coasts, this summer limit is reached at sites with comparatively warm winters. It is possible that *S. anglica* could endure colder winter conditions, comparable with those experienced by *S. alterniflora* at its northern limit, were summer conditions to be more favourable. On the other hand, there is evidence that severe winters are associated with death of *S. anglica* in northern Europe (Beeftink 1977*a,b*).

It is not clear what features of the biology of *S. anglica* make it so successful in the lower marsh in northern Europe. One factor which does confer advantage is its ability to withstand longer periods of submergence than its potential competitors, but the biological basis for this remains to be elucidated. One feature of *S. anglica* by which it differs from the major perennial low-marsh species in northern Europe (*Pu. maritima*) is in the development of a massive subterranean system of roots and rhizomes. It is possible that this system, although representing a diversion of resources from above-ground productivity, is the key to the species' success (Woolhouse 1981).

Even under warmer conditions, where the advantages of possessing the C_4 pathway should be fully realised, saltmarsh vegetation is made up of a mixture of C_3 and C_4 species. On arid coasts, where grasses may be limited in their abundance, the majority of saltmarsh biomass is concentrated in C_3 species.

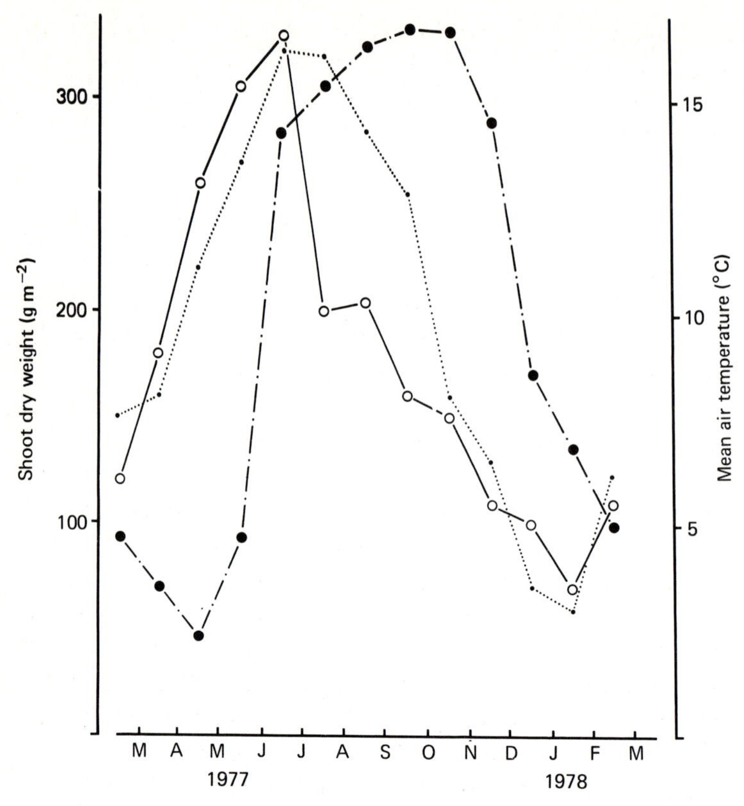

Fig. 4.4 Vegetative phenology of *Puccinella maritima* (C_3) and *Spartina anglica* (C_4) in north east Essex, UK. The shoot dry weights in monotypic stands are plotted against time. Note the earlier build-up of biomass in *Puccinellia*. ○ — ○ *Puccinellia*; ● – · – ● *Spartina*; ● · · · · ● mean air temperature. (Redrawn from Long 1983.)

The suggestion has been made that with increasing salinity, C_3 saltmarsh species may show features of C_4 species and that a switch between C_3 and C_4 mechanisms is possible, analogous to switches between C_3 and CAM. The evidence for this has been the demonstration of variation in the $\delta^{13}C$ values of plant tissue correlated with variation in salinity (Guy, Reid & Krouse 1980).

Farquhar *et al.* (1982) proposed a model in which shifts in $\delta^{13}C$ values could come about due to changes in stomatal conductance and internal CO_2 concentration without the necessity for a change in the biochemical pathway. Guy & Reid (1986) showed that this model was supported by studies on *Puccinellia nuttalliana* in which a marked shift in $\delta^{13}C$ occurred at increasing salinities without the C_4 pathway coming into operation. In *P.*

nuttalliana, increased salinity resulted in greatly decreased stomatal conductance with only slight effects on the photosynthetic capacity; in consequence, water-use efficiency was greatly increased. Currently, there is no convincing evidence for the existence of facultative C_4 plants.

Possible advantages of clonal growth

Perennial species predominate in most saltmarsh communities and, in many cases, flowering is intermittent. Jefferies (1972) and Jefferies, Davey & Rudmik (1979) have argued that sexual reproduction demands a diversion of resources from survival traits and is likely to be favoured only at times of lowered salinity. Although sexual reproduction occurs rarely, vegetative spread by means of rhizomes or stolons may produce extensive clones.

Salzman & Parker (1985) have shown that in *Ambrosia psilostachya*, which occurs in inland saline sites, transfer of photosynthate takes place in the rhizomes between ramets in non-saline microsites and those in saline microsites. In an heterogenous environment ramets in unfavourable sites may, thus, be buffered against their immediate surroundings by better-situated neighbours. Whether there are similar advantages to rhizomatous species in coastal saltmarshes is presently unknown. Saltmarsh environments show considerable heterogeneity at appropriate scales (Vestergaard 1982), but whether, in long-lived species, the rhizome connections between ramets remain functional over the life of ramets is doubtful.

How large, and how old, individual clones of saltmarsh species can become has not been investigated. In tall species with radial patterns of clonal growth, such as *S. anglica* (Caldwell 1957) or *J. maritimus*, easily visible discrete clones may be 10 m in diameter; the largest recognisable clones of *Triglochin maritima* may be one metre in diameter (Gimingham 1964; Heslop-Harrison & Heslop-Harrison 1957). In other cases, although vegetative growth is vigorous, the clone structure is not immediately visible and, in the absence of detailed investigation, clone size remains unknown. However, given the heterogeneity of saltmarsh environments, individual ramets of a single large clone may experience very different environmental conditions. It would be of interest to know whether the phenotypic plasticity of species capable of forming large clones is greater than that of species lacking this ability.

The possession of persistent underground storage organs allows carbohydrates and nitrogenous compounds to be stored during winter. The brief early season above-ground growth spurt shown by a number of north European perennials may largely reflect redistributon of these underground storage reserves (Jefferies & Rudmik 1984).

Coping with salinity

There is clear evidence that vascular plants differ in their tolerance to salinity but, nevertheless, there is a limit to tolerance, a salinity level beyond which no growth is possible. The salinity levels at which growth fails form a continuum from very low levels to concentrations well in excess of seawater. Is the physiological reason for growth failure the same throughout this continuum or do plants fall into a number of groups with differing adaptations to salinity and in which different mechanisms fail at different points along the continuum? The recent increase in interest in halophytes allows us to begin to hazard answers to these questions.

Historically, three reasons have been advanced for the adverse effects of salinity: directly toxic effects of sodium and chloride; interference with the uptake of essential nutrient ions; or the impact of lowered external water potential. Munns, Greenway & Kirst (1983) suggest that the most credible hypothesis is that the growth response of halophytes to increased salinity – both the growth stimulation shown by some species at moderate salinity and the decline in growth at high salinity shown by all species – is primarily a reflection of the impact of low external water potentials on the water relations of the species. Thus, the range of tolerance shown by halophytes indicates their varying ability to respond to lowered water potentials rather than the differential ability of their metabolism to withstand high ion levels *per se*. Nevertheless, despite this supremacy of water relations, the limits to tolerance appear to be reached when the generation of low internal water potentials results in toxic internal ion concentrations. While it is possible to develop a unified theory of salt tolerance in terms of plant water relations, Munns, Greenway & Kirst (1983) emphasise that at present it is still a hypothesis and that other hypotheses cannot be ruled out.

Before the plant's responses to a salinity-induced low water potential are discussed in detail, a number of instances of specific ion effects can be mentioned.

There are some glycophyte species in which growth reduction occurs at such low salt concentrations that adverse effects of lowered water potential seem implausible; Greenway & Munns (1980) cite the examples of soyabeans, avocado and grapevine where growth reduction occurs with exposure to 5–10 mol m^{-3} chloride. There are also demonstrations of much better growth in *Phaseolus* beans, maize and barley in polyethylene glycol than in isosmotic salt solutions (Greenway & Munns 1980). It is difficult in studies of this kind to determine whether chloride and/or sodium is directly responsible for the observed symptoms.

Termaat & Munns (1986) showed that growth of four species (wheat, barley and two species of *Trifolium*) was poorer in sodium chloride compared to that in concentrated macronutrient solution, suggesting a specific ion effect additional to any osmotic effect. They suggest transport of macronutrients from roots into the shoot is reduced when plants are exposed to high sodium chloride concentrations – competition between nitrate and chloride uptake and potassium and sodium may cause feedback control from the shoot of the uptake of other ions. However, as Termaat & Munns (1986) emphasise, it is not possible to distinguish between the possibility that reduced transport of ions from roots to shoot is the cause of shoot growth reduction or a response to feedback control from the shoot after growth rate has been reduced for other reasons.

Studies on these glycophytes suggest that specific ion effects are responsible for excluding them from saline habitats. The data, however, do not allow us to conclude that, for more tolerant species, differential tolerance is to be interpreted in terms of a spectrum of specific ion responses.

For halophytes growing at high salinities there is some evidence for specific ion effects but no generalisations can be made (see reviews by Flowers, Troke & Yeo 1977; Munns, Greenway & Kirst 1983). There is, however, little evidence for any specifically adverse effects of chloride or sodium. Growth in a range of salts is not better, or is often worse, than in the same concentrations of sodium chloride. Initially puzzling is the finding in a number of halophytes that growth is reduced more by high potassium than high sodium (Yeo & Flowers 1980), given that potassium is the cation present at highest concentration in most glycophytes. Munns, Greenway & Kirst (1983) suggest that, in algae at least, specific potassium toxicity may be a feature of halophytes, with glycophytic species not showing the same differential response with increased concentration. Preston & Critchley (1986) suggest that sensitivity to potassium may reflect the effects of this ion on oxygen evolution from chloroplasts. Oxygen evolution from isolated photosynthetic membranes of the mangrove *Av. marina* and the saltmarsh chenopod *S. quinqueflora* was lower in the presence of potassium chloride (and sulphate) than in sodium chloride and the degree of inhibition was related to the K^+/Na^+ ratio. No such response was shown by membranes isolated from spinach (which ancestrally is a coastal species with moderate salt tolerance). Preston & Critchley (1986) hypothesise that in halophytes there is differential ion compartmentation within cells so that the K^+/Na^+ ratio in the chloroplasts remains low.

These demonstrated effects of differential ion toxicity do not explain patterns of species distribution in coastal saltmarshes, where ionic conditions, although temporally variable in concentration, are basically similar throughout in terms of ionic ratios. Inland saline areas, however, are more varied in composition and ionic composition may exert a formative influence on the flora. Such areas differ in both cation and anion composition, and although there are a number of older studies (see summaries in Chapman, 1960*b*) suggesting that the dominant anion may be a major influence on the flora, there is little recent work on the effect of various anions. [Yeo & Flowers (1980) showed that for *Su. maritima* organic matter production was greatest in NaCl and less in Na_2SO_4 and $NaNo_3$ – interestingly dry weight was highest on Na_2SO_4 due to greater ion accumulation.]

The presence in the rhizosphere of high concentrations of sodium and chloride ions suggests that saltmarsh plants may have difficulty maintaining a balanced internal ionic composition. However, while there has been considerable discussion of this point, there has been little study of total ionic composition in halophytes and most attention has been paid to the effects of increased salinity on potassium status (Jeschke 1984).

Increasing salinity tends to lower tissue potassium concentration relative to sodium, although the decline is not always proportional to the increase in availability of sodium, suggesting differential selectivity in favour of potassium in some species. Foliar analyses suggest that in the field both the absolute and relative potassium concentrations are variable between species and it has been suggested that high potassium is a feature of the physiotype of saltmarsh monocotyledonous species (see p. 272). Interpretation of many experimental studies is difficult as two factors are frequently confounded – changes in total ionic concentration of the medium, and changes in the relative availability of sodium and potassium. The relevance to the field situation of experiments in which sodium chloride concentration in the medium is varied against a constant potassium concentration is not clear. The sodium/potassium ratio of the bulk soil solution will remain constant over a wide range of salinities, suggesting that such experiments pose unrealistic challenges. On the other hand, selectivity in uptake, coupled with possible excretion of sodium, may result in marked ionic changes in the immediate rhizosphere. Nevertheless, although failure to maintain an appropriate balance between ions may be an important factor preventing glycophytes from growing in saline sites, halophytes, despite the wide range of foliar potassium levels (both relative and absolute) which has been recorded, are not, in the field, potassium-deficient. While there is some evidence that the high potassium monocot physiotype is a feature of species found only in brackish to moderately saline, rather than highly saline sites, in general the habitat preferences of species are poorly correlated with their tissue potassium status.

A group of plants which might be expected to face particular problems in maintaining balanced mineral nutrition is parasites. Rozema *et al.* (1986) have made a comparison of the mineral content of two species, the root hemiparasite *Odontites verna* and the stem holoparasite *Cuscuta salina. Odontites* obtains water and mineral nutrients from the root xylem of its hosts. In the plants studied by Rozema *et al.* (1986) the majority of hosts were monocotyledons with relatively low xylem sodium concentrations. In the parasite the tissue sodium, calcium and magnesium concentrations were much higher than those in the host (by factors of 5–7); *Odontites* potassium levels were only slightly higher, or even lower, than those in the host.

C. salina is parasitic upon *S. pacifica* and other species on the Californian coast. *Cuscuta* obtains carbohydrate and mineral nutrients from the phloem of the host. Rozema *et al.* (1986) demonstrated that the tissue sodium concentration in *Cuscuta* was much lower than in the host and that there was no accumulation of potassium, calcium or magnesium in the parasite relative to the host. Although the host has very high tissue salt concentrations, the mineral content of the phloem may differ substantially from that in bulk tissue. Certain ions, including sodium, are largely excluded from phloem.

These differences in nutrient supply may partially explain the different distributions shown by the two types of parasite. Hemiparasites in the Scrophulariaceae (*Odontites*, *Euphrasia*, *Rhinanthus*) may be locally common in upper saltmarsh grasslands in northern temperate and boreal regions. Under prevailing climatic regimes, these habitats may experience high soil salinities only rarely. Although the hemiparasites will have much higher tissue sodium levels than their hosts, the generally low external salinity may mean that the levels found in the parasites are still relatively low compared with those in halophytes in other regions.

Holoparasites are relatively uncommon in saltmarshes. The few examples recorded (which include several *Cuscuta* spp. and *Cistanche lutea*) occur on saltmarshes where high soil salinities may prevail for much of the growing season. If the parasite fed on the xylem sap of the host, the resultant very high tissue salt concentrations could well prove toxic. However, utilisation of phloem does not result in potentially lethal tissue salt contents.

The mineral nutrition of parasitic fungi on the saltmarsh could be expected to vary similarly, depending on the particular host tissues attacked, but this remains to be confirmed.

Salinity and plant water relations

The water potential of plant cells is determined by two components (neglecting any possible contribution of a matrix component):

$$\Psi_i = \Psi_p + \Psi_\pi$$

where

Ψ_i = cell water potential; Ψ_p = turgor pressure potential; and Ψ_π = cell osmotic potential.

Turgor pressure is responsible for the maintenance of tissue rigidity and is also vital in sustaining growth (Hsiao 1973) – if plants are to grow, it is essential that turgidity be maintained.

For a glycophyte, the internal ionic concentration in the cytoplasm is in the range 100–200 mol m^{-3} which is mostly potassium salt (Dainty 1979). This would produce a Ψ_π of -0.5 to -1 MPa. If the cell were in equilibrium with a solution of zero water potential then Ψ_p would be $+0.5$ to $+1$ MPa, a turgor pressure adequate to sustain growth [this example is hypothetical – in nature no solution would be at zero water potential, but the soil solution in wet non-saline soil would be close to it].

For the coastal halophyte, the position is rather different. Seawater has a water potential of approximately -2.5 MPa; in order to survive Ψ_i of the roots of a halophyte must be below -2.5 MPa. If Ψ_p is maintained at a sufficient value for growth then Ψ_π of the root cells will be in the order of -3.5 MPa. For water to flow from the roots to the shoot there must be a gradient in water potential so that in a growing plant, the shoot Ψ_i and Ψ_π will be lower still.

A Ψ_π of -3.5 MPa would correspond to a salt concentration of about 700 mol m^{-3} NaCl (or osmotic equivalent). Many halophytes survive in habitats in which soil salinities regularly exceed that of seawater – these plants therefore, at least at times, have cellular water potentials below that of seawater.

Halophytes might, therefore, be defined as plants which attain internal water potentials sufficiently below that of the external saline soil solution to generate turgor pressures permitting growth without suffering metabolic impairment. [If an environmental distinction is to be drawn between glycophytes and halophytes, survival at salinities above 200 mol m^{-3} – corresponding to the internal cytoplasmic concentration within glycophytes – may be a more appropriate limiting value than some others which have been proposed.] It is possible to make direct measurements of turgor pressure, although the techniques are still in their infancy. Measurements have been made of turgor pressure in *Su. maritima* (Clipson *et al.* 1985). These show that turgor pressure in older leaves drops to values close to the minimum necessary to maintain tissue rigidity. In younger tissues turgor pressure is higher and clearly sufficient to drive cell expansion.

Three mechanisms, acting singly or in combination, can be suggested by which the concentration of solutes in plant cells might be increased in response to increased external salinity.

(1) Synthesis of organic solutes;
(2) Uptake of inorganic salts from the external environment;
(3) Dehydration.

The third approach, that of increasing solute concentration through dehydration might not, intuitively, seem to be inherently likely, particularly as it would involve some loss of turgor pressure (as a result of a decline in cell volume). However, it has been shown for the grass *Leptochloa* (*Diplachne*) *fusca* that considerable tissue dehydration occurs in response to imposed high salt treatments (Sandhu *et al.* 1981), and a similar, if less marked, response has also been demonstrated for *S. anglica* (Storey & Wyn Jones, 1978). *Eleocharis uniglumis* also showed dehydration when exposed to raised salinity levels (Gorham, Hughes & Wyn Jones 1981). It is suggested by Sandhu *et al.* (1981) that dehydration at high salinities may be a feature of monocots, in contrast to the salinity response shown by dicots (see also Glenn 1987). Glenn & O'Leary (1984), however, demonstrated considerable dehydration in a number of succulent dicotyledonous species when exposed to salinities in excess of 180 mol m^{-3} NaCl; Shennan *et al.* (1987*a,b*) also showed a decline in leaf fresh to dry weight ratio with increasing salinity in *A. tripolium* suggesting that reduction in water content may be an important factor in osmotic adjustment in a wide range of taxa.

If cell water potentials are lowered by accumulation of organic solutes, then the quantities required would represent a major sink for photosynthate. Greenway (1973) calculated that plant cells in equilibrium with an external salt solution of 100 mol m^{-3} NaCl would have to accumulate 30 g l^{-1} of hexose sugars or 60 g l^{-1} of disaccharides. This would represent about 20–40% sugar on a dry weight basis.

There are many demonstrations that halophytes growing under saline conditions accumulate large quantities of inorganic salts in their shoots (for reference to the earlier literature see Chapman 1960*b*). Salt content may exceed 50% of leaf dry weight in some species (for example, *Disphyma australe* – Neales & Sharkey 1981). This accumulation of salts by halophytes has been exploited by man in the past – the ash from burnt halophytes providing raw material for both glass making and soap manufacture (Bird 1981). In the case of coastal halophytes, the majority of the accumulated ions are Na^- and Cl^+ although other cations may be significant in some taxa (Flowers, Troke & Yeo 1977; Stewart & Ahmad 1983). In inland saline sites, other anions may be accumulated, for example, sulphate, while under certain circumstances nitrate can be present in large amounts (Smirnoff & Stewart 1985*b*). The balancing anion may be tissue-specific. Shennan *et al.* (1987*b*) showed that in *A. tripolium*, chloride was the major balancing anion in shoots at high salinity but in the roots chloride was never more than 38% of the sum of Na^+ and K^+.

Nitrate is accumulated in vacuoles but in a number of instances leaf nitrate concentrations have been shown to be reduced at higher salinities (Gorham *et al.* 1985), suggesting that the role of nitrate in osmotic adjustment in halophytes is likely to be limited. The effect of increased salinity on tissue nitrate levels appears to be specifically mediated by chloride; increased sulphate did not reduce leaf nitrate in *A. tripolium* although increased chloride had a marked effect (Stienstra 1986). The reduction in leaf nitrate may be related to both a reduction in uptake by the roots and reduced transport from roots to shoots. While nitrate may be reduced, the total nitrogen level may even be increased and there is no evidence to indicate that nitrogen deficiency is part of the syndrome of salt damage (Gorham *et al.* 1986). As Gorham *et al.* (1986) conclude

> 'The interactions between salinity and phosphorus or salinity and nitrogen seem to depend on the external nutrient concentrations, the nature of the applied salt treatment and the species being investigated'.

No simple unified model of the effects of salinity on anionic balance in plants has yet emerged.

In some cases, inorganic anions are insufficient to balance measured cation concentration and in this circumstance the difference must be made up by organic anions – representing a drain on photosynthate which might otherwise be devoted to growth.

On the basis of known salt contents of halophytes and their water content, estimates of the salt concentration within tissues indicate concentrations of an appropriate order of magnitude to generate the water potentials required for the maintenance of growth. It would thus seem that a hypothesis of salt uptake as the mechanism chiefly responsible for generation of low tissue-water potentials is upheld. Epstein (1980) suggests that there are four 'advantages' to plants in adapting to imposed salinity by salt uptake:

> Salt being responsible for the external low water potentials, ions of salts are the most readily available solutes to the plant. No new mechanisms are required for salt uptake which is one of the normal functions of roots.

Transport of ions from root to shoot is by the transpiration stream, a pathway which is metabolically inexpensive.

If inorganic ions are used for lowering water potential then photosynthate is not diverted to the production of organic osmotica.

If water potential in root cells were lowered by organic osmotica instead of by inorganic ions, then these osmotica (or their precursors) would need to be transported in the phloem from shoots to roots – a process which would be metabolically expensive.

Regulation of shoot salt content

If halophytes require to take up salts for osmotic reasons, there is also a need to regulate the internal salt content. Munns, Greenway & Kirst (1983) suggest that, at maximum, internal salt concentrations reach levels about three times external soil salinities. In the absence of mechanisms for regulating salt content, the uptake of salt via the transpiration stream into the shoot would eventually result in the shoot becoming, like Lot's wife, a pillar of salt.

A number of mechanisms have been reported by which salt levels in shoots might be regulated. The relative importance of different mechanisms varies between species and there is no single mechanism which appears to confer any special selective advantage on species possessing it.

At the present time, the number of species in which ionic regulation has been studied in detail is small; sweeping generalisations about ion regulation must therefore be treated with a degree of caution.

The mechanisms by which ionic regulation in the shoot might be achieved include:

Exclusion;
Growth;
Development of succulence;
Secretion;
Export from the shoot;
Leaf loss;
Reduction in transpiration.

Ion exclusion

The roots of vascular plants seem, in general, to be capable of a high degree of selectivity in ion uptake and halophytes do not appear to be different in this respect. At least under conditions of rapid transpiration it seems that all species show a marked exclusion of Na^+ and Cl^- from the xylem sap. The majority of measurements are from mangroves (which, being woody, are technically easier to study – see Scholander *et al.* 1962, 1966; Scholander 1968; Atkinson *et al.* 1967), although Rozema *et al.* (1981) have shown similar exclusion in a number of saltmarsh species and Flowers & Yeo (1986) report a high degree of exclusion in *Su. maritima*. Clough, Andrews & Cowan (1982) suggest that the evidence points to the low concentration of NaCl in the xylem sap being due to discrimination against sodium ions; uptake of chloride ions may not be directly controlled *per se*, rather the uptake of Cl^- is controlled by the need to maintain a balance between anions and cations in the sap. Unfortunately, there are few data on xylem sap concentrations of ions other than Na^+ and Cl^- in halophytes, but there is some evidence (Clough *et al.* 1982) that the concentration of K^+ may be higher in the xylem sap of some species than in seawater.

Despite the generality of exclusion, the reported sap concentrations indicate a wide variation between species (Scholander *et al.* 1962, 1966; Atkinson *et al.* 1967; Rozema *et al.* 1981; Munns, Greenway & Kirst 1983). Exclusion is most pronounced in those species lacking the capacity for excretion.

Low ion concentrations in the xylem are generally interpreted to indicate exclusion of ions at the endodermis. Munns *et al.* (1983) suggest that the data do not allow other explanations to be ruled out. In particular, they suggest two other possibilities:

> If there is a large active element in ion uptake not linked to water flow, the low concentrations may only pertain under conditions of rapid transpiration.
>
> Measurement of ion concentration in the xylem does not indicate the extent to which ions may be excluded at the endodermis. Ions may be removed from xylem sap, in either root or shoot (and sequestered, for example, in xylem parenchyma).

Nevertheless for most of the root system direct flow of ions and water into the stele via the apoplast is prevented by the suberised casparian strip of the endodermis. The endodermis is, however, absent towards the root tip, and in some cases also from the sites of lateral root formation. There is no proof that these regions do in fact provide an apoplastic pathway into the xylem but if there is an apoplastic path for water uptake, then, even if it is small in comparison to the symplastic path, under conditions of high external salinity it may provide an important route for salt uptake (Pitman 1977; Clough, Andrews & Cowan 1982; Munns *et al.* 1983; Yeo, Yeo & Flowers 1987). In the case of the mangrove *Rhizophora mucronata* the Cl^-

concentration in the xylem sap is about 3% that of seawater (Atkinson *et al.* 1967). Clough *et al.* (1982) calculated that this concentration could be supplied by an apoplastic route if 1–5% of the water uptake occurred via the apoplast.

The existence of such a pathway would be compatible with data presented by Scholander (1968), who showed that exposure of mangroves to low temperatures and to metabolic inhibitors had no effect on the ion concentrations in xylem sap. Scholander interpreted this finding as indicating that ion exclusion in halophytes was largely a result of physical properties of root cell membranes rather than of metabolically active ion pumps. There is no evidence as to whether the membranes of halophytes are physically different from those of glycophytes so the hypothesis remains an untested possibility – however, if there is a significant apoplastic pathway for ion uptake, it becomes unnecessary to postulate the existence of 'special' membranes in halophytes.

Using fluorescent tracer dyes, Moon *et al.* (1986) have shown convincingly that in undamaged roots of the mangrove *Av. marina* there is minimal apoplastic ion transport and even the low proportion of apoplastic transport postulated by Clough *et al.* (1982) does not occur. In mature roots the main barriers to apoplastic transport were in the periderm and at the exodermis. Moon *et al.* (1986) point out that if the endodermis provided the main barrier the cortex would be exposed to salt concentrations at least equal to those of the external soil solution (because of exclusion of ions at the endodermis, concentrations would increase with time). The very limited development of cortex in some saltmarsh species (only three cell layers in *Su. maritima* – Hajibagheri, Yeo & Flowers 1985; see also Waisel 1972) may be a response to the high salt levels external to the endodermis. Moon *et al.* (1986) also showed that only the finest roots were involved in ion uptake and that the large roots served to support the tree and to provide a connection between the absorbing roots and the aerial parts of the plant. Unfortunately, there is a dearth of information on what portions of the roots of saltmarsh plants are active in ion uptake.

Hajibagheri *et al.* (1985) showed that in *Su. maritima* the casparian strip differentiated much closer to the root tip in plants grown under saline conditions than under non-saline conditions and suggested that salinity may have a stimulatory effect on the differentiation of the endodermis of a number of species. These findings would suggest that the apoplastic pathway for ion movement into the xylem is minimised in at least some halophytes. In rice, a non-halophyte, Yeo, Yeo & Flowers (1987) suggest that apoplastic transport may make a significant contribution to sodium

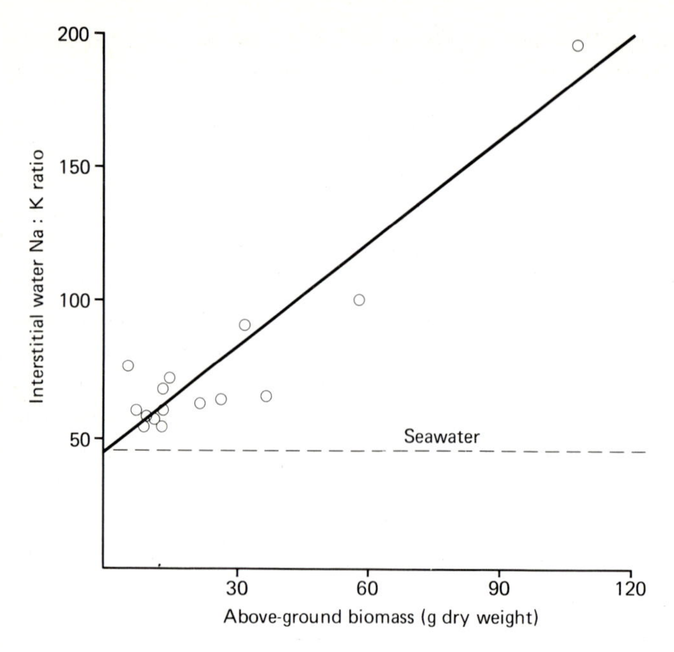

Fig. 4.5 The effect of above-ground growth of *Spartina alterniflora* on the Na : K ratio in the interstitial water in the sediment. (Redrawn from Smart 1982.)

uptake and that the proportion of the uptake attributed to the apoplastic route may increase with increasing 'stress' (which might include higher salinity). Investigation of a wider range of species is required to determine whether there is a consistent difference between halophytes and glycophytes in the contribution of the apoplastic pathway for sodium uptake.

The exclusion of ions by the roots of halophytes will change the soil solution adjacent to the roots. Not only will the concentration increase but if there is differential exclusion of ions the composition will change. Given the low hydraulic conductivity of saltmarsh soils (Clarke & Hannon 1967), the concentrations in the rhizosphere might rise considerably as has been demonstrated by Smart & Barko (1980) and Smart (1982) (Fig. 4.5). The effects of this upon rhizosphere microorganisms have been little studied. As far as the plant is concerned, it means that as ion uptake proceeds, there must be a further lowering of shoot water potential if the potential gradient between rhizosphere and shoot is to be maintained. Some comparisons of shoot and soil water potentials suggest that halophytes maintain a large potential gradient – such an assumption may be in error where the measured soil water potential is a bulk soil estimate; the effective gradient between rhizosphere and shoot may be much lower.

Although ion exclusion in halophytes is remarkably effective, under high external salinities appreciable quantities of ions will be carried into the shoots through the xylem. Indeed, if ion accumulation is the basis of turgor maintenance, it is essential that exclusion be only partial. The supply of ions to a leaf will be determined by two factors, the concentration of the xylem sap and the rate of transpiration. Ion concentrations in the xylem supplying different leaves have been shown to vary in some species, and it is possible that expanding leaves may be competitive sinks for ions (Flowers & Yeo 1986*a*). However, there is little evidence yet available to indicate that variation in xylem sap concentration is an important factor in ionic regulation of halophytes. Transpiration rate may decline with age of a leaf so leading to a declining rate of salt input. There is evidence from a number of non-halophytes for variation in transpiration rate with leaf age but no data are available from halophytes (Flowers & Yeo 1986*a*). While further study on these regulatory mechanisms at the individual leaf level is required there is good evidence that salt exclusion in halophytes is supplemented by additional mechanisms to regulate shoot ion content.

Growth and succulence

Ions will be carried into the aerial parts of a plant as it grows, so that the absolute ionic content will increase with age. However, the ionic concentration need not increase if the incoming ions can be 'diluted' either by growth or by increased succulence.

An example of dilution by growth is provided by the mangrove *R. mucronata* (Atkinson *et al.* 1967 – Table 4.2). In this species chloride and sodium concentrations remain almost constant for much of the life of a leaf. Leaf water content (on a % fresh-weight basis) also remains reasonably constant. Measurement of xylem sap concentration and the transpiration rate indicate an uptake of about 17 μmol m^{-3} Cl^- per leaf per day. In order for the observed leaf concentration to be maintained by growth, Atkinson *et al.* (1967) estimated that a mean dry weight increase of 3% per day was necessary. Similar dilution by growth was shown for the American saltmarsh composite, *Iva oraria* (= *I. frutescens*) by Steiner (1939). In *S. maritima* the rates of ion uptake and relative growth are balanced so that the tissue salt concentration in the foliage remains fairly constant (Yeo & Flowers 1986*b*). Flowers & Yeo (1986) suggest that for a range of species under steady-state salinity the sodium concentration of individual leaves remains constant over time. Flowers, Hajibagheri & Clipson (1987) conclude that ion input to the shoot controls growth.

Many plants on saltmarshes are succulent, as frequently are plants on seacliffs and sand-dunes exposed to high levels of salt spray (Waisel 1972; Jennings 1976; Saenger 1982; Munns, Greenway & Kirst 1983). There is,

Table 4.2. *Salt content of* Rhizophora mucronata

	Sample 1	Sample 2	Sample 3	Sample 4	Sample 5	Sample 6
Dry weight (g)	0.16	0.50	0.50	0.61	0.57	0.63
Water (% of fresh wt.)	56	65	66	65	67	69
Na^+ (μ-equiv./ml of leaf water)	305	313	431	435	461	461
K^+ (μ-equiv./ml of leaf water)	124	88	58.5	43.5	61	32
Cl^+ (μ-equiv./ml of leaf water)	370	562	522	530	515	522

Values are means for five leaf pairs on a shoot. Sample 1 was the unopened leaves at the tip and Sample 6 senescing leaves at the base of the shoot. In mature leaves (Samples 3–6), the sodium and chloride concentrations expressed relative to leaf water are reasonably constant. Potassium concentration declines with increasing leaf age, possibly indicating export from the leaves in the phloem. Data from Atkinson *et al.* (1967).

however, a lack of a clear definition of succulence which confuses comparison of various studies (Jennings 1976; Shennan *et al.* 1987*a*); Jennings (1976) proposes that water content or fresh weight per unit area are the most appropriate measures of succulence. Most of the succulent species are dicotyledons and the majority of saltmarsh monocotyledons (Gramineae and Cyperaceae) are never particularly succulent. Nevertheless, there are some halophytic succulent monocotyledons, (*contra* Munns *et al.* 1983) – the most widespread being *Triglochin* spp. Glenn (1987) showed that for a range of grasses maintenance of a water-potential gradient between shoots and the external soil solution was by co-ordination of sodium uptake and water loss. In all species studied by Glenn (1987), there was a reduction of water content at high salinity; Glenn (1987) suggests, however, that under non-saline conditions halophytes have lower water contents than glycophytes.

The degree of succulence in halophytes varies with external salinity and the succulence is an expression of the increased size of individual cells. Studies on *Atriplex hastata* (Black 1958) indicate that succulence is due to increased ion uptake. While the ion content of leaves increases throughout their life, increased succulence provides a greater volume of cells so that the concentration of ions may be controlled within fairly narrow limits.

Jennings (1976) and Handley & Jennings (1977) demonstrated that different ions may produce various degrees of succulence, with, for example, Cl^- promoting greater succulence than SO_4^{2-} in some instances (Jennings 1976). These results are physiologically intriguing, but, at least for coastal halophytes, are probably not of ecological significance.

Glenn & O'Leary (1984) suggest a distinction between species showing growth stimulation at moderate salinities (in their studies 180 mol m^{-3} NaCl), termed euhalophytes, and those with optimal growth under non-saline conditions but with the ability to survive at high salinities, termed miohalophytes. In euhalophytes, which accumulated more salt than miohalophytes, succulence may be increased in moderate salinities over the freshwater condition but does not necessarily increase at higher salinities, and, indeed, at high salinities cellular osmotic potentials may be maintained by dehydration. In miohalophytes increased water uptake (and hence increased succulence) in order to dilute salt when they are grown under high external salinities may be more general.

However, some halophytes show a great increase in succulence at comparatively low salinities suggesting that to view increased succulence primarily as a means of regulating ionic content of tissues may be inappropriate. [In addition, as Storey & Wyn Jones (1979) pointed out, the actual degree of dilution which might be achieved by increased succulence may be small compared with the increased salt concentration – in the case of *Suaeda monoica* a tripling of the fresh-to-dry weight ratio over a 50-fold increase in external salinity.] Storey & Wyn Jones (1979) suggested that increased succulence was simply the consequence of the accumulation of salts in the vacuole leading to increased turgor pressure. Under freshwater conditions insufficient ions may be taken up for generation of turgor pressure, thus explaining the increase in succulence at comparatively low salinities in some species (Storey & Wyn Jones 1979; Glenn & O'Leary 1984). Species showing growth stimulation at low to moderate salinities may have a particular ability to accumulate salts at low salinities, an ability also expressed by the high uptake of ions in some species even under non-saline conditions (Collander 1941; Flowers, Troke & Yeo 1977). Increased succulence may, through greater leaf expansion, be responsible for the increased growth shown by some species at moderate salinities. Increased succulence may also result in a reduction in stomatal frequency. The consequent decline in transpiration will reduce ion input into the leaves (Flowers 1985).

Secretion

Excretion of salt through salt glands and salt hairs has been the subject of considerable research over many years. The wealth of literature on salt secretion might perhaps encourage the view that secretion is practised by the majority of halophytes. This is not the case; although secretion is widespread taxonomically, only a minority of the halophytic flora possess salt glands.

Recent reviews discussing salt glands and salt hairs include those by Thomson (1975), Lüttge (1975), Hill & Hill (1976), Fahn (1979, 1988), Liphschitz & Waisel (1982) and Schirmer & Breckle (1982).

To date, salt glands have been reported from saltmarsh and mangrove dicotyledons in the Plumbaginaceae, Frankeniaceae, Primulaceae, Tamaricaceae, Acanthaceae, Malvaceae, Avicenniaceae, Verbenaceae, Convolvulaceae, Combretaceae and Myrsinaceae. In the Gramineae, salt glands have been recorded from a number of genera in the subfamilies Chloridoideae and Panicoideae (Liphschitz & Waisel 1974, 1982). Bladder hairs are a feature of a number of genera of the Chenopodiaceae but a salt-regulatory function has been demonstrated only in *Atriplex* (Osmond, Björkman & Anderson 1980; Schirmer & Breckle 1982) and *Halimione* (Baumeister & Kloos 1974).

Salt glands are variable in structure, but within a particular family are relatively uniform. Structurally the simplest glands are found in the grasses and consist of two cells (Lipschitz & Waisel 1982; Oross, Leonard & Thomson 1986). In the dicotyledons, salt glands are multicellular (see Fig. 4.6).

Salt glands excrete a salt solution to the outside of the leaf. This evaporates to produce salt crystals which may be easily visible to the naked eye and are subsequently washed off by rain or flooding tides.

Rates of salt excretion vary considerably between species (Rozema, Gude & Pollak 1981) and thus the ability of salt glands to maintain internal salt concentrations also varies. In the mangrove *Aegialitis annulata* and the saltmarsh species *S. anglica* and *L. vulgare* the salt glands seem well able to cope with the influx of salt experienced under field conditions (Atkinson *et al.* 1967; Rozema *et al.* 1981; van Diggelen *et al.* 1986); on the other hand, Rozema *et al.* (1981) found that the salt glands in *Arm. maritima* and *G. maritima* had a lower secretion efficiency. They suggest that this finding is ecologically significant in that *Spartina* and *Limonium* are exposed to higher salinities than *Armeria* and *Glaux*. However, their finding needs to be confirmed by studies on other populations; in many localities *Limonium* and *Armeria* co-exist in the same zone and may be exposed to high salinities in excess of those in seawater. Nevertheless, a lesser competence to cope with high salinity may partly explain the wider ecological amplitude of *Armeria* under the wetter oceanic climate of western Britain (Adam 1978, 1981*a*) compared with that on the east coast.

Salt glands show selectivity in the ionic concentration of the exudate, in general Na^+ being selectively excreted. However, there is variation between species in the degree and pattern of selectivity (Rozema *et al.* 1981).

In *G. maritima*, Rozema, Riphagen & Sminia (1977) demonstrated an increase in salt gland density with increased salinity. However, in *Samolus repens* (from the same family – Primulaceae) Adam & Wiecek (1983) were unable to show such an increase. Given the range and rate of fluctuations in saltmarsh soil salinity, it is unlikely that any such response would be of ecological significance, although it is possible that differences in gland density between populations in distinct subhabitats could occur.

The detailed physiology of salt glands is still imperfectly understood in spite of intense study. However, it is clear that excretion is an active metabolic process and that regulation of leaf salt content by salt glands involves the plant in expenditure of energy. Excretion rate is strongly dependent on temperature which may reflect variation in metabolic processes – however, temperature could also influence supply of salt to leaves through its influence on transpiration (Gorham 1987). Given that salt glands secrete a salt solution, this also involves a loss of water in excess of

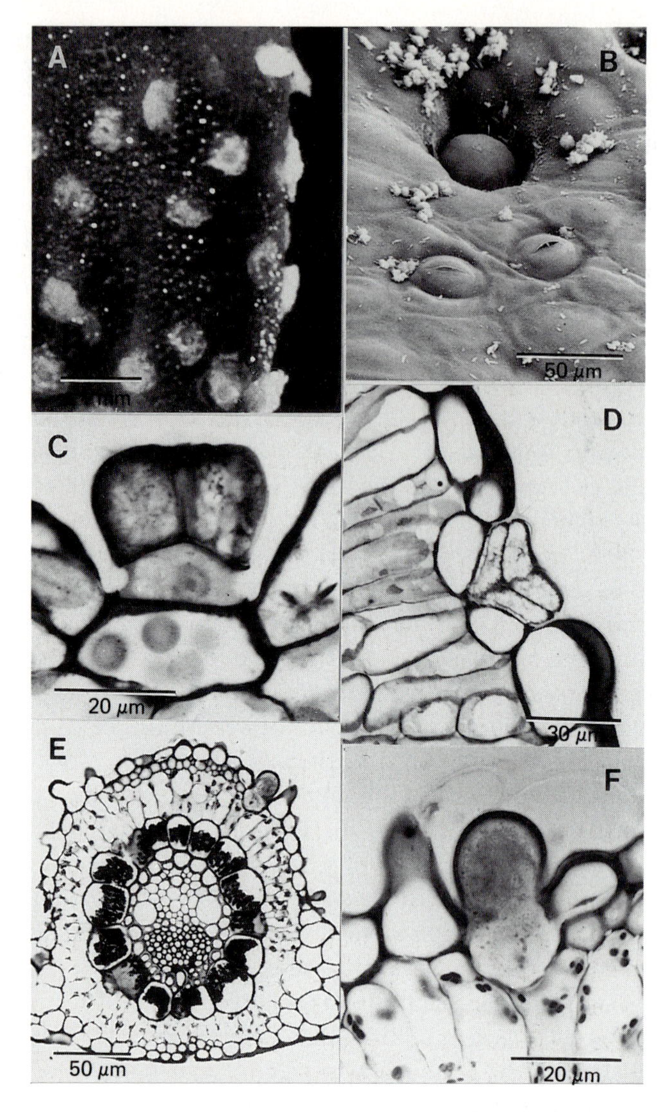

Fig. 4.6 Salt glands. (A) Portion of leaf of *Samolus repens* showing salt crystals over salt glands; scale bar = 1 mm. (B) SEM of surface of *Samolus repens* showing stomates and a salt gland; scale bar = 50 μm. (C) TS of salt gland of *Samolus repens*; scale bar = 20 μm. (D) Salt gland on leaf of *Frankenia pauciflora*; scale bar = 30 μm. (E) TS of leaf of *Sporobolus virginicus* showing Kranz anatomy characteristic of C_4 species; scale bar = 50 μm. (F) Salt gland of *Sporobolus*; scale bar = 20 μm. (Photographs: (A), L. Meier; (C), (D), B. M. Wiecek; (B), (E), (F), A. Fisher).

that transpired. Salt glands may thus be a disadvantage under drought stress – leaf death in early summer in upper marsh stands of *L. vulgare* during drought years may reflect this.

Gorham (1987) suggests that in the grass *L. fusca* excretion through glands prevents the build-up of salt in mature leaves, so prolonging their active photosynthetic life. Nevertheless, the main factor controlling leaf salt content in this species was exclusion of salt from the transpiration stream, rather than activity of the glands.

In the Chenopodiaceae the salt hairs are two-celled structures, consisting of a stalk cell and an upper bladder cell in which ions accumulate during its life. Unlike salt glands there is no secretion of a salt solution from the salt hairs. When the bladder cell reaches its limit of ion accumulation, the cell dies and is either shed from the leaf or bursts, allowing the salt to be washed off the leaf. The accumulation of dead hairs and salt on the leaf increases its reflectivity and may substantially reduce the heat load. Water loss from salt hairs is minimal and the majority of *Atriplex* species with salt hairs are arid-zone species occurring in an environment where salt glands would be a considerable disadvantage. The only major saltmarsh species to employ salt hairs appears to be *H. portulacoides* (Baumeister & Kloos 1974).

One matter on which few data are available is at what stage during the life of a plant salt excretory mechanisms become functional. Adam & Wiecek (1983) recorded the presence of salt glands on cotyledons of *Samolus repens* but had no data on their activity. Schirmer & Breckle (1982) report the absence of salt hairs on cotyledons on *Atriplex hortensis*. There appear to be no studies which have looked at salt excretion in young seedlings.

The wide taxonomic range of species from which salt glands have been reported and the diversity of glandular structures suggest that the ability to excrete salt has evolved independently a number of times. The similarity of structure between salt glands and other excretory glands (such as nectaries) may indicate that evolution of salt glands involved modification of pre-existing excretory structures within particular families. In the case of the grasses, Liphschitz & Waisel (1982) emphasise the restriction of salt glands to species with Kranz anatomy utilising the C_4 photosynthetic pathway. In the Chenopodiaceae, where, in the case of *Atriplex*, there are C_3 and C_4 members of the same genus, the possession of salt hairs does not appear to be linked to the C_4 pathway.

Shedding of parts and ion redistribution

In perennial plants older leaves and photosynthetic stems are replaced by new growth during the life of the plant. Even in annuals, older parts may be shed during the course of the season. In some species with no salt glands and little development of succulence, salt may be accumulated to high concentrations and be the determinant of the length of effective life of the organ. Shedding of salt-laden organs, and their replacement by new growth, may be regarded as a means of controlling salt content in such species, of which *Juncus* spp. may be the major example (Waisel 1972). The shedding of salt-laden material may increase the salt concentration in the litter and soil-surface layers. This is unlikely to be significant in intertidal saltmarsh, but in arid, non-tidal habitats it may produce conditions inimical to other species. The success of *M. crystallinum* as an alien species in both California and Australia is due in large measure to its ability to influence the distribution of salt in its local environment in this way (Vivrette & Muller 1977; Kloot 1983).

Internal redistribution of ions occurs in all plants, but may be particularly important in maintaining the required ionic balance in growing tissue of halophytes. Before old leaves are shed, some ions may be withdrawn and transferred to new growth. This seems likely to occur with potassium, as has been reported from mangroves by Atkinson *et al.* (1967) and from *Su. maritima* (Yeo 1981). Translocation of potassium from old leaves in *Suaeda* maintains the K/Na ratio in meristematic tissue (Gorham & Wyn Jones 1983).

The extent to which sodium is subject to redistribution has been discussed in the literature. Transport of sodium from leaves to roots, and subsequent release into the rhizosphere, have been reported (Ayadi & Hamza 1984; Kramer 1983, 1984; Lessani & Marschner 1978; Winter 1982). Differential salt tolerance between *Trifolium alexandrinum* (moderately tolerant) and the intolerant *T. pratense* is, in part, based on the ability of *T. alexandrinum* to export salt from its growing leaves (Winter & Läuchli 1982; Winter 1982*a*,*b*; Winter & Preston 1982). Winter (1982*a*) suggested that the advantage of retranslocation of sodium was that it prevented the development of possibly deleterious Na/K ratios in growing tissues. Sites on the root surface through which salt might be re-exported have not been identified.

Most species which have been claimed to export sodium from tissues have limited salt tolerance, and more tolerant species do not appear to do so (Gorham, Wyn Jones & McDonnell 1985; Flowers 1985; Flowers & Yeo 1986). Any retranslocation involves transport in the phloem and in general ionic composition and concentrations in phloem reflect the ionic status of cytoplasm (Raven 1977). If the hypothesis that halophytes sequester most of their intracellular sodium in vacuoles is correct, then little sodium would be expected in the phloem. Tissues in halophytes supplied by phloem rather than xylem (such as meristems and seeds) are characterised by high K^+/Na^+ ratios (Flowers & Yeo 1986), further suggesting that there is little retranslocation of sodium in halophytes. Even if sodium were phloem-mobile, retranslocation and excretion would probably incur a high energy cost, while any sodium excreted would increase the external concentration and immediately be liable for re-absorption. Nevertheless, Bhatti & Wieneke (1984) have demonstrated that in *L. fusca*, a grass species which could be considered as moderately halophytic, retranslocation and root efflux of both Na^+ and Cl^- occur and may be important processes in determining leaf salt levels.

Reduced transpiration

The higher the rate of transpiration, the greater the input of salt into the foliage. Reduction in the transpiration rate would, therefore, also reduce the rate of salt accession. Many saltmarsh plants are markedly xeromorphic, as was noted by Schimper (1903) and these xeromorphic features may act to lower transpiration rates. [Similarly, the xeromorphic nature of many freshwater bog plants has been suggested to be an adaptation to lower transpiration, so reducing the uptake of potentially toxic ions – Armstrong 1982.] Many saltmarsh grasses have thick cuticles and stomata concentrated at the base of grooves in the adaxial surface (Fig. 4.6). [In most species these leaves are also capable of rolling up.] The stem-succulent chenopods have minimised surface to volume, as have sedges and rushes such as *Baumea juncea* and *J. maritimus*. The thickening of the cuticle in *Su. maritima* under increased salinity may also serve to reduce water loss (Hajibagheri, Hall & Flowers 1983).

The use of the C_4 photosynthetic pathway would increase water-use efficiency compared with morphologically similar C_3 species (DeJong, Drake & Pearcy 1982). Thus C_4 photosynthesis may permit lower transpiration rates and so limit salt uptake. The mangrove *Rhizophora stylosa*, a C_3 species, shows a very high water-use efficiency compared to other trees (Andrews & Muller 1985). By permitting a reduction in transpiration, this reduces the salt loading to the leaves and this, in turn, may reduce the amount of energy expended in salt secretion (Andrews & Muller 1985). Whether C_3 halophytes in general show higher water-use efficiencies than glycophytes remains to be tested, but the shifts in isotopic carbon fractionation reported from several species in response to increased salinity (Farquhar *et al.* 1982; Guy, Reid & Krouse 1980; Guy & Reid 1986) may indicate a tendency to increased water-use efficiency (Farquhar & Richards 1984) at higher salinities.

The flux of water between roots and shoots will be influenced by changes in the hydraulic conductivity of the pathway. Oertli & Richardson (1968) suggested that increased salinity may reduce the hydraulic conductivity of roots. This was demonstrated by O'Leary (1969) in *Phaseolus vulgaris*, a glycophyte, and in the halophyte *Atriplex halimus* by Kaplan & Gale (1972). Ownbey & Mahall (1983) compared the effects of salinity on root conductivity in two species, *S. virginica* and *Raphanus sativus*. In both species, hydraulic conductivity declined with increased external salinity, but on transfer to distilled water the conductivity of *Raphanus* remained low, while in *Salicornia* the conductivity increased and, in some cases, was even enhanced over the initial pretreatment value. It is possible that such an enhancement could be of adaptive significance to a species growing in an environment where rapid, large changes in salinity may be of frequent occurrence. In *Raphanus*, conductivity appeared to be permanently impaired or, if recovery did occur, it was a slow process taking a longer period than the experimental investigation. Ownbey & Mahall (1983) suggest that the permanence of the impaired conductivity indicates an effect of increased ionic concentration rather than reduced water potential *per se* on the basis that recovery from a water-potential effect might be expected. However, evidence from a number of studies reviewed by Wainwright (1984) using organic osmotica strongly indicates that hydraulic conductivity declined primarily in response to a lowering of water potential, although these studies did not consider any possible recovery after salinisation. In addition, as Winter (1979) points out, it is difficult to be certain that flows and resistances measured experimentally in detopped roots bear any relationship to those in intact plants.

Many studies have shown a reduction in water uptake in a wide range of species exposed to increasing salinity (reviewed in Wainwright 1984). As well as possible reductions in hydraulic conductivity, there is much evidence to correlate this reduction in uptake with lower transpiration as a result of reduced stomatal conductance (Wainwright 1984; Winter 1979; Yeo, Caporn & Flowers 1985; Ball & Farquhar 1984*a*,*b*; Gorham, McDonnell & Wyn Jones 1984*b*; Gorham, McDonnell, Budrewicz & Wyn Jones 1985). If the closure of stomata were solely due to reduction in guard-cell turgor, reduction in transpiration might be a transient response to imposed salinity [reduction in stomatal conductance may be a rapid response to increasing salinity – Pezeshki, De Laune & Patrick 1987], with transpiration increasing after osmotic adjustment within the leaf had occurred. There is evidence that even after osmotic adjustment, transpiration rate may remain low (Gorham *et al.* 1984*b*). Ball & Farquhar (1984*a*,*b*) argue that in the mangrove *A. marina*, stomatal aperture remains controlled over a range of salinities, allowing minimal water loss relative to carbon uptake. Nevertheless, in some species stomatal resistance may vary little over a wide range of salinities [for example, *Salicornia* (*Arthrocnemum*) *fruticosa* – Abdulrahman & Williams 1981].

As the rate of carbon dioxide uptake is less affected by partial stomatal closure than water loss (Nobel 1974), water-use efficiency might be expected to increase under saline conditions. However, water-use efficiency is a ratio of net photosynthesis to water loss and the resultant changes in measured efficiency may reflect changes in both photosynthesis and transpiration. While increased salinity might result in lower transpiration, the water-use efficiency may still decline if the effects on photosynthesis are proportionally greater and, even if water-use efficiency increases, the increase may still be accompanied by substantial declines in photosynthesis. Thus water-use efficiency in *R. sativus* increases with increased external salinity, but the relative growth rate declines considerably, while in *S. virginica* the water-use efficiency and relative growth rate remain constant over a wide range of salinities (Ownbey & Mahall 1983). In *Spartina foliosa* which, as a C_4 species, has a comparatively high water-use efficiency at low to moderate salinities, both relative growth rate and water-use efficiency decline at higher salinities (Mahall & Park 1976*b*). Other environmental factors will also influence both photosynthesis and transpiration (for example, temperature – Gorham 1987) so that prediction of the response of plants in the field to increased salinity is very difficult. While net photosynthesis is the major determinant of growth, various parameters of growth may not vary linearly with net photosynthesis because of effects such as the influence of salinity on allocation of resources between organs (Pearcy & Ustin 1984).

While salinity may reduce photosynthesis simply through decreases in stomatal conductance, in some species reduction in photosynthesis occurs despite turgor maintenance (Beadle *et al.* 1985) and biochemical changes resulting in an increase in the mesophyll resistance to carbon dioxide uptake seem to be the immediate cause of a drop in the rate of carbon fixation. Beadle *et al.* (1985) suggest a number of ways by which salinity could affect photosynthesis. The water potentials in leaves of actively growing halophytes are frequently such that in glycophytes they would, in themselves, inhibit many processes associated with photosynthesis (Hsiao 1973). High salinity may result in reduced uptake (both proportionally and absolutely) of a number of ions which may be essential to high rates of photosynthesis. High levels of sodium chloride may disrupt a range of cellular processes, even though the osmotic potentials are not sufficiently low as to be totally inhibitory.

The relative importance of stomatal conductance and biochemical effects in the decline in photosynthesis with increased salinity varies between species. In the salt-sensitive rice, inhibition of net photosynthesis is mediated largely by the water deficit in the leaf cells due to apoplastic salt accumulation (particularly peristomatally), although there may also be some direct effects on mesophyll resistance (Yeo, Caporn & Flowers 1985). In sugar beet there appears to be little direct effect on the biochemical processes of photosynthesis, at least in short-term experiments, over a wide range of salinities (Papp, Ball & Terry 1983). In the grapevine, *Vitis vinifera*, which is regarded as being a salt-sensitive crop, the initial reduction in photosynthesis at low salinities was due to increased stomatal resistance, but at higher salinities photochemical efficiency was reduced and photorespiration increased (Walker *et al.* 1981). In the mangrove, *Av. marina*, both stomatal conductance and photosynthetic capacity decline with increasing salinity, so that water loss is minimised relative to carbon gain across the range of salinities (Ball & Farquhar 1984*a,b*).

Net photosynthesis may be effected, not only by changes in photosynthetic capacity, but also by changes in respiration or photorespiration rates, although there have been relatively few studies on respiration rates of halophytes under different salinity regimes. Such studies as have been reported do not provide evidence for any consistent effect of increased salinity on respiration rates, as changes appear to be largely species-specific (see Papp, Ball & Terry 1983). Walker *et al.* (1981) demonstrated an increase in photorespiration of grapevines at salinities above 150 mol m^{-3} NaCl, but in halophytes there is no evidence for any affects of salinity on photorespiration (Ball 1986).

It is possible that in some species salinity may induce anatomical changes in leaves which might, at least partially, compensate for the increased stomatal resistance to carbon dioxide uptake (Longstreth & Strain 1977; Longstreth & Nobel 1979). In some species, particularly succulents, stomatal density may be reduced by increased salinity (Bickenbach 1932; Waisel 1972).

There are too few data to assess whether, as a generalisation, it can be argued that halophytes differ from glycophytes in that halophytes can increase or maintain water-use efficiency while maintaining growth rates under saline conditions, but in glycophytes increases in water-use efficiency are accompanied by a marked reduction in growth rate. [C_4 species are clearly advantaged over C_3 species in water-use efficiency terms, but the C_4 pathway is not exclusive to, or even predominant amongst, halophytes.] However, even if such a generalisation could be supported, it would be uncertain as to whether it could be interpreted as a specific adaptation to saline conditions.

In a number of glycophytes, the selectivity of ion uptake is affected by the transpiration rate. In particular, the K^+/Na^+ of the xylem sap declines at higher transpiration rates (Pitman 1965; Jeschke 1984). If a similar effect occurs in halophytes, there would be a further advantage in reducing transpiration at high salinities.

Localisation of ions

At the level of whole organs, it can be argued that low osmotic potentials in halophytes are generated by the accumulation of inorganic ions – in the case of coastal halophytes principally sodium and chloride. However, in order to understand in more detail how halophytes cope with salt, it is necessary to enquire more closely into the exact location of the salt within tissues.

Water potential must be uniform throughout the cell, with no differences between any compartments in the cytoplasm or between cytoplasm and vacuole since hydraulic conductivity is relatively high. The simplest hypothesis, therefore, would be that salt is present in equal concentration in both cytoplasm and vacuole. If this were true, then the enzymes in the cytoplasm should be capable of functioning at salt concentrations in excess of 500 mol m^{-3}.

There is good evidence from studies on extreme halophilic bacteria that metabolism is not only possible at high ionic concentrations but, in fact, that there is a specific requirement for high salt conditions (see review in Hochachka & Somero 1984). In the case of halophytes, however, there is no convincing evidence to suggest that, as a generalisation, the enzymes of halophytes differ in their *in vitro* response to increased salt (in most experimental systems, NaCl). The degree of salt inhibition of enzyme function *in vitro* varies with the ion concerned but such variation appears to be a physicochemical effect rather than to be due to any specific properties of particular ions. The order of effectiveness of ions in inhibiting enzymes follows the Hofmeister lyotropic series – $K^+ < NH_4^+ < Na^+ < Mg^{2+} < Ca^{2+}$ for cations, and for anions $Cl^- < NO_3^- < Br^- < I^-$ (Wyn Jones *et al.* 1977; Wyn Jones & Pollard 1983; Stewart & Ahmad 1983) and enzyme function is generally impaired at high salt concentrations [the literature is reviewed by Jennings 1976; Flowers, Yeo & Troke 1977; Munns, Greenway & Kirst 1983; Wyn Jones & Pollard 1983]. The degree of salt inhibition may be altered by extraction procedures, or by substrate concentrations in the assay (Greenway & Sims 1974; Shomer-Ilam, Moualem-Beno & Waisel 1985). While enzyme properties *in vitro* may differ from those in the cytoplasm, the salt inhibition of enzyme activity has been reported from so many enzyme systems from different species as to make it unlikely that the effect is simply an artefact.

This viewpoint, although widely accepted, does not gain universal support. From studies on the effect of sodium chloride on phosphoenolpyruvate carboxylase, Shomer-Ilan, Moualem-Beno & Waisel (1985) and Shomer-Ilan & Waisel (1986) showed that resistance to salt is considerably enhanced by high substrate (phosphoenolpyruvate) levels. If such phosphoenolpyruvate levels did pertain *in vitro* then halophyte phosphoenolpyruvate carboxylase activity may be little reduced by high salinity, but the effective PEP concentration in halophytes has not been investigated. Shomer-Ilan & Waisel (1986) suggest that salt-tolerance might not only depend on salt exclusion and compartmentation but could also be achieved by increasing enzyme substrate concentrations and the concentration of various protective compounds.

In some species, differential sensitivity to salt has been demonstrated for different enzymes. Gettys, Hancock & Cavalieri (1980) showed for the grasses *S. alterniflora* and *S. patens* that malate dehydrogenase was much more salt-sensitive than leucine aminopeptidase or peroxidase; similar results were obtained by Daines & Gould (1985) from tissue cultures of *D. spicata*. It would seem from most plants studied that malate dehydrogenase (MDH) is a particularly salt-sensitive enzyme. Studies by Cavalieri & Huang (1977) on a number of American saltmarsh species show, however, that root MDH was more salt-tolerant than that in the shoots. Cavalieri & Huang (1977) also report an exception to the general rule of MDH sensitivity. In *Borrichia frutescens*, a shrubby composite, shoot MDH showed optimal activity at 250 mol m^{-3} NaCl while in the roots the optimum was at 500 mol m^{-3}. Furthermore, MDH from the cytosol had greater salt tolerance than that from the mitochondria.

The demonstration of different sensitivities to salt by different enzymes and the report of differences between cytosolic and mitochondrial MDH in *Borrichia* suggest that there may be some internal compartmentation of ions within the cytoplasm. However, the reports of high chloride concentrations in chloroplasts of some halophytes appear to be in error (Ball 1986) and there is no evidence to suggest that chloride levels in halophyte chloroplasts are significantly different from those of glycophytes.

Kaiser, Weber & Sauer (1983) showed that *Spinacia oleracea*, a moderately salt-tolerant species (Coughlan & Wyn Jones 1980), accumulates chloride in its chloroplasts when grown under conditions of low salinity. As external salinity levels were increased, the chloride level in the chloroplasts did not show a concurrent rise but stabilised at a constant value. Similar results were reported from the more salt-tolerant *M. crystallinum* by Demmig & Winter (1986). However, it is probable that the Na^+/K^+ ratio in halophyte chloroplasts is higher than that in non-halophytes (Kaiser *et al.* 1983; Robinson, Downton & Milhouse 1983; Preston & Critchley 1986; Ball 1986).

Notwithstanding demonstrations of high salt optima for some enzymes and the possibility of high salt compartments within the cytoplasm, the *in vitro* salt sensitivity of the majority of halophyte enzymes studied supports the inference that high salt concentrations do not prevail throughout the cytoplasm of halophytes.

There have been a number of attempts to localise ions and measure their concentrations within cells. The techniques used are histochemical, by X-ray microanalysis, and by compartmental flux analysis. Flowers & Läuchli (1983), Stewart & Ahmad (1983) and Sacher & Staples (1984) review approaches to the localisation of ions in plant cells.

For the results of histochemical and X-ray microanalysis studies to be interpretable, it is necessary that internal ion redistribution should not occur during specimen preparation and that ion migration be prevented during sectioning. Both requirements are technically difficult to meet and Jennings (1976), after reviewing earlier studies, suggests that little faith could be placed upon the results.

Harvey *et al.* (1981) using an improved X-ray microanalysis technique (transmission analytical electron microscopy) provide data indicating that *Su. maritima*, grown under saline conditions, accumulates the majority of the sodium and chloride ions within the vacuole. However, this pattern was not observed in all cells and it was suggested that this result was genuine and that different cells in the leaf tissue may differ physiologically in their patterns of ion distribution. Gorham & Wyn Jones (1983) express doubt as to whether Harvey *et al.*'s (1981) technique provides an accurate picture of the distribution of ions *in vivo*. However, their own data (Gorham & Wyn Jones 1983) demonstrating higher K/Na ratios in non-vacuolated tissues of *S. maritima* compared with those in vacuolated cells, also strongly suggest that sodium is excluded from cytoplasm. Harvey *et al.* (1985) discuss the problems associated with various approaches to X-ray analysis and the interpretation of results.

The technique of compartmental flux analysis has been applied only rarely to halophytic higher plants. Studies on *Su. maritima* (Yeo 1981) and on the roots of *T. maritima* (Jefferies 1973) yielded results which are consistent with an active sodium transport from cytoplasm to vacuole and accumulation of sodium in the vacuole. A contradictory result, however, was obtained by Shepherd & Bowling (1979) from the roots of *Eleocharis uniglumis* where active sodium transport from vacuole to cytoplasm is suggested. *E. uniglumis* is a species of brackish habitats and would probably not experience salinities comparable to those to which *Suaeda* and *Triglochin* would be exposed. There is an absence of data to allow assessment of the significance of this difference between *Eleocharis* and the more halophytic species; it is possible that those species characteristic of brackish conditions exhibit a particular syndrome of physiological adaptations.

At the present time, the main evidence for compartmentalisation of ions between cytoplasm and vacuole is indirect, being based on studies of effects of salt on enzymes. Evidence from other sources is too sparse to permit generalisation although some of the attempts to localise ions do support the inferences from enzyme studies.

In considering the location of ions within tissue, cytoplasm and vacuole are not the only options. Particularly in leaves the possibility arises that ions may accumulate in cell walls. If this were the case, they would not contribute to the lowering of cellular water potential. Oertli (1968) has suggested that accumulation of ions in cell walls could be responsible through loss of turgor for the adverse effects of salinity on some plants. The apoplastic compartment of most leaves is a small proportion of the total leaf volume; thus, if there is a failure to take up ions into cells the apoplastic concentration would rise rapidly (Flowers and Yeo 1986). Flowers and Yeo (1986) suggest that the decline in turgor of older leaves of *Su. maritima* may be correlated with a build-up of salt in the apoplast. Yeo and Flowers (1986*a*) argue that, in rice, tissue damage in salt-exposed plants is not directly due to metabolically toxic effects of salt but to water stress in cells caused by accumulation of salt in cell walls.

If it is assumed that, at high external salinities, the salt accumulated by halophytes is largely sequestered in the vacuoles of cells this raises some questions about whether membrane properties of halophytes are different from those of glycophytes. It also becomes necessary to postulate the occurrence in the cytoplasm of some non-toxic solute(s) at sufficient concentration to osmotically balance the salts in the vacuole.

Membrane properties in halophytes

In glycophytes it is argued that, at least for cations, there are small concentration gradients between cytoplasm and vacuole (Dainty 1979), and that little metabolic energy need be expended in maintaining ionic distribution within the cell. Much more energy may be involved in ion transport across the plasmalemma.

If the model developed in the preceding sections is correct then the situation in halophytes may be somewhat different. As far as the requirement for energy expenditure at the plasmalemma is concerned, then halophytes and glycophytes are probably not dissimilar but there will be a considerable concentration gradient for sodium and probably chloride between cytoplasm and vacuole. The establishment and maintenance of this concentration gradient may require considerable expenditure of energy. In non-halophytes, the tonoplast appears to be much more permeable than the plasmalemma (Dainty 1979). If this were also the case in halophytes, then it would add to the energy required to maintain ionic distribution between cytoplasm and vacuole.

It seems probable that if there are any qualitative differences between glycophytic and halophytic species then they may reside in the tonoplast. In halophytes the tonoplast may be relatively less permeable than in glycophytes and must contain the ionic pumps necessary to maintain the appropriate ion gradients. Erdei, Stuiver & Kuiper (1980) showed that the lipid composition of *Plantago* spp. may change in response to salinity, suggesting that membrane structure alters with increasing salinity.

In glycophytes, sodium is one of the ions most readily leached from leaves (Tukey 1970); it is possible that much of any sodium leached may be extracellular, residing in the cell walls. Waisel (1972) has suggested that leaching may provide one means by which the salt concentration in halophyte tissues may be regulated. There are few measures of leaching from halophytes but if sodium is actively accumulated in vacuoles, it may not be readily leachable. Indeed leaching of even apoplastic salt from halophyte leaves could be potentially harmful in that it could lead to the development of excessive turgor. Hajibagheri, Hall & Flowers (1983) have shown a considerable increase in cuticular thickness in *Su. maritima* in response to increased salinity – this is possibly an adaptation to prevent leaching occurring.

It is likely that in most coastal halophytes any accumulation of sodium ions is balanced by chloride. In inland saline areas where the dominant available anion may be one other than chloride, it is probable that other anions are accumulated preferentially.

Cytoplasmic osmotica

In most mature plant cells, the vacuole occupies 90–95% of the cell volume so that the absolute quantities of any such cytoplasmic solutes could be small, particularly if they are of low molecular weight.

A wide variety of organic compounds have been shown to accumulate in plants exposed to high salt concentrations. This accumulation may be a consequence of salt stress but need not necessarily be of adaptive significance – it could be a manifestation of metabolic disturbance induced by salinity rather than the means by which metabolism is protected from damage. Strogonov (1964) has argued that the build-up of certain amines (for example, putrescine) in some species exposed to high salinity is a major factor in salt toxicity. In more recent years, the accumulation of a range of other compounds has been claimed as being of adaptive significance in the evolution of salt tolerance. It is difficult, however, to establish the validity of such claims. Wyn Jones (1980) has suggested four criteria to be considered in assessing possible adaptive significance

> The quantitative distribution of the compound in plants collected from the wild should be correlated with particular habitats and, possibly, with families associated with these habitats;
> The compound should be constitutively accumulated in tolerant but not in sensitive species;
> Accumulation should be enhanced by a moderate stress;
> Exogenous application of the compound might promote *in vivo* the tolerance of the whole organism or *in vitro* enhance the tolerance of specific metabolic processes.

With respect to this last criterion, there have been few cases where compounds have been applied externally to whole plants with the aim of conferring salt tolerance, although Wyn Jones (1980), Wyn Jones & Storey (1981) and Ahmad, Wyn Jones & Jeschke (1987) suggest in certain circumstances that this may be possible.

Although the number of postulated osmotically active cytoplasmic solutes continues to increase, the evidence suggests that in the majority of plants the compounds accumulated belong to relatively few molecular types. Indeed, there is evidence to suggest that not only in plants but over all living organisms a similar small range of families of molecules is involved in generating low cytoplasmic water potentials (Yancey *et al.* 1982).

The classes of compounds for which evidence exists that they may act as cytoplasmic osmolytic solutes in halophytes are the imino acid proline, onium compounds (principally quaternary ammonium compounds but possibly also tertiary sulphonium compounds), and soluble carbohydrates [polyhydric alcohols (polyols) and sugars]. The claims that these compounds are of adaptive significance in halophytes have not been judged against all the criteria proposed by Wyn Jones (1980) in every case, but examples exist for all the classes of compounds which meet at least some of the criteria.

Nitrogen-containing compounds; proline and quaternary ammonium compounds

There is considerable evidence to indicate that free proline levels increase in tissues of plants subject to drought stress (Hsiao 1973; Aspinall & Paleg 1981; Wyn Jones 1980; Wyn Jones & Storey 1981; see also Stewart & Lee 1974; Cavalieri & Huang 1979). Although there is evidence in some cases to suggest that ability to accumulate proline is associated with drought tolerance, in general it would seem that proline accumulation is a consequence of drought stress rather than an adaptation to withstand drought.

There are data to indicate that proline may be a major component of the aminoacid pool in halophytes and that accumulation of proline occurs in response to increased salinity. However, in some species accumulation of proline occurs only at very high salinities (Cavalieri & Huang 1979 – Fig. 4.7) when indications of impaired growth are apparent. This would suggest that in these species proline accumulation is associated with toxicity rather than tolerance and that high levels of salinity, and associated low water potential, impose what is effectively drought stress on plants.

However, such an interpretation is not appropriate in all instances. Stewart & Lee (1974) showed that in *T. maritima* (Fig. 4.8), *Ruppia maritima*, and a coastal population of *Arm. maritima*, proline accumulation was stimulated by low salinities and proline content increased with increasing salinity. In *T. maritima* free proline could constitute up to 10–20% of the shoot dry weight. Jefferies (1980, 1981) confirmed these high levels of proline in *Triglochin* exposed to saline conditions. Stewart, Larher, Ahmad & Lee (1979) demonstrated similar accumulation of proline in *Pu. maritima*, and Cavalieri & Huang (1979) provided evidence that in *L. carolinianum* and *J. roemerianus* proline accumulation was stimulated at fairly low salinities.

Cavalieri & Huang (1979) showed that for some other species proline accumulation appeared to have a threshold salinity of about 500 mol m^{-3} NaCl, roughly equivalent to seawater. This pattern of accumulation occurred in the grasses *S. alterniflora*, *S. patens* and *D. spicata*, and has also been reported in *S. anglica* (Storey & Wyn Jones 1978; Wyn Jones & Storey 1978) and *L. fusca* (Sandhu *et al.* 1981).

For those species with a low salinity threshold for proline accumulation, it seems probable that accumulation is of adaptive significance in developing salt tolerance. The role of proline accumulation in these grasses with a much higher salinity threshold is less certain and may indicate the onset of stress rather than being of adaptive significance.

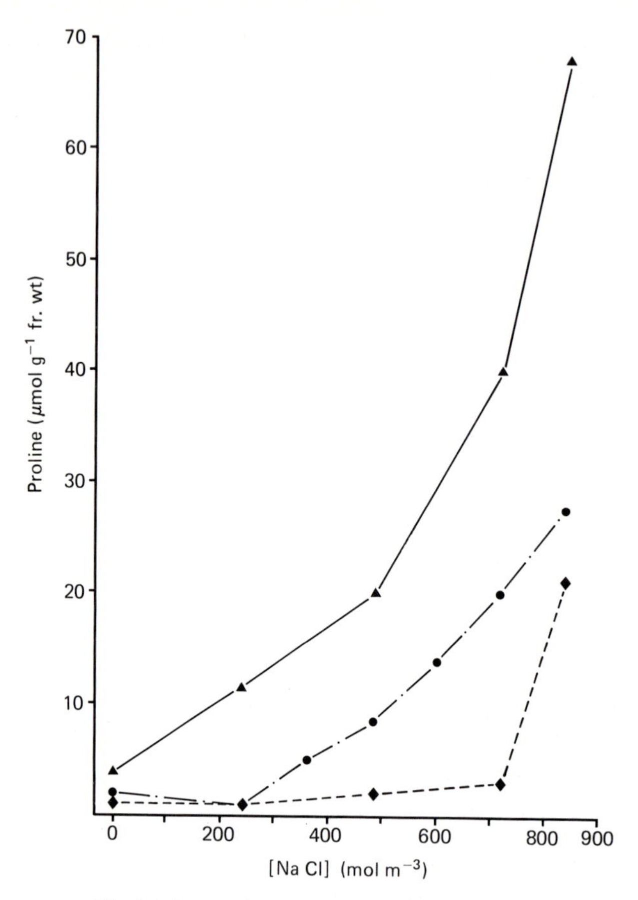

Fig. 4.7 Contrasting proline accumulation patterns in shoots of halophytes from the east coast of the USA. ▲ – ▲ *Limonium carolinianum*; ● – · – ● *Spartina alterniflora*; ◆ – – – ◆ *Borrichia frutescens*. Note the high threshold salinities for accumulation in *Spartina* and *Borrichia*. (Redrawn from Cavalieri & Huang 1979.)

Flowers & Hall (1978) found a marked increase in proline, as a proportion of total aminoacids, in some plants of *Su. maritima* exposed to high salinities. However, the total aminoacid pool remained constant and the concentrations were such that the contribution to lowering cytoplasmic water potential would have been small. As this increase in proline was not shown by all plants, Flowers & Hall (1978) suggest that it may have reflected an inadvertent water stress (additional to the salt stress) imposed on some of the experimental plants.

The first demonstration of salt-induced accumulation of glycinebetaine

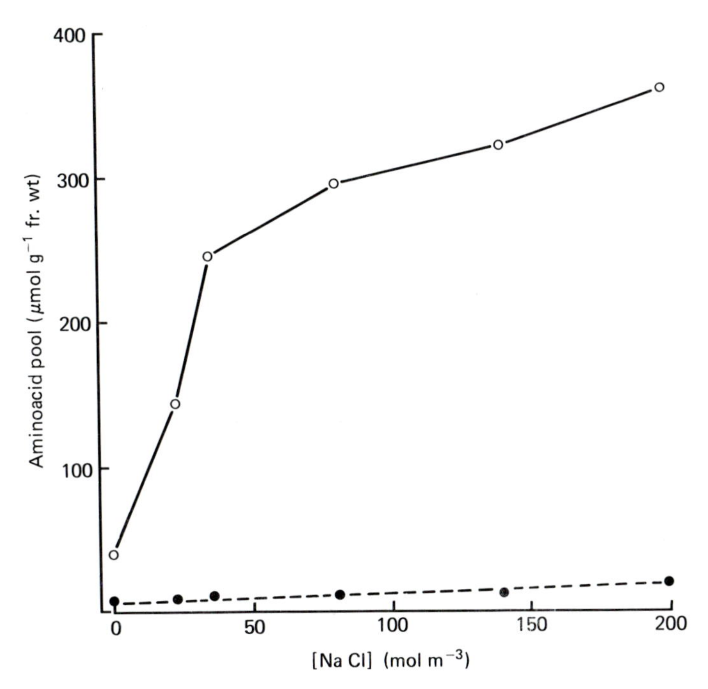

Fig. 4.8 Aminoacid and proline accumulation in shoots of *Triglochin maritima* grown at different salinities. Cuttings were grown in non-saline media for two weeks prior to being transferred to the saline medium. Shoot tissue was harvested for analysis after 10 days of saline treatment. ○ —— ○ proline; ● – – ● aminoacid pool less proline. (Redrawn from Stewart & Lee 1974.)

was provided by Storey & Wyn Jones (1975), subsequent papers reporting this compound from a number of halophytes (Storey & Wyn Jones 1977, 1979; Storey, Ahmad & Wyn Jones 1977; Wyn Jones & Storey 1981). [The trivial nomenclature of ammonio compounds is discussed by Wyn Jones & Storey (1981) – N,N,N-trimethylglycine has been referred to as 'betaine' in the literature but Wyn Jones and Storey reject this terminology in favour of the trivial name, glycinebetaine.]

It was suggested that proline accumulation was a general plant response to water stress, but that in some species it also had adaptive significance in the development of salt tolerance. Accumulation of glycinebetaine, however, is not taxonomically widespread (Storey & Wyn Jones 1977; Wyn Jones & Storey 1981). Glycinebetaine accumulation is a feature of a fairly small group of plants, in which members of the Chenopodiaceae and Gramineae are particularly prominent (Wyn Jones & Storey 1981). The majority of demonstrations of accumulation of glycinebetaine have been in

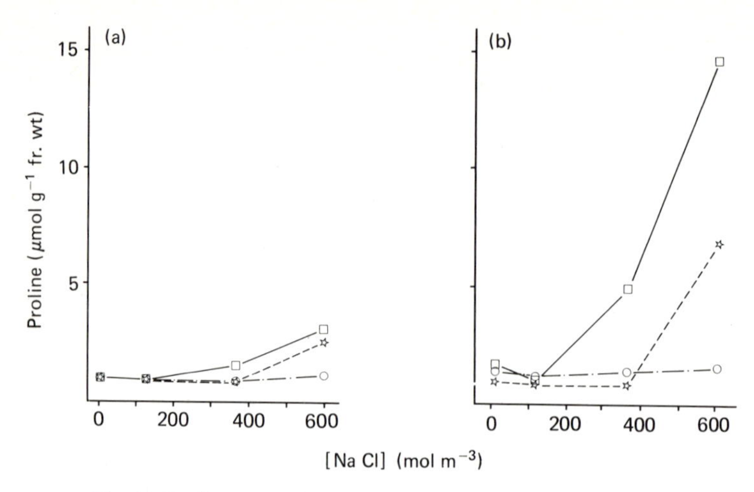

Fig. 4.9 Proline accumulation by roots (a) and leaves (b) of *Spartina alterniflora* grown hydroponically in NH_4Cl and NaCl. Plants were grown under experimental conditions for two months prior to harvesting. Each point is the average of four measurements. □ — □ 14.0 mg l^{-1} available nitrogen; ☆ -- ☆ 1.4 mg l^{-1}N; ○ – ○ 0.14 mg l^{-1}N. (Redrawn from Cavalieri 1983.)

response to increased salt concentrations, but there is evidence to suggest that accumulation may also be induced by non-saline applications of low water potential (for example, polyethylene glycol treatments) (Wyn Jones & Storey 1981; Jefferies, Rudmik & Dillon 1979).

It may thus be appropriate to refer to both proline and glycinebetaine (as well as to other organic solutes accumulated in some halophytes) as 'stress metabolites' (Smirnoff & Stewart 1985*a*), recognizing that a range of environmental factors may induce similar responses from plants (Steponkus 1980).

A number of halophytes accumulate both proline and glycinebetaine (Stewart, Larher, Ahmad & Lee 1979; Jefferies, Rudmik & Dillon 1977; Storey, Ahmad & Wyn Jones 1977; Storey & Wyn Jones 1978; Cavalieri 1983. Figs. 4.9, 4.10). In these species, glycinebetaine is present in high concentration, even at low salinities, and the increase in concentration as salinity is raised is not proportional to the applied salinity increase (i.e. the species are osmoregulators, see p. 270). At high external salinities there is a rapid increase in free proline; the threshold salinity at which proline accumulation commences is influenced by nitrogen availability and is lowered with higher nitrogen supplies (Cavalieri 1983; van Diggelen *et al.* 1986).

A major difference between glycinebetaine and proline is their behaviour after plants are transferred from a high to a low salinity environment.

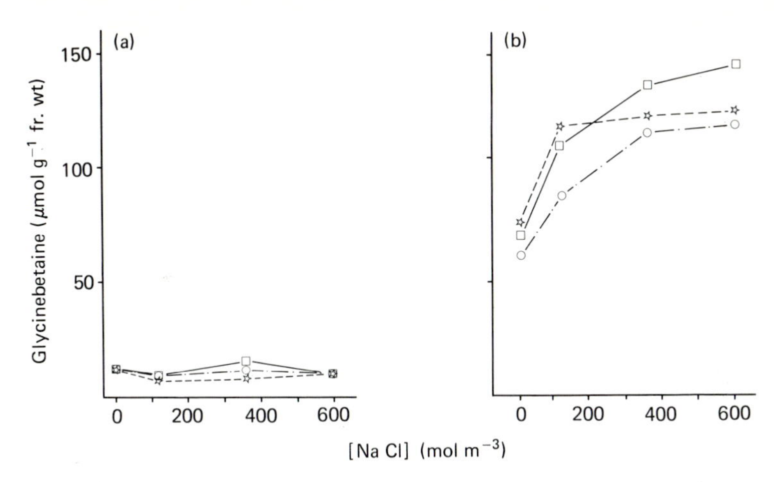

Fig. 4.10 Glycinebetaine accumulation in roots (a) and leaves (b) of *Spartina alterniflora*. Conditions and symbols as in Fig. 4.9. (Redrawn from Cavalieri 1983.)

Proline levels decrease rapidly (Wyn Jones & Storey 1981; Cavalieri 1983) whereas glycinebetaine concentrations remain almost constant (Hansen & Nelson 1978; Wyn Jones & Storey 1981; Cavalieri 1983 – see Fig. 4.11). This difference has important implications for the nitrogen economy of halophytes.

Glycinebetaine appears on current evidence to be the most widely distributed ammonio compound amongst halophytes. A number of other ammonio compounds have been recorded from plants (Wyn Jones & Storey 1981). In *L. vulgare* β-alaninebetaine is accumulated in sufficient concentrations for it to be a major cytoplasmic solute (Larher & Hamelin 1975).

The sulphonio analogues of ammonio compounds have been reported from a number of vascular plants and a range of algae (Wyn Jones & Storey 1981). Amongst halophytes, β-3-dimethylsulphoniopropionate (DMSP) has been recorded from *S. anglica* by Larher, Hamelin & Stewart (1977) although glycinebetaine is the major cytoplasmic solute in this species (Storey & Wyn Jones 1978). van Diggelen *et al.* (1986) demonstrated that shoot levels of DMSP did not increase in response to increased external salinity and suggested that it was unlikely to act as a cytoplasmic osmoticum (see also p. 295).

Sorbitol and other polyols

Polyhydric alcohols (polyols), at high cytoplasmic concentrations, are recorded from a wide range of organisms (Yancey *et al.* 1982). Glycerol is particularly prominent in unicellular marine algae (Brown & Simpson

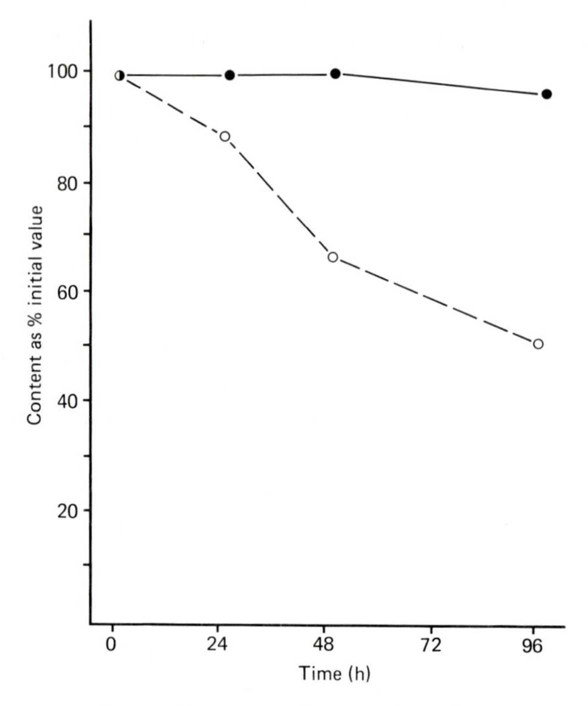

Fig. 4.11 Time course of decrease in proline and glycinebetaine content in leaves of *Spartina alterniflora* after transfer from saline conditions. Plants were grown hydroponically in nutrient solutions containing 28 mg l^{-1}N and 500 mol m^{-3} NaCl for a month and then transferred to a nutrient solution lacking NaCl at time 0. ● —— ● glycinebetaine; ○ – – ○ proline. (Redrawn from Cavalieri 1983.)

1972; Hellebust 1976). Amongst saltmarsh plants, salt-induced accumulation of sorbitol has been demonstrated in *Pl. maritima* (Ahmad, Larher & Stewart 1979; Jefferies, Rudmik & Dillon 1979) and *Pl. coronopus* (Gorham, Hughes & Wyn Jones 1980; Briens & Larher 1983; Smirnoff & Stewart 1985*a*; Kuiper 1984). Pinitol is accumulated in *Spergularia media* in sufficient quantity to suggest that it may be an important cytoplasmic solute (Gorham *et al.* 1980). Other species reported by Briens & Larher (1982) to accumulate some form of polyol are *A. tripolium* (in which myoinositol was recorded by Albert & Popp 1978), *J. maritimus* and *Phragmites australis* (although the salt tolerance of this species is limited).

Pinitol is found at relatively high concentrations in the upper saltmarsh sedge, *Carex extensa*, but accumulation did not increase with increased salinity. However, in the strand-line species *Honkenya peploides* pinitol may be involved in cytoplasmic osmotic adjustment (Gorham, Hughes & Wyn

Jones 1981). Gorham, McDonnell and Wyn Jones (1984*a*) demonstrated accumulation of pinitol in salt-stressed plants of *Sesbania aculeata*, a member of the Leguminosae tolerant of both salinity and flooding but localisation within the cell has not yet proved possible so its role in salt tolerance is still a matter of assumption. A number of Leguminosae occur in the upper zones of temperate saltmarshes but the nature of any of stress metabolites in these species is unknown.

While polyol accumulation has been recorded from a number of unrelated species, it would appear on present evidence that the role of these compounds as cytoplasmic solutes in vascular halophytes from saltmarshes is limited in comparison with their role in the algae.

However, the distribution of types of cytoplasmic solutes in mangroves may be rather different from that in saltmarsh species. Studies on Australian mangroves (Popp 1984*a*,*b*; Popp, Larher & Weigel 1984,1985) suggest that pinitol and mannitol are the most widely distributed putative cytoplasmic solutes and that proline and methylated quaternary ammonium compounds are found in only a few species.

Sugars

The concentrations of a number of sugars have been shown to rise in a range of plants exposed to high salt – this tendency being particularly marked in monocots (Albert & Popp 1978; Gorham *et al.* 1980; Briens & Larher 1982). The importance of these compounds as osmotically active cytoplasmic solutes is not clear. Briens & Larher (1982) suggest that sucrose may play a significant role in *J. maritimus*, *Phragmites australis* and *Scirpus maritimus*, while Gorham *et al.* (1980) speculate that hexoses may be a more important component of salt tolerance in halophytes than sucrose. In a number of species examined by Gorham *et al.* (1980) the concentration of total soluble carbohydrates was more than double the sum of individually measured free sugars; the nature of the compounds making up the difference is unknown – however, if they are high molecular weight compounds their contribution to the osmotic potential would be small.

From the data of Briens & Larher (1982), it would appear that patterns of sugar accumulation, both in terms of the particular sugars and the spatial distribution within the plant, vary between species. Outstanding amongst their observations was the accumulation of rhamnose (up to 33% dry weight) in the roots of *Pl. maritima*.

Evidence for the role of compounds as compatible solutes

Demonstration of the accumulation of a range of compounds in plants in response to exposure to high salinity serves to define a list of substances which might serve as cytoplasmic solutes, osmotically balancing salts in the vacuole. In order to confirm this hypothesis, it is necessary to localise the distribution of the compounds within cells and to demonstrate that at high concentrations they do not have a deleterious affect on cell metabolism.

Hall, Harvey & Flowers (1978) have provided histochemical evidence that in *Su. maritima* glycinebetaine is largely restricted to the cytoplasm. Unfortunately, while there is a histochemical procedure for localisation of quaternary ammonium compounds there are no such procedures for the other compounds of interest. Other, more indirect evidence for these solutes being cytoplasmic rather than vacuolar is also, however, largely restricted to glycinebetaine; for the other compounds restriction to the cytoplasm remains a plausible hypothesis.

Meristematic cells lack large vacuoles, so any compounds present in high concentrations in meristematic tissue are likely to be present in the cytoplasm. In *Atriplex spongiosa* glycinebetaine concentrations in apices are much higher than in mature vacuolated tissue (Wyn Jones & Storey 1981). Separation of whole intact vacuoles from cells indicates that glycinebetaine is partially excluded from vacuoles (Wyn Jones *et al.* 1977; Wyn Jones 1980). Higher glycinebetaine concentrations in younger, compared with older, tissues have also been reported in *Su. maritima* (Gorham & Wyn Jones 1983).

The major lines of circumstantial evidence for preferential cytoplasmic location remain the inference derived from studies on the effects of salt on individual enzyme or metabolic processes (Brady *et al.* 1984; Gibson, Speirs & Brady 1984) that metabolism and high salt concentrations are incompatible, and the necessity for osmotic balance between cytoplasm and vacuole.

Nevertheless, although osmotica may be preferentially located in the cytoplasm, there is evidence which suggests that at times high concentrations may occur in the vacuole.

Jefferies, Davy & Rudmik (1979) showed that in three species (*T. maritima*, *H. portulacoides* and *L. vulgare*) total soluble nitrogen levels in the foliage dropped between early and late summer even though the soil salinity rose. They argued that a possible interpretation of these findings is that early in the growth season soluble nitrogen-containing compounds are distributed between cytoplasm and vacuole and represent a pool of both nitrogen and organic skeletons available for a variety of purposes. During the summer, part of this resource is utilised in growth but the requirements for exclusion of ions from the cytoplasm mean that a certain amount is retained in the cytoplasm to provide balancing osmotica. Evidence from reported tissue contents of potential osmotica also suggests that at low external salinities, a large part of the total may be in vacuoles – conversion of the tissue contents to concentrations on the basis of restriction to cytoplasm would imply very low water potentials. The high glycinebetaine levels reported from plants with osmoregulatory behaviour at low external salinities may be a case in point. Leigh, Ahmad & Wyn Jones (1981) demonstrated that in *Beta*, isolated vacuoles may contain appreciable quantities of glycinebetaine and proline.

Jefferies (1980, 1981) argued that proline stored in vacuoles would provide a store of nitrogen for use in growth and development as well as the potential for rapid osmotic adjustment to lowered salinities by accumulating ions in the vacuole and exporting proline to the cytoplasm. It is possible also to envisage a pool of soluble carbohydrates being used for both growth and osmotic adjustment in a similar fashion. However, the apparent stability of tissue glycinebetaine levels once accumulated suggests that while transfer between vacuole and cytoplasm may occur, diversion for growth is less likely.

Internal recycling of nitrogen compounds between older leaves and new growth will permit maximum use of limited resources. In many perennial species underground storage organs allow resources synthesised in one season to be used for rapid growth in the next spring (Jefferies, Davy & Rudmik 1979).

Within the cytoplasm there may be differences in the concentration of osmotica between organelles. Study of isolated chloroplasts from *M. crystallinum* by Demmig & Winter (1986) showed that proline did not make a major contribution to osmotic adjustment although proline levels in leaf sap were high under conditions of high salinity.

The evidence for the putative cytoplasmic solutes being non-toxic is more extensive than that for their localisation.

Sorbitol is widely used as an osmoticum in the isolation of cell organelles so that it is unlikely that it has adverse effects on metabolism at the concentrations in which it would occur in plant cells.

There are a number of demonstrations that high concentrations of proline and glycinebetaine do not affect metabolism adversely (including Stewart & Lee 1974; Flowers, Hall & Ward 1978; Cavelieri 1983; see review by Paleg, Stewart & Starr 1985). Larkum & Wyn Jones (1979) compared proline and glycinebetaine as osmotica for extracted spinach chloroplasts and found glycinebetaine to be superior. [Robinson & Jones (1986) showed that *in vivo*, glycinebetaine accumulated in chloroplasts of spinach exposed to high salt concentrations while Hansen *et al.* (1985) indicate that the chloroplast may in fact be the site of glycinebetaine synthesis.] Glycinebetaine may also confer some positive protection against the inhibitory effects of salt on metabolism (Pollard & Wyn Jones 1979). Ahmad *et al.* (1982) demonstrated that glycinebetaine, proline and sorbitol all stabilised the enzyme glutamine synthetase (form GS_{II}) from *T. maritima.* Jolivet, Hamelin & Larher (1983) demonstrated that glycinebetaine may protect membranes from the destabilising effect of oxalate (which particularly in the Chenopodiaceae may be present at high concentration). Ahmad, Wyn Jones & Jeschke (1987) suggest that glycinebetaine may modify the transport characteristics of membranes and this may be a significant factor in the maintenance of compartmentation of ions within cells.

Until recently, studies of potential compatible solutes have carried the implicit assumption that the findings could be generalised across taxa. However, Manetas, Petropoulou & Karabourniotis (1986) demonstrated that proline inhibited phosphoenol pyruvate carboxylase (PEPCase) activity in *Salsola* spp. and provided no protection against sodium chloride whereas in the grasses *Cynodon dactylon* and *Sporobolus pungens* proline was not inhibitory and provided protection against inhibition by sodium chloride. Glycinebetaine was compatible with PEPCase in all species.

The effects of high sugar concentrations on metabolism have not been as well investigated; however, Rozema, Buizer & Fabritius (1978) indicate that sugars may have considerable inhibitory effect and would appear to be less desirable cytoplasmic solutes than proline or glycinebetaine. While a number of species accumulate sugars in response to salinity, this strategy may only be advantageous at low or moderate salinities; the majority of species shown to accumulate sugars tend to predominate in brackish rather than fully saline sites.

Smirnoff & Stewart (1985*a*) suggested that stress metabolites may serve to afford protection from environmental hazards other than salinity. They pointed out that accumulation of stress metabolites has been shown to occur in a number of coastal species (from shingle and sand-dune habitats) at low salinities and hypothesised that this accumulation may confer greater heat resistance on these species; Paleg *et al.* (1981) demonstrated that a number of stress metabolites, including proline and glycinebetaine, did confer protection to enzymes from heat damage. Even in temperate latitudes, shingle and sand-dune plants may be exposed to high temperatures, as clearly may halophytes in arid-zone habitats; however, the ecological significance of greater heat resistance to coastal saltmarsh species is not so readily apparent. Sorbitol and mannitol may be more effective than nitrogenous compounds in preventing heat damage to enzymes (Smirnoff &

Stewart 1985*a*). Stewart & Popp (1987) speculate that the accumulation of polyols in mangroves may be related to the provision of protection from heat damage as leaf temperatures up to 15° C above ambient air temperature have been recorded in some mangrove species (Clough, Andrews & Cowan 1982).

Osmotic adjustment and nitrogen availability

The widespread utilisation of nitrogen-containing compounds as cytoplasmic osmotica necessitates a review of nitrogen availability and use in saltmarshes.

Although total nitrogen levels in saltmarsh soils may be comparable with those in many fertile inland soils, the amounts of biologically available forms of nitrogen are low. The majority of the soil nitrogen is in the form of organic nitrogen compounds which break down very slowly (Abd.Aziz & Nedwell 1979). As well as being present in small quantity, the amounts of inorganic nitrogen in the soil vary considerably over the growing season (Henriksen & Jensen 1979; Jefferies 1977*b*). Jefferies (1977*b*) found that the quantities of nitrogen available to plants over the growing season did not differ between the upper- and lower-marsh. However, there was more available nitrogen in soils immediately adjacent to creeks – the tendency of *H. portulacoides* to favour creekbank sites, although probably largely a response to better conditions of aeration (Armstrong *et al.* 1985), may also reflect improved nutrient supply given the interaction between salinity-stimulated growth and available nitrogen as that reported by Jensen (1985*b*).

The form of inorganic nitrogen will be dependent upon the state of aeration of the soil. In anaerobic conditions, the predominant form will be as ammonium ions. However, while ammonium predominates in the bulk soil solution in anaerobic soils, the extent of oxidation in the rhizosphere is unknown. Mendelssohn (1979*a*,*b*) showed that the short form of *S. alterniflora* was apparently nitrogen-limited, even though interstitial ammonium concentrations were higher than those in sites of the tall form. Morris & Dacey (1984) demonstrated that uptake of ammonium in *S. alterniflora* was dependent on rhizosphere oxygen concentration. These two studies both suggested that rhizosphere conditions may be an important additional influence upon nitrogen availability. Despite the likely predominance of ammonium over much of the marsh, measurements of nitrate reductase in a number of species by Stewart, Lee & Orebamjo (1972, 1973) indicate that nitrate is the major available nitrogen source – either the heterogeneity of the environment, itself, provides aerobic microsites with available nitrate or rhizosphere oxidation mitigates the prevailing ammonium. Stewart *et al.* (1972, 1973) found a declining gradient in both nitrogen reductase and tissue nitrogen concentrations from seaward to landward in *S. maritima* and other species on a saltmarsh in North Wales. However, nitrogen availability in the strandline was high.

The anaerobic conditions in saltmarsh soils will tend to promote denitrification.

A number of studies have shown improved growth, particularly of the above-ground parts of plants, in response to addition of nitrogen in various inorganic forms to saltmarshes (see Smart 1982; Valiela & Teal 1974; Long & Mason 1983; Jensen 1985*b*). These demonstrations of probable nitrogen limitation to growth of saltmarsh plants suggest they may be particularly demanding of nitrogen.

This view is strengthened by the finding of Smart & Barko (1980) that the mass ratio of nitrogen to phosphorus under conditions when minimum requirements for growth are just met in *D. spicata* and *S. alterniflora* was 16.1:1 compared with 7.5:1 in the freshwater sedge *Cyperus esculentus* (Barko & Smart 1978). The extra requirement for nitrogen may reflect the need for synthesis of organic nitrogen compounds for maintaining osmotic balance between cytoplasm and vacuole.

Glycinebetaine and proline may make up a large proportion of the total organic nitrogen in halophytes (Stewart & Lee 1974; Larher *et al.* 1982). Although the highest levels recorded may represent, in part, a labile pool of available nitrogen to be utilised for both growth and osmotic regulation (Jefferies 1980, 1981), the proportion of the total soluble nitrogen pool in the form of cytoplasmic osmotica remains high throughout the growing season (Jefferies, Davy & Rudmik 1979).

The accumulation of cytoplasmic osmotica is an essential requirement for the survival of halophytes. However, not all species utilise nitrogen-containing compounds for this purpose. Given the scarcity of available nitrogen in the saltmarsh environment, it is perhaps surprising that nitrogen compounds are used by so many species. [Although the tissue nitrogen levels and nitrogen reductase activity in lower-marsh *Su. maritima* (Stewart, Lee & Orebamjo 1972) suggest that nitrogen availability is not limiting for that species in that habitat.] Jefferies & Rudmik (1984) propose that availability of fixed carbon may impose greater constraints on the particular compound used as cytoplasmic osmotica than nitrogen supply; species using carbohydrates do not appear to have competitive advantage over species accumulating glycinebetaine or proline. It is possible that these nitrogen-containing compounds confer other benefits as yet unknown.

Heydemann (1981) suggests that, despite the general tendency of consumers to favour tissues of high nutrient (particularly nitrogen) status, high proline levels appear to deter some phytophagous insects.

Despite the frequent reports of growth stimulation following nitrogen-fertiliser application, it is not always the response recorded. Mendelssohn (1979*a*,*b*) suggested that short *S. alterniflora* in a North Carolina saltmarsh might be unable to assimilate available ammonium. Jefferies (Jefferies 1977*b*; Jefferies & Perkins 1977; Jefferies, Davy & Rudmik 1979) has shown that in some species there may be population differentiation in the ability to respond to additional nitrogen. On the north Norfolk coast the upper-marsh predictably experiences high soil salinities in the early summer, while in the lower-marsh there is less variation in salinity (Jefferies 1977*b*). Upper-marsh species demonstrate a genetically determined growth pattern with slow growth in early summer and a growth spurt later in the growing season when lower salinities prevail; lower-marsh populations grow throughout the summer. The low growth rates of upper-marsh populations in the early summer are not affected by additions of nitrogen (Jefferies & Perkins 1977), although growth in the late summer is stimulated by such additions. Lower-marsh populations respond to added nitrogen during the early summer. To argue teleologically, it could be suggested that, were upper-marsh plants to grow in the early summer their nitrogen requirement would be greater than could be met by the environment – not only would nitrogen be required for growth *per se*, but for every extra growth increment additional nitrogen would be needed for osmotica in the young tissues. In these circumstances, an inherently low growth rate, insensitive to stimulation, would, intuitively, be seen as advantageous. It would be of interest to see whether similar temporal responses to nitrogen exist in other parts of the world and, also, whether species in less-predictable habitats (as far as salinity regime is concerned) exhibit different strategies.

In studies on *S. anglica*, van Diggelen *et al.* (1986) found that stimulation of growth by increased nitrate only occurred at low salinities. Tissue proline and glycinebetaine levels varied in response to nitrogen availability at higher salinities but increased levels of these putative cytoplasmic osmotica did not influence the salinity level at which growth inhibition occurred.

In interpreting the results of field experiments, it is important to recognise that application of nitrogen fertilisers may have a range of effects on the environment. As pointed out by Smart & Barko (1980) and Smart (1982) such additions may influence microbial populations, affecting rates of decomposition and mineralisation, and also soil pH. These other changes may also affect the growth of vascular plants.

Smart (1982) has suggested that the response of *S. alterniflora* to added nitrogen involves interactions and feedbacks between several factors.

Under low salinity conditions, the supply of nitrogen (as ammonium) to the roots is governed by the mass flow of water generated by transpiration. Growth stimulation results in greater transpiration, while ion exclusion by the roots results in an increased salinity in the rhizosphere. The increased salinity results in lower transpiration rates and, thus, lower nitrogen transport into the rhizosphere. Fertilisation eliminates the dependence of nitrogen supply on transpiration. However, the increased growth increases transpiration, promoting higher salinities in the rhizosphere. Tidal flushing, by reducing rhizosphere salinities, may also promote growth (see Fig. 4.12).

The availability of nitrogen also influences the internal allocation of resources by vascular plants. Under low nitrogen conditions, a greater proportion of biomass may be allocated to below-ground organs (Smart & Barko 1980). With increased nitrogen above-ground, growth is stimulated to a greater extent than below-ground. Production of flowers in some long-lived perennials is also favoured by increased nitrogen availability (Jefferies & Perkins 1977; Jefferies, Davy & Rudmik 1979).

In addition to inorganic nitrogen released during the decay of organic matter and any inputs brought in by the tide, it is possible that fixation of atmospheric nitrogen may be an important source of available nitrogen in saltmarshes.

Cyanobacteria (blue-green algae) may occur abundantly on mud surfaces in creeks and pans and, particularly, in the lower-marsh on the sediment surface underneath vascular plants. These cyanobacteria may be active in nitrogen fixation (Jones 1974; Mann 1979).

Nitrogen-fixing bacteria may occur both within the rhizosphere and cortical cells of the roots of *S. alterniflora* (Patriquin 1978; Patriquin & Keddy 1978; Patriquin & McClung 1978; Mann 1979), and probably also occur in association with other species. Acetylene reduction assays suggest that these bacteria actively fix nitrogen in these circumstances. However, whether much of the nitrogen fixed either on marsh surface or in the rhizosphere becomes available to vascular plants is debatable. Any nitrogen-fixing activity may well be counteracted by the activity of denitrifying bacteria.

Legumes may be locally common in upper-saltmarsh communities, being particularly prevalent in some of the grazed communities in north-west Europe. In general, the Leguminosae are not regarded as being particularly salt tolerant, but amongst the species studied there is a range of

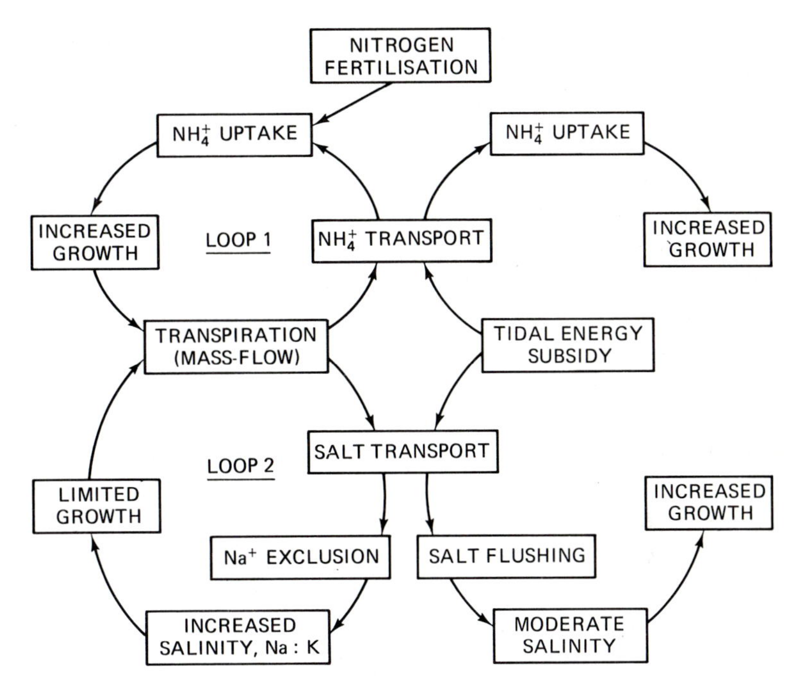

Fig. 4.12 Postulated feedback loops linking nitrogen availability to *Spartina alterniflora* growth. (Redrawn from Smart 1982.)

salt tolerance, although most species of potential agricultural importance show seriously impaired growth at low salinities (Läuchli 1984). Salinity causes a major reduction in acetylene reduction activity of *Trifolium subterraneum* (Hopmans, Douglas & Chalk 1984), while in soyabeans (*Glycine*) Singleton & Bohlool (1984) have shown that the earliest stages of nodulation initiation failed at very low salinities. Whether saltmarsh legumes are not active in nitrogen fixation or whether their symbiotic bacteria are adapted, nodulate and function under saline conditions has not been investigated. Reddell, Foster & Bowen (1986) have shown that in the actinomycete *Frankia*, which forms a symbiotic relationship with the genus *Casuarina*, there is genotypic variation with respect to the effects of salinity on nitrogen-fixing activity.

While nitrogen availability in saltmarshes is low, the levels of extractable phosphorus are frequently also low (Long & Mason 1983). However, although there are some reports of increased growth of saltmarshes species following additions of phosphorus (Pigott 1969; Tyler 1967), in most cases addition of phosphorus alone has little effect, but combined application of nitrogen and phosphorus may produce a significant synergistic interaction. For some crop plants, increased salinity may promote phosphate uptake, even to the extent of inducing phosphorus-toxicity symptoms (reviewed in Gorham *et al.* 1986). Such an effect has not been reported with any saltmarsh species.

Okusanya & Fawole (1985) suggest that in the case of the seacliff species *Lavatera arborea* salt tolerance is improved when the plants receive additional phosphate. They suggest that this treatment produces increased root growth at the higher salinities, thus providing increased root area for nutrient uptake, the potassium and calcium content of shoots being considerably increased compared with plants exposed to seawater treatments without additional phosphate. An increased root growth under saline conditions after the provision of extra phosphate has also been reported from *Spergularia* spp. (Okusanya & Ungar 1984). Such an effect is potentially important to seacliff species, where locally nesting birds may provide very high phosphate levels. Saltmarsh sites used as high-tide roosts by birds may exhibit higher soil phosphate levels, but these areas are not preferentially occupied by particular species, nor does growth in general appear markedly more luxuriant than elsewhere on the marsh. Areas of the extreme upper-marsh where domestic stock congregate at high tide are often markedly different in their flora from the rest of the marsh – (in northern Europe species such as *Urtica dioica*, *Cirsium arvense*, *Stellaria media* and *Potentilla anserina* may be prominent – Adam 1981*a*) – but while these floristic differences may be a response to added nutrients, the salinity of such sites is low and a nutrient/salinity interaction is unlikely to be particularly important.

Osmoregulators and osmoconformers

Wyn Jones (see Storey & Wyn Jones 1979; Wyn Jones 1980) has suggested that there are two strategies of osmotic adjustment shown by plants which, to borrow terminology from animal physiology, may be divided into osmoconformers and osmoregulators. Osmoconformers maintain an almost constant gradient of osmotic potential between the external solution and the plant as the external potential is lowered. In osmoregulators, the osmotic potential of leaf tissue drops very rapidly on imposition of salinity and is then maintained approximately constant over a wide range of external salinities (Fig. 4.10). Stewart & Ahmad (1983) indicate that the distinction between the categories is not sharp and that some species are somewhat intermediate in their behaviour. Nevertheless, a number of physiological studies suggest that a broad division of halophytes into the two classes is possible.

Little is known, however, about the ecological consequences of the two strategies. Osmoregulation might appear to be a more appropriate strategy in the face of rapidly fluctuating salinity. Jefferies (1980) suggested that there was a correlation between rooting depth and strategy. Deep-rooted plants are likely to be osmoconformers. Changes in salinity at depth are likely to be slow and essentially seasonal in nature. Osmotic adjustment by the plant can, therefore, be gradual. Shallow-rooted plants, however, are likely to experience rapid and repeated changes in salinity. Response to these changes by alteration in the osmotic potential of cells may be inefficient energetically and insufficiently rapid to cope with the external changes. Osmoregulators may expend less energy by 'adjusting' to an external salinity close to the highest likely to be experienced. Little is known about the active rooting depth of many halophytes, but it would be of interest to test this hypothesis.

However, there is no close association between either strategy and taxonomy. The family Chenopodiaceae contains both osmoconformers (including *Atriplex spongiosa* (Storey & Wyn Jones 1977) and *S. europaea* (Guy, Warne & Reid 1984) and osmoregulators (*Su. monoica* – Storey & Wyn Jones 1979 and *Su. torreana* – Glenn & O'Leary 1984). Thus, both strategies are found in species in which glycinebetaine is the major cytoplasmic solute, and this suggests that, to some extent, the recognition of the strategies may be an artefact of experimentation. When external salinity is steadily increased, the glycinebetaine content of an osmoconformer increases in step. However, as was discussed on p. 261, glycinebetaine is not metabolised if the external salinity is subsequently reduced (see Fig. 4.11). Thus, in the field, the glycinebetaine content of an osmoconforming species is likely to reach a maximum consequent upon the highest salinity experienced and would then be maintained for the rest of the growing season. If glycinebetaine levels were monitored in field-collected samples, it is probable that, in samples from after the salinity peak, there would be little apparent difference between socalled osmoconformers and osmoregulators.

In environments of rapid salinity change, proline accumulation may be a more appropriate adaptation than increasing glycinebetaine content. At times of low salinity, the leaf tissues may maintain a low osmotic potential because of the persistent high concentration of glycinebetaine in the cytoplasm (balanced by the retention of ions in the vacuole). In these circumstances, equilibrium between internal and external water potentials would demand an increase in turgor, imposing considerable stresses in the cell wall. [Adjustment could be achieved, possibly at greater energetic expense, through export of ions from the shoot to the root (and, hence, the soil) – but see p.244 – and redistribution of glycinebetaine between cytoplasm and vacuole, assuming the tonoplast is appropriately permeable.] In a proline-accumulating species, adjustment to higher external water potential (i.e. decreasing salinity) by raising the cell osmotic potential with no, or only slight, change in turgor pressure is possible. It would be of interest to measure the correlation, if any, between the salinity variations expected in particular marsh subhabitats and the proportion of the flora accumulating different cytoplasmic-compatible solutes.

It is appropriate to draw attention to Reed's (1984) criticism of much of the terminology applied to discussions of adjustments in cellular solute concentration accompanying changes in external water potential. Reed argued that a number of widely used terms, most particularly osmoregulation, are both inappropriate and misleading. For general usage, he proposed the use of osmoacclimation, with a variety of other terms reserved for particular contexts. The categories osmoconformer and osmoregulator are permissible as classes but, as Reed points out, the term osmoregulation has been used non-specifically in reference to osmotic responses regardless of whether the organisms were osmoconformers or osmoregulators.

Physiotypes

The development of a greater understanding of salt tolerance permits new criteria to be used in the classification of halophytes. In an earlier section division into categories of osmoregulation and osmoconformers was discussed. In this section the concept of physiotypes is introduced.

Albert (see Albert 1975; Albert & Popp 1977, 1978) has used the term physiotype to describe groupings of halophytes made on the basis of the chemical composition of their foliage. A precise definition of physiotype is not provided in Albert's accounts but Gorham, Hughes & Wyn Jones (1980) suggest that the concept implies the existence of 'a defined spectrum of physiological traits associated with a specific taxon and related to the tolerance of members of that taxon to a specific environment'. Storey, Ahmad & Wyn Jones (1977), and Gorham, Hughes & Wyn Jones (1980, 1981) have examined the relationships between the distribution of cytoplasmic organic osmotica and patterns of ionic composition in various saltmarsh and other coastal species.

In general, there is agreement between the results of Albert from species from an inland saltmarsh in Austria and those of Wyn Jones from coastal species in Wales. There are, however, some differences: Albert (1975) classified *Pl. maritima* as belonging to an SO_4^{2-}-accumulating physiotype; Storey, Ahmad & Wyn Jones (1977) found the major inorganic anion to be Cl^-. Such a difference could indicate ecotypic differentiation in a geographically very wide-ranging species or reflect variation in the availability of sulphate in the substrate at the two collecting sites.

Monocotyledonous species in general conform to a 'high potassium physiotype' with a K^+/Na^+ ratio greater than unity (Gorham *et al.* 1980). In addition, this physiotype is associated with free sugars as the major organic osmotica (Albert & Popp 1978; Gorham *et al.* 1980). However, not all monocots conform to this pattern. In *S. anglica* and *T. maritima* the K^+/Na^+ ratio is below 0.5; *Triglochin* and *Pu. maritima* are proline accumulators, while *Spartina* accumulates glycinebetaine (Gorham *et al.* 1980).

The Chenopodiaceae, or at least the halophytic members of the family, are characterised by the accumulation of sodium and high levels of glycinebetaine (Gorham *et al.* 1980). In some members of the family, large quantities of oxalate are found in the foliage (Osmond 1963; Osmond, Björkman & Anderson 1980; Karimi & Ungar 1986). This accumulation of oxalate may be the result of the need to maintain cellular pH, nitrate reduction in leaves generates hydroxyl ions which are removed in the synthesis of oxalate (see Raven 1985; Smirnoff & Stewart 1985*b*). For arid-zone chenopods, nitrate may be readily available (Osmond *et al.* 1980), although this is less likely to be the case in coastal saltmarshes.

Other betaines seem to be characteristic of particular families (Wyn Jones & Storey 1981) but are not necessarily associated with salt tolerance.

In the case of other dicotyledonous species studied, the concordance between ionic physiotypes and taxonomy and between ionic composition and major organic osmotica is less clear, although some possible groupings are recognisable (Gorham *et al.* 1980).

Such physiotypes as can be recognised are neither strictly taxonomic nor ecological groupings. While there are strong taxonomic overtones to the physiotype groups, there are exceptions to the generalisations that can be made and, as more species are studied, the more exceptions emerge. For example, *T. maritima*, *Pu. maritima* and *S. anglica* are exceptions to the monocotyledonous physiotype proposed by Albert (Albert 1975; Albert & Popp 1978). While all very salt-tolerant Chenopodiaceae, when growing under saline conditions, appear to conform to a high sodium-glycinebetaine physiotype, this is not so for less-tolerant species such as *Spinacia oleracea* (Coughlan & Wyn Jones 1980).

Were physiotypes to be restricted to particular taxonomic groups, then they might reflect sets of conservative characters which had remained linked during evolution. As such, physiotypes might be useful in indicating the relatedness of taxa, but provide little value in the assessment of adaptations to the environment. Were particular physiotypes to be closely associated with particular environmental conditions, then understanding the value of a particular physiotype under those conditions would be of considerable physiological and ecological interest. Unfortunately, no ecological pattern in the distribution of physiotypes is readily apparent, although this may simply reflect the relatively small number of species investigated and the lack of detailed characterisation of the environment. However, it would appear that the majority of monocotyledons, with a high potassium-free sugars physiotype do not occur in habitats exposed to very high salinities (Jefferies & Rudmik 1984), while the exceptions to this physiotype (such as *Triglochin*, *Puccinellia* and *Spartina*) are ecologically more versatile. It is possible that more detailed comparison of physiotypes with specific environmental features may be of value in elucidating the possible adaptive significance of particular physiological traits.

Glycophytic and halophytic species

In the light of current knowledge of halophyte physiology, it is apropriate to re-examine the differences between halophytes and glycophytes.

For halophytic species growth in saline conditions is accompanied by considerable uptake of ions and the available evidence supports the hypothesis that these ions are mostly sequestered in the vacuole, while maintaining osmotic balance between cytoplasm and vacuole involves accumulation of organic osmotica in the cytoplasm.

However, as was emphasised in Chapter 3, the flora of many saltmarshes may contain species conventionally regarded as glycophytic. How do these species cope with saline conditions?

The evidence would suggest that, in general, salt-tolerance in glycophytes is achieved through the exclusion of ions, at least from the foliage (Greenway 1973; Greenway & Munns 1980; Hannon & Barber 1972; Yeo 1983; Noble, Halloran & West 1984). This exclusion is seen not only between salt-sensitive and tolerant varieties of crops where the limits to tolerance are still low, but also between salt-tolerant and salt-sensitive ecotypes of species such as *F. rubra* in which the tolerant forms may be exposed to salinities higher than the growth optima of many halophytic species (Hannon & Barber 1972).

However, not all moderately salt-tolerant glycophytes exclude salt from the foliage. Marschner, Kylin & Kuiper (1981) demonstrated that the most salt-tolerant of the three genotypes of sugar beet they studied accumulated more salt under higher salinity treatments, although the salinities involved (up to 150 mol m^{-3} NaCl) were low. This finding raises again the problem of categorising plants as either halophytes or glycophytes. Given that the wild progenitors of sugar beet probably came from coastal habitats and that *Beta* is a member of the Chenopodiaceae, a family to which many very salt-tolerant species belong, it would be possible to regard sugar beet not as a glycophyte with some salt-tolerance, but as a halophyte with very limited salt-tolerance. A species which might more confidently be classified as a glycophyte, but in which limited salt-tolerance is associated with salt accumulation, is *Lupinus luteus*; tolerance in other members of the Leguminosae is associated with exclusion (Läuchli 1984).

Although most glycophytes exclude ions from their leaves, turgor is maintained so that osmotic adjustment must rely on organic solutes. Ahmad, Wainwright & Stewart (1981) showed that a saltmarsh ecotype of *A. stolonifera* accumulated more organic solutes and fewer ions than an inland provenance when cultivated under saline conditions. Greenway (1973) suggests that salt-tolerant glycophytes, although adjusting to imposed moderate salinities, may be particularly susceptible to additional drought stress.

Exclusion of salt from the leaves may involve total exclusion from the roots, excretion by the roots of much of the influx of salt, or reabsorption of ions from the xylem along the transpiration pathway (Yeo *et al.* 1977; Yeo 1983; Kramer 1983; Jeschke 1984; Pitman 1984; Hannon & Barber 1972). However, whatever the mechanism, it would appear that at some critical level of external salinity (which varies between species) the exclusion breaks down and a large influx of ions into the aerial parts occurs (Fig. 4.13). The deleterious effects of this may be due to a failure of compartmentation and consequent adverse effects of high salt levels in the cytoplasm, specific ion effects, or adverse effects on water relations (possibly because of salt accumulation in the apoplast).

The exclusion strategy does not appear to be practised by any species from the most saline parts of a saltmarsh. It is possible that none of the mechanisms by which exclusion is achieved is capable of evolving to exclude ions in the face of salinities in excess of 3–400 mol m^{-3}. However, even if such exclusion were possible, the diversion of photosynthate into osmotica would be considerable and would cause considerable reduction in growth. Species with such physiology would be at a disadvantage in competition

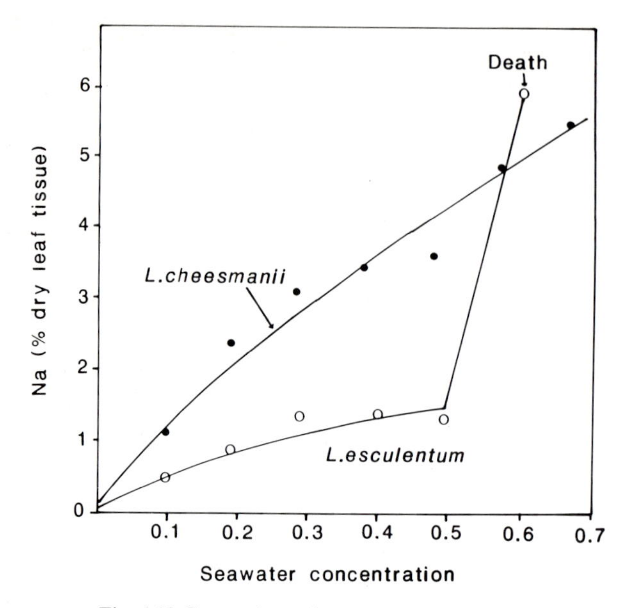

Fig. 4.13 Comparison of two tomato species grown under conditions of different salinities. *Lycopersicon cheesmanii* is a relatively salt-tolerant species and tissue salt increases with external salinity. *L. esculentum* excludes sodium from the shoots until the external salinity exceeds 50% seawater. (After Epstein 1980.)

with species with a metabolically 'cheaper' solution to the problems of growth at high salinity.

If the exclusion of glycophytes from saline sites can be explained by their lack of salt-tolerance, how is the exclusion of halophytic species from non-saline habitats to be explained in view of the lack of any obligate requirement for high salt levels? Most halophytic species grow well under freshwater conditions and many show optimal growth in these circumstances, but even so their growth rates are likely to be lower than those of glycophytes. Halophytes are, thus, at a competitive disadvantage sufficient to exclude them from favourable freshwater sites (Barbour 1970, 1978; Gray & Scott 1977*b*; Szwarcbaum & Waisel 1973).

As was discussed in Chapter 2, a number of European halophytes have bimodal distributions also occurring at restricted inland sites. This bimodality is a comparatively recent phenomenon; in the late Glacial and early post-Glacial periods species such as *Arm. maritima* and *Pl. maritima* were widespread. Adam (1977*a*) argued that a unifying feature of the present inland sites of these species was the high probability that they had remained open and unwooded throughout the post-Glacial period; the decline of the species across the broader landscape reflecting competition and shading.

The particular cytoplasmic osmotica found in a species may partially determine its competitive ability. Jefferies & Rudmik (1984) pointed out that the majority of the species restricted, or almost restricted, to saltmarsh habitats in northern Europe accumulate nitrogenous compounds. On the other hand, most of the glycophytic element in marsh floras accumulate soluble carbohydrates. Although some nitrogenous compounds may be divertible to growth, this is less likely in the case of glycinebetaine. For osmoregulatory species in which accumulation of osmotica is stimulated at low salinities, the resultant diversion of resources from growth may reduce the competitive advantage over glycophytes; the osmoregulators may, thus, be restricted to saltmarshes today by their low competitive ability in brackish conditions. The glycophytic element, utilising carbohydrate, may be better able to switch resources between osmotic adjustment and growth, and so be favoured in upper saltmarshes in which high salinities are transient [*A. stolonifera* and *F. rubra*, which accumulate nitrogenous compounds to some extent are an apparent exception (Jefferies & Rudmik 1984). However, the competitive ability of these ecotypes relative to inland strains under non-saline conditions has not been fully investigated.]

For at least part of the growing season the salinity conditions in saltmarshes are likely to excede even the highest growth optima of any halophytic species. Thus, most plants on saltmarshes will experience long periods of suboptimal growth conditions. Thus, competition is between species which are all disadvantaged to some degree – clearly a case of 'in the land of the blind, the one-eyed man is king'.

The question still arises as to how salt-tolerant glycophytes can assume an important role in saltmarsh vegetation. Growth under saline conditions is likely to involve a reduction in growth rates, while salinities in the habitats in which glycophytes are abundant are likely to be close to the growth optimum of those species showing salt-stimulated growth for at least part of the growing season. *A priori*, it is difficult to envisage salt-tolerant glycophytes ever having a competitive advantage over halophytic species under saline conditions. However, in Europe, salt-tolerant glycophytes are particularly abundant at sites which experience heavy grazing pressure (Chapter 3; Adam 1978). It is probable that differential tolerance to grazing determines the relative success of halophytic species and glycophytes. [However, it needs to be borne in mind that in northern Europe cessation of grazing is followed by spread of another salt-tolerant glycophyte – *F. rubra*.] It is also possible that the glycophytes are able to respond more effectively to very low salinities, which might occur at some seasons, than halophytic species and so gain an advantage which is not totally lost in subsequent periods of high salinity. [Despite the greater competitive ability of glycophytes at low salinities, once established, halophytic species may persist for many years in conditions of low salinity, as in reclaimed saltmarshes used only for grazing.]

Halophytic species appear, in general, well able to expand into less-saline areas, provided competition is reduced. Glycophytes in upper saltmarshes are frequently widespread species inland, for which saltmarshes are but one of a large number of possible habitats. There is, however, a category of species which appears far more restricted in its range. These are those species which, at least at the present, are restricted to sites which are permanently brackish (where they grow mixed with salt-tolerant glycophytes and some halophytes). In northern Europe, examples would be provided by *Eleocharis uniglumis* and *Blysmus rufus*. Further south, *Carex divisa* and *Alopecurus bulbosus* show similar characteristics. Whether there are certain physiological features which limit these species to a narrow range of habitats is unknown. However, brackish habitats have been particularly prone to disturbance and destruction and, in many areas, both habitats and the species they support are becoming increasingly rare.

Agricultural potential of halophytes

Salinisation and consequent loss of land to agriculture is a major problem, not only in the Third World, but also in developed nations such as the United States and Australia. In addition, extensive areas in arid and semi-arid climates are naturally saline and this salinity is a considerable constraint upon development in these regions.

The possibility that by understanding how wild halophytes cope with high salinity it might be possible to develop new crops for saline areas, or breed salt-tolerant traits into existing crops, has provided the motivation for funding much of the recent research on halophyte physiology.

It is unlikely that intertidal saltmarshes will be used for other than grazing, haymaking or local collection of some native species for human consumption (see p. 364), although the possibility of using *S. anglica* for papermaking has been investigated (Ranwell 1967). However, it is possible that coastal saltmarsh species may be utilised in non-tidal saline areas.

The potential for development of salt-tolerant plants of agricultural utility has been the subject of several reviews (see, for example, Hollaender *et al.* 1979; Staples & Toenniessen 1984; Pasternak & San Pietro 1985; Barrett-Lennard *et al.* 1986; Epstein & Rains 1987). Amongst plants currently of economic importance there is a wide spectrum of salt-tolerance. Even amongst crops which are very sensitive to salt, for example, citrus fruit, there are some strains which do show a degree of tolerance (Walker *et al.* 1984). The occurrence of this variation suggests that there is scope through conventional breeding programmes for increasing salt-tolerance of many crops. However, the degree of tolerance which might be achieved in this way may not be significant when set against the prevailing salinities in the field situations for which potential crop species are required. Greater tolerance might be achieved by introducing genes from closely related species which are much more tolerant than the current crop species. Wyn Jones, Gorham & McDonnell (1984), Gorham, McDonnell & Wyn Jones (1984*b*) and Gorham *et al.* (1985, 1986) have shown that a number of wild members of the grass tribe Triticeae (which also includes the cultivated wheats and related cereals) possess considerably more salt-tolerance than wheat. Shah *et al.* (1987) showed that the D genome from *Aegilops squarrosa* had considerable influence on cation selectivity in hexaploid wheats exposed to saline conditions and suggest that there is potential for using this genome in breeding salt-tolerant cereals. On the other hand, Jefferies & Rudmik (1984) point out that a number of extremely important families from an agricultural viewpoint, for example, the Rosaceae, do not have any markedly halophytic members.

Even if the ability of crop species to survive at high salinities could be increased, it is probable in many cases that it would be at the expense of those attributes which originally made the species desirable. Tolerance of salinity in halophytes involves diversion of resources between organs and away from growth. If tolerance is achieved by exclusion of salt from foliage (which for many crop species would appear to be the direction in which selection from currently available more-tolerant strains would lead), then the requirement for synthesis of osmotica could be high (and would certainly be higher than in salt accumulators). In addition to having lower growth rates, most halophytes are perennial and sexual reproduction may be dependent on amelioration of salinity conditions and/or a more readily available source of nitrogen (Jefferies & Perkins 1977; Jefferies, Davey & Rudmik 1979; Jefferies & Rudmik 1981). Many crop plants are grown for their fruit or seed, the yields of which might be considerably reduced in highly salt-tolerant strains. The major seed-producing crops are annuals and in saltmarshes annuals are of restricted occurrence. Natural selection has favoured vegetative growth and survival traits over frequent sexual reproduction – artificial selection for high seed yield under highly saline conditions may be severely constrained by the limited availability of resources for allocation to reproduction. While these constraints might be overcome by provision of additional nitrogen, this may not be feasible on economic grounds.

At the present time, it seems unlikely that improvement of existing crops will allow other than marginally salinised land to be reclaimed. It is possible, however, that amongst the wild halophytic flora there are species with potential for development as novel crops. O'Leary (1984) suggests that the saltmarsh grass *Distichlis* might have some potential as a source of grain for human consumption, while O'Leary, Glenn & Watson (1985) point to the potential of several species as oilseed crops. Fruits and seeds tend to have a low sodium content as they are supplied with solutes via phloem (O'Leary, Glenn & Watson 1985). Epstein (1985) also suggests that for development of crops for human consumption attention should be concentrated on fruit-bearing species. If the aim is to provide an economic return from salinised land rather than simply to grow food crops, then species could be grown for other markets; O'Leary (1984) suggests that the sea lavenders (*Limonium* spp.) have horticultural value (some *Limonium* spp. are grown at present to supply the cut-flower trade, but not under saline conditions); other species may also be of horticultural use. As with many plants, halophytes may contain secondary metabolites of potential value, but this has not been investigated in detail (O'Leary 1984).

Utilisation of halophytes for grazing or as fodder crops would seem more immediately practical than attempting to grow crops for human consumption on saline sites. O'Leary (1984) and O'Leary, Glenn & Watson (1985) suggest that the high salt content of halophyte foliage may be a constraint on such usage. However, domestic livestock grazing on European saltmarshes do not appear to be troubled by the salt levels in their food. Haymaking has been practised extensively on both European and North American marshes and the salt level of the hay has not been reported as a major problem. Hubbard & Ranwell (1966) conducted trials on silage making using *S. anglica*; the nutritional value of the silage was reasonable and the product was acceptable to livestock. Salt contents in grasses are relatively low compared with those in chenopods which dominate many inland saline areas.

The vegetation composition of saltmarshes may be considerably altered by grazing and some halophytes are intolerant of even light grazing pressure. If halophytes are introduced to salinised soils with the intention of providing pasture, then it is important that appropriate grazing-resistant species or ecotypes be chosen.

It has been proposed that halophytes could be grown to reduce the salt content of soils. Accumulation of salt in foliage and its subsequent harvest and removal could result in a lowering of soil salt concentrations. O'Leary (1984) suggested that, while this might occur under non-irrigated conditions, continued supply of new salt would prevent such an approach succeeding in irrigated areas.

For the most-saline areas, use for grazing or fodder crops would appear to be the only economical approach to utilisation for the foreseeable future. Engineering solutions may permit lowering of salinity and the growing of a wider range of crops, but the cost of such works may be prohibitive in many of the poorer countries affected by land salinisation. It is, however, important that any new irrigation developments be designed so as to minimise the risks of future salinisation. Irrigation with seawater could be a viable option under appropriate circumstances (O'Leary 1984; O'Leary, Glenn & Watson 1985). Severe technological problems remain to be solved before seawater irrigation can be a practical possibility on a large scale (O'Leary 1984) and economic considerations may limit the opportunities for applying the technology, even assuming it is developed.

One of the problems associated with the development of more salt-tolerant crops is uncertainty as to what attributes should be selected for (Shannon 1985). Salt-tolerance in halophytes involves integration of the physiology of the whole plant and is likely to involve control by numerous genes [even though the genetic basis of salt-tolerance is poorly understood, Tal (1985)]. While attention has been given to studies of salt-tolerance at the cellular level (Lerner 1985; Spiegel-Roy & Ben-Hayyim 1985), the nature of salt-tolerance is likely to involve selection at the whole plant level (Yeo & Flowers 1986*a*; Passioura 1986). In addition to salt-tolerance involving interaction between organs, Yeo & Flowers (1986*a*) point out that in the whole plant the effective salt concentration to which cells are exposed may be very different from that in the outside medium. It would be difficult, therefore, to relate the salt-tolerance of cell cultures to that of whole plants.

Richards (1983) has argued that a further constraint to breeding programmes is imposed by the heterogeneity of the environments in which any salt-adapted crops would be utilised. However, heterogeneity is a feature of most environments and even heavily cultivated non-saline soils are likely to show local heterogeneity in nutrient availability. Most crops are exposed to considerable environmental heterogeneity, even if tolerance of such conditions was not specifically tested in their breeding, and variation in soil salinity might not prove to be insurmountable once the basic problems of breeding for yield (be it of total biomass or some specific component) have been overcome.

Consequences of tidal flooding

Tidal flooding has, through its alteration of soil aeration and chemistry, considerable effects on plant growth. However, it may have other more direct effects which have, unfortunately, been little studied.

During spring tides, the surface of the entire marsh may be covered with water. Even on the highest tides, the vegetation of the upper-marsh may be partially emergent, but the lowest vegetative zones will be completely submerged for several hours.

The extent to which saltmarsh plants are capable of photosynthesising while submerged has been little studied. Hubbard (1969) recorded that plants of *S. anglica* could survive for at least 4.5 months fully submerged in clear seawater. The seawater flooding tidal marshes is rarely clear, but is often extremely turbid. Frequently, the effect of flooding is, therefore, not simply to reduce the available light, but to cut off light altogether and alter the photoperiod. Even if the flooding water were clear, the quality of light (including importantly the red/far red ratio) reaching plants would be altered. Many morphogenetic processes in plants are controlled by light quality or photoperiod. Given that the pattern of tidal submergence at any particular elevation on the shore varies through the year in a predictable fashion, it is possible that the timing of developmental processes in saltmarshes plants could be cued by tide.

This topic has been little studied and the only indication of some involvement by tides in the control of development comes from Hubbard's (1969) work on *S. anglica* in Poole Harbour. Despite its ability to withstand prolonged submergence under experimental conditions, in the field *S. anglica* is never completely submerged for long periods. At its lowest limit in Poole Harbour, it may be submerged for up to 23.5 h/day during November neap tides. At its lowest point, *S. anglica* is exposed to the air during daylight for an average only 2.3 h/day during November. Hubbard (1970) suggested that November was a particularly critical month as the shoots for the next growing season were initiated then, and this process might be photoperiodically controlled. Interestingly, *S. anglica* close to its northern distribution limit in Denmark also seems incapable of growing where it would not be exposed to daylight for several hours a day during November (Christiansen & Møller 1983). However, despite the suggestive nature of these field observations, experimental physiological confirmation that there is a critical minimum daylight requirement in November for *Spartina* growth is lacking.

There is considerable variation from site to site in the relative elevation of the lower *S. anglica* limit. In Bridgwater Bay, Somerset (slightly north and west of Poole Harbour, but with a broadly similar climate), *S. anglica* at its lower limit was immersed for only a third of the time prevailing at Poole Harbour (Morley 1973). Morley (1973) suggested that the failure of *S. anglica* to extend further seaward was due to the exposed nature of the site and that wave action may cause stunting of growth and loss of vigour. The tidal range at Bridgwater Bay is much greater than at Poole Harbour and although *Spartina* is flooded for a short time the depth of flooding at Bridgwater is much greater than at Poole Harbour.

It is interesting to note that the pattern of interruption of photoperiod by tidal flooding will vary from site to site. At Bridgwater Bay, the highest spring tides occur around 0700 and 1900 hours, so that the marsh is always exposed to daylight at midday. The reduction in light input due to flooding will be less, therefore, than at sites where the major spring tides occur around midday and midnight (for example, Morecambe Bay) where the lower-marsh may lose a high proportion of potential daylight towards the beginning and end of the growing season. Other factors being equal (wave action, substrate, grazing pressure, etc.), it might be expected that there would be variation in the elevation limits of species depending on the time of day spring tides occur – this hypothesis remains to be investigated.

The sediment which is responsible for interrupting the photoperiod may also be deposited on the foliage of marsh plants as the tide recedes. Frequently, the position of the last tidemark can be seen easily in the contrast between the green foliage of unflooded areas and the grey silt-covered leaves lower down the marsh. During the summer, in the absence of rain, the foliage may remain silt covered for prolonged periods. The effects of this silt layer on light capture by the leaves and on leaf gas exchange have not been investigated.

As well as altering light regimes, the flooding tide may cause rapid and large changes in temperature. In summer, the flooding water may be much cooler than the air. In the winter, a frozen marsh surface may be rapidly immersed in water more than 10° C warmer. Such rapid changes in temperature may well be physiologically disruptive, but any such effects have not been studied. Temperature change might also interact with variation in photoperiod in controlling morphogenetic processes.

Mechanical effects of tidal flooding

As tides rise and fall across a marsh, the plants are exposed to mechanical forces from the tidal current. The velocity of these currents will vary with tidal range.

Within established vegetation, the velocity of the tidal current will be much reduced and is unlikely to exert significant effect. However, in open low-marsh stands, the full force of the current is felt by the vegetation. Mature plants with well developed root or rhizome systems appear able to withstand currents, although when the latter are accompanied by strong wave action local erosion can occur. Germinating seedlings and young plants with poorly developed and shallow root systems are less well placed. It has been argued that inability to withstand strong currents as seedlings is a major factor limiting the seaward extension of some species. Wiehe (1935) studied the survival of *S. europaea* (s.l.) seedlings along a transect on a saltmarsh in the Dovey estuary and showed that at lower elevations few seedlings survived from germination in spring to early summer, and the few plants surviving showed signs of having been dragged by the tide (Fig. 4.14). Wiehe (1935) suggested that there was a threshold period of 2–3 days during which time the seedlings developed sufficient roots to withstand subsequent tides. Longer periods between submergence were not associated with lower mortality rates. Clapham, Pearsall & Richards (1942) (see also Chapman 1960*b*) suggested that for *A. tripolium* the threshold period free from tidal submergence required for successful establishment was five days.

The extent to which the length of threshold period varies with sediment type, tidal range (which will influence current velocity), or climatic conditions (growth rates might vary with latitude and between years) has not been investigated, nor have threshold times been calculated for many other species. Nevertheless, it is possible that variation in threshold time between species is a factor maintaining the species zonation patterns apparent in most saltmarshes.

Although flooding tides do not uproot species in the upper-marsh, they may cause taller species to lodge. The resultant tangled mat of vegetation may suppress lower-growing species and this may be a major factor in the development of the species-poor upper-marsh communities characteristic of ungrazed marshes in northern Europe.

Tides are also responsible for carrying flotsam and jetsam into marshes and redistributing it over the marsh surface. Although the major accumulation of litter is normally along the drift line of the highest tides, lower tides may deposit significant amounts lower in the marsh where it may remain for weeks or even months. If sufficiently abundant, this litter may cause death of the underlying vegetation, providing localised sites for short-term successional sequences or, possibly in some cases, the initiation of new pans.

Salt entry directly into leaves

Most of the studies on the effects of salt on the growth of plants have concentrated on salt treatments applied to roots. Plants growing on

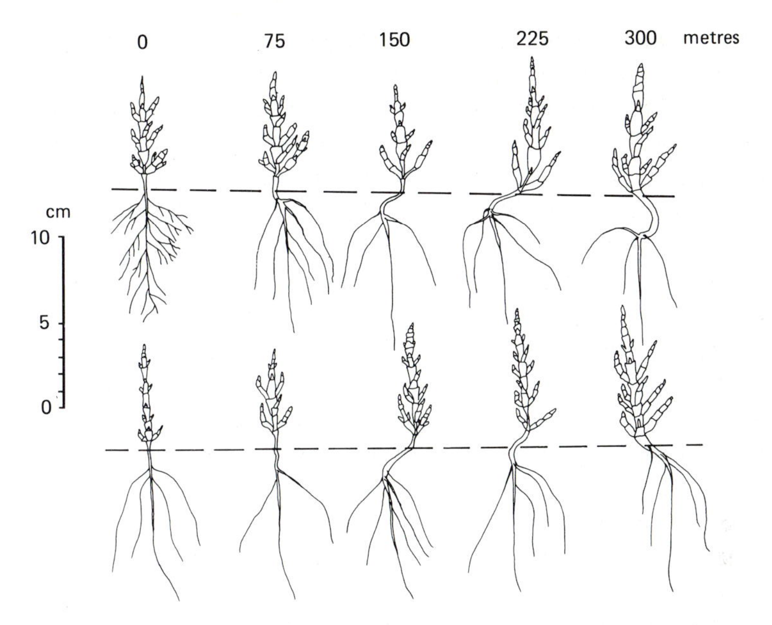

Fig. 4.14 Scale drawings of *Salicornia* plants sampled randomly from five positions along a transect from landward (0 m) to seaward in the Dovey estuary in July. Plants seaward of mean high water neap tides (150 m) show the effects of exposure to tidal currents. (Redrawn from Wiehe 1935.)

saltmarshes are certainly exposed to high salt levels in their root environment. However, they must also endure periods of partial or total immersion in saline water.

The effects of this on the ionic balance of plants has not been given much consideration. Immersion could lead to leaching of salts from tissues (loss of apoplastic salt could lead to very high turgor pressures) but also could permit the ingress of salts directly into leaves. Absorption of salt is the cause of damage to a number of sensitive crop species which are affected adversely by sprinkler-irrigation systems (as distinct from dripper systems) using saline water (Meiri & Plaut 1985; Maas 1985). Given the range of susceptibilities shown by crop plants, it is possible that native saltmarsh species are adapted to minimise any potential for leaf absorbance, but this remains to be tested. However, the thick cuticle possessed by many species may be as much an adaptation to minimise salt entry as it is to minimise transpiration.

In addition to exposure to flooding tides, the foliage of saltmarsh species is also impacted by windborne salt (Rozema *et al.* 1983); this might limit the ability of some glycophytes to survive in the upper-marsh and, again, a thick cuticle may confer protection from damage.

Waterlogging

There have been relatively few studies on the effects of waterlogged soil conditions upon the growth of saltmarsh plants, although it has frequently been argued that salinity and waterlogging are the two major factors controlling the distribution of species on saltmarshes.

Anaerobic conditions are of frequent occurrence in saltmarsh soils and these conditions are accompanied by considerable changes in soil chemistry. It can be suggested that plants growing on waterlogged soil are faced with two potentially inimical factors: a direct lack of oxygen, and the presence of phytotoxins. The topic of plant growth in waterlogged soils has been reviewed in Hook & Crawford (1978), Kozlowski (1984*a*,*b*), Armstrong (1982), and Crawford (1982,1983,1987).

Lack of oxygen

There has been considerable debate as to whether or not roots of plants growing in waterlogged soil exhibit metabolic adaptations. Crawford and his colleagues (see general reviews by Crawford 1978, 1982, 1983) have argued that wetland plants differ from species not tolerant of flooding in possessing an ability to accumulate malate (or other organic acids) under flooding conditions.

Their hypothesis can be outlined as follows (see McMannon & Crawford 1971): when intolerant species are flooded and aerobic respiration ceases, glycolysis proceeds to produce acetaldehyde and ethanol. Acetaldehyde induces alcohol dehydrogenase activity which accelerates glycolysis. Carboxylation of phosphoenolpyruvate can lead to formation of malate, but this is rapidly decarboxylated to form pyruvate, which contributes further to acetaldehyde and ethanol production. Eventually, acetaldehyde and ethanol accumulation reaches toxic levels. In wetland species, acetaldehyde and ethanol accumulate, but acetaldehyde fails to induce alcohol dehydrogenase, so glycolysis does not accelerate. Malate formed by carboxylation of phosphoenolpyruvate is not decarboxylated to pyruvate, owing to lack of malic enzyme, and so accumulates.

Attractive as this hypothesis is, the evidence supporting it is weak.

The hypothesis would indicate that ethanol toxicity is the prime cause of damage to non-tolerant species. However, normally ethanol would be lost readily by flooded plants (Kozlowski 1984*a*). Ethanol applied to roots in culture solutions at concentrations 100 times those in flooded soil are non-injurious (Jackson, Herman & Goodenough 1982). It seems unlikely that the amount of ethanol in, or close to, flooded plants would ever be sufficient to cause injury. Monk, Crawford & Brändle (1984) suggest that under anaerobic conditions rhizomatous monocotyledonous species from wetlands retain less ethanol in their rhizomes than non-wetland species, despite relatively high rates of ethanol synthesis in the wetland species. Loss of ethanol from roots and rhizomes would represent a drain of carbon which could otherwise be utilised for growth.

The hypothesis demands the absence of malic enzyme to decarboxylate malate to pyruvate. However, Davies, Nascimento & Patil (1974) have demonstrated the presence of malic enzyme in flood-tolerant species, including several studied by McMannon & Crawford (1971). Crawford (1983) argues that in flood-tolerant species malic enzyme is inhibited by flooding so that malate accumulation would still occur as in the original hypothesis. However, there is little confirmatory evidence, and the malate hypothesis has been strongly rejected by Keeley (1978), Smith & ap Rees (1979), ap Rees & Wilson (1984), and ap Rees *et al.* (1987). Smith & ap Rees (1979) studied the anaerobic root metabolism of a number of wetland species and showed convincingly that malate did not accumulate and the overall pattern of metabolism was the same as that in pea roots, which are intolerant of flooding.

A fundamental drawback of the malate accumulation model is that little, if any, ATP is produced. [If the glycolytic substrate is free, rather than phosphorylated, hexoses then the net ATP yield is zero.] Even alcoholic fermentation would yield substantially less ATP than aerobic respiration. This low energy yield could be of significance to halophytes where regulation of ion uptake could impose a very high demand for ATP.

If there is no convincing evidence for consistent metabolic differences between growing roots of wetland and non-wetland species, how are differences in tolerance between species to be explained?

For a root growing in non-waterlogged soil, the oxygen demand (at least of the cortex) will be met by oxygen entering the root from the soil (Armstrong 1982). This is not possible under anaerobic conditions. However, there is a network of intercellular spaces which connects the tissues of the root to the above-ground parts and, ultimately, through the stomata to the atmosphere. In non-wetland plants the porosity of the ground tissue is normally low (2–7% by volume), insufficient to provide an adequate diffusion path to the roots (Armstrong 1978, 1979, 1982). In wetland species up to 60% of the plant body is pore space. These air spaces may result from loose packing of cortical parenchyma or the existence of a system of more organised airspaces. Not only do these air spaces reduce the resistance to gaseous diffusion, but they also result in the respiration rate per unit volume of tissue being lower than in tissue of lower porosity (Armstrong 1982).

There is variation in the ability of plants to form aerenchyma, depending on species and degree of soil anaerobiosis. Plants which cannot form aerenchyma appear to be intolerant of waterlogged habitats (Armstrong 1982). This is well seen in the behaviour of *H. portulacoides* which is usually restricted to parts of saltmarshes which have well aerated soils, at least during the growing season. *Halimione* does not develop aerenchyma and root porosity is low, not exceeding 5% (Armstrong *et al.* 1985). [Although *Halimione* is most frequently found in well drained sites, there are occasions where the species dominates extensive areas of marsh extending from creek levees into waterlogged intercreek basins – how these plants cope with the anaerobic soil remains to be explored.]

There is considerable variation between saltmarsh species in the degree of porosity of their root and rhizome systems (see Rozema, Luppes & Broekman 1985). Some species have very extensive aerenchyma, for example, *S. alterniflora*, *D. spicata* and *J. roemerianus* (Anderson 1974). Other species have much lower porosities, while some have very low porosities, similar to those of terrestrial mesophytes (for example, *E. pycnanthus* and *H. portulacoides* – Armstrong *et al.* 1985; Rozema *et al.* 1985).

Studies of the root anatomy of *Su. maritima* by Hajibagheri, Yeo & Flowers (1985) showed that the cortex contained larger intercellular air spaces in plants grown under saline conditions than in plants not exposed to salt. While this response was to salt, in a saltmarsh where plants are exposed both to anaerobiosis and salinity such an increase in airspace development, although slight in absolute terms, may be advantageous.

Species in the upper-saltmarsh may have most of their shoots above water, even on the highest tides. Provided there is an adequate aerenchyma connection between shoot and roots and rhizomes, these species may face no special hazard from tidal flooding. For species at mid- and low-marsh levels, tidal flooding results in total immersion for periods of from a few minutes to several hours. Even in species with extensive aerenchyma, the ability of the aerenchyma to provide a reservoir of oxygen (Williams & Barber 1961) is limited (Armstrong 1982; Crawford 1983). However, at least on some tides, the aerenchyma may be sufficient to maintain aerobic respiration in some species. The time of flooding may have considerable influence; if the lower part of the plant is flooded during the day, photosynthesis in the upper part of the plant may produce sufficient oxygen to maintain aerobic respiration. Flooding at night may result in transient anoxia. This effect has been reported in *S. alterniflora* by Gleason & Zieman (1981).

Whether the aerenchyma of saltmarsh plants is sufficient to maintain aerobic root respiration in the particular marsh subhabitats occupied by each species remains uncertain. For *S. alterniflora* the evidence of diffusion of oxygen from roots (Teal & Kanwisher 1966) suggests that the aerenchyma should supply sufficient oxygen to support root respiration. Nevertheless, Mendelssohn, McKee & Patrick (1981) argued that the aerenchyma only supplied sufficient oxygen in moderately, but not highly, reduced soils. Under continuously flooded conditions plants respired anaerobically. Burdick & Mendelssohn (1987) suggest that in *Spartina patens* aerenchyma could be regarded as only a partial adaptation to waterlogging and that anaerobic metabolism remained important in water-logged soils.

Curran, Cole & Allaway (1986) have shown that in the mangrove *Av. marina*, the aerenchyma of the root system contains sufficient oxygen to maintain aerobic respiration during tidal flooding. *Avicennia* possesses pneumatophores and other mangroves have other root modifications (knee roots, stilt roots) which may assist in root aeration. No comparable structures are found in any saltmarsh species. Possibly the low porosity of tree trunks and the large oxygen demands of an extensive root system have led to the selection of these features in mangroves, while in saltmarsh species the direct aerenchyma connection between roots and shoots renders such root modifications unnecessary.

While growth of plants in wetland soils seems to be based on the ability of species to restrict development of anaerobic conditions in root tissue through the provision of an internal pathway for oxygen diffusion rather than metabolic adaptation, there are occasions when tolerance of anaerobiosis may be essential for survival. In some species, die-back of aerial parts may limit the potential pathways for oxygen supply to roots and rhizomes outside the growing season. Under these conditions, maintenance metabolism of the underground portions of the plant may depend on anaerobic respiration. Plants in this situation exhibit forms of anaerobic metabolism in which the intermediate substrates and accumulated end products of respiration are relatively non-toxic. For example in *Iris pseudacorus* rhizomes there may be a pathway utilising shikimic acid (Boulter, Coult & Henshaw 1963). The low retention of ethanol in rhizomes of wetland species reported by Monk, Crawford & Brändle (1984) may be a factor in permitting survival under anaerobic conditions.

Another situation where anaerobic respiration may be essential is in germinating seeds. Many species have impermeable testas so, prior to rupture of the testa, the early stages of germination depend on anaerobic metabolism with an accumulation of malate (Crawford 1983). For a number of halophytes, the period of anaerobiosis during germination may be more extended. Clarke & Hannon (1970) reported the ability of a number of Australian saltmarsh species to germinate while submerged in up to 5 cm of water.

Drainage conditions may change during the life of a plant. Deterioration of drainage and ponding-up of water may result in the death of some species but others may be capable of responding by production of new roots with a higher proportion of aerenchyma. This has been demonstrated for *Sp. virginicus* by Donovan & Gallagher (1984).

It is tempting to see in the range of root porosities the basis for the ecological segregation of species into different zones and microhabitats in saltmarshes. The separation in distribution of *S. alterniflora* and *S. patens* can be explained in terms of the more effective development of aerenchyma in *S. alterniflora* (Gleason & Zieman 1981; Burdick & Mendelssohn 1987); more work is required on a wider range of species to see how far this explanation might be generalised.

Toxic substances

The reduced state of several metal ions is more toxic to plants than the higher valency form. Many of the products of microbial metabolism under anaerobic conditions are potentially toxic to higher plants. To what extent does adaptation to wetland soils involve avoidance or tolerance of these toxins.

Detoxification

Potentially, toxins may be nullified by oxidation which could be achieved in three ways: by diffusion of oxygen from roots (radial oxygen loss); by enzymatic oxidation, either on the surface or within the root; or by microbial oxidation close to the root (Armstrong 1982).

Roots growing in waterlogged soil frequently have deposits of ferric compounds along part of their length (Armstrong 1982). Such deposits may frequently be observed, for example, associated with the roots of *S. anglica*. Ferric deposits suggest that plants may have the ability to exclude iron from roots by oxidation of ferrous ions. Other inorganic toxins may also be oxidised.

Oxygen flux from roots can be measured by polarographic techniques (Armstrong 1979, 1982). The loss from wetland species is higher than from non-wetland species. Teal & Kanwisher (1966) argued that the dominance of *S. alterniflora* in reduced mud reflected possession of an aerenchyma system sufficient to meet both the needs of root respiration and the oxygen demand of the sediment. The ability of *S. alterniflora* to oxidise a variety of compounds, both in the rhizosphere and in root tissue, has been demonstrated by a number of workers (Mendelssohn & Postek 1982; Howarth 1979; Carlson & Forrest 1982). However, Morris & Dacey (1984) were unable to confirm Teal & Kanwisher's (1966) finding of net oxygen loss, and Mendelssohn, McKee & Patrick (1981) suggest that under permanently waterlogged conditions aerenchyma is unable to maintain aerobic conditions within the roots.

Armstrong (1982) has developed a model to predict the diameter of the oxygenated sheath around roots which indicates that the thickness of the sheath may be many times root diameter, provided the porosity of the root is sufficient to deliver adequate oxygen to the root surface.

However, it would appear that direct oxidation by diffusing oxygen accounts for only a proportion of the oxidising capacity of a root system. Enzyme-mediated oxidation may account for the major part of the oxidising activity (Armstrong 1982) in some freshwater wetland species. The extent of such activities in saltmarsh species remains to be explored.

It has been shown that the bacterium *Beggiatoa* may protect rice from sulphide toxicity (see Armstrong 1982). *Beggiatoa* carries out the intracellular oxidation of hydrogen sulphide to sulphur. To prevent damage from the hydrogen peroxide produced, the bacterium requires an external supply of catalase. Catalase may be released from rice roots, permitting an effective symbiosis between rice and *Beggiatoa* – whether similar systems operate in saltmarshes is unknown.

Tolerance

Notwithstanding the ability of many species to effect an oxidation of toxic substances in the rhizosphere, differential tolerance/susceptibility of species to various toxins may be a factor in determining their distribution on saltmarshes.

It has been shown that growth of a number of saltmarsh species under waterlogged conditions is accompanied by increased tissue iron and manganese content (Cooper 1982; Rozema & Blum 1977; Rozema, Buizer & Fabritius 1978) – in the case of *A. stolonifera* and *J. gerardi*, the greater increase was in the roots, although the shoot concentrations also increased. In *G. maritima*, shoot and root concentrations increased to the same extent (Rozema & Blum 1977; Rozema *et al.* 1978). However, Gorham & Gorham (1955) showed that the iron and manganese content of a number of halophytes was, in general, lower than that of inland species.

Singer & Havill (1985) confirmed the toxicity of manganous ions to a number of species. There was some differentiation between species in tolerance to low concentrations, but tolerance was not correlated with

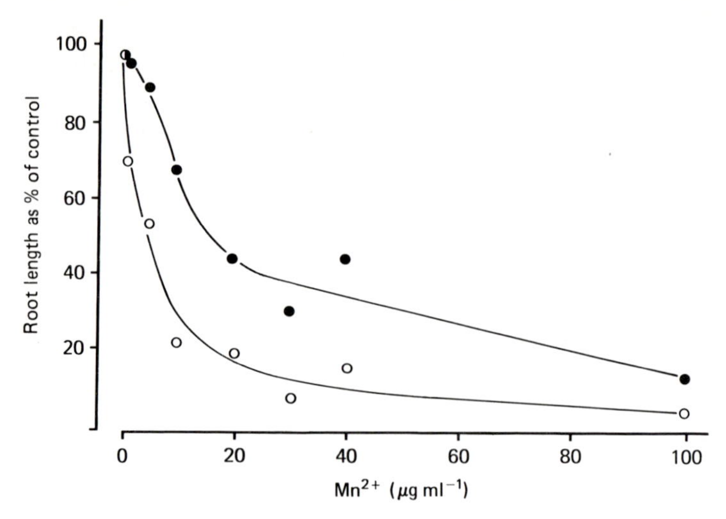

Fig. 4.15 Response of root growth in a saltmarsh (● — ●) and inland (○ – ○) ecotype of *Festuca rubra* to a range of Mn^{2+} concentrations. (Redrawn from Singer & Havill 1985.)

position in the marsh. However, there was evidence that saltmarsh populations of *F. rubra*, *Arm. maritima* and *Pl. maritima* were slightly more tolerant of manganese than inland populations (Fig. 4.15). If the experiments were carried out in the presence of sodium chloride, the sensitivity of the species to manganese was much less, due to reduction in manganese uptake by roots (Fig. 4.16). The physiological mechanism for antagonism of manganese uptake by sodium remains to be elucidated. Rozema, Luppes & Broekman (1985) suggest that the toxicity of manganese is a function of the iron concentration of the tissue, a Mn/Fe ratio in excess of 10–30 being required for manifestation of toxic symptoms. In the presence of sodium, uptake of manganese is insufficient to exceed the toxicity threshold.

Despite the relatively high concentrations of manganous ions in saltmarsh soils (Fig. 4.17), Singer & Havill (1985) concluded that differential manganese toxicity was not an important factor in saltmarsh ecology.

Adams (1963) suggested, on the basis of hydroponic studies, that *S. alterniflora* was restricted to waterlogged muds by a requirement for high levels of ferrous iron. However, this hypothesis, although supported by correlation of field distribution with ferrous ion concentration in the soil, has not been subject to detailed experimental investigation. Haines & Dunn (1976) did, however, demonstrate tolerance of high iron levels in hydroponic culture. Smart (1982) reviewed a number of studies and pointed out that iron deficiency had been reported only from plants grown in aerated

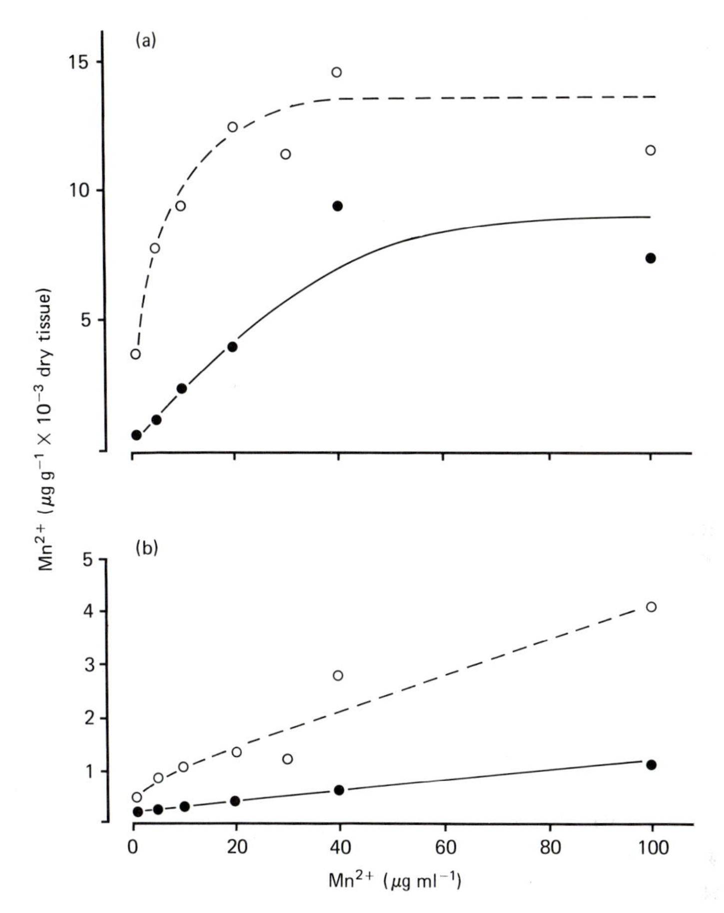

Fig. 4.16 The effect of NaCl on Mn^{2+} uptake by *Festuca rubra* roots (a) and shoots (b). Manganese content of tissue of plants grown in the presence ● — ●, and absence ○ - - - ○ of NaCl. (Redrawn from Singer & Havill 1985.)

culture solutions and had not been observed under anaerobic culture conditions.

Armstrong (1982) reviewed the effects of waterlogging and suggested that differential tolerance of manganese and iron occurred in species from a number of habitats. However, while there is differential tolerance amongst saltmarsh species (Rozema *et al.* 1985), the significance of this in controlling the distribution of species is not clear. In general, variation in iron availability appears, as with manganese availability, to be relatively unimportant in saltmarsh ecology (Cooper 1982; Rozema & Blum 1977; Rozema, Luppes & Broekman 1985).

High levels of sulphide can be found in saltmarsh soils. High concentra-

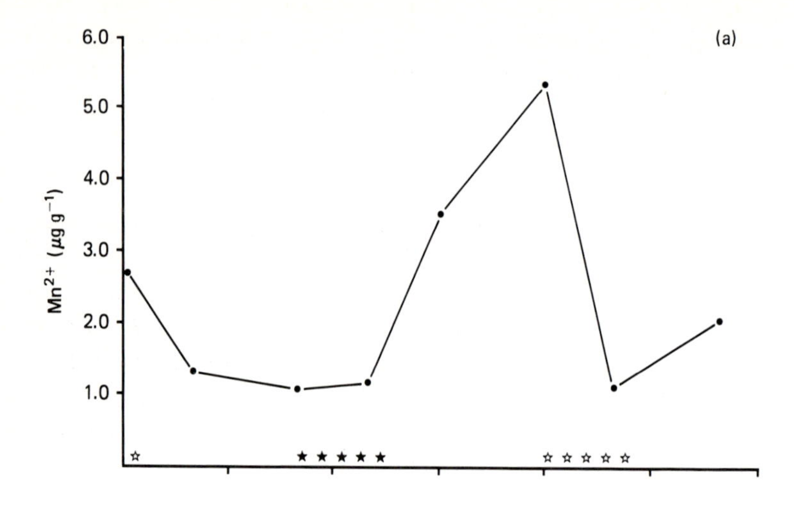

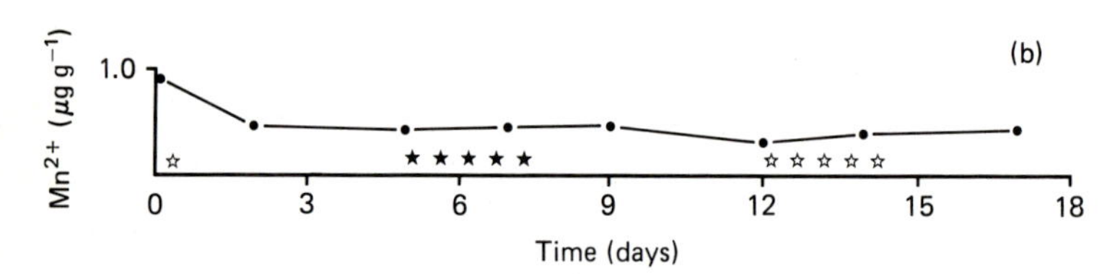

Fig. 4.17 Soil Mn^{2+} concentration over a fortnight tidal cycle, ★★ spring tides, ☆☆ neap tides. Note the increase in manganese concentration following spring tides in the upper-marsh (a) and the relative constancy in the lower-marsh (b). (Redrawn from Singer & Havill 1985.)

tions of dissolved sulphide in sediment have been correlated with reduced growth in *S. anglica* (Goodman & Williams 1961) and in *S. alterniflora* (DeLaune, Buresh & Patrick 1979; Mendelssohn, McKee & Patrick 1981; King *et al.* 1982). However, these studies do not provide direct evidence for sulphide toxicity at the concentrations prevailing in the field.

The possible importance of sulphide toxicity to saltmarsh plants has been reviewed by Ingold & Havill (1984) and Havill, Ingold & Pearson (1985).

The mechanisms of sulphide toxicity are not clear, but three possible effects in anaerobic soils can be considered (Havill *et al.* 1985):

Soluble sulphide may, in itself, be directly toxic;
Precipitation of sulphides may reduce the availability of sulphate, with consequent sulphur deficiency;

> Formation of insoluble sulphides may reduce the availability of Fe, Mn, Cu, Zn and other transition metals. These metals are essential co-factors for enzymes so a wide range of metabolic functions may be adversely affected.

Growth studies of halophytes in cultures demonstrated the toxicity of soluble sulphide to plant growth and differential tolerance between species (Ingold & Havill 1984; Havill *et al.* 1985). *S. europaea* and *A. tripolium* were tolerant of high sulphide levels; *Pu. maritima* moderately tolerant (tolerance increasing towards the end of the growing season); while *F. rubra* and *Atriplex patula* were intolerant. However, the suggestion (Ingold & Havill 1984) that *S. europaea* is the most sulphide-tolerant species of those studied is difficult to equate with Brereton's (1971) finding that *S. europaea* was favoured by well-drained (but saline) conditions.

Havill *et al.* (1985) and Ingold & Havill (1984) argue that sulphide tolerance is an important factor controlling species distribution. However, Ingold & Havill (1984) demonstrated a lack of correlation between sulphide concentration and redox potential and a heterogeneous distribution of sulphide. A number of authors have reported that *Pu. maritima* shows improved growth under waterlogged conditions (Cooper 1982; Brereton 1971; Gray & Scott 1977*b*; Rozema *et al.* 1985). It is possible, however, in view of Ingold & Havill's (1984) results, that *Puccinellia* does not occur at waterlogged sites with high sulphide content.

The suggestion that *S. alterniflora* is affected adversely by high sulphide concentrations and the demonstration of benefit from improved aeration under conditions of high nutrient supply (Linthurst & Seneca 1981) are, apparently, at variance with Adams' (1963) suggestion of a limitation of *S. alterniflora* to waterlogged soils with high available ferrous ion concentrations. However, the anomaly may be resolved if Ingold & Havill's (1984) finding of the local distribution of high sulphide concentrations is more widely applicable.

In addition to the detoxification of sulphides by oxidation, there is a possible internal mechanism for assimilation of excess sulphide. Havill *et al.* (1985) and Rozema *et al.* (1985) speculate that uptake of sulphide and its incorporation in compounds such as β-3-dimethylsulphonioproprionate (DMSP) may be an important mechanism in sulphide tolerance.

Van Diggelen *et al.* (1986) showed that in *S. anglica* there was an increase in DMSP when plants were grown in nutrient solutions augmented with high levels of sodium sulphide. No such response was demonstrated when sodium sulphate was substituted for sulphide. Sulphate levels in shoots increased as the external sulphide concentration was raised, indicating the ability of the plant to accomplish oxidation of reduced sulphur compounds. The sulphate is probably accumulated in vacuoles (Rennenberg 1984). Van Diggelen *et al.* (1986) suggest that, when ability to oxidise sulphide is saturated, excess sulphide is metabolised to DMSP, and that this increase in DMSP may be of adaptive significance in decreasing sulphide toxicity for plants growing on highly anaerobic sediment.

In some species, excess sulphur can be lost as volatile compounds, such as hydrogen sulphide (Rennenberg 1984). Breakdown of DMSP to dimethylsulphide gas could be a mechanism for loss of sulphur in *S. anglica*. However, van Diggelen *et al.* (1986) were unable to detect such losses from leaves, although appreciable dimethylsulphide production was recorded from germinating seeds.

Further research on sulphur metabolism of saltmarsh plants would be of interest, not only for understanding mechanisms for tolerance of anaerobic conditions, but also to quantify processes involved in the biogeochemical cycling of sulphur compounds.

There have been few studies on possible toxic effects of the organic products of anaerobic microbial activity on plants. While a number of these compounds are theoretically harmful, little is known about their effective concentration in the rhizosphere. Ethylene has been implicated in some of the anatomical and morphological changes occurring in flooded plants, but if it is a controlling factor, endogenous production may be more significant than exogenous levels.

The available evidence suggests that the main effects of flooding on plants arise directly from a decrease in oxygen and that tolerance is a function of the ability of species to provide an internal pathway for oxygen diffusion to root tissue. Toxicity effects do not appear to be major factors in the ecology of saltmarsh species, except in the case of sulphide.

Interactions between salinity and anaerobiosis have been poorly studied. Sites subjected to continuous waterlogging are likely to experience relatively constant salinity conditions, while well-drained, frequently aerobic soils, may experience large fluctuations in salinity.

Ahmad & Wainwright (1977) showed that a saltmarsh ecotype of *A. stolonifera* was more tolerant of waterlogging, as well as being more salt-tolerant, than populations from a seacliff and an inland site. Similar linking of tolerance to salinity and anaerobic conditions is also suggested by Singer & Havill's results (1985), showing greater manganese tolerance in saltmarsh populations of *F. rubra*, *Pl. maritima* and *Arm. maritima*.

Anaerobic rooting conditions might limit the uptake of ions. This could have important consequences for halophytes, in which accumulation of ions is an important factor in generating low tissue-water potentials. Karimi & Ungar (1986) showed that hydroponically grown *Atriplex triangularis* accumulated higher sodium and chloride concentrations in leaves under aerated conditions than in an unaerated treatment. On the other hand Barrett-Lennard (1986) showed that in glycophytic crop plants waterlogging led to increased salt uptake and aggravated the adverse effects of salinity. Possibly glycophytes and halophytes in general differ in their response to saline waterlogging.

Morris (1980, 1982, 1984) has investigated nitrogen uptake by *S. alterniflora*. Growth in the field appears to be nitrogen-limited even though soil nitrogen levels are relatively high. Morris (1980) suggested that nitrogen uptake could be inhibited by lack of oxygen, sulphide toxicity or increased salinity. These various factors would not vary independently. Under experimental conditions, Morris (1984) demonstrated that following the onset of anaerobic conditions in the rhizosphere of *S. alterniflora* or *S. patens*, uptake of ammonium declined considerably. Under aerobic conditions increasing salinity reduced ammonium uptake in both species. Under anaerobic conditions the same increase in salinity had no effect on uptake by *S. alterniflora* but caused a considerable further reduction in uptake in *S. patens*. *S. patens* is likely to be at considerable competitive disadvantage under conditions which are both anoxic and saline.

Mahall & Park (1976*a*,*b*,*c*), after an investigation of the zonation of species on a Californian saltmarsh suggested that salinity, rather than waterlogging, was the major determining factor. Parrondo, Gosselink & Hopkinson (1978) studied grasses in Louisiana and reached the opposite conclusion; flooding and a degree of drainage were more significant than salinity in affecting distribution. The general lack of correlation between salinity tolerance of species and position in saltmarsh zonation, and the demonstration of a wide range of root porosities in saltmarsh plants, suggest that differential tolerance to waterlogging deserves greater attention as an explanation for the local distribution of species on saltmarshes.

'Costs' of living in saltmarshes

The ability of plants to grow in the saltmarsh environment is dependent upon the possession of at least some of a number of physiological traits. These physiological features may demand utilisation of limited resources which might otherwise be devoted to growth or reproduction. To an extent, therefore, growth in saltmarshes may be more 'expensive' than growth in a non-saline, non-tidal habitat. However, the diversion of resources has not been an evolutionary option – if higher plants are to grow in the saltmarsh environment then certain physiological requirements must be met. It may be that certain combinations of physiological features are more 'expensive' than others, at least in some circumstances, so permitting selection of the most economical physiological traits for particular environmental regimes, but compared with terrestrial plants, growth and survival in any saltmarsh habitat will require diversion of resources from growth. The measure of these diverted resources might be regarded as the 'cost' of being a saltmarsh plant. Unfortunately, 'costs' are often viewed as being disadvantageous, whereas, in this case they are essential for occupation of the habitat. Some of the 'costs' may be incurred constitutively rather than in response to environmental triggers (patterns of allocations of resources between photosynthetic and non-photosynthetic tissue, ion accumulation at low salinities); in a non-saline environment such 'costs' would disadvantage plants incurring them and, presumably, underlie the low competitive ability of halophytes as against glycophytes at low salinities.

The costs of being a halophyte have been considered by a number of authors (Greenway 1973; Greenway, Munns & Wolfe 1983; Greenway & Munns 1983; Osmond 1980; Yeo 1983; Gale & Zeroni 1985; Raven 1985; McCree 1986). The assessment of the cost of possession of a particular physiology depends on the resource being considered. Accumulation of cytoplasmic osmotica has a cost which could be measured either in terms of the carbon diverted to the osmotica or in terms of the photons required for synthesis. Whatever osmoticum is utilised in a particular species, then a cost will be incurred in either currency. However, osmotica may also represent a cost in terms of nitrogen but if, for example, sorbitol were the major cytoplasmic osmotic solute then the nitrogen cost would be zero. Use of nitrogenous compounds also involves additional energy expenditure in nitrogen assimilation; this cost will vary according to the nitrogen source with ammonium being an appreciably 'cheaper' resource than nitrate (Raven 1985). The magnitude of the cost associated with the accumulation of organic solutes will be dependent on the intracellular location of the solutes, the costs if solutes are restricted to the cytoplasm being relatively low compared with those if the solutes are also in the vacuole (Greenway 1973; Greenway, Munns & Wolfe 1983; Greenway & Munns 1983; Yeo 1983; Raven 1985). Salt-tolerant glycophytes which exclude ions must incur considerable costs in osmotic adjustment.

The nitrogen costs for use of nitrogenous compounds for osmotic adjustment, even if these are restricted to the cytoplasm, are proportionally much higher than the cost in terms of photons (Raven 1985). Jefferies &

Rudmik (1984) suggest that under saline conditions reduction in photosynthesis due to increased stomatal resistance may result in carbon being the limiting resource for growth, so favouring use of nitrogenous compounds for osmotic adjustment. However, the growth response of halophytes to additional nitrogen (p. 267) would appear to run counter to this hypothesis by demonstrating nitrogen limitation. The preponderance of non-nitrogenous cytoplasmic osmotica amongst ion-excluding monocotyledons in brackish to moderately saline habitats may also reflect the relative costs and availability of carbon and nitrogen. The costs of ion transport in halophytes may be higher than those of plants in non-saline habitats. Although transport in the xylem is inexpensive, accumulation within cells may require a large expenditure of energy (Yeo 1983). If ions are accumulated in vacuoles and excluded from the cytoplasm, there will be large electrochemical gradients between cytoplasm and vacuole. Maintenance of these gradients may involve considerable expenditure of energy, although little is known about the properties of the tonoplast in halophytes (Dainty 1979). Salt excretion will also be associated with metabolic costs – little is known, however, about the relative energetic costs of different means of regulating tissue-salt levels.

Reduction in transpiration may be of value in reducing salt input into foliage, but increased stomatal resistance will also reduce photosynthesis; any saving in energy costs from adopting a 'passive' method of salt regulation will involve a trade-off of growth reduction due to lower carbon dioxide uptake.

Drawing up a balance sheet of the costs and benefits of various physiological syndromes exhibited by saltmarsh plants in different habitats within the marsh might provide a better understanding of the reasons for the distribution of species within saltmarshes than is presently possible.

Discussion

The majority of physiological studies of halophytes have either addressed specific physiological questions or have been aimed at detecting differences between the ends of the halophyte–glycophyte spectrum. Few studies have been designed to provide explanations for the distribution of species within a marsh or to account for the partitioning of resources underlying the co-existence of species in communities. While the numerous physiological studies discussed in this chapter shed light on various aspects of these latter questions, the data do not as yet provide for a coherent synthesis although, if approached with appropriate questions in mind they may yield testable hypotheses.

However, there is a firmer basis for discussion of the halophyte–glycophyte continuum. The evidence available suggests that it is possible to explain both the absence of many terrestrial species from saltmarshes and the failure of halophytes to invade non-saline habitats by reference to physiological traits. Lawton, Todd & Naidoo (1981) argued that 'Modern work reviewed by Flowers, Troke & Yeo (1977) has failed to correlate any one aspect of the physiology of halophytes in general with a reasonable hypothesis of salt tolerance'. This conclusion is unnecessarily pessimistic. It can be argued that the ability of halophytes (both halophytic and glycophytic elements) to grow under saline conditions is based on the regulation of concentration and distribution of ions so as to permit turgor maintenance while avoiding harmful effects of either specific ions or total ionic concentration. In the case of species from the glycophytic element, salt-tolerance appears, in most cases, to demand salt exclusion and the generation of low tissue-water potentials by organic solutes. Amongst halophytic species low water potentials are generated largely by ion accumulation, but the ions are probably sequestered in vacuoles and are balanced in the cytoplasm by organic solutes. While the exact mechanism by which salt tolerance is achieved varies between species (there is no single unique strategy of achieving tolerance) the various mechanisms do result in the majority of the enzymatic systems, in at least the aerial part of plants, being spatially separated from high salt concentrations. Contrary to Lawton *et al.* (1981), it seems reasonable to argue that halophytes segregate enzymes from salt and maintain turgor in the face of high external salinity. While there are several 'solutions' to both of these problems, possession of the physiological mechanism to achieve at least one solution to each is a *sine qua non* of being a halophyte.

The compartmentation of ions and organic solutes necessary for salt tolerance occurs at cellular, tissue and organ levels of organisation. As such, salt tolerance is a function of the whole plant rather than of isolated cells or tissues (Pitman 1984). While some aspects of salt tolerance are manifest in tissue cultures of certain species (Smith & McComb 1981; Warren & Gould 1982; Daines & Gould 1985; Tal 1984; Stavarek & Rains 1984), in other instances, cell cultures of species known to be extremely salt tolerant show very low tolerance (Smith & McComb 1981). While screening cell cultures for salt tolerance may provide a rapid assessment technique in some species, it would not be appropriate for an initial survey of a wide range of species whose properties were unknown.

The processes by which salt tolerance is achieved are not, as far as is known, unique to halophytes; halophytes differ quantitatively rather than qualitatively from other plants. Ion transport and regulation occur in all plants. What is different about halophytes is the control of these processes. None of at least the major classes of putative cytoplasmic osmotica are unique to halophytes. In terms of response to salt, higher plants exhibit a complete spectrum from species intolerant of low salinities to those tolerant of high salinities – along this spectrum there is variation in the efficiency of ionic regulation and compartmentation and the mechanisms by which it is achieved. The existence of this spectrum of responses and the quantitative nature of physiological differences suggest that it is unlikely that any higher plant is an obligate halophyte. However, Yeo & Flowers (1986*b*) point out that, while it is clear that the quantities of ions transported by halophytes and their distribution within tissues differ from those in non-halophytes, most investigations have reported the end result of transport and compartmentation processes rather than details of the processes themselves. The transport pathway provides many potential sites at which regulation of ion movement could occur (Gorham, Wyn Jones & McDonnell 1985; see Fig. 4.18). It is possible that detailed examination of the processes and pathways of ion transport in halophytes may yet reveal unique features of halophytes which will allow them to be characterised as qualitatively different from non-halophytes.

Although investigations of physiological traits have revealed a variety of mechanisms associated with salt tolerance, the ecological significance of this variety remains obscure. The distribution of species on the marsh and the composition of communities do not seem well correlated with particular physiological traits. If particular syndromes of traits have adaptive significance in restricted subhabitats, then it is not immediately obvious. Patterns of variation in physiology (as partly recognised in the physiotype concept) are more readily correlated with taxonomy than with environmental features. This lack of correlation with the environment may reflect inadequate definition of environmental conditions or that other physiological traits remain to be discovered. It is possible that the distribution of species is determined at critical stages in their life cycle and, provided the plant is salt tolerant at other stages, the actual mechanisms of achieving tolerance are not of significant adaptive value. Variation in germination responses of halophytes, however, also appears inadequate to explain the distribution of species (pp. 316–334). Few studies have been conducted on seedlings and it is possible that differences in the requirements for seedling establishment provide the mechanism for differentiating between the distribution of species.

Although some species show growth stimulation at relatively low

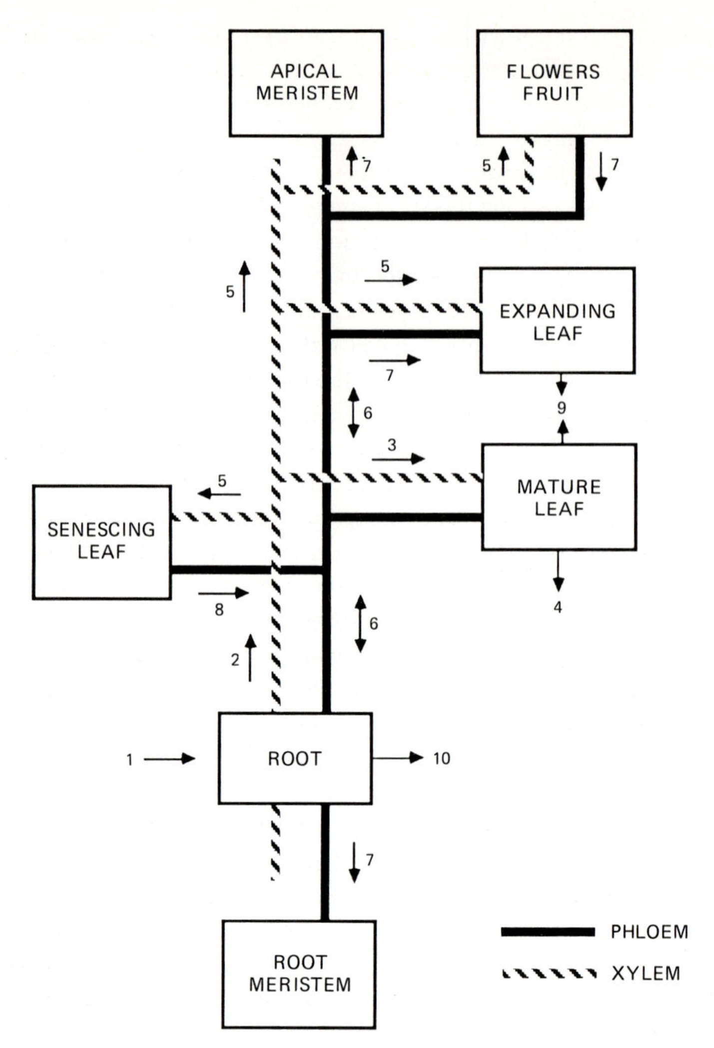

Fig. 4.18 Water and ion transport pathways through the whole plant: (1–4) main pathway of water movement through the plant; (5) xylem supply to other organs; (6) selective solute redistribution via phloem; (7) phloem supply to developing tissues and regions of low transpiration; (8) selective recovery of nutrients (for example K) from senescing leaves; (9) salt excretion from leaves; and (10) salt excretion from roots.
There are many opportunities along this pathway for both quantitative and qualitative control of ionic content. (Redrawn from Gorham, Wyn Jones & McDonnell 1985.)

salinities, the conditions under which these species are found in the field would normally be at salinities well above the growth optimum. Whether the stimulation shown by these species results in any advantage is unclear. Species not showing growth stimulation may be ecologically as versatile and, apparently, as successful as species with growth optima at low salinities. It is possible that growth stimulation is a consequence of the set of traits conferring salt tolerance in these species rather than being the trait which, in itself, has been selected. Growth stimulation appears to be primarily a turgor-mediated response. Uptake of ions at low salinities increases turgor pressure, driving cell wall expansion and increasing the area of photosynthetic tissue. The uptake of ions is, however, essential for maintenance of water relations at the higher salinities which will apply for the majority of the time. The decline in growth at the higher salinities may reflect many factors which vary between species (direct effects of high ionic concentrations on photosynthetic efficiency, diversion of resources to osmotica or non-assimilating tissues such as roots, or reduced stomatal conductance, possibly resulting from the effects of high apoplastic salt accumulation). Growth under optimal conditions is likely to be an extremely transient event in the life of most halophytes. It seems likely that natural selection has favoured survival traits rather than growth stimulation under conditions which might never occur in the field. Unfortunately, there has been less interest in physiological studies on any traits which might be more directly related to survival characteristics than to growth.

While species which show growth stimulation might be regarded as nutrient-deficient under non-saline conditions (Yeo & Flowers 1980, 1986*b*), these conditions do not prevail in the natural habitats of the species. However, if growth stimulation is regarded as the consequence of the uptake of ions which permit survival at higher salinities, then it is not obvious why other species which also accumulate ions do not show any stimulation.

A number of aspects of the physiological response of halophytes to their environment still require investigation. While under steady-state conditions, halophytes can generate tissue water potentials lower than those in the rooting medium, in the field, salinities are variable. How rapidly do plants respond to such changes? Uptake of ions and the synthesis of osmotica will not be instantaneous. Flowers, Troke & Yeo (1977) reported that *Su. maritima* leaves adjusted to the imposition of a seawater treatment over 48 h, which required an increase in total ion concentration of about 200 mol m^{-3} day^{-1} (see also Clipson 1987). In the case of synthesis of osmotica, several studies report a lag of 10–12 h before synthesis commences and that full adjustment takes place over two to three days (see, for example, Cavalieri & Huang 1979; Daines & Gould 1985; Fig. 4.19). Although many changes in the external medium could take place over a similar time scale (increasing concentrations as a result of evaporation, dilution following rainfall) much more rapid changes might take place after tidal flooding. Unfortunately, little is known about the rate of change in saltmarsh soils – the low rates of infiltration reported by Clarke & Hannon (1967) may mean that effective rates of change are relatively slow. However, assuming that rapid change does occur, then plants may experience several days of water stress after each such event. Dainty (1979) has suggested that lowering of tissue-water potential for short periods could be achieved by partial loss of turgor pressure; depending on the elastic properties of the cell wall, this could take place with only a slight reduction in cell volume. This possibility has not been subjected to detailed investigation, although the advent of techniques for direct measurement of turgor pressure (Clipson *et al.* 1985) renders the topic amenable to experimental study. Further study of rates of osmotic adjustment in species of the mid- and upper-marsh where the largest and most rapid changes in external salinity might be expected would be of interest; a species with more rapid adjustment might gain considerable advantage over other species.

It is possible also that the effects of a rapid increase in salinity are experienced only on the first occasion that it occurs. If the plant has already accumulated large quantities of ions, then it may be effectively buffered against further salinity fluctuations (see also discussion on osmoregulators and osmoconformers, pp. 270–272). The high rates of ion uptake, even at low salinities, exhibited by many halophytes (Collander 1941) may be of adaptive significance in that they provide a means of pre-empting the effects of increases in external salinity.

The majority of physiological studies on the effects of salinity have investigated changes in leaves. Nevertheless, it is the roots which are in direct contact with the saline medium. Although there are opportunities to regulate salt input into the leaves at various points along the transpiration stream (Gorham, Wyn Jones & McDonnell 1985), the root system must play a key role in the regulation of input (both specificity and quantity) and throughput. How are the activities of roots and shoots integrated? Flowers, Hajibagheri & Clipson (1986) argue that the evidence suggests that in halophytes, shoot growth is governed by input of ions from the roots and that shoot growth itself does not provide a feedback which regulates ion uptake by roots. Relatively little is known about the characteristics of ion uptake in halophytes (Yeo & Flowers 1986*b*; Pitman 1984) and it is as yet

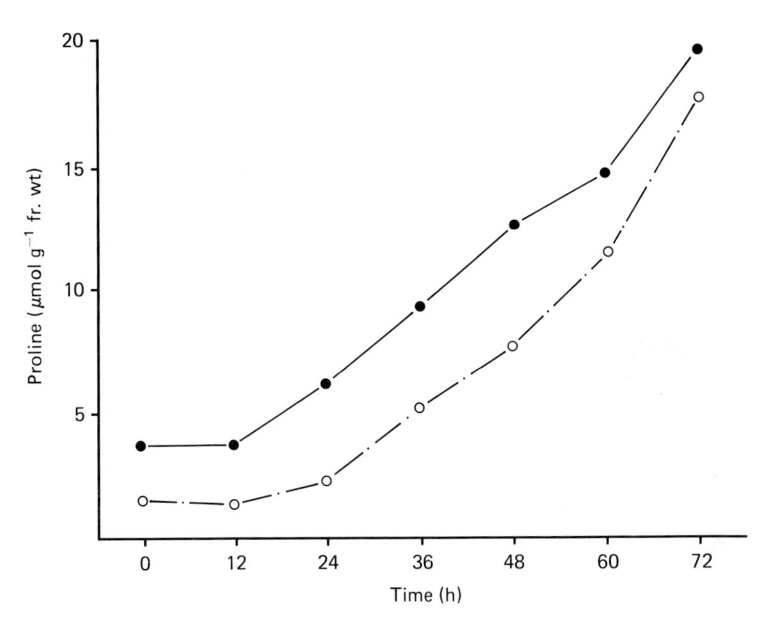

Fig. 4.19 Build up of proline in shoots of *Limonium carolinianum* after exposure to 500 mol m^{-3} NaCl at time 0. ○ – • – ○ control plants with no saline pre-treatment; ● —— ● pre-treatment with 200 mol m^{-3} NaCl for five days. (Redrawn from Cavalieri & Huang 1978.)

unclear whether the processes in halophytes are qualitatively different from those in non-halophytes. Erdei & Kuiper (1979) argue that in the genus *Plantago*, the key to salinity tolerance resides in efficient sodium chloride transport from root to shoots, which prevents ionic overloading of root cells, but in salt-tolerant glycophytes which exclude ions from the shoots, sodium and chloride concentrations in the roots may be much higher than in the shoot (Hannon & Barber 1972). If leaf metabolism cannot be maintained if salt accumulation occurs (for whatever proximal cause – specific ion effects, failures of compartmentation, or water stress consequent on apoplastic salt accumulation), how does root metabolism continue unimpaired? What causes the eventual breakdown of the exclusion process? Unfortunately, these topics have been poorly studied. The salt relationships of root cells are little known – are ions segregated between cytoplasm and vacuole in the same way as in leaves? Investigations of a number of species suggest that the quantities of putative cytoplasmic osmotica in roots are much lower than those in leaves (Stewart & Lee 1974; Storey, Ahmad & Wyn Jones 1977; Cavalieri 1983) and that the quantities change little with varying external concentrations (Cavalieri 1983; see Figs.

4.9, 4.10) except possibly at very high external salinities (Storey *et al.* 1977). It is possible that the osmoticum utilised in a particular species might vary between root and shoot (although within above-ground tissues the same compound is distributed throughout – Stumf & O'Leary 1985). Any osmoticum (or its biochemical precursors) utilised in roots would, ultimately, be derived from the shoot. Osmotic adjustment in the root would, therefore, be a drain on shoot reserves. If compartmentation of ions in roots were less complete than in shoots, then there would be a reduction in the demands for cytoplasmic osmotica. The consequence, however, would be higher cytoplasmic salt concentrations and studies of shoot enzymes (see p. 249) indicate that most higher-plant enzyme systems are salt sensitive. Some indication that root enzymes may be considerably less sensitive to salt was provided by Cavalieri & Huang (1977). However, if plants have the ability to produce more tolerant enzymes in roots, it is difficult to see why the same salt-tolerant enzymes are not also found in shoots as this would reduce the need for cytoplasmic osmotica and release resources to be diverted to growth. A species with more salt-tolerant enzymes would thus gain competitive advantage. However, Cavalieri & Huang's (1977) results suggest that the different forms of the enzymes are segregated within the plant; the adaptive significance, if any, of this distribution remains obscure.

Further work on roots is required to determine whether the compartmentation of ions is less effective than in shoots and whether root enzymes are generally more salt tolerant than those in shoots. It is possible, however, that variation in physiological processes in roots may be the key to explaining the distribution of species within a saltmarsh – an explanation which for most species cannot be provided on the basis of shoot properties.

The other major factor in saltmarsh ecology which has been subject to physiological investigation is waterlogging. While the patterns of distribution of some species may be explained by their relative tolerance of anaerobic conditions (for example, the poor internal ventilation of *H. portulacoides* (Armstrong *et al.* 1985) and the resultant intolerance to anaerobic conditions explains the field restriction of this species to better drained substrates and habitats), the general picture is obscure. Distribution of species within sites appears only partly correlated with possession of features assumed to confer tolerance to anaerobiosis, although it must be stressed that few species have been investigated in detail. Germination responses of species to waterlogging also seem inadequate to explain distribution patterns (Rozema 1975, p. 324). The interactions between salinity and waterlogging have rarely been studied, but may reveal more than studies of responses to salinity or waterlogging singly.

The direct effects of tidal flooding have been considered rarely, but may serve to differentiate between species. Seedling root-growth rate is an important factor determining the lower limit of species distribution in the lower-marsh, but in the upper-marsh, where tidal flooding is infrequent, earlier development of resistance to tidal currents is likely to be of lesser importance.

There is clearly a wide range of combination of various traits concerned with both growth and survival which permit species to flourish in saltmarsh environments. No single trait confers overwhelming competitive advantage on any species. It is likely that in any one community, species will differ in the particular traits they possess. Most studies report on species from different communities (or sites) and given ecotypic and temporal variation in a species' physiological responses it would not be possible to assemble disparate accounts of individual species into an assessment of community organisation. However, one of the few studies of the comparative physiology of the major species of a single community does indicate that the component species do not show identical physiological responses to salinity (Schat & Scholten 1986). The four characteristic species of the Centaurio–Saginetum moniliformis (*Plantago coronopus*, *Samolus valerandi*, *Centaurium littorale* and *Sagina nodosa* var. *moniliformis*) show marked interspecific differences in growth, patterns of change in root/shoot ratio, ionic content, and osmotic adjustment in response to increased salinity (Schat & Scholten 1986). However, increased salinity prolonged the vegetative phase in each species. Schat & Scholten (1986) suggest that the distribution of the four species within the community reflects microtopographic variation (with consequent heterogeneity in a number of environmental factors) but that the persistence of the apparently more salt-sensitive species (*Centaurium* and *Sagina*) depended on their ability to endure short periods of very high salinities.

Much of patterning of species within a marsh is the outcome of competitive interactions between species rather than a reflection of absolute physiological control of species' distributions. While competitive ability can be studied experimentally it is much harder to investigate its underlying basis. Under a particular set of conditions what physiological characteristics contribute to competitive success?

The study of the ecophysiology of halophytes has made considerable advances. What is needed now is for the response of plants to variation in the total environment, rather than to variation in easily manipulated factors treated individually, to be investigated, and for responses to be interpreted in terms of long-term adaptation and fitness (rather than simply growth) of particular populations (rather than species as a whole being characterised by the behaviour of a few individuals).

Ernst (1985) has posed many questions which need consideration if an understanding of the control of species' distribution is to be achieved. Integration of the answers to these questions will be no easy task;

> 'will ecologists ever be able to comprehend the complexity of the processes which are governing the physiology of an individual, the dynamics and genetics of a population or the composition of vegetation? Or will they restrict their activities to easy measurable parameters without relevant ecological information . . .?' (Ernst 1985).

5

Plant life history studies

Despite long-term successional processes, plant communities on most saltmarshes appear to be stable over many years. The little information on the longevity of saltmarsh plants suggests that, at many sites, communities may be stable for periods longer than the lifespan of their individual components. Any understanding of the processes involved, both in the maintenance of communities and the transition between communities, will require knowledge of the regeneration niche (Grubb 1977, 1986) of saltmarsh species. The concept of the regeneration niche encompasses 'the requirements for effective seed set, characteristics of dispersal in space and time, and requirements for germination, establishment and onward growth that have to do not only with gap shape and size, but also with weather, pests and diseases.' (Grubb 1986).

At the present time, it would not be possible to provide a fully comprehensive account of the regeneration niche of any saltmarsh species, although data on *Salicornia* spp. (see pp.330–334) are sufficient to define several components of the niche. The regeneration niche may be unique to genotypes, and so is not necessarily constant across the whole range of a species. Given the great intraspecific genetic diversity within species (pp.107–131), life history characteristics are likely to be as much subject to variation as other traits. In the case of *A. tripolium*, within a single marsh there can be variation from long-lived perennials in the low marsh to short-lived perennials (or even annuals) in the upper-marsh (pp.112–114).

Pollination

The pollination biology of saltmarsh species has been little studied. A particular difficulty facing saltmarsh plants is of tidal flooding occurring during the flowering period. While taller-growing species may hold their inflorescences above the highest water-level, lower-growing plants and plants in the lower-marsh zones may be completely submerged. To what extent the timing of flowering is synchronised with periods free from flooding is unknown.

In a number of north European species, there is a sequence in the time of flowering from seaward to landward on the marsh. As Jefferies (1977*b*) pointed out, such a sequence may serve to provide genetic isolation between populations of species on different parts of the marsh. However, Marks & Truscott (1985) suggest that late flowering in the upper-marsh in northern Europe may limit subsequent seed set, as climatic conditions in early autumn may reduce the ability of species to divert resources to seed filling.

It has been generally assumed that many saltmarsh species are wind pollinated as this is the characteristic pollination mode in the major saltmarsh families Gramineae, Cyperaceae, Juncaceae and Chenopodiaceae, but the extent to which this is true remains to be tested, although a number of reports have shown that pollen collection by insects from assumed anemophilous saltmarsh species is widespread. Pojar (1973) described pollen collection by bumble bees from a number of saltmarsh species in western Canada. Keighery (1979) has suggested that honey bees and syrphid flies may transfer pollen in *Su. australis* and that insect pollination may be of significance in at least some populations of this species. Adam, Fisher & Anderson (1987) reported pollen collection by honey bees from *S. quinqueflora*. Leereveld, Meeuse & Stelleman (1981) showed that *S. maritimus* flowers were regularly visited by syrphid flies and that, at some times, these flies may feed almost exclusively on *Scirpus* pollen. There is no conclusive evidence that the flies effect pollination.

There are a number of saltmarsh herbs with visually attractive flowers which, *a priori*, would be expected to be insect pollinated. Eisikowitch & Woodell (1975) investigated pollination of *Arm. maritima* at Scolt Head Island (North Norfolk), where populations occur on the saltmarsh, sand-dunes and shingle beach. Although some plants are in flower at most times of the year, the main flowering season is early summer. During this period, inflorescences may be submerged on spring tides. Eisikowitch & Woodell (1975) showed that stigmatic and pollen viability were maintained following submergence in seawater (and more concentrated solutions) in plants of saltmarsh origin, and this was in contrast to plants of inland origin which were adversely affected by seawater treatment. However, spraying the flowers with distilled water adversely affected pollen tube production (this treatment may mimic the effect of rainfall).

Although bees were regular visitors to sand-dune plants of *Armeria*, visits to saltmarsh plants were rare. Eisikowitch & Woodell (1975) suggest that this was due primarily to the fact that the weight of a bee was sufficient to bend the inflorescence stalk and deposit the visitor on the wet soil surface or its algal covering. The wetting was regarded by Eisikowitch & Woodell (1975) as the major factor discouraging bees. The stronger stems of sand-dune plants did not bend as much when the inflorescences were visited by bees. After immersion, the nectar in the flowers may be saline, which may be another factor discouraging bee visitors.

Arm. maritima has dimorphic flowers and is self-incompatible (Clapham, Tutin & Warburg 1962). If seed is to be set, then cross-pollination must be achieved. What the effective pollination agent is in saltmarsh populations has not been fully determined. Eisikowitch & Woodell (1975) found that most flowers from saltmarsh plants contained small members of the Thysanoptera, which were tolerant of submersion in seawater and which often had pollen on their bodies and these insects may act as pollinators.

L. vulgare, which is also self-incompatible, but whose main flowering season is slightly later than that of *Armeria* and which has, in general, a longer inflorescence stalk, was observed by Woodell (1978) to be visited by bees (*Bombus lucorum* and *B. lapidarius*).

The study of pollination mechanisms in saltmarsh plants is an open field. Plants of the upper-marsh, whose inflorescences are borne well above the highest tides, presumably pose few unusual challenges to potential pollinators. Casual observation suggests that these plants are visited by a wide range of insects. For plants of the lower-marsh, tidal immersion during the flowering period would seem to pose difficulties for both wind and insect pollination.

Little is known about the compatibility characteristics of saltmarsh plants, with the exception of the Plumbaginaceae (Baker 1966; Clapham *et al.* 1962). In the Plumbaginaceae, many of whose members are saltmarsh species, a heteromorphic incompatibility syndrome is well developed (Baker 1966). In dimorphic species (which include *Arm. maritima* and *L. vulgare*) there are two pollen types (A and B) and two stigma types (cob and papillate – see Fig. 5.1). Plants with cob stigmas have type A pollen, those with papillate stigmas have B type pollen. The cob stigmas are only pollinated by B pollen, papillate stigmas by A pollen (Baker 1966; Clapham *et al.* 1962). Not all members of the Plumbaginaceae are dimorphic; even in the genus *Limonium* some species are monomorphic and self-compatible (for example, *L. humile*), while others are monomorphic and apomictic (including *L. binervosum* agg.).

The incidence of autogamy and apomixis in saltmarsh floras is poorly documented. It has been suggested that *G. maritima* is self-pollinated (Clapham *et al.* 1962) (although Faegri & van der Pijl (1979) report that

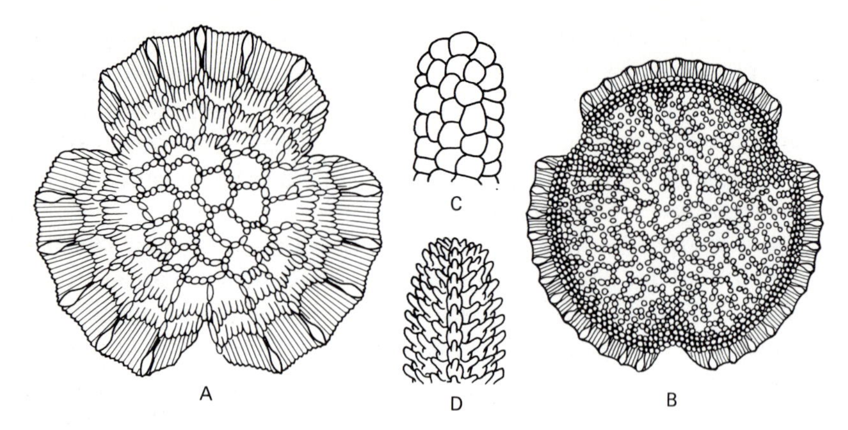

Fig. 5.1 Pollen and stigmas of dimorphic *Limonium*. A, B, the two pollen types; C, 'cob' stigma; D, 'papillate' stigma. A and B, × *c*.800 and C and D, × *c*.200. [After H. G. Baker, from Clapham, Tutin & Warburg 1962.]

Glaux may be one of the very few species pollinated by ants). *Glaux* has relatively inconspicuous flowers born in the leaf axils and, unlike most saltmarsh plants in which the flowers are borne on erect stems, the shoots are often prostrate. The flowers are thus more likely to be subject to complete submergence than those in most other saltmarsh species. *S. europaea* agg. exhibits cleistogamy (Ball & Tutin 1959; Dalby 1962; Ball & Brown 1970) which ensures the maintenance of a genetic constitution adapted to habitat conditions and avoids the uncertainties of outbreeding. Silander (1985) has suggested that patterns of clonal growth may influence the spatial pattern of genetic heterogeneity and breeding systems, but these hypotheses remain to be investigated in saltmarsh species.

The majority of saltmarsh species studied have perfect (hermaphroditic) flowers. Eleuterius (1974, 1978, 1984) has demonstrated that *J. roemerianus* is gynodioecious (plants having either perfect or female flowers), the only report of this condition amongst rushes. The grass *D. spicata* is dioecious, and at both inland sites (Freeman, Klikoff & Harper 1976) and coastal saltmarshes (Bertness, Wise & Ellison 1987) the two sexes are spatially separated. Freeman *et al.* (1976) showed that female plants were more abundant in sites with low salinity. Bertness *et al.* (1987) found male plants at lower elevation than females which may reflect a similar underlying response.

Vivipary

Viviparous seedlings occur in a number of mangrove species and have long attracted attention (Schimper 1903). In these species, normal fertilisation is followed by embryo development but there is no dormant period and the embryo continues to grow into the seedling while still attached to the mother plant. Although there has been considerable speculation on the adaptive significance of viviparity, not all mangroves show this mode of reproduction, and it is not possible to make dogmatic assertions as to the advantages (Saenger 1982; Tomlinson 1986).

True vivipary has not been reported for any saltmarsh species. In some species, in wet seasons, germination may occur in the infructescence, but this is not a normal pattern of development. Inflorescence proliferation (formation of plantlets in place of spikelets) has been reported from a number of saltmarsh grasses. This may be a response to particular climatic conditions or a consequence of genetic abnormalities (Smith-White 1984); it occurs rarely and cannot be regarded as characteristic of either particular species or particular regions.

Dispersal

Range extension of individual species and the establishment of new saltmarshes on habitats created by human activity or natural geomorphological processes demonstrates the successful dispersal of saltmarsh species.

There is little information on the effective dispersal agencies in most cases. Seeds and fruits of halophytes may be carried internally or externally by birds (de Vlaming & Proctor 1968; Proctor 1968; Olney 1963; Siira 1970), which may thus provide a potential vector for long-distance dispersal.

The seeds of most saltmarsh plants retain viability during submersion in seawater and so dispersal by flotation, or by being moved by bottom currents, is feasible. Long-distance waterborne dispersal is likely to be effective only rarely but may, nevertheless, be a significant factor over geological time periods. It may be the most important mechanism permitting colonisation of new sites within an estuarine system. Many, perhaps most, of the propagules dispersed by water are likely to be deposited in environments which, even if they support germination, do not permit establishment. [Propagules may arrive either in a non-saltmarsh environment or be carried across the marsh to the driftline – drift litter may contain large numbers of viable seeds (Boorman 1968; Bakker, Dijkstra & Russchen 1985) – equally, saltmarshes may well be sinks for propagules of non-saltmarsh species.]

The role of vegetative propagules in dispersal and establishment remains to be quantified.

Large patches of marsh surface may be eroded and be transported either within a single marsh or between systems. At high latitudes the major agency is probably ice scour (Roberts & Robertson 1986; Richard 1978).

Elsewhere, erosion is a consequence of wave and current activity. Some of these eroded blocks may be deposited in sites where new marsh growth is possible, but how often, if ever, they initiate marsh establishment remains to be discovered.

On marshes grazed heavily by sheep, tufts of grass pulled up by stock may be moved by the tide. In northern Europe, a narrow band of tufts of *Pu. maritima* frequently marks the limit of the most recent tide. Many of these tufts can be established in cultivation – in the field, establishment would require protection from tidal scour and some agency (possibly trampling) for pushing the tuft into the sediment. While this may be an infrequent occurrence, it could be a mechanism promoting small-scale genetic diversity in species spreading extensively by vegetative growth.

Seedlings (at the cotyledon stage) of annual *Salicornia* spp. may occasionally be observed floating on a rising tide in some abundance. Most of these seedlings are probably stranded in mature closed vegetation and perish. However, given the high proportion of seed deposited close to the parent plant (Watkinson & Davy 1985), dispersal of seedlings may provide a mechanism for the colonisation of new sites. Ellison (1987*a*) demonstrated that the hairs on the surface of *Salicornia* seeds could 'latch onto the passing wrack in the water much as Velcro hooks bind to fabric'. As deposits of drift litter would create openings in perennial vegetation, this might ensure that some *Salicornia* seeds are transported to colonisable gaps.

The seed bank in saltmarsh soils

Seeds become incorporated into the soil in most vegetation types. Depending on various characteristics of the seed, viability may be retained for many years. The proportion of seeds which will ever germinate in the soil is probably small unless some major disturbance occurs which creates conditions favourable for germination. The presence of viable seed in the soil may confer resilience on communities in the face of disturbances which cause clearance of existing vegetation.

The soil seed bank is difficult to study. Seeds can be extracted and then identified or germination can be promoted and the seedlings identified (Numata 1984). Extraction techniques are appropriate for medium- to large-sized seeds, but not for very small seeds (such as those of *Juncus* spp.). In addition, use of chemicals to disaggregate soil and separate soil from seeds may affect the viability of seed extracted so that estimates of the size of the viable seed bank may be difficult. Germination tests can reveal the viable seeds, but the germination requirements of each species must be met, so there will be a tendency for results to be biased in favour of the easily germinable species. Tests may need to be repeated over long periods in order to detect seeds with innate dormancy, and seedlings may need to be grown on until they can be identified.

With the exception of studies on the seed bank in *Salicornia* communities (p. 332), there have been few studies on coastal saltmarsh seed banks (Milton 1939; Hopkins & Parker 1984; Jerling 1983; Bakker 1985; Ungar 1987*a*). Both Milton (1939), who studied a Welsh marsh, and Hopkins & Parker (1984), who investigated a marsh in California, reported both low densities of seeds and low diversity. However, in Hopkins & Parker's study there was a contrast between a relatively undisturbed site and marshes subject to various forms of disturbance which had much higher seed densities and a higher diversity (although most of the increase in diversity was due to annual species introduced to California). The seed densities from saltmarshes (Hopkins & Parker 1984) were much lower than those from freshwater tidal marsh (Parker & Leck 1985), and inland freshwater marsh (van der Valk & Davis 1978).

Knowledge of the composition of the seed bank compared to that of the vegetation is important if the potential of the buried seed to re-establish the vegetation cover is to be assessed. Milton (1939), in general, found a lower diversity in the seed bank than in the vegetation; although in the case of a *J. maritimus* stand, the buried seed showed greater diversity than the vegetation and contained several species not recorded from anywhere on the marsh. Hopkins & Parker (1984) found general concordance between the seed bank and the vegetation, except that *S. foliosa*, a major component of the vegetation, was not represented by buried seed. Jerling (1983) emphasised the spatial heterogeneity in the distribution of seeds; there was some correlation between representation of species in the seed bank and in the vegetation, but the match was not very exact.

Bakker (1985) found that a large number of saltmarsh species were only represented in the seed bank if present in the canopy, suggesting that dispersal was very limited. Seeds of other species, including some from terrestrial habitats, could be found distributed throughout the marsh regardless of their presence or absence in the community.

The poor concordance between vegetation and seed bank need not indicate a poor ability of the vegetation to recover from disturbance. Disturbance of existing vegetation sufficient to provide habitats for germination may well create a set of environmental conditions more similar to those of lower zones in the marsh and not those immediately favouring species in the pre-existing vegetation.

The longevity of seeds in saltmarsh soils is unknown, as is the importance of various agents in the mortality of seed.

Germination and seedling establishment

Although many perennial species of saltmarsh spread vegetatively and flower and set seed infrequently, establishment from seed is the main means by which new sites are colonised. Annual species clearly must develop from seed.

The germination behaviour of halophytes has been extensively reviewed by Ungar (1978, 1982, 1987*a*). In general, the germination response of both halophytes and glycophytes under saline conditions is similar, with a reduction in the percentage of seeds germinating and a delay in the onset of germination. While some halophytes can germinate in saline conditions, germination in most species is maximal in freshwater. The percentage of seeds germinating is negatively correlated with salinity, although the salinity at which germination ceases varies between species. For many saltmarsh species, germination ceases at salinities far lower than those prevailing in the field, although a few species can germinate in full-strength seawater (Ungar 1978, 1982; Woodell 1985). Species of *Salicornia* are particularly tolerant of high salinity and a number of species may show significant germination at salinities well in excess of seawater (Ungar 1974, 1978; Chapman 1974*b*).

A feature of halophytes, which Ungar (1978) suggests may distinguish them from glycophytes, is that although, in most cases, germination does not occur under saline conditions, the seeds of halophytes may survive prolonged immersion in seawater (and in even more saline conditions – Woodell 1985). On transfer to freshwater, high germination rates may occur and, in a number of cases, there is 'salt stimulation' – a higher germination percentage than in freshwater alone (Boorman 1968; Woodell 1985; see Fig. 5.2). For species without stimulation, germination after salt treatment may be similar to that of the freshwater treatment, or it may be reduced depending on species (Woodell 1985; see Fig. 5.2). In *L. vulgare*, Boorman (1968) showed that the extent of the salt stimulation of germination was increased for longer periods of immersion and that seeds survived at least 21 weeks' immersion in seawater.

These results suggest that, at least for halophytes, the effects of salinity on germination are mediated primarily through the water potential of the medium rather than via specific ion effects, which might be expected to be irreversible. [While immersion in seawater may destroy the viability of many glycophytes, there are other species in which viability is retained after prolonged periods in the sea – the classic example being the coconut, *Cocos nucifera*. Ungar's (1978) distinction between glycophytes and halophytes on the basis of behaviour following immersion may not be of general validity.]

The inhibition of germination by salinity indicates that in most species germination in the wild would require a reduction in the soil surface salinity, such as that brought about by rain. The high germination percentage of *Salicornia* spp., even at high salinities, can be seen to be of potential adaptive value in view of the environments in which these species occur. In the low-marsh, frequent tidal flooding is likely to limit the lowering of soil-surface salinities, except after exceptional rainstorms (Hill 1909); for an annual species, successful occupation of the low-marsh of necessity involves the ability to germinate successfully at high salinities.

Ungar (1978) suggested that, on transfer to freshwater after saline treatment, the germination of halophytes may be speeded up. Such an effect could be of adaptive significance in that low salinity periods are likely to be variable in extent, but frequently brief. While such an effect occurs in some species, it is certainly not general and is shown only in the data for a few of the European species studied by Woodell (1985).

Ungar (1978) suggested that the failure of most halophyte species to germinate under saline conditions could be of adaptive advantage in that it would prevent germination of floating seeds, which might not reach a stable environment soon enough to survive. However, seeds capable of germinating at high salinities may germinate while floating, and Ungar (1978) records this as occurring in *S. europaea*. What proportions of such seedlings subsequently become established are unknown, but most may become stranded in sites where successful establishment is unlikely.

The poor drainage characteristics of many saltmarshes result in shallow depressions being flooded for long periods. If the water in the depression is highly saline, germination in most species would not occur but, particularly in the upper-marsh, heavy rain may result in brackish, or even fresh, water accumulating. A number of species may germinate under water, and Rozema (1975) showed, in several *Juncus* species, germination was promoted by inundation. Clarke & Hannon (1970) demonstrated that several Australian saltmarsh species could germinate under several centimetres of water.

A wide range of other factors, particularly temperature and light, will influence the germination of halophytes as they do for non-halophytes and may determine the habitat range which can be occupied by a particular species. Temperature exerts considerable control on germination, both on the rate and the final germination percentage (see Fig. 5.3). There may be an interaction between the effects of salinity and temperature; above a certain

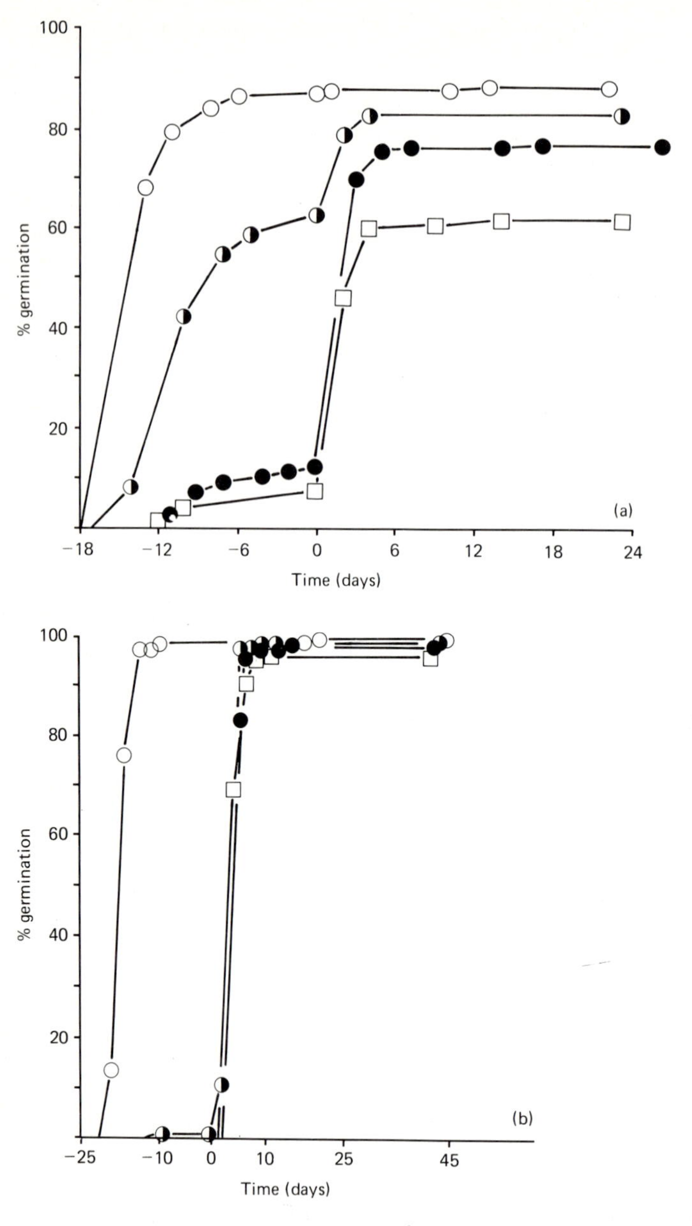

Fig. 5.2 Germination in relation to salinity. Seeds were placed on filter paper in Petri dishes and watered with solutions of various salinities. Dishes were weighed and salinities were maintained during the experiment by adding distilled water to return weight to the original value. The temperature for the duration of the experiments was 20 ± 3° C. After 18–25 days the seeds were transferred to new filter papers in new dishes and moistened with distilled water. ○ —— ○ freshwater; ◑ —— ◑ half strength seawater; ● —— ● full strength seawater; □ —— □ one and a half strength seawater. Day 0 is the day of transfer to distilled water. (a) *Salsola kali*, a strandline species.

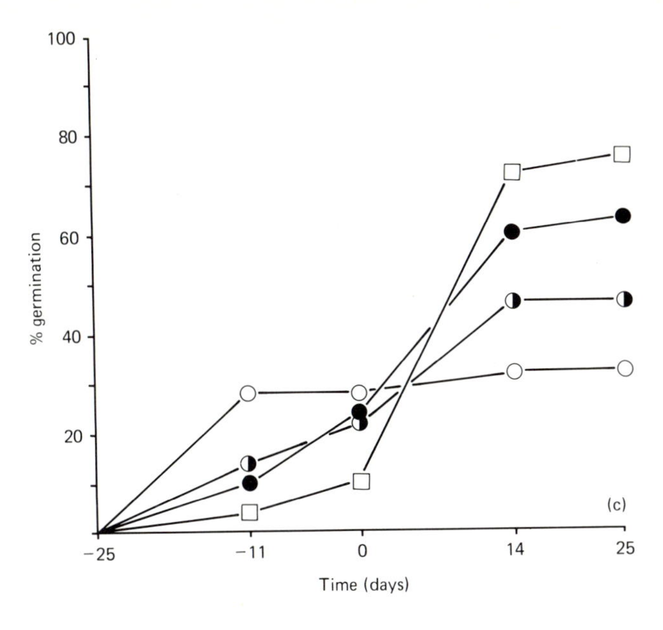

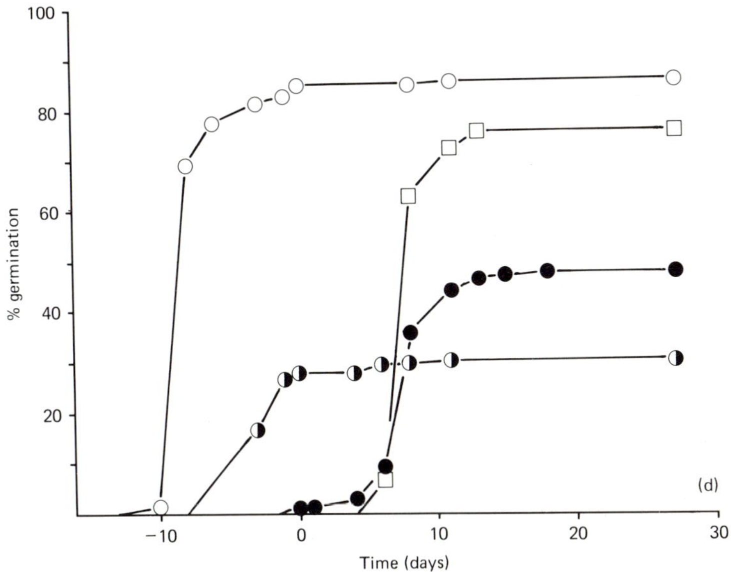

Germination is inversely correlated with salinity: after transfer to distilled water the final percentage germination is also inversely correlated with salinity.
(b) *Rumex crispus*. No germination in saline conditions; final germination in distilled water not impaired by prior saline treatment.
(c) *Limonium vulgare*. Low germination in salinity treatments but 'salt stimulation' exhibited following transfer to distilled water.
(d) *Juncus maritimus*. Optimal germination in distilled water, but, after transfer of treated seeds to distilled water, germination is correlated with treatment salinity. (Redrawn from Woodell 1985.)

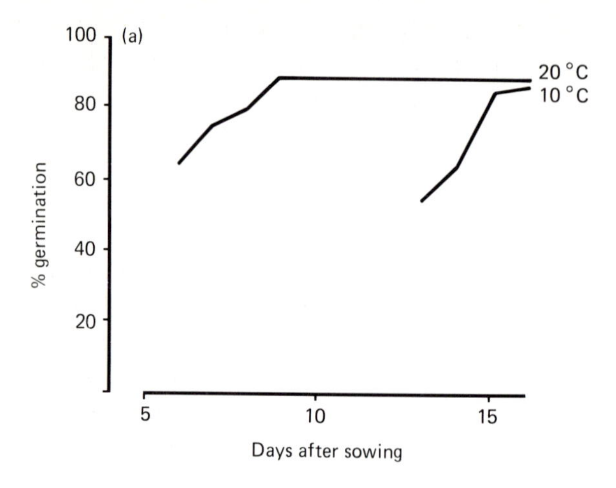

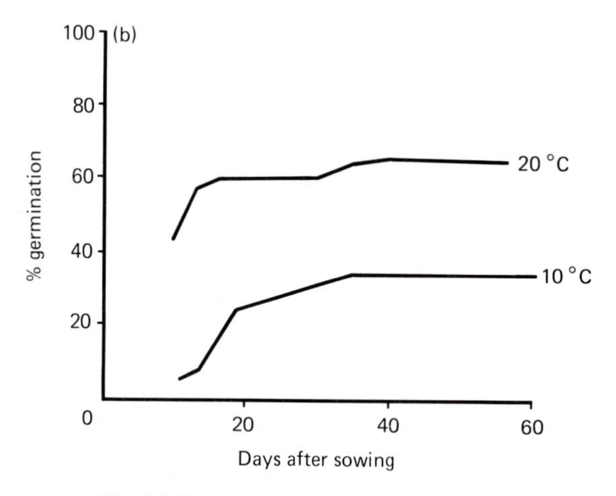

Fig. 5.3 Germination of (a) *Samolus valerandi* and (b) *Oenanthe lachenalii*. Seeds sown on moist filter paper in Petri dishes (4 replicates, each of 100 seeds for *Samolus*, 25 fruits for *Oenanthe*). Petri dishes maintained in constant temperature rooms with a 12-h day/night cycle. In *Samolus* germination was rapid and almost complete but the onset of germination was delayed at the lower temperature. In *Oenanthe* the onset of germination was not affected by temperature but the percentage germination was lower at the lower temperature. Germination continued until the experiment was terminated at 180 days (reaching 70% at 20° C and 40% at 10° C). (Adam, unpublished data.)

temperature (varying between species) germination under saline conditions may be reduced with increasing temperature (Ungar 1978). Stratification, a period of low temperature experienced by imbibed seed, may promote subsequent germination as, for example, in both *S. alterniflora* (Seneca, 1974) and *S. anglica* (Hubbard 1970; Marks & Truscott 1985). [This response may be part of the reason for the failure of *S. anglica* to become established in tropical and subtropical regions – Ranwell 1967.]

For a number of glycophytes, germination is enhanced by fluctuating temperatures (Fenner 1985). Binet (1965) has recorded a similar effect in *G. maritima*, while Okusanya & Ungar (1983) demonstrated optimal germination in *Spergularia marina* under 5/15 °C alternating temperatures. *S. marina* is unusual, however, in that germination is enhanced by a preceding period of storage of imbibed seeds in darkness. Such a response to alternating temperatures would promote germination of seeds on, or very close to, the soil surface; at greater depths of burial, considerable damping of the diurnal variation in temperature would occur.

Light is required for the germination of some species and this would also promote germination on the soil surface. A light requirement is almost universal for the germination of very small seeds and is exhibited by saltmarsh *Juncus* species (Rozema 1975) and *S. marina* (Okusanya & Ungar 1983). As *Juncus* spp. are normally a feature of upper marshes rather than the pioneer zone, successful establishment from seed may require the creation of small open microsites in otherwise continuous swards; *S. marina* is characteristic of relatively open disturbed sites, often experiencing very large salinity fluctuations. Hubbard (1970) suggested that germination of *S. anglica* was promoted by darkness. This would ensure that germination of buried seed would occur, with the emerging seedling being better protected against being washed away by tidal currents than seeds germinating on the surface. Marks & Truscott (1985) were unable to confirm this finding.

A number of saltmarsh species show a synchronous burst of germination, as in *Samolus valerandi* (Schat & Scholten 1985; Adam, unpublished data), but in other species germination is prolonged over a long period (see Fig. 5.3). An extended germination period 'spreads the risk' for the newly germinated seedlings in a changing environment. Nevertheless, amongst the relatively few saltmarsh species studied, rapid full germination is not uncommon in laboratory studies. It would be of interest to monitor germination in the field and to assess the relative safeness of the germination sites compared with those of species with extended germination.

Several species of halophyte, particularly amongst the Chenopodiaceae, have dimorphic seeds, with the different seed types having different germination requirements (Ungar 1979, 1982, 1987*b*). This variation in germination behaviour may serve to extend the germination period.

Physiological requirements for germination may also vary between morphologically similar seeds as has been shown in *Spergularia marina* by Okusanya & Ungar (1983). *S. marina* has a long flowering period and seeds are set over a correspondingly long season. Optimal germination requirements vary for seeds produced at different times of the season; effectively populations produce physiologically dimorphic seeds, with one type germinating in autumn and the other in spring after overwintering.

Little is known about the fine-scale horizontal patterns of distribution of halophyte seeds and any effects this might have, on both germination and subsequent seedling establishment. Thompson (1986) has demonstrated significant clumping, at a fine-scale, of the seed bank in acid grassland and, in reviewing the literature, emphasised that most previous studies have been inadequate to detect such patterning. Linhart (1976) investigated density dependency of germination of seeds sown in clumps of different sizes (under experimental rather than field conditions) and suggested that a positive density-dependent response was characteristic of species of closed habitats. If Linhart's findings can be extrapolated to the field and be generalised across a wider range of habitats, then it might be predicted that many saltmarsh species would show a positive density-dependent germination response.

Waite & Hutchings (1978) showed that in *Pl. coronopus* germination and salinity tolerance were enhanced when seeds were sown in clumps. The degree of enhancement varied with substrate, but occurred on sand and soil as well as on filter paper, suggesting that it may be a significant phenomenon under field conditions. Seeds of *Plantago* are strongly mucilaginous, which may tend to limit dispersal and promote aggregation of seeds from the same inflorescence.

Although *Pl. coronopus* can occur in closed vegetation (as did the population providing the seed for Waite & Hutchings' (1978) study), it is also found in relatively open disturbed habitats, so does not provide a good test for Linhart's (1976) hypothesis in the saltmarsh context. More work is required on a wider range of species before the hypothesis can be assessed.

Germination behaviour as an explanation of field distribution

The degree of both salinity and flooding tolerance for germination varies between species. In some cases there is intra-specific variation in response which can be correlated with the maternal microhabitat (Bülow-Olsen 1983) but in general Rozema's (1975) conclusions that 'germination tolerance to salinity and flooding are not clear cut parameters to be used in distinguishing between halophytes and non-halophytes, or between hygrophytes and non-hygrophytes' and 'plant zonation can only partly be explained by the factors salinity and flooding regarding germination' are amply supported. It is possible that investigation of other factors affecting germination would indicate a closer relationship between the preferred habitat of mature plants and germination requirements, but it is also likely that the early stages of seedling growth are a particularly sensitive stage.

Although germination in many halophytes may be promoted by low salinities, such conditions are likely to be short-lived and the newly germinated seedlings may be exposed to high salinities. Little is known about the physiological capabilities of young halophyte seedlings, but it is possible that exposure to high salinities before they have developed physiological competence to adjust to such conditions may be a major cause of mortality.

On the other hand, Stumpf *et al.* (1986) suggest that seedling establishment in *Salicornia bigelovii* requires moderate to high salinity. They argue that the generation of sufficient turgor to facilitate cotyledon expansion requires ion uptake. At very low salinities, growth is impaired by low turgor. As developing seeds are supplied with ions via the phloem, they may have an ionic composition different from that of the rest of the plant and, in general, may also have an ionic content much lower than that in foliage (O'Leary 1984). This would suggest an early need for ion uptake so that the results reported by Stumpf *et al.* (1986) may apply to more species.

Many tidal incursions which do not completely cover adult plants will submerge tiny seedlings under several centimetres of water. While flooding tolerance of adult plants appears only partially correlated with their position in the zonation, that of the seedling may be more critical in determining survivable sites. Ability to withstand tidal currents may set limits on the proportion of the intertidal zone occupied by a species (see pp.283–285).

Grazing and plant reproduction

Grazing may have considerable impact on plant reproductive processes. Grazing mammals have been reported to consume inflorescences of several species selectively (Rowan 1913; Jefferies & Perkins 1977; Jerling & Andersson 1982; Jensen 1985*a*). On marshes subject to very heavy grazing pressure by domestic livestock, flowering of any species may be rare and, indeed, such grazing may impose selective pressure in favour of prostrate, vegetatively reproducing ecotypes. However, grazing animals may create microsites favourable for germination and seedling establishment (Jerling & Andersson 1982 – see pp.327–331) so that, in the long term grazing of some species may increase population densities.

Mature seed may be consumed in large quantities by birds, both wildfowl

(Ogilvie 1978; Joenje 1985) and passerines (Fuller 1982). The effects of such consumption on the maintenance of plant populations in saltmarshes is not known. On the other hand, it is probable that seeds of saltmarsh species may be an essential resource for the survival of some bird species. For example, Yugovic (1984) suggests that the survival of the rare Australian seed-eating Orange-bellied Parrot (*Neophema chrysogaster*) may be dependent on relatively limited areas of *Halosarcia halocnemoides* on Victorian saltmarshes.

On most saltmarshes, the major group of grazers affecting plant reproduction (by consuming either flowers or seeds) is likely to be insects. However, with the exception of the detailed study by Bertness, Wise & Ellison (1987), there has been little investigation of the effects of insects on seed set in saltmarshes. Bertness *et al.* (1987) monitored the effects of insects on reproduction of the four major perennial plant species on a saltmarsh on Rhode Island. Insects were responsible for major reductions in seed set and Bertness *et al.* (1987) suggest that selective pressure imposed by consumers may have influenced both phenology and reproductive resource allocation. The most important insect consumer on the marsh was the grasshopper *Conocephalus spartinae* which is a generalist feeder. Some loss in potential seed set was also caused by more host-specific species, including planthoppers and microlepidoteran and dipteran larva. *Conocephalus* numbers increased during the early summer to a peak in July. During the summer there is also a shift in abundance from high- to low-marsh. *J. gerardi* at the study site flowered very early in the season, with most plants setting seed by mid July. As a result, *J. gerardi* largely escapes predation by grasshoppers. Its capsules are predated by microlepidopteran larvae (*Coleophora* spp.) – microlepidopteran larvae attack many species of *Juncus* and both *J. gerardi* and *J. maritimus* on European marshes are often heavily infected.

S. alterniflora was the last species to commence flowering (Fig. 5.4) and, with most flowering occurring after the peak in grasshopper abundance, predation was again reduced.

S. patens and *D. spicata* had very low seed output and Bertness *et al.* (1987) suggest that 'heavy flower consumer pressure has selected for reduced sexual reproductive effect in these species'.

Bertness *et al.* (1987) conclude that: 'By limiting seed production, consumers may minimise the importance of seedlings in marsh plant dynamics, amplify the importance of adult clonal interactions in determining the success of marsh plants and the interspecific spatial patterns of marsh plant communities and reduce the genetic diversity of marsh plant populations.' It would be of interest to have comparative studies on a wide range of marsh types.

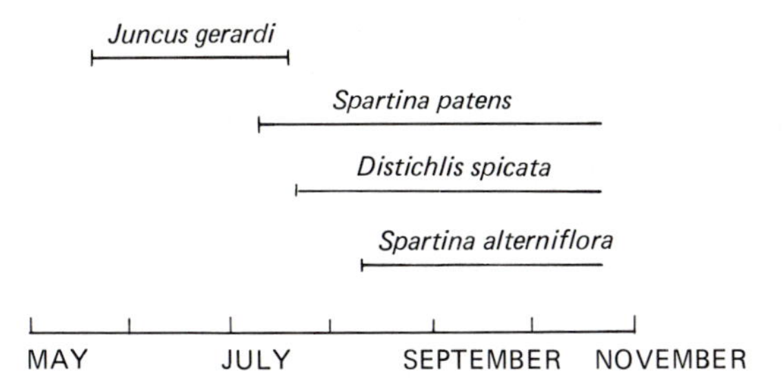

Fig. 5.4 Flowering periods of the species studied by Bertness *et al.* (1987). (Redrawn from Bertness *et al.*)

Potential flower and seed predators have been recorded from many marshes, but there is little evidence as to their quantitative significance. However, Jefferies & Rudmik (1984) report substantial reductions in seed output of *Salicornia* caused by a small beetle and a lepidopteran larva.

Even if seed set is successful, there may be predation of shed seed or fruit. Crabs may consume or damage large numbers of mangrove fruit or viviparous seedlings, but there is no evidence that they are significant consumers of the much smaller propagules of saltmarsh species.

Clonal growth

Many saltmarsh plants have the capacity to spread by vegetative means. Given the apparently low rate of successful establishment from seed in many species, it is possible that the long-term maintenance of many communities is dependent on survival of clonal populations.

Clonal growth offers the potential for physiological integration between ramets, providing a means of minimising possibly adverse effects of spatial heterogeneity in the environment (Pitelka & Ashmun, 1985 – see also p.225). There are virtually no data on the effectiveness and longevity of the connections between ramets in saltmarsh species. Lytle & Hull (1980*a*,*b*) have demonstrated that carbohydrate can be transferred between ramets of *S. alterniflora*. Lytle & Hull (1980*a*) showed that secondary tillers in *S. alterniflora* competed with developing flowers for carbohydrate. Vigorous vegetative growth may, therefore, lower the potential for successful sexual reproduction, although this is not a general pattern of resource allocation in clonal grasses (see review in Pitelka & Ashmun 1985).

In addition to transfer of carbohydrates (via the phloem), there is also evidence for movement of water (in the xylem) and mineral ions (possibly involving both xylem and phloem) between ramets (see Pitelka & Ashmun 1985). Such exchanges could be of particular value where there was fine-scale heterogeneity in soil-water potential, a situation likely to prevail in many saltmarshes. Integration of ramets within a clone may also be advantageous in permitting species to overcome short-term temporal changes in environmental conditions.

Many clonal plants are potentially very long lived (Cook 1985), although there are no data for saltmarsh species. Many saltmarsh species are widespread across a marsh and, although attaining maximal abundance in a particular zone, are found commonly in several zones. If the species can tolerate and keep pace with sedimentation, then it is possible that the same 'individual' may persist from one zone to another as the marsh develops. If this does occur, do the species concerned display high levels of phenotypic plasticity in order to accommodate the wide range of environmental conditions experienced?

As well as being potentially long lived, clonally spreading plants have the ability to extend over large areas (Cook 1985; Silander 1985). It is possible for a single clone to dominate a large area, although Silander (1985) suggests that such behaviour is the exception rather than the rule, and that in many swards of clonal species there may be considerable genetic heterogeneity and numerous clones growing intermingled. The only study of the genetic constitution of a saltmarsh sward dominant, that of *Pu. maritima* by Gray, Parsell & Scott (1979, see p.119), demonstrated small-scale heterogeneity, but it would be unwise to base generalisations on a single study. Silander (1985) has suggested that the spatial distribution of genotypes may influence the breeding behaviour of species, but this has not been investigated in saltmarshes.

There is considerable variation in the patterns of growth of saltmarsh species, and both ends of the spectrum between phalanx and guerilla patterns (Harper 1985; Silander 1985) are represented. There are few data on the rates of clonal growth and on the detailed morphological architecture of individual species. Such data are required before it will be possible to explore patterns of resource capture and competitive interactions in saltmarsh communities. Studies of clonal growth pattern in halophytes include those of Caldwell (1957) in *S. townsendii* (s.l.), Reidenbaugh (1983), Metcalfe, Ellison & Bertness (1986) in *S. alterniflora*, Tadmor, Koller & Rawitz (1958) in *J. maritimus* and Eleuterius & Caldwell (1981) in *J. roemerianus*. In this last study, three different populations were studied and, although the basic pattern of clonal growth was the same in all, shoot growth, kinetics and longevity were very different between populations (Eleuterius & Caldwell 1981). Given the genetic variation shown in many other characters (see pp.107–131), it is likely that parameters of clonal growth will vary similarly.

Metcalfe *et al.* (1986) showed that *S. alterniflora* seedlings did not compete successfully with adult plants; seedlings were common in mature swards, but did not survive a season's growth. Survivorship and seedling growth increased with size of patch in bare areas. The rapid growth of seedlings in cleared patches suggests that if disturbance (for example, by ice scour or deposition of rocks) creates large bare areas, then recolonisation by seed and rapid establishment of new clones within the clearing may pre-empt re-establishment by rhizome growth from the periphery. Over time, such a pattern of sward restoration would promote genetic diversity within the sward.

Case studies

Aspects of life histories of two widespread northern hemisphere species are discussed in this section: *Pl. maritima*, a perennial species, and annual *Salicornia*.

Plantago maritima

Pl. maritima is one of the most widespread of northern hemisphere saltmarsh species. Jerling (1984, 1985) and Jerling & Liljelund (1984) have studied the population dynamics of a population of *Pl. maritima* on a Swedish saltmarsh; this provides an example of the variation in population dynamics of a single species over a small area.

The marsh studied was not exposed to regular tidal flooding. The water level in the Baltic varies seasonally and duration of flooding on the marsh also varies markedly between years (Fig. 5.5).

The distribution of seedlings and adults of *Pl. maritima* was recorded along a transect running from a low area (zone 1) with open vegetation to a higher area with closed vegetation and accumulated litter (zone 4). Along this gradient, maximum density of adults and seedling *Pl. maritima* occurred towards the centre (Figs. 5.6, 5.7) but with considerable variation in absolute density from year to year. In the case of the adults, there was a considerable drop in abundance after 1979 – a year of much greater submergence than usual. The prolonged flooding appears to have caused considerable mortality and a landwards retreat of the density peak.

Pl. maritima produces seeds which are strongly mucilaginous when wet, the stickiness of the seed limiting its dispersability. Jerling & Liljelund (1984) suggested that most seed was deposited very close to the parent. None, or very little seed, is contributed to the seed bank in the soil (Jerling 1983), so that the seedling number in any one year is dependent on the previous year's flowering success.

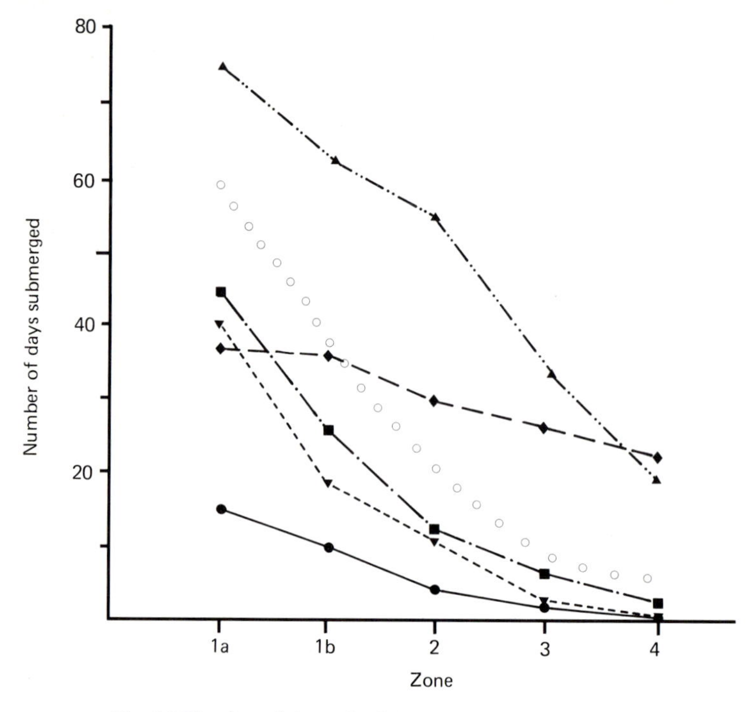

Fig. 5.5 Number of days of submergence in the different vegetation zones along the transect studied by Jerling between 1977 and 1981. ■ – · – ■ 1977; ◆ – – ◆ 1978; ▲ – · · – · · ▲ 1979; ● ——— ● 1980; ▼ – – – – ▼ 1981; ○ ○ ○ ○ Average. (Redrawn from Jerling 1984.)

Flowering varied, both between years and between zones (Fig. 5.8). Rosettes flowered more frequently in the lowest zone, but in the highest those plants flowering bore more flower spikes. Seed production (Fig. 5.9) was dependent not just on numbers of inflorescences but also on inflorescence size. Reproduction was achieved at the expense of growth, and flowering plants showed less vegetative production than non-flowering specimens (Jerling & Liljelund 1984 – see also Jefferies 1972).

Germination of *Pl. maritima* requires light (Arnold 1973). In dense vegetation, seeds may either not make contact with the substrate because of existing litter or be smothered by litter before germination. Seedling establishment and survivorship varied between zones and years. In the lowest zones, flooding was the major cause of mortality, although, because of the open nature of the vegetation, drought was also important when submergence was alleviated (Jerling 1981, 1984); in the higher zones, shading from the vegetation and build-up of litter prevented success.

In the lower zones, years of high water level caused death of existing

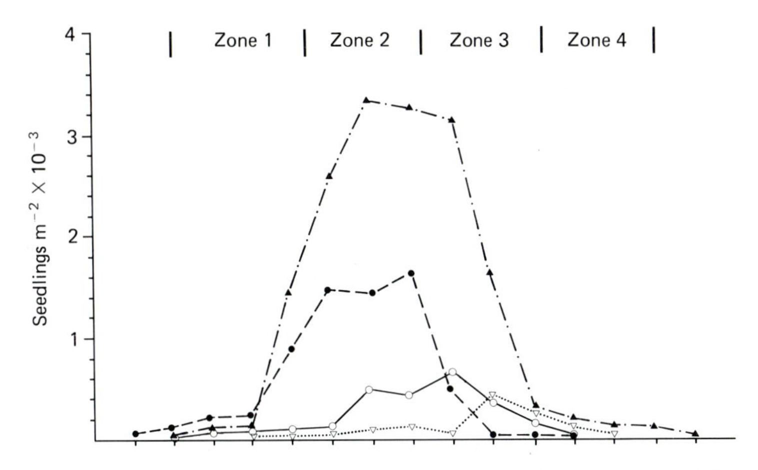

Fig. 5.6 Seedling numbers of *Plantago maritima* along the transect. (The intervals represent sampling points which are 3 m apart.) ●– – ● 1978; ▲ – · – ▲ 1979; ○ ——— ○ 1980; ▽· · · · · · ▽ 1981. (Redrawn from Jerling 1984.)

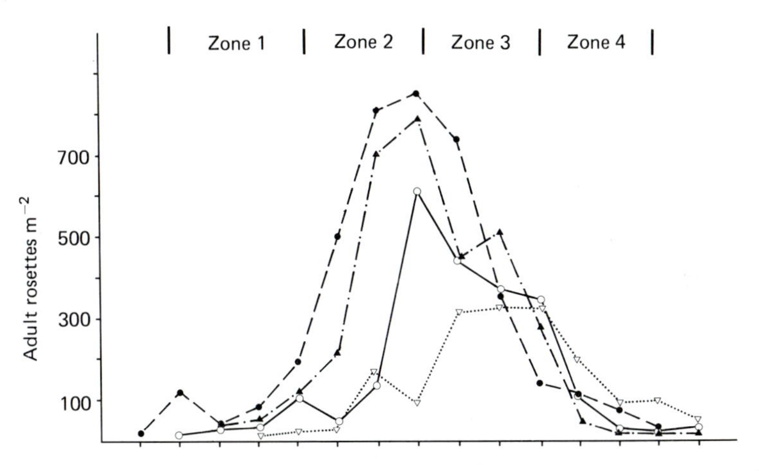

Fig. 5.7 Number of adult rosettes of *Pl. maritima* along the transect; symbols as in Fig. 5.6. (Redrawn from Jerling & Liljelund 1984.)

plants and reduced vigour of survivors. However, these adverse effects also created appropriate open sites for subsequent seedling establishment. In this zone, the risks of death by flooding are 'compensated for' by greater reproductive output.

Grazing has an important influence on the population size of *Pl. maritima* (Jerling & Andersson 1982). Grazing cattle remove many flowering spikes and substantially lower seed output. However, grazing

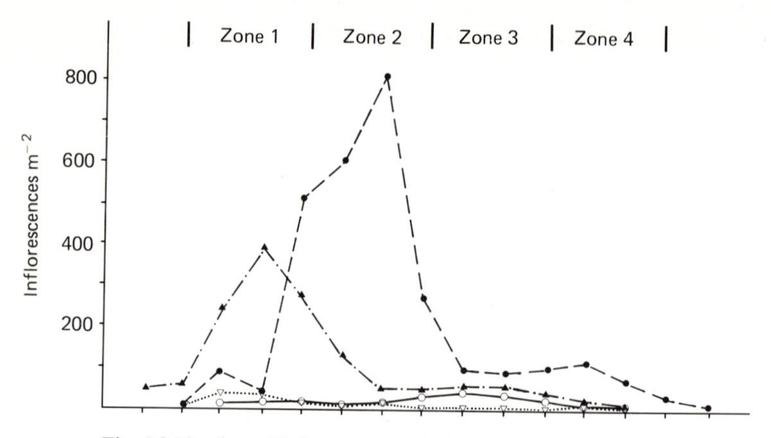

Fig. 5.8 Number of inflorescences of *Pl. maritima* along the transect; symbols as in Fig. 5.6. (Redrawn from Jerling & Liljelund 1984.)

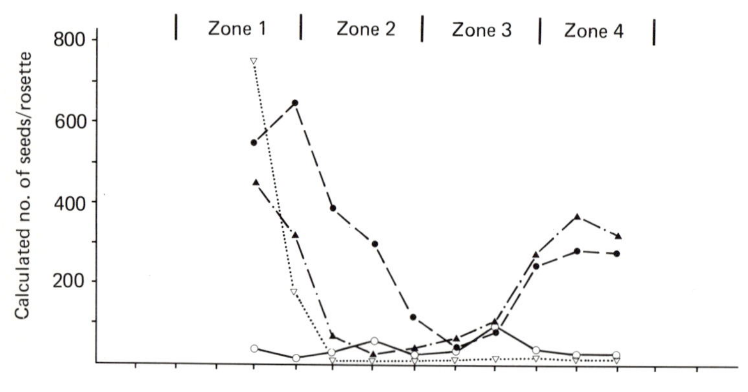

Fig. 5.9 Calculated number of seeds per rosette along the transect. Symbols as in Fig. 5.6. (Redrawn from Jerling & Liljelund 1984.)

keeps the vegetation low and permits light to reach seedlings so that survival is increased. Jerling & Andersson (1982) suggest that, despite the loss of reproductive output to cattle, a reduction in grazing would not be followed by an increase in the *Pl. maritima* population. The growth of taller vegetation as a result of lack of grazing would reduce, and possibly eliminate, successful establishment of seedlings.

A summary of the population characteristics is given in Table 5.1.

Salicornia spp.

Pl. maritima is a perennial and, while fluctuations in the adult members occur, failure of successful reproduction in any particular year does not presage the immediate demise of a population.

Table 5.1 Importance of characteristics of *Pl. maritima* in the different zones along the transect. (From Jerling & Liljelund 1984.)

	zone 1a	zone 1b	2	3	4
ROSETTES					
density (no. m^{-2})	low number, strong variation between years	low number, strong variation between years	intermediate to high number, variation between years	high number, almost constant	low number, some variation between years
mortality	mortality, mainly due to flooding, almost density-independent	mortality due to flooding, almost density-independent mortality	variable mortality due to flooding but also due to competition if periods between high waters are long	low mortality mostly due to competition, i.e. mainly density-dependent	mortality varies, mainly due to competition, i.e. density-dependent
FLOWERS & SEEDS					
flowering frequency	high	high	high	low	low
flowers m^{-2}	few - strong variation	few - strong variation	few to many depending on momentary number of rosettes - strong variation	many - less variation	few
reproductive effort	high	high	low to high - variable	low to intermediate	intermediate
factor limiting flowering	low vigour due to flooding	low vigour due to flooding	low vigour due to flooding plus maybe lack of nutrients	lack of light	lack of light
no. of seeds m^{-2}	low to intermediate	low to intermediate	high to intermediate	intermediate	intermediate to low
causes of seed losses between flower and seedling	heavy losses in connection with flooding - also grazing	heavy losses in connection with flooding - also grazing	losses mainly due to lack of light	losses mainly due to lack of light	heavy losses, mainly due to lack of light
SEEDLINGS					
no. m^{-2}	low	low to intermediate	intermediate to high	low to intermediate	low
establishment	bad to very good depending on water regime	bad due to lack of light and frequent flooding	bad to good depending on flooding frequency and light conditions at the moment	bad - limitations mainly due to lack of light	bad to intermediate - limitations mainly due to lack of light
ALLOCATION OF BIOMASS	–	–	high proportion to production (i.e. leaves and stem) and reproduction	high proportion to production (i.e. roots, leaves and stem) but low to reproduction	high proportion to production (i.e. roots, stem and leaves) but low to reproduction
GROWTH	–	–	constant throughout summer	constant throughout summer	decreasing in late summer
PHENOLOGY	–	–	early flowering	early flowering	later flowering

For many terrestrial annuals, the seeds produced in any one year may enter the soil seed bank and remain viable for many years and germinate only when environmental conditions permit (Fenner 1985). The soil seed bank provides a buffer against environmental fluctuations and allows a species to maintain its geographical distribution over time.

The majority of the saltmarsh flora is perennial, but in the northern hemisphere annuals occur in three differing marsh subhabitats: in the low-marsh zone where *Salicornia* spp. and *Suaeda* spp. may constitute the bulk of vegetation; on drift lines; and in openings in otherwise closed upper-marsh communities (such openings may be created by smothering by drift litter, ice rafting, grazing and trampling by domestic stock or by human activities such as turf cutting). Few of these annuals have been studied, but *Salicornia* spp. have been the subject of several investigations (Jefferies, Davy & Rudmik 1981; Jefferies, Jensen & Bazely 1984; Jensen & Jefferies 1984; Davy & Jefferies 1981; Davy & Smith 1985; Watkinson & Davy 1985; Beeftink 1985*a*; Ungar 1987*b*; Ellison 1987*a,b*).

In temperate coastal marshes there is no long-term accumulation of a seed bank in the soil; seed shed in autumn is almost entirely used up by the following summer (Jefferies, Davy & Rudmik 1981; Davy & Jefferies 1981; Beeftink 1985*a*; Jensen & Jefferies 1984).

Such behaviour seems, at first sight, inappropriate given the heterogeneity of saltmarsh environments. However, while there is considerable spatial heterogeneity within a zone, the pattern of temporal change is, in many respects, highly predictable (Jefferies 1977*b*; Jefferies, Davy & Rudmik 1979). The very changeability of the environment provides the necessary conditions for regeneration on an annual basis. A number of north temperate sand-dune annuals have similar reproductive behaviour which results in each generation being a discrete entity (Watkinson & Davy 1985; Watkinson & Harper 1978). The main difference between saltmarshes and sand-dunes as habitats for annuals is that the saltmarsh environment favours summer annuals, while most sand-dune annuals are winter annuals.

It is possible that in less temporally predictable environments *Salicornia* spp. may build up a substantial seed bank. Ungar (1982) suggests that populations in an inland saline habitat in Ohio do maintain viable seed in the soil and that this may buffer the local populations against environmental fluctuation. *Salicornia* spp. produce dimorphic seeds and it is the smaller seeds which constitute the seed bank (Ungar 1987*a,b*). At the northern limit of its distribution, the production of ripe seed is low compared with temperate latitudes and in a cold, wet summer fecundity is reduced further (Jefferies, Jensen & Bazeley 1983). Selective advantage would accrue to populations which could compensate for bad summers by drawing on a buried seed bank in the following year. Jefferies *et al.* (1983) suggested that a seed bank was present in the sediment of the Hudson Bay marsh they studied. Unfortunately, the viability of these buried seeds is unknown.

Although temperate saltmarshes provide predictable environments, chance events can, and do, occur which might prevent the annual seed bank being replenished. The overwintering seed bank of *Salicornia* contains seed produced locally, but also 'imported' seeds carried in by the tide. Although many seeds are deposited close to the parent plants (Watkinson & Davy 1985; Ellison 1987*a*), there is significant import and export of seeds

(Beeftink 1985*a*). If a population of *Salicornia* is destroyed in the summer prior to seed set, then import of seed may permit seedlings to establish on the site in the following year. The size of the catchment supplying seed to any particular site is rarely known but, presumably, depends largely on the geomorphological setting. Should the cause of mortality be widespread in effect (such as a major oil pollution incident), then the supply of seed for a wide area may be lost. In general, the dispersion of a proportion of the seed out of *Salicornia* stands by the tide will serve as a buffer against environmental unpredictability. For inland sites, such as those studied by Ungar (1982), while dispersion around the site is possible, movement between separate sites is more unlikely. In non-tidal environments, a long-term seed bank may be of greater selective advantage than in coastal habitats. Nevertheless, even in the coastal marsh, a small number of seeds in the soil do not germinate in the year following their production and so may constitute a small persistent seed bank (Beeftink 1985*a*). However, the viability and biological significance of this small number of seeds are unknown.

Different species of *Salicornia* occupy different habitats within the marsh. Jensen & Jefferies (1984), Davy & Smith (1985) and Beeftink (1985*a*) have shown that there are significant differences in life-history characteristics between populations in different habitats. Reciprocal transplantation experiments by Davy & Smith (1985) showed that the fitness of each population (assessed by the ability of individuals to leave descendants) was reduced considerably when grown outside the original habitat. The life-history characteristics shown by each population thus have adaptive value (Davy & Smith 1985).

Detailed studies by Beeftink (1985*a*) have allowed quantification of the life-cycle stages of two populations of *S. procumbens* agg. (one on the open mudflat habitat and one on closed saltmarsh) and one of *S. europaea* in the upper-marsh. The mortality at different stages in the life cycle was very different between the two *S. procumbens* populations (Table 5.2).

On the mudflats, there was much lower seed set per flower than in the marsh, possibly because the frequent tidal immersion affects pollination adversely. However, this poor seed set is compensated for by a high local seed rain and higher probability of successful incorporation (through accretion) into the soil on the mudflat. On the marsh, with much lower sedimentation, more seeds remain on the soil surface and are not incorporated.

Germination and establishment of the young seedling is less successful on the mudflats. Beeftink (1985*a*) suggests that low rates of root growth

Table 5.2. *Probabilities of one flower or seed of* Salicornia *producing an established seed or a mature plant.*

	Bergen op Zoom			North Norfolk	
	S. procumbens		*S. europaea*	*S. europaea*	
Period	zone 2	zone 3–5	zone 5	lower-marsh	upper-marsh
Flower stage to established seedling	0.004	0.002	0.013		
Mature seed to established seedling	0.026	0.004	0.021	0.040–0.086	0.131–0.401
Established seedling to mature plant	0.670	0.090	0.300	0.044–0.081	0.291–0.639
Flower stage to mature plant	0.003	0.0002	0.004		
Mature seed to mature plant	0.018	0.0003	0.006	0.001–0.028	0.035–0.083

S. procumbens and *S. europaea* at Bergen op Zoom in the Netherlands: zone 2, open pioneer *Spartina anglica* – *Salicornia* vegetation, zones 3–5, closed vegetation in low- to mid-marsh. From Beeftink (1985*a*). Data for north Norfolk from Jefferies, Davy & Rudmik (1981).

under cool spring conditions expose seedlings to the risk of uprooting by tidal currents (the importance of tidal currents in limiting *Salicornia* spp. was first emphasised by Wiehe 1935).

The overall probability of a *Salicornia* seed producing a mature plant is very low although the probabilities reported by Jefferies, Davy & Rudmik (1981) for *S. europaea* in Norfolk are much higher than those for either *S. procumbens* or *S. europaea* in the Netherlands (Beeftink 1985*a*) – see Table 5.2. Low as the probabilities are, they are not very different for those of a range of other annual species (Beeftink 1985*a*).

Density of *Salicornia* seedlings may be very high (10 000 plants m^{-2}) but no study has reported density-dependent self-thinning (Ungar 1987*b*; Ellison 1987*a,b*). Ellison (1987*b*) suggests that at high densities the morphology of plants, essentially unbranched cylinders, results in growth being only in height. Plant volume increases as a linear function of height, not exponentially; the area of ground covered by a plant remains constant and the conditions associated with self-thinning in other species are not attained.

Salicornia spp. are the only saltmarsh annuals for which detailed quantitative studies of life histories have been reported. Whether the characteristics of *Salicornia*, particularly the non-persistent seed bank of temperate populations, are general features of saltmarsh annuals remains to be investigated.

6

Saltmarshes as ecosystems

Other chapters in this book have been concerned with plant communities on saltmarshes and have stressed the variability in community composition both within and between marshes. While the physiological basis for the survival of saltmarsh plants in a saline, waterlogged environment is now established, the differentiation between species expressed in distribution patterns of species and communities is less clearly understood. Concentration on organisation at the species and community level is appropriate for testing hypotheses of a biogeographic or ecophysiological nature and may provide input towards an eventual synthesis of saltmarsh ecology but any such synthesis will also demand an understanding of processes operating at the ecosystem level. Knowledge of how saltmarshes function as ecosystems will be necessary for long-term management and for full understanding of the linkages between estuaries and saltmarshes. As Mann (1982) has argued 'we shall never make good predictions about ecosystems unless we learn to observe ecosystems, and make testable hypotheses about them'.

The study of saltmarshes as ecosystems is still in its early stages. Compared to terrestrial systems the relative species paucity of many saltmarshes may simplify ecosystem studies; on the other hand, the intertidal nature of the habitat presents many practical problems.

While there have been many studies of particular processes, there have been few investigations which have adopted an integrative approach to the whole system. Many of the generalisations about saltmarsh ecosystems are based not on complete studies but on assumptions developed to fill the gaps between studies on particular processes. Nixon (1980) analysed much of the widely accepted dogma about saltmarsh ecosystems and revealed the shaky foundations underlying the edifice of conventional wisdom. Premature extrapolation has hindered the development of an appropriate conceptual framework for understanding saltmarsh processes.

If ecosystem studies are to provide the basis for prediction, it is essential that they provide generalisations which can be extrapolated between sites.

Given the great variations between marshes in both the physical environment and the biota, it is likely that these generalisations will be about the nature of processes operating within the ecosystems rather than the outcome of those processes. Quantitative estimations of the magnitude (and even of net direction) of fluxes of energy and nutrients at any particular site are likely to require data collection at that site although knowledge of processes and pathways derived from other studies should enable these data to be recorded in an economical manner.

The most detailed ecosystem study of a saltmarsh is that on Sapelo Island reported in Pomeroy & Wiegert (1981). Earlier studies are discussed by Nixon (1980) and Long & Mason (1983), and an extensive review of estuarine studies is provided by Knox (1986). This chapter provides an introduction to only some of the topics which could be covered under the heading 'Saltmarshes as ecosystems'. The various processes discussed in more detail by Pomeroy & Weigert (1981) operate in all marshes but the quantitative details of energy and nutrient pathways are likely to be site-specific.

Primary production

Saltmarshes differ from most terrestrial ecosystems in that, at least potentially, *in situ* production by plants is not the only major carbon and energy source for heterotrophs. An alternative supply is provided by material washed into the marsh by the tide. In sites studied so far, primary production within the marsh is the major carbon source (Long & Mason 1983) but the balance between the two sources is likely to vary depending on local hydrological and geomorphological conditions.

It has been frequently asserted that saltmarshes are amongst the world's most productive ecosystems (Whittaker 1975; Howes, Dacey & Goehringer 1986). Some studies of saltmarshes have reported estimated productivities which are, in comparison with those from other communities, high, but various problems associated with productivity measurement make it difficult to accept the reliability of these estimates (Long & Woolhouse 1979; Long & Mason 1983).

Most productivity studies refer to *S. alterniflora* stands in eastern America (reviewed by Turner, 1976). Virtually all provide estimates of above-ground production and ignore below-ground components. Long & Mason (1983) and Groenendijk & Vink-Lievaert (1987) reviewed available estimates and showed that below-ground biomass often exceeded that of above-ground material. While the relationship between biomass and productivity is likely to vary between species, the limited data available indicate that below-ground productivity is high (Long & Mason 1983). Also, as no studies have attempted to measure material excreted or leached from roots and rhizomes (Long & Mason 1983) all estimates of below-ground production are likely to be serious underestimates.

The majority of estimates of above-ground productivity may also be underestimates, as they fail to account for all losses during the growing season (Long & Mason 1983).

Most studies of above-ground productivity involve measurements of very limited areas, the results then being extrapolated to the whole marsh. However, saltmarshes display considerable heterogeneity in physico-chemical factors even in the lower-marsh where regular tidal flooding might intuitively be thought of as creating uniform conditions. Even when this environmental variability is not reflected by changes in species composition it is likely to influence growth rates of individual species. Single estimates of productivity from a site are therefore a poor basis for generalisations about whole marshes. Relationships between productivity and tidal amplitude or site latitude (Turner 1976) are unproven; a large part of the range of values supporting these claimed relationships can be found between different parts of a single marsh. [*A priori* it is probable that there is a strong relationship between latitude and productivity as a result of the effect of climate on productivity.]

As well as varying spatially, productivity will also vary from year to year (Long & Mason 1983; Hussey & Long 1982) depending on climatic and other factors.

In considering the relationship between vegetation and other organisms, the best estimate of the availability of energy and material for circulation is provided by measures of net primary production (the balance between gross primary production and respiration). However, in order to understand the factors controlling the magnitude of net primary production, both gross primary production and respiration must be studied (Long & Woolhouse 1979; Long & Mason 1983). There are few data on these topics but such information as is available indicates that, at least for *Spartina*, respiration may account for up to 70% of gross primary production (Long & Mason 1983). Compared with terrestrial communities this is a high proportion, but may be explained by the maintenance of the large below-ground biomass and the respiratory demands of mechanisms for maintaining salt balance (Long & Mason 1983). Despite the apparently high net primary production, the relatively small difference between gross production and respiration suggests that any environmental change which results in only slight changes in the balance between assimilation and respiration could cause a large drop in net primary production (Long & Mason 1983). If the relationship between gross and net primary production demonstrated in *Spartina* is similar for other marsh types, then managers need to be aware that high net primary productivity may not provide a buffer against environmental change.

Algal productivity

Algae contribute considerably towards the species diversity of saltmarsh vegetation; to what extent do they contribute to marsh productivity? Most studies of marsh above-ground net primary productivity refer only to vascular plants, indeed most of the techniques used in these studies would be inappropriate to the investigation of algal productivity, although the larger marsh fucoids may be studied by similar methods.

Algae are found in a number of different habitats in the saltmarsh, as phytoplankton in creeks and pans, as epibenthos on creek sides and pan bottoms, on the sediment surface below vascular plants, and epiphytically on the stems of plants. Microalgae on mud may have the ability to migrate vertically from a depth of several millimetres in the sediment to the surface. This permits survival during sedimentation and may also occur during tidal flooding (Pomeroy *et al.* 1981). In addition to microalgae, mats of cyanobacteria may be extensive.

Epibenthic algae differ from vascular plants in several ways. The standing crop is almost always low compared with that of higher plants, but, because of a much faster turnover rate, differences in productivity may be much smaller. Unlike vascular plants, where the majority of the biomass may be below-ground, the whole algal biomass is photosynthetic. It is suggested from various studies that despite seasonal changes in species composition, the seasonal variation in algal production may be less than for vascular plants – in temperate regions algal production may be the major component of total marsh production during autumn and winter (Pomeroy *et al.* 1981; Zedler, Winfield & Mauriello 1978).

Techniques for measuring epibenthic productivity are discussed by Pomeroy *et al.* (1981). The majority of published studies are from eastern America and, after some allowance for spatial heterogeneity, suggest that algae may contribute 20–25% of total production. How representative this figure is of saltmarsh in general is unknown. Data from the Tijuana estuary in California (Zedler, Winfield & Mauriello 1978) show a much higher contribution from algae (40–50%), with, somewhat surprisingly, particularly high contributions at high elevations on the marsh.

Herbivory

The role of grazing by domestic livestock in modifying the species composition of saltmarsh vegetation is discussed on pp. 356–361. Such grazing must also represent the consumption of a significant proportion of marsh primary production but there do not appear to have been any studies which have quantified the impact of grazing practices on saltmarsh productivity and nutrient and energy cycles.

The role of herbivory in marshes not subject to direct human utilization has also been poorly studied and the few quantitative data available refer to *Spartina* marshes. Although there are some marshes in which vertebrates clearly play a major role, in most cases the dominant terrestrial herbivores are insects (including both sap suckers and leaf chewers) (Davies & Gray 1966; Pfeiffer & Weigert 1981). Studies on saltmarshes elsewhere indicate similar diversity and abundance of insects to that in the USA (Paviour-Smith 1956; Davis & Gray 1966; Dijkema 1984*b*) but whether consumption rates are also similar is unknown. Nevertheless, at least for *Spartina* marshes it seems probable that Teal's (1962) conclusion that only a small proportion of primary production is consumed *in situ* is valid.

The low incidence of herbivory may be related to the poor nutritional status of the vegetation. Caswell *et al.* (1973) have suggested that herbivory may be lower on C_4 than C_3 species as a consequence of the lower nutrient status (particularly of nitrogen) of C_4 species. Vince, Valiela & Teal (1981) showed that application of fertiliser to plots of *S. alternifolia* on a Massachusetts marsh resulted in an increase in herbivore standing crop with greater growth of individual animals and improved survival and fecundity. These effects were greatest in the treatment resulting in the highest foliage nitrogen level. As well as changes in biomass there were also, in low-marsh plots, some changes in relative abundance of individual species. It would be of interest to repeat this experiment in other parts of the world where saltmarshes are dominated by C_3 species to see whether this apparent limitation of herbivores by nutrient status is a general feature. It should be borne in mind that many saltmarsh plants have relatively high levels of low molecular weight soluble nitrogen compounds which are utilised to generate low cytoplasmic-water potentials and so might be expected to be relatively nutritious. Parsons & de la Cruz (1980) showed that conocephaline grasshoppers grazed a portion of leaves of *J. roemerianus* selectively about 10–15 cm below the tip. This segment of the leaf had the highest proline concentration. The authors suggested that this high intake of proline was important to the flight mechanisms of the grasshoppers. Insects are able to oxidise proline to glutamic acid which can be deaminated to α-ketoglutaric acid, a Krebs cycle acid. Heydemann (1981) suggested however, that there is a significant negative correlation between the presence of some aphid species and the plant proline level so that high foliar proline levels may not be advantageous to all insects.

The impact of insect herbivory on host productivity and allocation of photosynthate has been little studied. Consumption is apparently small compared to production but the interactions between plants and herbivores can be complex and subtle. In the saltmarsh environment, herbivory may pose particular hazards to plants. Given the importance of nitrogenous compounds to survival at high salinity consumption of these compounds by sap-sucking insects might be expected to have a considerable impact on productivity. The effects on plant-water relationships of the presence of damaged tissues, which presumably represent sites of water loss, have not been quantified. Herbivorous insects can act as vectors for disease agents but there is very little information on pathogens in saltmarsh species. Indirect losses following insect grazing may be greater than actual consumption. With grasshoppers on *J. roemerianus*, grazing is restricted to a segment of the leaf and the portion distal to the grazed area suffers premature death (Parsons & de la Cruz 1980). The impacts of grazing on reproductive effort are discussed in Chapter 5.

Grazing by insects is spread across the marsh and there are no reports of localised (either in space and time) heavy grazing, although there are reports of episodes of defoliation in mangrove stands (Whitten & Damanik 1986). Other grazing animals may cause more obviously visible effects on marsh vegetation.

The most obvious grazing impacts are caused by vertebrates (mammals and birds) although grazing intensities may vary from year to year and may be limited in area. The crab *Sesarma reticulatum* has been recorded as being capable of reducing tall creekbank stands of *S. alterniflora* in Delaware to stubble (Kraeuter & Wolf 1974). Although the same *Sesarma* species also occurs in Georgia, similar intensive grazing has not been observed in that state (Pfeiffer & Weigert 1981).

In Arctic and north temperate regions the major grazers on saltmarshes not exploited by domestic livestock are wildfowl (chiefly geese but also ducks). These birds can have a dramatic impact on saltmarsh vegetation.

Geese consume large quantities of plant material which passes through the gut very quickly (Sibly 1981). It has been suggested that this rapid passage was maximised at the expense of efficient digestion (Sibly 1981) but

Buchsbaum, Wilson & Valiela (1986) argue that the high throughput is related to the low protein availability in the forage. The digestibility of protein by geese is high and Buchsbaum *et al.* (1986) suggest that 'maximising protein intake (by high turnover rates of food and efficient protein digestion) is probably more important to geese than increasing the efficiency of carbohydrate digestion (by retaining foods for longer periods of time)'.

On the east coast of the United States, snow geese *Anser caerulescens* impose very heavy pressure locally on vegetation and are responsible for 'eatouts', where biomass is reduced to a very low level over areas which may be from several hectares to square kilometres in extent (Lynch, O'Neill & Lay 1947; Smith & Odum 1981; Smith 1983; Daiber 1982).

Snow geese feed mainly on the rhizomes of grasses and sedges, to a depth of 20–25 cm. Although the vegetation over large areas may be completely uprooted, only a small proportion of the available food material is actually consumed (Lynch *et al.* 1947). Flocks of geese may be very large, up to 100 000 birds. Smith (Smith & Odum 1981; Smith 1983) studied the effects of snow geese on several saltmarshes in North Carolina and demonstrated significant differences in responses between sites. Nevertheless, at all sites, plant cover was reduced in the spring after grazing. *Scirpus robustus* recovered more rapidly than other species (Smith & Odum 1981) and in some circumstances may rapidly invade grazed *Spartina* marsh (Smith 1983). *S. robustus* is able to reproduce from seed and rhizomes but is advantaged over *Spartina* when grazed in that a larger proportion of the reproductive effort is invested in seeds. Vegetative reproduction in *Spartina* is depressed by the consumption of rhizomes and this cannot be compensated for by germination of seed. Smith & Odum (1981) suggested that under a goose-grazing regime *S. alterniflora*-dominated marsh would be unstable and that over a period of time *Spartina* should be eliminated from grazed areas. This hypothesis was not supported by a longer-term study, *Spartina* persisted although biomass was much reduced (Smith 1983). Smith suggests that this finding is compatible with optimal foraging theory in that once biomass was reduced to a critical level it was energetically unprofitable for snow geese to continue grazing so that the flock moved either to another part of the marsh or to a different marsh.

Other species of geese and duck also graze extensively during the winter on saltmarshes in the north temperate region. These species generally consume above-ground plant material but on occasion uprooting of below-ground biomass occurs.

During the summer, these wildfowl migrate to higher latitudes. The effects of snow geese on the vegetation of a Hudson Bay marsh have been studied intensively (Jefferies, Jensen & Abraham 1979; Cargill & Jefferies 1984*b*; Bazely & Jefferies 1985, 1986).

These Arctic marshes have a net above-ground primary production which is much lower than that of *S. alterniflora* marsh at low latitudes. However, although geese consumed approximately 80% of the above-ground production, productivity on grazed marshes was significantly higher than on ungrazed sites (Cargill & Jefferies 1984*b*). In the early part of the growing season productivity was similar on both ungrazed and grazed sites but in the later part of the season it was sustained on grazed sites while productivity on ungrazed areas declined. Cargill & Jefferies (1984*b*) and Bazely & Jefferies (1985) suggest that the stimulation of production under grazing is provided by accelerated nitrogen cycling, as a result of nitrogen in goose faeces being more readily available for uptake.

Bazely & Jefferies (1986) demonstrated that grazing by geese also affected the composition of the vegetation, promoting dominance by *Pu. phryganodes*. If grazing was excluded, there was a rapid change in species composition, vegetation structure and a build-up of litter. While grazing may retard succession, it does not arrest it (Jefferies, Jensen & Abraham 1979).

It seems clear that heavy grazing by geese during the summer both promotes vegetation favoured by the herbivores and affects nutrient cycling within the marsh so as to promote productivity. Grazing on temperate marshes during the winter occurs when growth of the vascular plants is minimal. The vegetation may not be able to respond to any increase in availability of nitrogen which might, in consequence, be lost to the system. In view of the critical importance of nitrogen availability to saltmarsh plants, does sustained grazing by wildfowl in winter have long-term effects on nutrient cycling and productivity of saltmarshes? [Grazing by domestic stock on temperate marshes is normally reduced during the winter so might not have the same effect.]

Geese are relatively unselective feeders but are influenced in their choice of site by vegetation structure and possibly by nutritional quality. On both wintering and breeding grounds wildfowl tend to avoid areas which have previously been ungrazed (Cadwalladr & Morley 1974; Bazely & Jefferies 1986). This response raises the question as to how conditions favourable to wildfowl were maintained prior to utilisation of saltmarshes by livestock. In the Arctic, grazing is concentrated during the short growing season. For most of the time when wildfowl are absent, no growth is possible. Provided grazing intensity is not reduced between seasons, the vegetation structure favoured by birds should be maintained. On the wintering grounds, the vegetation is subject to highest pressure during the dormant season. Today the vegetation is kept short during the summer by livestock, and if this summer grazing is reduced, wildfowl shun the subsequent rank vegetation. In prehistoric times, was the proportional area of saltmarsh available to wildfowl much reduced or did other herbivores have a more significant impact in summer than has been acknowledged?

Large numbers of grazing wildfowl also have a significant impact on saltmarshes through trampling and puddling the substrate. Where rhizomes are grubbed up, the resulting disturbed areas may be enlarged and deepened to become pans (Jefferies, Jensen & Abraham 1979). Wildfowl may also consume a large proportion of the seed production of saltmarsh plants (Joenje 1985). Other specialist seed-eating birds are also found on saltmarshes (see p.324) but their impact on vegetation has not been quantified.

The impact of wild mammalian herbivores on saltmarsh vegetation has rarely been studied. In the southern United States, muskrats *Ondatra zibethicus* produce eatouts similar to those of snow geese (Lynch, O'Neill & Lay 1947; Daiber 1982). Rabbits can graze saltmarsh vegetation to a very low sward (Fig. 6.1). In general, however, mammalian grazing, other than by livestock, seems to be localised and of fairly minor impact.

Consumption of below-ground material

The below-ground standing crop (in the form of roots and rhizomes) is often far greater than that of above-ground material. However, if the effects of herbivory on above-ground material are poorly quantified, those below-ground remain virtually unknown.

As discussed above, snow geese may consume a part of the below-ground material on some marshes but they also bring to the surface far more material than is immediately eaten and this excess may become input into detrital food chains (Lynch, O'Neill & Lay 1947; Smith & Odum 1981). de la Cruz & Hackney (1977) have suggested that crabs may also be important in bringing root and rhizome fragments to the sediment surface and that they may be an important agency by which underground biomass is made available for recycling.

The quantitative significance of direct underground consumption of plant material is unknown. Root aphids occur on a number of species (p.143), but the feeding habits of most invertebrates living in saltmarsh soil remain to be studied. It is possible that the major losses of below-ground material are in the form of exudates.

Fig. 6.1 The impact of grazing by rabbits on saltmarsh vegetation. Outside the exclosure the sparse *Puccinellia–Festuca* sward is only 1–2 cm high. (Holy Island, Northumberland, UK)

The fate of algae

There have been few studies on grazers of algae. However, algal production may be a significant proportion of the marsh total; in addition algae may be of higher quality nutritionally than vascular plants. It is probable that a higher proportion of algal production is grazed than would be the case for vascular plants (Pomeroy *et al.* 1981). Herbivorous fish, such as mullet, snails, fiddler crabs and various insects may all graze on algae.

While algal grazing may have significant impacts on some marshes, and may be responsible for the maintenance of some communities, it would seem that in most marshes only a relatively small proportion of production is directly consumed by herbivores and that large amounts of material are available to enter detrital food chains.

The breakdown of detritus

The breakdown of saltmarsh detritus is reviewed in Pomeroy & Weigert (1981) and Long & Mason (1983). The breakdown process is summarised in Fig. 6.2. One matter which has been of considerable concern has been the location of the sites where most decay occurs – is most material

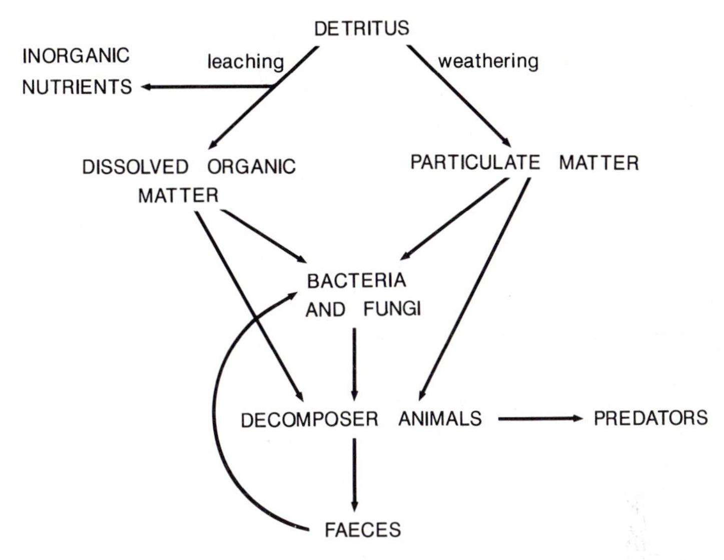

Fig. 6.2 The breakdown of detritus. (Redrawn from Long & Mason 1983.)

washed out of marshes prior to decay or does most processing of detritus occur *in situ* in the marsh? In addition to material produced in the marsh, material washed in includes some of marine and estuarine origin (algae and seagrasses), litter from terrestrial vegetation (leaves and branches washed down stream from the catchment), and the non-biodegradable products of civilisation.

The composition and structure of algae may result in proportionally higher consumption than for vascular plants. These same factors may also influence the rates of breakdown of dead material although most studies of decomposition have concentrated on vascular plant material. The spatial distribution of algal and vascular plant debris may differ with much of the algal material being deposited in the lower marsh (Beeftink 1977a).

Figure 6.2 identifies two major initial pathways in the decay process, one involving the leaching of material and the other, fragmentation of detritus. The term dissolved organic matter – DOM (or dissolved organic carbon) – is a convenient term to disguise our ignorance of its nature. Attempts to characterise DOM have been only partially successful (Pomeroy & Imberger 1981; Kennish 1986). While a large proportion of DOM may be readily utilised by micro-organisms a significant fraction is more refractory. It is thought that the major components of this refractory DOM are humates and fulvates (Pomeroy & Imberger 1981). These are complex polyphenolic compounds which cannot be regarded as potential substrates for micro-organisms (Pomeroy & Imberger 1981). As well as being leached from dead plant material DOM may be lost from both above- and below-ground parts of living plants.

Particulate organic matter (POM) is also chemically very complex. Not only will there be variation depending on the initial source but changes will occur during decay. Various pollutants may become concentrated in decaying litter (Breteler *et al.* 1981; Drifmeyer & Rublee 1981); the extent to which this process amplifies environmental hazards remains to be assessed. The changes which occur during decomposition mean that at different stages in the process, decaying litter as a food source for detritivores varies in both energy content and nutritional value. Valiela & Rietsma (1984) have shown that various chemical factors determine the palatability of saltmarsh detritus.

Studies on the rate of decomposition of litter are reviewed by Long & Mason (1983). The rates of decomposition vary between species, and in the case of the most studied species, *S. alterniflora*, with position in the marsh, material decaying more quickly in the lower-marsh. Studies on a number of European species by Buth & de Wolf (1985) similarly showed variations in rates dependent on species and habitat within the marsh. Buth & de Wolf (1985) emphasised the value of long-term studies in understanding the factors affecting decomposition. Initial rates of decay are frequently much faster than later ones, reflecting the early utilisation of easily metabolised components (Odum & de la Cruz 1967; Long & Mason 1983; Buth & de Wolf 1985). Temperature has a considerable effect on decay rates and in temperate latitudes decomposition in winter is very slow (Buth & de Wolf 1985).

After the early loss of DOM the remaining material is broken down into smaller fragments, both by physical factors such as movement by tides and by animals. As breakdown proceeds, the smaller particles are, by virtue of the increasing surface-to-volume ratio, more exposed to the activities of micro-organisms. In most accounts, the major microbial activity is ascribed to bacteria, and fungi are generally considered to be of relatively minor importance. However, in the early stages of decay, with standing dead material and larger fragments, fungi may assume greater importance (Torzilli & Andrykovitch 1986).

The surface of detritus is rapidly invaded by micro-organisms and these decomposers are subject to grazing by protozoa. Detritus is also ingested by larger animals. The extent to which the detritus itself is a major food source, or whether the detritivores are assimilating the surface micro-organisms remains uncertain (Long & Mason 1983). Even if detritivores do not directly utilise detritus and consume micro-organisms their activities may, nevertheless, stimulate the rate of decay. Many detritivores comminute detritus, increasing the surface area available for microbial colonisation. In addition, the activities of detrivores may alter the structure of the microbial community, promoting bacteria at the expense of fungi (Long & Mason 1983). Nitrogen and phosphorus excreted by invertebrates may be taken up rapidly by bacteria. Lopez, Levinton & Slobodkin (1977) developed a model of a simple detritus-based system involving the amphipod *Orchestia grillus*, bacteria and *S. alterniflora* litter which emphasised the feedback between excretion by *Orchestia* and the stimulation of the bacterial population.

The activities of detritivores are of considerable importance in structuring the saltmarsh environment. Many species ingest a mixture of detritus and sediment, and the faecal pellets represent aggregations of smaller sedimentary particles which may form a more-consolidated substrate for marsh development. The burrows of many species serve to increase the regions of aerobic sediment, which may provide habitats for meiofauna and locally stimulate root growth.

Despite the activities of burrowing animals and radial oxygen loss from roots much of the subsurface sediment in most saltmarshes is anaerobic; even in a largely aerobic sediment there will be numerous anaerobic microniches. Anoxic decomposition may represent a major pathway for energy flow in saltmarshes, particularly in view of the large standing crop of below-ground biomass recorded from many sites. Anaerobic processes in saltmarsh sediments are reviewed by Wiebe *et al.* (1981) who identified four major processes – fermentation, dissimilatory nitrogenous oxide reduction, dissimilatory sulphate reduction, and methanogenesis (see Fig. 6.3). Except under virtually freshwater conditions, such as might prevail in upper estuarine sites, methanogenesis is likely to be of minor importance in saltmarshes owing to inhibition of methanogenic bacteria by sulphate reducers.

Sulphate reducing bacteria utilise relatively few substrates (Wiebe *et al.* 1981), most of which, such as lactate and acetate, are products of fermentation. Sulphate, which is abundant in seawater, acts as electron acceptor, being reduced to sulphide while the carbon source is oxidised to carbon dioxide. Only a few genera of bacteria (*Desulfovibrio*, *Desulfuromonas* and *Desulfomaculum*) are known to be capable of this process (Weibe *et al.* 1981). The sulphide may react with iron to be precipitated as ferrous sulphide, eventually (sometimes rapidly) forming pyrite (Howarth 1979; Howarth & Teal 1979). Only a proportion of the sulphide is precipitated

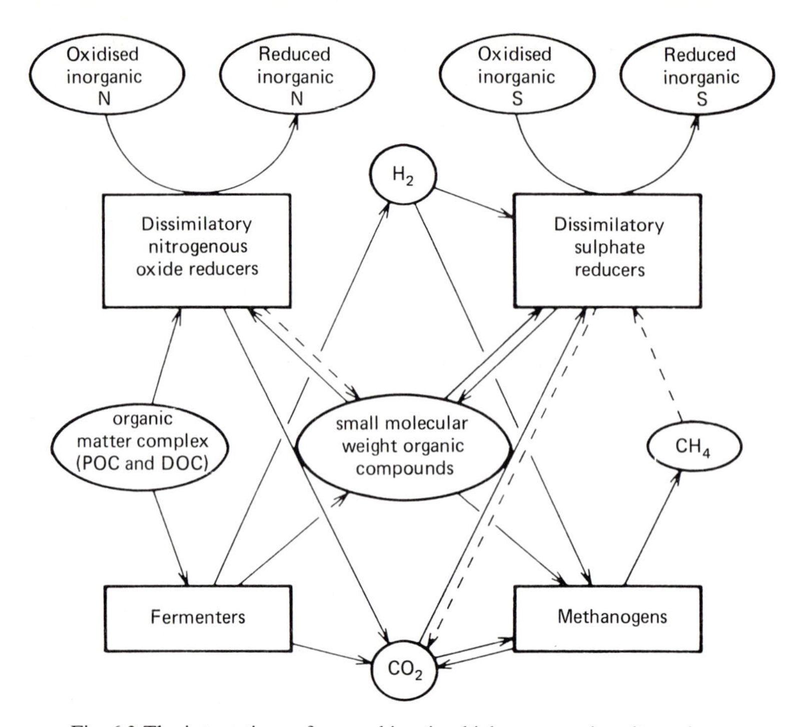

Fig. 6.3 The interactions of anaerobic microbial processes in saltmarsh sediments. (Redrawn from Wiebe *et al.* 1981.)

and the rest may diffuse into aerobic sediment where it is oxidised by bacteria (such as *Thiobacillus* spp.). These sulphur-oxidising bacteria are potentially available to be grazed by detritivores (Odum, McIvor & Smith 1982). Peterson, Howarth & Garritt (1986) were not able to demonstrate that these bacteria were important food sources for the consumers they studied, but the pathway may be important to the meiofauna.

Sulphate reduction appears to account for a major part of the available organic carbon sources in some marshes (Jørgensen 1977; Howarth & Teal 1979, 1980; Skyring, Oshrain & Wiebe 1979). Wiebe *et al.* (1981) warn that sulphate reduction may have been overestimated and that further research and development of methods to measure sulphate reduction are required.

The proportion of sulphide which is immobilised as pyrite may vary between sites, but there have been too few studies to venture generalisations (Wiebe *et al.* 1981).

The linkage between saltmarshes and adjacent waters

Most of the studies on decomposition concentrate on events occurring within marshes. There remains the possibility that detritus is exported from marshes and utilisation by detritivores occurs outside the saltmarsh. In addition to the export of carbon, the possibility of export of other materials, notably nitrogen and phosphorus, has also been discussed in the literature.

The concept of saltmarshes as 'exporting' ecosystems was developed in the early 1960s, rapidly became accepted dogma (Nixon 1980), and is still firmly entrenched in many quarters. As the results of more detailed studies become available, it is clear that generalisations about the role of all saltmarshes as either sources or sinks of material cannot be sustained.

The arguments presented in the early ecosystem studies of saltmarshes (for example, Teal 1962) were logically appealing, and the conclusions, that large amounts of detritus were exported and support estuarine and offshore detritus-based food chains, suggested that saltmarshes were of direct economic value at a time when in many parts of the world they were under threat of destruction. However, there was little confirmation from direct measurements of movement of material in or out of saltmarshes. Certainly export of detritus may be readily observed as rafts of material drifting on estuarine waters, while import may equally be seen in the heterogeneous assemblage of material forming the driftline. Export of large detrital particles, although visible, appears to account for only a small proportion of saltmarsh production (Pomeroy & Imberger 1981; Dame 1982).

The measurement of fluxes in and out of marshes is technically difficult. In addition, any results obtained cannot easily be extrapolated to other areas or time periods. Reed, Stoddart & Bayliss-Smith (1985) argued that for sediment particles, net flux was affected by tidal amplitude and climate, and as the asymmetry of flows between ebb and flood in creeks varied with amplitude, measured fluxes on tides not flooding the marsh could not be extrapolated to high spring tides. Odum, Fisher & Pickral (1979) have argued that fluxes during storms may be larger and in different directions from those prevailing under average conditions. However, there are few measurements made under these conditions. The equipment to monitor materials moving in creeks is expensive and can rarely be left to record continuously over long periods, particularly in situations where storms may cause damage or loss. Extrapolations from short recording periods, even if accounting for seasonal changes, may give a totally misleading result on the net direction and magnitude of fluxes if storm periods are not sampled.

In addition to tidal and climatic variables, Gallagher *et al.* (1980) have pointed out that the export characteristics of any particular marsh will also depend on vegetation type, physiography and geomorphology and the element being considered.

Accepting the problems of measurement and extrapolation, the majority

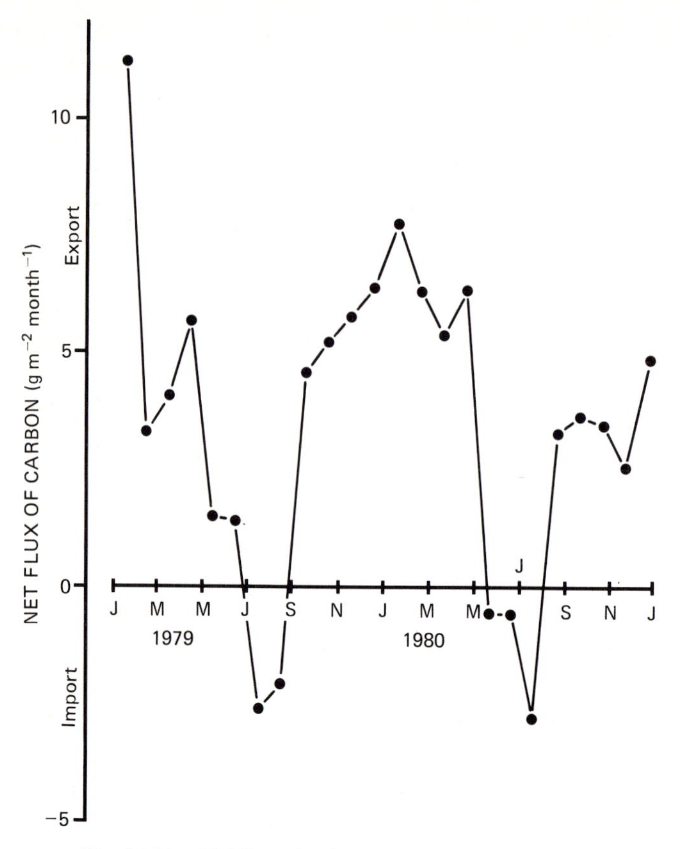

Fig. 6.4 Net tidal flux of carbon derived from saltmarsh vegetation at Seafield Bay, Suffolk. Over the year (months indicated are January, March, May, July, September and November) there is net export although brief periods of import occur. The majority of the carbon is derived from *Spartina anglica*. (Redrawn from Jackson, Long & Mason 1986.)

of studies to date do not suggest that a major fraction of marsh production is exported as POM (see reviews by Nixon 1980; Long & Mason 1983). Situations of net export, approximate balance and net import have been recorded. The highest estimates of net export amount to about 30% of above-ground production. Even where, over a year, a marsh can be shown to export POM there may be periods of net import. In a study of a marsh in Suffolk, Jackson, Long & Mason (1986) showed that 15–20% of above-ground annual net production was exported (70% to the estuary and the rest of the drift line) but that during the summer there was a period of net import (Fig. 6.4).

Calculation of nitrogen budgets for saltmarshes is made more complex by the number of different forms of the element and by interconversions between various forms (Whitney *et al.* 1981). Since these interconversions are mediated by bacteria, environmental variations (particularly of temperature and sediment redox potential) will result in significant temporal and spatial variation in the dominant reactions in the nitrogen cycle. As in the case of POM, export, import and balanced nitrogen budgets have been reported from different studies, Woodwell *et al.* (1979) suggested that nitrogen fixation on the marsh surface contributed towards the Flax Pond marsh being a small net export of inorganic nitrogen to coastal waters. Valiela & Teal (1979*a*,*b*) found that the Great Sippewissett marsh in Massachusetts exported rather larger quantities of nitrogen to coastal waters. At this site there was an appreciable input of nitrate in groundwater and effectively the marsh converted this nitrate input into ammonium and particulate organic nitrogen and exported both. Little is known about groundwater inputs in many marshes; however, the effects of fertiliser use on groundwater quality suggests that if there is a groundwater influence on a marsh, then the importance of this to the nitrogen budget in developed countries may have changed in recent years. The Sapelo Island marsh in Georgia is a net importer of nitrogen, most probably as nitrate from flooding waters (Whitney *et al.* 1981). In an Essex saltmarsh Abd. Aziz and Nedwell (1986*a*,*b*) reported an approximately balanced budget but point out

> 'Even if there is a net annual balance in the exchange of nitrogen between marsh and estuary, this does not mean that saltmarshes are unimportant to the estuarine processors of nitrogen. Nitrate is removed from tidal water but organic −N, which may be used by the estuarine fauna, is exported. Again, the export or import of the different forms of nitrogen may change seasonally, leading to a nitrogen-buffering effect of the saltmarsh upon the availability of nitrogen in the waters of the adjacent estuaries.'

Various studies have reported high rates of nitrogen fixation in saltmarshes. This occurs in two ways, either by mats of cyanobacteria which are likely to be spatially heterogenous in their distribution across the marsh, or by rhizosphere bacteria associated with particular species of higher plants. Although there will be seasonal variation in the intensity of fixation and denitrification, the available data suggest that on an annual budget basis, losses through denitrification far exceed gains from fixation (Whitney *et al.* 1981; Valiela & Teal 1979*a*,*b*).

The phosphorus cycle in saltmarshes also involves a number of different chemical forms (Long & Mason 1983). Although considerable exports of phosphorus from marshes have been claimed, detailed studies indicate that on an annual basis the phosphorus budget is likely to be approximately balanced (Nixon 1980; Long & Mason 1983; Whitney *et al.* 1981), although there may be significant seasonal variation in fluxes (Woodwell *et al.* 1979); or they show saltmarshes to be phosphorus sinks. As with nitrogen, transformations of phosphorus within the saltmarsh may be of importance to the maintenance of ecosystems of adjacent waters – 'The remobilization and loss of some of the phosphorus from sediments buried on the marsh may, in fact, make marshes a source of reactive and organic phosphorus for adjacent waters' (Nixon 1980).

In discussing other materials, particularly pollutants, the role of saltmarshes as 'sinks' rather than as 'exporters' has been stressed (see Nixon 1980).

Even if there were more data on exports and imports from saltmarshes such information does not permit a direct assessment of the importance of saltmarsh to the maintenance of off-shore ecosystems – 'studies which demonstrate that marshes export or import organic matter do not tell us whether or not this organic matter is used by consumers either in the marsh or off shore' (Peterson, Howarth & Garritt 1986).

One approach both to demonstrating whether export occurs and to identifying pathways of energy flow is provided by studies on the distribution of stable isotopes in both potential food sources and heterotrophs.

The most widely used technique utilises differences in abundance of the carbon isotopes, ^{13}C and ^{12}C (Rounick & Winterbourn 1986). The ratio of these isotopes varies between plant groups (algae, C_3 and C_4 plants), and varies little in heterotrophic food chains based on the different sources (Rounick & Winterbourn 1986). The ratio of $^{13}C/^{12}C$ is generally expressed relative to a standard (the international standard is prepared from a fossilised belemnite from the Pee Dee formation in South Carolina) as $\delta^{13}C_{PDB}$ such that

$$\delta^{13}C_{PDB} = \frac{[^{13}C/^{12}C \text{ sample} - {}^{13}C/^{12}C_{PDB} \text{ standard}]}{^{13}C/^{12}C_{PDB} \text{ standard}} \times 10^{-3}$$

(Rounick & Winterbourn 1986).

The technique has been used to identify the source of particulate organic matter in tidal creeks in Georgia by Haines (Haines 1976, 1977, 1979; Haines & Montague 1979). Prior to these studies, the expectation was that this material would have originated primarily from the adjacent saltmarshes. As the saltmarshes were overwhelmingly dominated by *S.*

alterniflora, a C_4 species, then the $\delta^{13}C$ of the POM should reflect this. Haines however, found that this was not the case and the $\delta^{13}C$ indicated that phytoplankton production was far greater than had been assumed.

Jackson *et al.* (1986) have taken advantage of the C_4 photosynthesis pathway in *S. anglica* to use the $\delta^{13}C$ technique as a means of identifying the contribution of saltmarsh detritus to the diet of saltmarsh macro-invertebrates in Suffolk.

Nevertheless, despite the advantages of the $\delta^{13}C$ approach, it does have several drawbacks. First, the $\delta^{13}C$ of a heterotroph provides no information on the trophic status of the organism. Second, the technique can allow clear-cut identification of the ultimate food sources only if they are overwhelmingly dominated by one type with an identifiable $\delta^{13}C$ signature. Many detritivores are likely to ingest a complex cocktail of detritus from several original sources – the composition of the cocktail possibly varying both in space and time. The $^{13}C/^{12}C$ of the detritivore will integrate $^{13}C/^{12}C$ of the various sources and resolution into dietary components will not be possible. Peterson, Howarth & Garritt (1985, 1986) have suggested that simultaneous studies of several stable isotopic ratios (carbon, sulphur and nitrogen) may increase the resolution in identifying the ultimate sources of organic detritus although detailed analysis of food webs still requires direct observation. Using $\delta^{13}C$ and $\delta^{34}S$ Petersen *et al.* (1986) showed that most of the macrofauna they studied were ingesting organic matter originally derived from both *Spartina* and phytoplankton and that the species could be arranged along an axis from mainly *Spartina* to mainly phytoplankton. The technique of multiple isotope studies appears to have considerable potential.

The evidence from both direct measurements of exports/imports and isotope ratio studies suggests that the model (Fig. 6.5) for saltmarsh functioning developed from the work of Teal (1962) requires modification and that the model advanced by Haines (1979) (Fig. 6.6) may be more appropriate. Haines (1979) emphasised the importance of algal production in maintenance of estuarine systems. Saltmarshes remain important components of the model but as sites for the transformation of various materials and in the provision of refuges, feeding habitats, and nursery grounds for various estuarine animals. Under Haines' (1979) model a major export from saltmarshes is in the form of living organisms (juvenile fish, invertebrate larvae). Pomeroy & Imberger (1981) similarly suggest that for the saltmarshes of the Duplin River, Georgia, the significant exports are organisms and refractory DOM.

Research on a wider range of marsh types is required to validate this

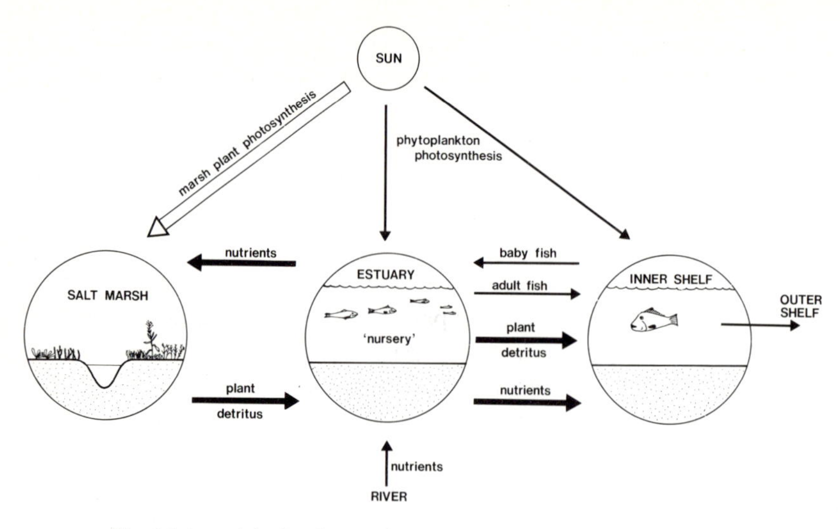

Fig. 6.5 A model of exchanges between saltmarshes, estuaries, and coastal waters developed from studies in the 1960s and early 1970s. (Redrawn from Haines 1979.)

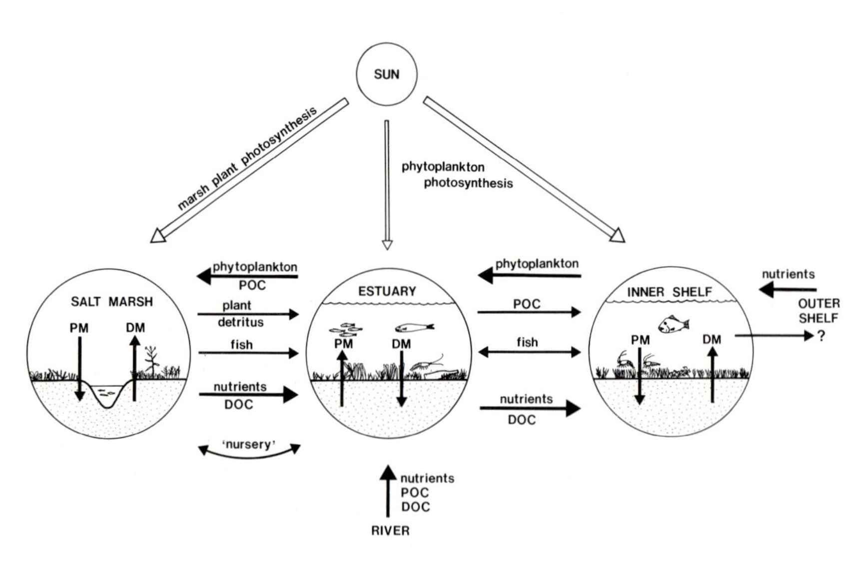

Fig. 6.6 A model of exchanges between saltmarshes, estuaries and coastal waters, based on more recent studies, developed for Georgia, but probably more generally applicable. (Redrawn from Haines 1979.)

model. However, even if most marshes do act as transformers of nitrogen and phosphorus compounds, and as nurseries for commercially important invertebrate and fish species, the question that engineers and planners would then put to ecologists – what would be the impact on commercial stocks of modification/destruction of a particular marsh? – is one which remains difficult to answer and for which any quantitative response is likely to require very detailed site-specific information. Even if the necessary studies were conducted it is unlikely that the resultant data could be assessed in isolation. The value of a particular site as a nursery may depend on the distribution of other marsh areas in a region. If these other areas are modified, it is possible that the value (both absolutely and relatively) of a particular site could change. The importance of determining an appropriate context for assessment cannot be stressed too highly. At this stage it may be more valuable to impress on those who require the reassurance of apparently quantitative predictions, the complexity of ecosystems and to suggest that the question reflects an inappropriate approach to estuarine management. Until more is known about the nexus between saltmarshes and adjacent waters, decisions which pre-empt future management options should be avoided.

Wiegert & Pomeroy (1981) stress that 'there is no single answer to the question about the flow of materials that will be valid for all salt marshes'. Nevertheless, this conclusion should not be viewed as an argument against ecosystem studies of saltmarshes. Detailed studies similar to those reported in Pomeroy & Wiegert (1981) are urgently required for a range of marshes. Such studies will give an indication of the range of 'answers' which might be expected and will allow assessment of whether some marsh types are more strongly linked to estuarine and offshore waters than others. The search for correlation between variations in patterns of fluxes and variation in the physical environment may generate testable hypotheses on the control of ecosystem processes. These studies will also serve to identify processes universal to saltmarshes. Modelling of these processes will allow assessment of their sensitivity to particular environmental changes thus indicating those aspects of development proposals which should be given greatest attention during environmental-impact studies.

7

Modification, management and conservation

Many of the world's major cities are sited on estuaries. In consequence, saltmarsh often provides the only extensive areas of apparently natural vegetation close to major conurbations. Saltmarshes are almost always dominated by native species and look very different from the farmland which may fringe the city – however, there are many human impacts on saltmarshes and the species and community composition of many sites may be a consequence of these impacts. If these areas are to be managed, it is important that decision makers are aware of past impacts and the sensitivity of the ecosystem to changes in the pattern of human impacts.

Grazing

Livestock have been grazed on saltmarshes for centuries. At the present time, grazing is an important use of saltmarshes in northern Europe (Dijkema 1984*d*), eastern Canada (Roberts & Robertson 1986) and Japan (Ishizuka 1974), and occurs widely on marshes from high latitudes to the tropics. In Europe, grazing is most intensive in northern and central regions and is less intensive in the south (Beeftink 1977*a*). Grazing may, however, be an important ecological factor even on sites where it is carried out on an irregular casual basis.

Cattle and sheep are probably the two species most widely grazed on saltmarshes, although in some regions horses are important. Pigs are allowed access to some marshes and may have been more important in the past. Domestic geese graze marshes in both Europe and Asia (Dijkema 1983; Chung 1982) and in the past may have been important grazers at many sites (Chadwick 1982).

In some regions saltmarsh grazing is a useful bonus to essentially inland agriculture but in others it may provide the major livelihood to its practitioners.

Gray (1972) provides a discussion of grazing practices in north-west England where sheep are the main species pastured on saltmarsh. Virtually all the saltmarshes around Morecambe Bay are grazed and for some farmers in the area, saltmarsh provides the majority of the grazing with only small areas of non-tidal land available as holding paddocks during spring tides. Compared with upland grazings, saltmarshes could support high stocking rates, were largely free of parasites and did not suffer from nutrient deficiencies. Disadvantages included the need to remove stock from the marshes on the highest tides and occasional stock losses through drowning. However, despite the productivity of marsh grazing in Britain and throughout northern Europe (Beeftink 1977*b*), the utilisation of marshes for agriculture has declined considerably over the last decade in consequence of general changes affecting European agriculture. If this trend persists, and is not reversed, then there are likely to be major changes in European marsh vegetation.

The grazing regime is a major factor controlling diversity in flora and vegetation between sites. Not only are there major differences between grazed and ungrazed sites (for example, see pp.181–203) but grazed sites themselves differ considerably. The nature of the grazing stock, stocking rate, and timing of grazing periods are all likely to have impacts on the vegetation. Not only will there be differences in effects between species of grazer (Jensen 1985*a*) but even different breeds of the same species may vary in their grazing selectivity. The various components of the grazing regime will have changed continually over time so that every site will have a grazing history that in detail will be unique. Nevertheless, it is possible to make some generalisations about the effects of grazing in northern Europe although it is not clear how generally applicable they are.

Grazing may have direct effects on species composition of vegetation through the susceptibility of some species to grazing and the avoidance by grazing animals of others. It is probable that grazing tolerance and palatability are not absolute attributes of any species but vary with genotype and environmental factors. For example, in North America *T. maritima* is regarded as toxic to stock (Miller & Egler 1950) but in Europe *Triglochin* may be selectively grazed by sheep (Jensen 1985*a*).

Grazing reduces the build-up of litter and according to Grime's (1979) model relating plant strategies and environmental factors to species-richness, this should lead to increased species diversity. In northern European saltmarshes the communities of grazed mid- and upper-marshes (Armerion communities – principally Juncetum) frequently show much greater diversity than communities on ungrazed marshes at the same elevations. In the low-marsh, diversity on grazed sites is not increased and may be reduced.

To the direct effects of grazing must be added the impact of trampling – both through damage to plants and from effects on the substrate. Shrubby plants and succulents are particularly susceptible to direct damage from trampling. *H. portulacoides* is an example of a species sensitive to this type of damage. The degree of disturbance to the substrate will depend on the nature of the stock and the substrate. Cattle and horses are likely to cause more poaching of the surface, especially on softer substrates, than sheep while pigs will cause considerable damage by trampling and rooting. Unless the grazing pressure is very high sheep cause little surface disturbance except at favoured creek-crossing points or around water troughs. However, even if the soil surface remains unbroken there will be a compaction effect associated with high grazing pressure. Bakker & Ruyter (1981) recorded a change towards a more porous upper soil layer after grazing ceased.

The response of the flora to trampling damage depends, in part, upon the zone affected. In the lower-marsh in northern Europe persistently poached areas provide a niche for *Salicornia* spp. and *Su. maritima*, species common in the pioneer zone and on creek sides. At higher levels relatively large areas with deep holes produced by livestock provide a habitat for a distinct group of communities of the alliance Puccinellio–Spergularion, characterised by *Pu. distans* and *S. marina*. If such trampled areas are subject to freshwater flushing, *A. stolonifera* and *Alopecurus geniculatus* may form a lush grassland community. Grazing may also result in a diffuse pattern of smaller openings within the sward, either directly or as a result of trampling. These openings may provide high-level sites for low-marsh species (Bakker 1985; Bakker, Dijkstra & Russchen 1985) or small annual species of the Saginetea.

As well as influencing the fine-scale patterning of marsh vegetation, grazing may influence the whole process of succession. Reduction in standing crop on grazed marshes will reduce the effectiveness of the vegetation in trapping and stabilising sediment (Randerson 1979); coupled with any compaction effect, this will reduce the rate of accretion and hence vegetation change. As well as slowing succession, grazing may also facilitate the process. Ranwell (1961) has demonstrated that grazing may open out *S. anglica* communities and permit the invasion of species such as *Pu. maritima* at lower elevations than have been recorded from ungrazed marshes.

In general, in northern Europe, grazing appears to have been responsible for an increase in both species and community diversity, at least in mid- and upper-marshes. There have been few studies in other regions which have addressed the effects of grazing on marsh floristics.

In North America many marshes have been subject to stock grazing (primarily by cattle) since the early days of European settlement (see Teal & Teal 1969). This grazing pressure does not seem to have had the same effect on species diversity as in Europe. The great diversity of glycophytic species in grazed Juncetum in Europe does not have an equivalent in North America. This may reflect the nature of the dominant *Spartina* species or absence of suitable upland pasture species to invade newly created niches.

Reimold, Linthurst & Wolf (1975) showed that, in a *S. alterniflora* marsh, grazing lowered the primary productivity and suggested that, in the interests of conservation, grazing pressure should be reduced. On the other hand, the economic value of marsh grazing conferred some protection on marshes which might otherwise be destroyed for development.

While Japanese marshes share many species with European and the arrangement of communities is also similar, the available accounts do not indicate that grazed marshes develop communities equivalent to the grazed variants of the Juncetum although Ishizuka (1974) reported that grazing by horses has led to extensive trampling of saltmarshes on Hokkaido.

In Australia native hooved animals are absent and both Bridgewater (1982) and Kirkpatrick & Glasby (1981) suggest that cattle (and sheep) grazing of saltmarshes has had a number of adverse effects. Trampling is particularly deleterious to the various succulent chenopod shrubs, and on grazed marshes openings are created between the bushes. In the upper-marsh in temperate regions, these are invaded by alien species, particularly the grasses *Polypogon monspeliensis* and *Parapholis incurva*. Grazing thus increases the species diversity of the marshes but this increase is by introduction of growth forms previously absent. Grazing may also be responsible for an increase in the native grass *Pu. stricta*. In Tasmania, Kirkpatrick & Glasby (1981) noted that *L. australe* is absent from grazed sites, with the implication that it is grazing-sensitive. Grazing by rabbits has also had some effects on Australian marsh vegetation, similarly increasing the ratio of grasses and herbs to succulent shrubs (Bridgewater 1982).

The majority of studies discussing grazing have examined changes to floristics, rather than impacts on fauna or ecological processes. Nevertheless, it is likely that grazing has considerable affects on all aspects of saltmarsh ecology.

The diversity of the terrestrial invertebrate fauna is likely to be highest on ungrazed marshes. Reduction in the litter layer and structural diversity of the vegetation reduces the niches available for fauna (see Fig. 7.1). The size

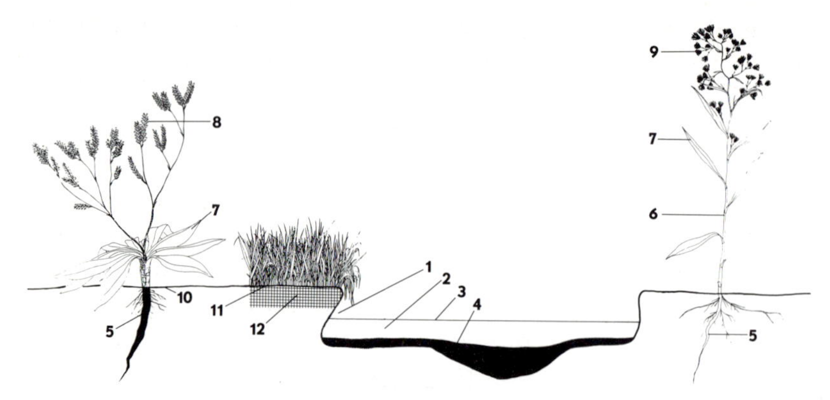

Fig. 7.1 Microhabitats in the saltmarsh: (1) underneath the overhanging margin of a pan; (2) permanent water (micro-pelagial); (3) water surface (epineustal); (4) organic mud on bottom of pan; (5) roots and rhizosphere; (6) plant stems; (7) leaves; (8) flowers; (9) seeds; (10) soil surface under relatively open vegetation; (11) soil surface under dense vegetation; (12) major rooting zone. (Redrawn from Dijkema 1984*b*.)

of the reduction is likely to vary with grazing intensity and nature of the livestock; sheep produce a more uniform, shorter sward than cattle.

The impact of grazing on nesting birds will also depend on intensity. High grazing pressure may result in considerable disturbance of nesting birds and actual trampling on nests (Doody, Langslow & Stubbs 1984) but light or moderate grazing may produce a habitat favourable to breeding birds (Dijkema 1984*d*).

In the northern hemisphere saltmarshes are heavily utilised by geese and ducks – during the breeding season at high latitudes and over winter further south. The direct impact of this grazing is likely to be considerable; at high latitudes the growth season is short and productivity relatively low while utilisation of the wintering grounds occurs when plants have either ceased growth or will be showing very low growth rates. For several wildfowl species, it has been shown that short, grazed swards are favoured over rank ungrazed marshes in both breeding and wintering grounds (Cadwalladr *et al.* 1972; Cadwalladr & Morley 1973, 1974; Ebinge, Canters & Drent 1975; Bazely & Jefferies 1986). It is likely therefore that the present numbers of wildfowl are, at least in part, a reflection of saltmarsh management practices. The decline in stock grazing on marshes is likely to result in a reduction in plant species diversity and the development of rank communities unfavourable to wildfowl (Westhoff 1969; Westhoff & Sykora 1979; Bakker

1978; Bakker & Ruyter 1981; Cadwalladr *et al.* 1972; Cadwalladr & Morley 1974). If bird populations are to be maintained, it may be necessary for conservation authorities in Europe to continue, or re-institute, summer stock grazing as a management technique.

Grazing reduces the above-ground standing biomass but the effects of grazing on the allocation of resources to below-ground structures have not been studied. While biomass is reduced, it does not necessarily follow that above-ground productivity will decline similarly; light to moderate grazing may stimulate productivity in some grassland species. Cargill & Jefferies (1984*a*,*b*) suggest that summer grazing by geese in a subarctic marsh may increase productivity by increasing nitrogen availability. While consumption involves a net loss of nitrogen from the system, goose droppings may, nevertheless, provide a local source of nitrogen more readily available to plants than the largely refractory organic sources which dominate the nitrogen pool in an ungrazed marsh.

The effects of grazing on nitrogen availability in temperate marshes do not appear to have been quantified. Given the importance of nitrogen to the growth of many halophytes, and the relatively low rate of transformation of organic nitrogen to phyto-available inorganic forms (Abd.Aziz & Nedwell 1979), any increase in available nitrogen might be expected to stimulate production at least in the lower-marsh; mid- to upper-marsh communities may be less responsive (Jefferies & Perkins 1977). Grazing may also increase the spatial heterogeneity in nitrogen availability throughout a marsh. In addition, at high tide, stock retreat to the upper-marsh fringe or to higher knolls. This often results in local eutrophication and the development of a 'weed'-rich flora.

Utilisation of saltmarshes by grazing stock represents a pathway for export of marsh primary production to the local market. This export inevitably means that less of the marsh production is potentially available for export to estuaries. One approach to studying the interconnections of saltmarshes and adjacent waters may be a comparison of estuaries fringed by grazed marshes and those with an ungrazed border.

Cessation of grazing

In northern Europe traditional practices of saltmarsh grazing have been declining over the last two decades, even though many marshes continue to be grazed. A number of studies on the vegetation of formerly grazed saltmarshes suggest that 'An abandoned salt marsh is characterised by litter accumulation, dominance of a single species and a subsequent small species diversity and coarse grained vegetation mosaic' (Bakker 1985) (see Westhoff 1969; Westhoff & Sykora 1979; Bakker 1978, 1983, 1985; Bakker & Ruyter 1981; Cadwalladr *et al.* 1972; Cadwalladr & Morley 1974).

This response is similar to that reported from terrestrial grasslands after the end of grazing and accords with the theory advanced by Grime (1979). In saltmarshes it is best displayed in the mid- and upper-marsh. In the lowest zones reduction in grazing results in taller vegetation but either no change in composition or an increase in richness. The species which become dominant in the higher zones vary with circumstances but include *F. rubra*, *E. pycnanthus* and, in brackish sites, *Phragmites australis*.

While the development of dominance is well attested an actual decline in the number of species present is less certain. As Rice & Westoby (1983) point out, if species diversity is recorded in fixed-sized plots, it is almost inevitable that diversity will appear to decline as a few species become more abundant. However, it is possible that if the whole marsh is considered, then the species which are no longer recorded in the plots are still present, albeit as very scattered individuals. If species are actually totally lost from the site then this has implications for the diversity which might develop if grazing were re-introduced. There are few data on this point although Bakker (1985) indicates that while single-species dominance is reached within five years of the cessation of grazing the decline in total species number occurs over a much longer time interval (more than ten years).

Harvesting

There has been a long history of exploitation of various saltmarsh resources. While at least in the northern hemisphere, many of these traditional marsh uses have declined their impacts will have played a major role in moulding present-day marsh community composition.

Plants

Haymaking

Haymaking was once practised extensively on upper marshes in northern Europe and on *S. patens* upper-marsh communities in the north east USA (Teal & Teal 1969). On both sides of the Atlantic haymaking has declined considerably during the present century but is still carried out on a number of European marshes (Bakker 1983; Dijkema 1984*d*). In Britain there is evidence for past haymaking on a number of marshes (see Parkinson 1985) but there has been little recent extensive cutting.

There is little evidence on the effects of haymaking on marsh vegetation although Bakker (1978) demonstrated that under an experimental mowing regime species composition could change rapidly. In Europe, haymaking is presumably a factor contributing to the diversity of the upper-marsh. In America, the upper-marshes remain predominantly *S. patens* despite the long period of exploitation.

Traditional haymaking was restricted to the upper marsh. Experimental harvesting of lower-marsh *S. anglica* showed that it was possible to make silage from it (Hubbard & Ranwell 1966); this has not been developed as a practical management technique. It has also been demonstrated that *S. anglica* could be utilised in paper manufacture (Ranwell 1967).

Reed cutting

Phragmites australis is an almost cosmopolitan species with a wide variety of uses. The species is widespread in both fresh and brackish sites and it is probable that many coastal reed beds have been harvested to fulfil local needs.

In Britain, any such harvesting has been on the decline for most of the present century. However, in 1975, commercial mechanical harvesting of the largest intertidal reed beds in Britain (on the Firth of Tay) commenced. The long-term impact of this is uncertain but to date, there has been an increase in stem density and a reduction in stature (Ingram *et al.* 1980).

One of the major uses of *Phragmites* would be for thatching. *Juncus* spp. and *Scirpus* (*sensu lato*) spp. could provide a lower quality alternative to reed for this purpose and may well have been harvested but such uses of marsh vegetation are poorly documented.

In the Arctic, the straw of *Pu. vaginata* was used by Eskimos for basket making and Sørensen (1953) has suggested that the current distribution of this species may have been determined by past exploitation.

Turf cutting

One form of marsh exploitation, which on a commercial scale is probably restricted to north west England, is the cutting of turf for specialist uses such as bowling greens. *F. rubra*, or *Festuca-Agrostis*, turves are favoured and are prepared for cutting by a programme of selective weed-killing and mowing (Gray 1972). The turf is cut mechanically, large areas often being lifted at one time. Although there is some re-seeding with *F. rubra* seed, recolonisation is essentially a natural process. Successional sequences are very diverse and early chance arrivals may spread rapidly and pre-empt the niche. Ephemerals and bryophytes are more conspicuous in the early stages of succession and turf-cutting areas provide important habitats for species such as *Sagina maritima*, *Centaurium pulchellum*, *C. littorale*, *P. heimii* and *Barbula* spp. *Bryum* spp., *Trichostomum brachydontium* and *Tortella flavovirens* may also be important elements in the early stages of recolonisation. Although turf cutting appears at first sight to be a major disruption of the marsh, as traditionally practised it does permit an increase in the richness and diversity of the flora.

Cutting of turves of saltmarsh vegetation to face earth embankments continues to be a widespread practice.

Edible plants

The exploitation of saltmarsh plants for human food is poorly documented but appears to have been a widespread practice.

A number of the succulent saltmarsh species are edible and have been used as vegetables. In Britain the main species so harvested was *S. europaea* agg., the tradition of samphire gathering was well established in south-east England and elsewhere. [The name samphire also applied to the seacliff and shingle-beach species *Crithmum maritimum* the collection of which was the 'dreadful trade' referred to in '*King Lear*', and *Inula crithmoides* found on seacliffs and some saltmarshes. In early accounts one cannot be certain which species is being discussed.] *Cochlearia* spp., referred to as scurvy grass, were also used occasionally as vegetables. *A. tripolium* was used as a vegetable in northern Europe although not apparently in Britain. Extensive use of such species is a thing of the past but local collection still occurs. As Beeftink *et al.* (1982) point out, the main users of these plants today are elderly, but for these people saltmarsh species may still be a regular component of the diet.

Given the polluted nature of many estuaries, and the propensity of saltmarsh species for ion uptake, Beeftink *et al.* (1982) suggest that there may be a health risk associated with the consumption of *Salicornia* and *Aster* and that in some areas the practice should be discouraged. [Concern does not seem to have been expressed about the consequences to domestic stock grazing on saltmarshes.]

In Australia, Cribb & Cribb (1975) suggest that *S. quinqueflora* and *Su. australis* are both edible and could be used in the same way as samphire; whether aborigines made extensive use of these species is unknown. Succulent chenopods elsewhere in the world are also potentially edible and may have been used as human food.

Despite the probable widespread harvesting of some groups of saltmarsh species, it is unlikely that the exploitation of marsh resources for human food was anywhere a major impact on marsh vegetation.

A striking feature of the European coastal flora is the number of species which are either the ancestors of, or are closely related to, major vegetables (including the *Brassica* group, *Crambe*, *Raphanus*, *Daucus carota*, *Foeniculum vulgare*, and *Asparagus officinalis* – see also Hepburn 1952). The majority of these species are from sandy or shingle beaches, or the spray zone on cliffs rather than saltmarshes although they may occasionally be found on saltmarsh drift lines. The most widespread species in this category in north Europe is *Beta vulgaris*, cultivated forms of which have developed storage taproots, while the leaves can be used as a form of spinach.

The occurrence of edible coastal species is not restricted to northern Europe. In Australasia, the New Zealand spinach (*Tetragonia tetragonioides*), a strand-line species which also occurs on saltmarsh driftlines, was one of the few native species regarded as edible by the early white settlers (Cribb & Cribb 1975). The pigface, *Carpobrotus* spp., which occurs in similar habitats in Australia, provided both edible fruit and foliage to aborigines and early explorers (Cribb & Cribb 1975).

The use of strand-line species as vegetables was widespread but there is no indication that it caused any long-term change in the vegetation.

A number of species, particularly those of brackish marshes, store substantial carbohydrate reserves in below-ground organs. If these were utilised as food resources by humans, it is possible that the disturbance involved in uprooting the plants could have had a major effect on the vegetation. There is, however, no evidence to suggest that this occurred on a wide scale but the possibility should be considered when trying to explain local anomalies in species' distributions.

In addition to providing food, native vegetation may provide medicinally useful species. In Europe, a number of marginal saltmarsh species (such as *Althaea officinalis*) were valued by herbalists but little evidence is available elsewhere in the world for the medicinal value of saltmarsh species. It is, however, unlikely that collection of material for medicinal reasons would seriously influence species distribution or vegetation composition.

Animals

Wildfowl

The vast flocks of wildfowl which seasonally frequent some saltmarshes would have been harvested throughout history. At the present-day they are not exploited as a major food source but for sport. Wildfowlers have been a major, and often successful, pressure group promoting the conservation of saltmarshes.

Fish and shellfish

Fish and shellfish in saltmarsh creeks will have long been harvested although this may have had little impact on the marsh itself.

Currently there is considerable interest in various forms of aquaculture. It is possible that the provision of facilities for this may involve destruction of saltmarsh habitat – already aquaculture pond construction has involved destruction of mangroves in some parts of the world (Terchunian *et al.* 1986).

Digging of annelids for bait by anglers can be a major disturbance in saltmarsh creeks and in the pioneer zones of marshes although whether this has long-term consequences for marsh development has not been recorded.

Fur

Muskrat *Ondata zibethicus* which occurs on brackish marshes and upper saltmarshes in the southern USA has been trapped for its fur over a long period, first by Red Indians and then by European trappers. Management of marshes, particularly by burning, to favour muskrat has been intensive and may have had considerable impact on vegetation and community structure (Hackney & de la Cruz 1981; Daiber 1982). Burning was carried out both to promote new growth of species favoured by muskrat and to make the marshes more accessible to trappers and hunters.

It is likely that any medium- to large-sized mammalian herbivores visiting saltmarsh would have been considered 'fair game' by local peoples and would have been hunted, either opportunistically or systematically, depending on circumstance.

Sediment extraction

In the late nineteenth century large quantities of fine clay were extracted from saltmarshes in the Medway estuary for use in brick and Portland cement manufacture (Kirkby 1984). The industry declined rapidly in the present century although the last cargo of clay was taken as recently as 1965. The process caused a great reduction in marsh area and the extremely complex local topography of the remaining marshes is probably a legacy of clay winning.

Extraction of clay for local purposes has probably occurred on many marshes while the construction of embankments at the expense of borrow-pits in the marsh is very widespread.

Salt and chemical production

Salt has long been a major trade commodity. In arid or semi-arid areas, the simplest means of production is by evaporation of shallow, natural or artificial, seawater pans, these pans often being situated in saltmarshes. Production of salt in this way is still carried on, often on a very large scale, and construction of these facilities may result in destruction of extensive marshes.

In Europe, salt production was widely carried out on both coastal and inland marshes. In England, place names referring to salt (as in Saltcotes) are common for farms and fields adjacent to saltmarsh. Throughout the Medieval period, there are numerous references to sites used for salt production in England (Parkinson 1985) and Scotland (Proctor *et al.* 1983). Two methods of production appear to have been employed; boiling of seawater in pots over open fires, or evaporation in shallow pans. It is possible that at some sites complex systems of hollows in the upper-marsh may relate to salt manufacture.

Many saltmarsh plants contain large quantities of mineral ions (see p.231) and on burning, leave an alkaline salt-rich ash.

Prior to the development of the modern chemical industry and cheap mass production of alkalis in the latter half of the nineteenth century, the major source of alkali was plant ash and coastal plants were particularly valued as sources of supply.

In Europe, the two major supplies were from seaweed (kelp) in France, Ireland and Scotland (see Rymer 1974) and barilla (*Salsola kali* and *S. soda*) from Spain. *Salsola* species are very widespread, being found in the drift-line and in hypersaline upper-marsh communities. In the eighteenth and early nineteenth centuries barilla was a valuable export from Spain and the marshes (barillars) were cultivated to favour *Salsola* production. In addition, the trade was protected by a ban on export of *Salsola* seed (Bird 1978, 1981).

In the early days of the Australian colonies, soap was scarce but one of the ingredients, tallow, was available in abundance. Exploitation of native vegetation for ash led to the development of a soap manufacturing industry which provided one of the first valuable exports of the new colonies. In the absence of barilla, the major species exploited were mangroves and the clearance of mangroves for burning may have had long-term effects on their distribution at their southern limit (Bird 1978, 1981). In the absence of mangroves other coastal species, notably *Atriplex cinerea*, *Rhagodia* spp.

and succulent-stemmed chenopods were exploited. In Tasmania, saltmarsh species provided a major part of the ash (Whinray 1981). In the short term this undoubtedly led to particular marshes being 'cut out' but the documentation is not adequate to allow us to determine whether there have been long-term effects of the soap industry.

Presumably the burning of marsh species for alkaline ash was a widespread practice in many countries.

In addition to its use in soap manufacture, alkaline ash is also a raw ingredient in glass manufacture; the vernacular name glasswort for *Salicornia* spp. refers to this use. The existence of the name suggests that use of plant ash in glass making was widespread over a long period, but the industry is poorly documented.

Manipulation of species abundance

Management of saltmarshes, either to increase the value of the habitat for specific purposes, or to control species regarded as harmful to human interests has had a long history (other management practices such as grazing also profoundly alter species composition in a manner which promotes continuing use).

Fertilising

As Gray (1972) pointed out, one of the advantages of saltmarshes for grazing purposes is that there is little need for pasture improvement. Nevertheless, upper saltmarshes have been fertilised to increase their value for grazing and haymaking and this still occurs on some European marshes (Bakker 1983).

Burning

Saltmarshes would not generally be regarded as particularly flammable. However, fires have been recorded from many localities: many fires are deliberately lit – either for management or arson but cases of fires caused by lightning and other natural combustion sources have been recorded (Hackney & de la Cruz 1981).

In North America, burning of marsh vegetation has long been practised (Teal & Teal 1969; Hackney & de la Cruz 1981; Daiber 1982). In the Gulf States, burning of brackish habitats to improve conditions for muskrat has been carried out over a long period while on the east coast *S. patens* and *J. roemerianus* communities were burnt on occasion. Miller & Egler (1950) suggest that burning may have been carried out by Indians prior to white settlement and may have been a factor determining the present nature of the marsh vegetation. Hackney & de la Cruz (1981) have reviewed the information on the effects of burning and emphasise that little is known about many aspects of the topic.

In the fire-prone Australian environment, burning of upper-marsh vegetation is a frequent occurrence. Kirkpatrick & Glasby (1981) suggest that fire, followed by grazing, may eliminate the shrubby, succulent chenopod *Sclerostegia arbuscula* from Tasmanian marshes. The succulent nature of *Sclerostegia* makes fire very rare; the much more frequently burnt *Gahnia* and *Stipa* tussock communities recover well after fire.

In northern Queensland, extensive *Sp. virginicus* meadows are burnt regularly to promote new growth which is more palatable to grazing cattle (Anning 1980).

In Europe, fire in saltmarsh would be rare but does occur, mainly as a result of accidental ignition by humans. *J. maritimus* stands have been burnt sometimes in attempts, usually unsuccessful, to control rush spread.

Control of individual species

Plants

Spartina

The species which, at least in this century, has been subject to the most intensive control measures is *S. anglica*. The ability of this species to colonise mudflats at elevations previously unavailable to native flora has caused locally substantial reductions in intertidal mudflat, and loss of amenity to seaside resorts as well as causing problems for small-boat navigation. Digging out of plants has been attempted at several sites but this approach is labour-intensive and while successful in the earliest stages of colonisation is impractical in established stands. Herbicides have been used at a number of sites in Britain and in New Zealand. The consensus from various studies is that the most effective agent to date is dalapon (a sodium salt of dichloroproprionic acid) (see Truscott 1984). Large areas can be treated by aerial spraying but the cost is substantial [Truscott 1984 estimated £79 ha^{-1} at 1979 prices].

While dalapon has low toxicity towards a number of marine invertebrates and fish (Truscott 1984) the effects on non-target plant species have been less well studied; if aerial spraying is used then it is inevitable that application cannot be precisely controlled.

Edwards & Davis (1974) have reviewed the effects of several herbicides, applied to *S. alterniflora*, on the saltmarsh ecosystem.

Other species

Gray (1972) reports the use of herbicides to kill dicotyledonous species selectively in the preparation of pure grass swards before turf cutting.

A range of other species has probably been subject to local control at a number of sites. In Britain, spread of *J. maritimus* may reduce the value of upper saltmarsh for grazing; in the past, it may have been controlled by cutting (Kay & Rojanavipart 1977). This is now uneconomic but there are various small-scale local control measures (often carried out on an opportunistic unplanned basis) – digging out, burning and use of herbicides.

Insects

Saltmarshes provide a habitat for a number of insects which have considerable nuisance value, and at times constitute a real health risk, to humans. These include mosquitoes, biting midges and biting flies (tabanids).

The main control of mosquitoes has been by habitat modification or by spraying.

Drainage to destroy mosquito breeding sites has been practiced in many parts of the world (Daiber 1982; Ranwell 1972; Géhu 1984*c*; Dale *et al.* 1986). The practice has also been subject to considerable criticism, on the grounds of long-term ineffectiveness as well as adversely modifying the marsh environment. Miller & Egler (1950) condemned mosquito ditching in the following terms:

> 'Existing marshes have been lacerated with ditches with that admirable thoroughness and pseudo-foresightedness with which mankind is apt to treat the lands of his heritage. In this instance, the advent of modern methods of insect control calls into critical question the present necessity of such ditching, at best a violent activity which, though it destroys the mosquitoes, also destroys the permanent pools so valuable to wildlife, completely rearranges the mosaic of natural plant communities, and eventually produces other pools of the same kind that the ditches were designed to eliminate'.

The spoil heaps created from ditching of US marshes are often colonised by the shrubby composites *Baccharis halimifolia* and *Iva frutescens*, species limited to the upper-marsh fringe in unditched marshes. Ditching reduces the value of the marsh for grazing or haymaking (Ranwell 1972).

As an alternative or addition to ditching, spraying with pesticides has been employed to control both mosquitoes and biting midges. However, these 'modern methods of insect control' are not without problems. Early approaches using chemicals such as DDT resulted in considerable mortality of non-target organisms such as fish and various crustacea (see Ranwell 1972). Although more recent pesticides (such as Abate used to control biting midges in northern New South Wales and southern Queensland – Watson & Watson 1984–) may be less environmentally damaging, there has been little research on the wider impacts of the use of pesticides in the saltmarsh environment or of the consequences of reducing insect populations.

Continuing pressure to control insects is inevitable and where health hazards are involved control measures may be essential. However, the problems associated with insects should be a constraint against the siting of new developments close to saltmarsh or mangrove shores.

Reclamation

The reclamation of saltmarsh has a very long history. Major areas have been reclaimed either for agriculture or development in all parts of the world (see, for examples, Wagret 1968; van Veen 1955; Doody 1984*a*,*b*; Gosselink & Baumann 1980; Nichols *et al.* 1986).

Considerable concern has been expressed about the destruction of habitat associated with reclamation. While such concern is amply justified, it has to be recognised that the effects of historical reclamations have been very varied.

Distinction must be drawn between reclamation for various forms of development and for agriculture.

Reclamation involving infilling of sites to raise their surface level entails loss of existing habitat with few opportunities for compensatory development of new habitat types. Around the world, major urban, industrial and port areas (Coughlan 1979) have been developed on former saltmarsh and mangrove areas. In most of these cases, accretion of new marsh areas adjacent to the development would be undesirable and construction would be engineered so as to prevent this. In addition to the direct effects, these developments are likely to result in pollution sources affecting the whole estuary.

Close to major urban areas, there may be a shortage of sites for the disposal of fill (domestic garbage, industrial waste, rubble, etc.). Saltmarshes have been seen as cheap sites for this purpose. After infilling, such sites are normally sealed with a layer of topsoil and used either for recreational purposes (sports fields or public parks) or as building land. Leaching from sites used for waste disposal may constitute a chronic pollution problem.

In the developed world, pressures from these traditional agents of marsh destruction may be somewhat less than in the past; however, new threats come from canal-estate developments and facilities, such as marinas, associated with recreational developments. In the third world, however, construction of an urban and industrial base may put great pressure on estuarine foreshores which are seen as providing relatively low-cost development sites.

Reclamation for agricultural land has resulted in the creation of some of the most valuable productive farmland in the world, as around the Wash (Gray 1976). Not all reclamations have been so successful – embanking of very sandy marshes as around Morecambe Bay (Gray & Adam 1974) was financially unrewarding and the reclaimed land little more than poor grazing land. Drainage and subsequent oxidation of sulphide-rich clays has caused problems with soil acidity in several parts of the world.

Embankment of marshes frequently creates conditions for accelerated accretion in front of the new seawall (Kestner 1979). This may permit subsequent reclamation of further marsh and the progression of successively younger reclaimed areas seawards. This is well seen in the Wash (Gray 1976 – Fig. 7.2). The process can be speeded up by the construction of an artificial drainage system and brushwood groynes extending through the marsh on to the mudflats to seaward; use of such sedimentation fields has been widespread on north European coasts (Verhoeven 1983; Dijkema 1983).

It is not possible to total the area of reclaimed saltmarsh to estimate the area of marsh which might have existed at some particular time in the past (Doody 1984*b*). Within an estuary, marsh area reflects an equilibrium between erosion and accretion – after reclamation, the new marsh is developing in a new hydrological and geomorphological regime. The area which develops may be greater or less than that reclaimed. In Morecambe Bay, much saltmarsh was reclaimed as a consequence of railway construction rather than specifically for agriculture in the 1850s. The marshes that developed subsequently were considerably larger than those reclaimed (Gray 1972; Gray & Adam 1974). Doody's (1984*b*) estimate of a former 1000 km^2 of saltmarsh in Britain may thus be excessive. While the distribution of marshes will have changed over time and there may have been local net losses, it is probable that at any one time, the area will have been of the same order of magnitude as that at present.

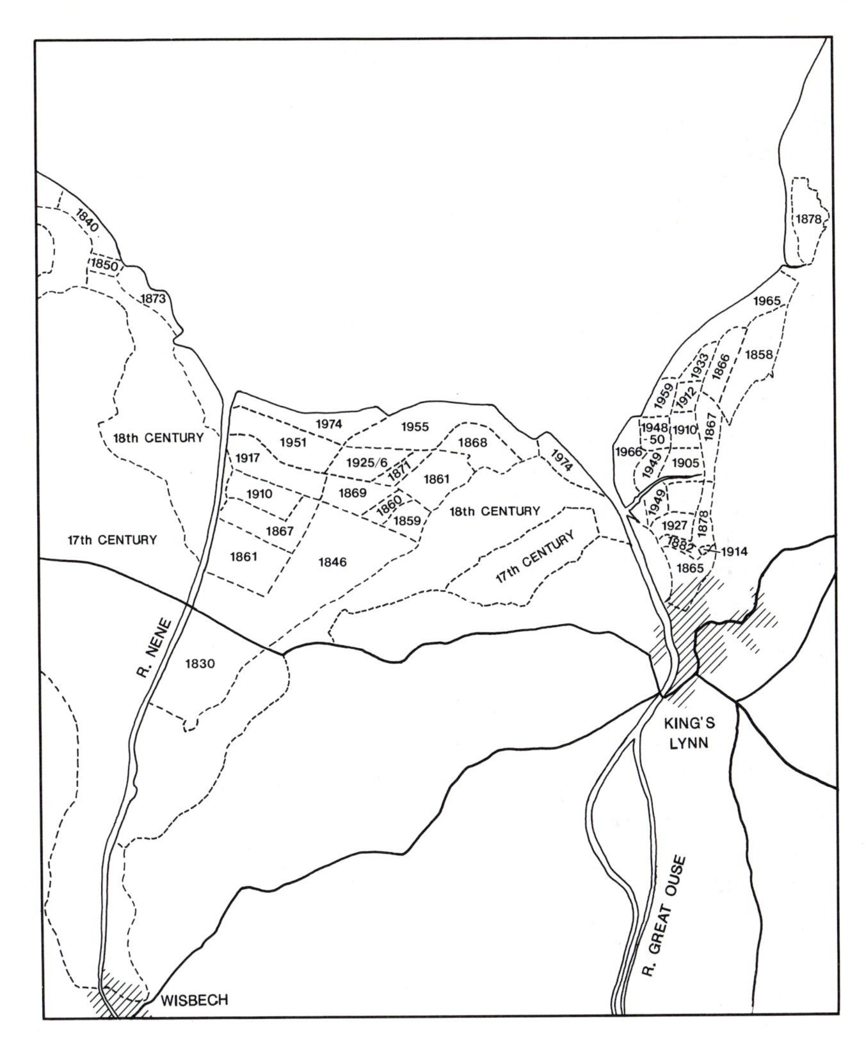

Fig. 7.2 The reclamation history of saltmarsh and intertidal flats on the southern shores of the Wash, UK. (Redrawn from Gray 1976.)

This is not to suggest that reclamation for agriculture is to be regarded as environmentally neutral. While new marsh is created, it is not necessarily identical (either in species or community composition) to that it replaces. The length of time before reclamation may be insufficient for the development of a mosaic of upper-marsh communities. The extremely low species and community diversity of the Wash marshes may reflect this. In addition, the imposition of an embankment in the upper-marsh reduces the diversity of potential high-marsh habitats – the complex local variation associated with local shingle areas or freshwater flushes may be completely lost. Expansion of new marsh following reclamation is at the expense of intertidal mud- or sand-flats. These are important bird habitats and the cumulative impact of their loss may be considerable. Gray (1979*a*) has cautioned that the regeneration of saltmarsh to replenish that lost by reclamation cannot operate indefinitely. Sediment supply may eventually become limiting or marsh development will not occur because of extension into environments where high erosive forces operate. The effects of scale also require consideration. Historically reclamation has been a piecemeal process so that other nearby sites were available to supply propagules for colonisation. However, the potential for much larger-scale reclamations may result in a failure of colonisation.

Modern technology offers the prospect of total alteration of whole estuaries in which saltmarshes would be only one of the ecosystems affected. Tidal power-generation schemes, or estuarine freshwater-storage barrages (Taylor 1979) are proposals with such potential. In the current climate of public opinion, proposals for such major developments (at least in the western world) would be accompanied by detailed ecological studies. The very size of the required engineering works and the high capital costs are likely to lead to a very prolonged evaluation. The various water storage schemes discussed by Taylor (1979), although investigated for many years, remain unbuilt, largely because earlier estimates of water demand have been revised downwards making provision of new water supply schemes uneconomic.

Reclaimed land in itself may provide a series of habitats of considerable interest – seawalls, brackish dykes and rough pasture (Gray 1970, 1977, 1979*a*,*b*; Beeftink 1975; Williams & Hall 1987). In regions where natural brackish marshes have been modified, reclaimed land may support the major populations of some species restricted to brackish habitats. The long-term conservation values of these sites are lessened by changes in agriculture practice. More efficient drainage and a change from grazing to cropping might reduce the diversity of these lands. The very rapid decline in grazing marsh in Essex, largely because of conversion to arable between 1938 and 1987, is documented by Williams & Hall (1987).

The embankments protecting reclaimed saltmarsh are vulnerable to damage by storms and high tides. Over the centuries there have been numerous examples of breaches to seawalls with inundation of the low-lying land behind them (Wagret 1968; van Veen 1955). In some instances, repair was not immediately possible so that halophytic vegetation was temporarily re-established. Even without storm damage, seawalls require regular upkeep. If, for economic or other reasons, maintenance is not carried out, then walls can become derelict with subsequent re-establishment of saltmarsh (Kirkby 1984). Areas of reclaimed land which have subsequently reverted to saltmarsh may be of interest because of the long survival of unusual mixed assemblages of halophytic and glycophytic species.

While reclamation in various forms has been responsible for substantial saltmarsh losses (even if compensated for in some instances by new marsh growth), losses also arise from natural processes of storm damage, relative sea-level rise, or channel changes (Gosselink & Baumann 1980). The relative importance of different agencies in saltmarsh loss is poorly documented. Human modification of estuaries can alter these natural processes. For example, stabilisation of channels by training walls can cause accelerated erosion in some parts of an estuary while promoting accretion and marsh development elsewhere (Gray & Adam 1974).

Saltmarshes might also be affected by development far removed from the estuary. Many rivers have been substantially modified by water abstraction and flood-mitigation works. These have the effect of reducing both total freshwater inputs and the temporal variability of inputs (see Nichols *et al.* 1986). The resulting modified salinity regime is likely to be reflected in changes in the estuarine biota; whether these are direct consequences in saltmarshes remains to be investigated.

Pollution

Saltmarshes are exposed to both chronic and acute pollution. The effects of some acute incidents have been studied but there have been few studies of the effects of chronic pollution. Even when experimental toxicological studies have been conducted, it is difficult to extrapolate results from the laboratory to the field and even harder to predict from data on one or two species the effects of pollution on the whole ecosystem. The behaviour of pollutants under conditions of changing salinity may be particularly difficult to predict.

As many estuaries are heavily industrialised, the sources of pollution are often local to the saltmarsh but pollutants may also be carried into estuaries from upstream and from out to sea. Indeed, one of the major potential 'threats' to saltmarsh is posed by one of the most diffuse global pollutants of all. If increased atmospheric carbon dioxide concentrations caused by fossil fuel combustion cause a global warming then many saltmarshes may be inundated by the consequent rise in sea level.

Although saltmarshes are natural sediment sinks, in many cases increased sedimentation as a result of human activity could be regarded as a form of pollution. Catchment clearance may result in substantial inputs of sediment into estuaries (Froomer 1980; Nichols *et al.* 1986). Clearing and development close to saltmarsh may lead to direct sedimentation in the upper zones of the marsh. While increased sedimentation in the lower-marsh may result in marsh expansion, very high rates of sedimentation may smother vegetation (Chung 1982). Deposition of large amounts of sediment in the upper-marsh would normally result in either death or reduced growth of the vegetation.

Freshwater can be viewed as a pollutant in some circumstances. Urbanisation of areas landward of saltmarsh may be accompanied by the construction of stormwater drains discharging directly into the marsh. These drains may introduce large amounts of sediment, and oil and heavy metals washed from road surfaces. The direct input of water may lead to lowering of soil salinities permitting invasion of species of brackish habitats. In southwest Western Australia, extensive invasion by *Typha orientalis* and *Bulboschoenus caldwellii* displacing saltmarsh species as a result of flushing by stormwater drainage has been reported (Pen 1987). In California, Beare & Zedler (1987) suggest that changed estuarine salinity conditions resulting from the timing of releases of freshwater from upstream reservoirs may permit invasion of saltmarshes by *Typha domingensis*.

In addition to sediment, the tides also wash into marshes drift litter and a great variety of non-degradable products of affluent society. Considerable accumulations of plastic material may build up in the drift line.

Although most public concern has been concentrated on industrial and domestic sources of pollution, agricultural run-off is the major diffuse source of pollutants for most estuaries (see Nichols *et al.* 1986). The effects of herbicide and pesticide residues, on either the estuary or its fringing saltmarsh, have been little studied. However, increased nutrient loadings have promoted algal growth. Probably the best studied example is the Peel-Harvey estuary in Western Australia (Hodgkin *et al.* 1985). In this case, phosphate fertiliser use in the catchment over many years led to an increase in the nutrient content of the estuarine waters and blooms of algae. Masses of algae washed on to the shore create a public nuisance as they decay and may smother saltmarsh. Géhu (1984*c*) records similar smothering of saltmarshes in France by algal growth stimulated by piggery effluent. Adverse effects of algal growth on saltmarsh have also been recorded in southern England (Ranwell 1972, 1981*b*). Domestic sewage contributes to raising many nutrient levels in estuaries.

Many industrial processes release heavy metals into the environment. Saltmarsh and mangrove soils appear to act as sinks for heavy metal ions which under anaerobic conditions may be precipitated as insoluble sulphides as well as being bound by cation exchange on clay minerals. Not all these ions become inert and a proportion is taken up by plants, thus providing the potential for accumulation at higher levels in the food chain. The uptake of ions and their distribution in the plant is species-specific and Beeftink *et al.* (1982) suggest that European saltmarsh species can be grouped into several categories on the basis of their accumulation characteristics. The behaviour of heavy metal ions in the saltmarsh environment is reviewed by Giblin *et al.* (1980), Ragsdale & Thorhaug (1980) and Beeftink *et al.* (1982).

Although high levels of heavy metal ions have been recorded from saltmarsh sediments there have been no reports of acute toxicity symptoms in plants. Whether there are sublethal effects has not been investigated. Giblin *et al.* (1980) demonstrated elevated levels of some ions in a number of animals after a saltmarsh was experimentally treated with sewage sludge to which heavy metals had been added but there have been few studies on the effects of heavy metals on saltmarsh fauna.

Saltmarshes appear to perform a 'filtering' role in removing metal ions from circulation. However, since many of the ions are precipitated as sulphides, any disturbance causing oxidation may result in re-solution.

Most pollutants are not readily visible. Oil, however, is highly conspicuous. In most navigable estuaries, there are regular small spillages of petroleum products, while in commercial ports larger spills are to be expected (for example, the mangroves and saltmarshes in Botany Bay, in New South Wales, have been the recipients of several spills over recent years – Allaway 1982; Anink *et al.* 1985). Oily discharge from refineries and other industrial sites may cause chronic pollution of saltmarshes. Major oil spills from shipwrecks have affected some marshes. [For example, oil from the 'Torrey Canyon' and 'Amoco Cadiz' was washed onto marshes in Britain and France while saltmarshes in the Magellan Straits were polluted by oil from the 'Metula']. Fortunately, the number of tanker wrecks is small, but since accidents could occur almost anywhere on the seas, virtually all marshes are vulnerable to major oil-pollution incidents.

The effects of oil on saltmarsh vegetation have been reviewed by Baker (1979, 1983). It is difficult to generalise about impacts as the susceptibility of species varies, the toxicity of different crude oils and oil products varies considerably (the lighter volatile fractions appear to be the most acutely toxic), and the time of year pollution occurs may influence plant response, while the amount of oil reaching the marsh surface will depend on both the amount spilt and the tides at the time.

Some species are killed by only one oiling. In northern Europe annual

Salicornia spp. fall into this category (Baker 1979). As these species do not support a soil seed bank bridging between generations it is possible that oil spills could cause temporary local extinction. Other species are much more tolerant. The most tolerant in Baker's (1979) studies was the umbellifer *Oenanthe lachenalii*, a species normally associated with *J. maritimus* in the upper saltmarsh. *J. maritimus* is a susceptible species but with frequent oiling *O. lachenalii* increases in abundance at the expense of the dying *Juncus*. Other species lie on a continuum of susceptibility between *Salicornia* and *Oenanthe*.

The actual toxic components of oil and the mechanisms by which they act are poorly understood. In some cases, death may be an indirect consequence of oiling. *S. anglica* is tolerant of single oilings but succumbs to more frequent pollution (Baker 1979). Baker (1979) suggested that oil may interfere with the transfer of oxygen from the atmosphere to the rhizosphere via aerenchyma. Ferrell, Seneca & Linthurst (1984) similarly suggest that death of *S. alterniflora* and *S. cynosuroides* following oiling is not due to direct toxicity but to interference with gas transfer and through effects on the substrate.

Old weathered oil may stimulate growth of a number of species (Baker 1979, 1983).

Once oil has been deposited on a marsh, there may be pressure by authorities for some form of clean-up treatment. While oil is unsightly and may be hazardous to bird life, and on grazed marshes may interfere with farmers' livelihood, currently available clean-up procedures – use of chemical dispersant, burning, cutting, sediment stripping – may cause more long-term damage than allowing the oil to weather without treatment (Baker 1979, 1983). However, in the case of very heavily polluted marshes, sediment stripping may be required but as far as possible this should be followed by replanting to stabilise the surface and to promote accretion.

As far as possible, attempts should be made to prevent oil reaching marshes through improved procedures at oil berths and the rapid deployment of booms after spills. However, despite the development of contingency plans to deal with oil pollution, accidents rarely conform to forecasts.

Recreational and educational use

Saltmarshes are not suitable for many recreational uses. The largest recreational user group is probably wildfowlers, who have been active in securing the conservation of many marsh areas. Ornithologists are also attracted to saltmarshes. The growing popularity of ornithology as a hobby makes bird watchers an active pressure group for conservation. In some localities visiting ornithologists make an appreciable contribution to the local economy. Others interested in natural history are also attracted to saltmarsh.

Saltmarshes are also places of great beauty. The scene constantly changes with the ebb and flood of the tide and seasonally with the phenological cycle of the dominant species changes summarised by:

> 'A more desolate region can scarce be conceived, and yet it is not without beauty. In summer, the thrift mantles the marsh with shot satin, passing through all gradations of tint from maiden's blush to lily white. Thereafter a purple glow steals over the waste, as the sea lavender bursts into flower and simultaneously every creek and pool is royally fringed with sea aster. A little later the glass-wort, that shot up green and transparent as emerald glass in the early spring, turns to every tinge of carmine.'
>
> *Mehalah – a story of the salt marshes*
>
> S. Baring-Gould (1880)

Even when relatively close to urban or industrial development, being on a large saltmarsh can provide a wilderness experience, often heightened by dramatic skyscapes and the vast flocks of estuarine birds. While not everyone recognises the beauty of the saltmarsh scene, it has been captured by many artists.

In various parts of the world, increasing recreational use of off-road vehicles (trail-bikes and 4-wheel drives) poses problems to marsh management. Vehicular use may cause immediate destruction of vegetation, with the areas affected being vulnerable to erosion. Although wheel tracks may be only a few centimetres deep, their presence may considerably alter local hydrology. Tracks may act as artificial creeks and promote improved drainage of previously waterlogged areas; conversely under other circumstances they can remain water-filled and convert previously well-drained areas into waterlogged ones (Godfrey, Leatherman & Buckley 1978). As one set of tracks becomes deeper, it tends to be abandoned in favour of a new trail so that very rapidly large areas can be converted into a quagmire. Even with very infrequent use vehicle tracks can create a new environment, often providing a niche for lower-marsh species to invade the upper-marsh, which may survive for many years (see Fig. 7.3). Succulent species may be particularly prone to vehicular damage (Pen 1983).

While perhaps not a recreational use of marshland, a practice causing similarly destructive damage is recorded by Géhu (1984*c*). In western France, ploughshares are de-rusted by being used in saltmarshes.

Fig. 7.3 Wheel ruts in upper saltmarsh on Holy Island (Northumberland, UK) – these provide a habitat for the low-marsh species *Spartina anglica* and modify the local drainage pattern.

Saltmarshes have been used for educational visits for many years. Amongst the earliest formal fieldcourses were those organised by Professor F. W. Oliver to the saltmarshes of the Bouche D'Erquy and recorded in early volumes of the *New Phytologist*.

Saltmarshes are ideal for demonstrating and studying zonation patterns while the relatively low species diversity of many communities makes them appropriate for exercises of vegetation description and sampling. There is an almost unlimited potential for individual projects involving aspects of saltmarsh ecology. Care must be taken in planning educational exercises on saltmarshes to avoid overuse of particular sites. Trampling may cause as much lasting damage as vehicles and regularly visited quadrats may soon become bare areas linked by beaten tracks.

Conservation

Wetland conservation has become a matter of international concern (Maltby 1986) as more and more of the wetland areas of the world have been drained or modified. For a long time the general public attitude towards wetlands has been best expressed as 'wetlands are wastelands'. Wetland modification has been seen as a public-spirited enterprise, an attitude fostered by local and national governments and by entrepreneurial developers. This attitude even finds expression in the term reclamation with its connotations of restoring to some earlier (and better) state. These views are still widely held but are increasingly questioned. Not only are the values of wetlands being championed but detailed analysis of various developments involving wetland modification suggests that often the supposed advantages, in either economic or environmental terms, failed to materialise.

In the developed world, increasing environmental consciousness and public pressure have led to various mechanisms to allow consideration of the impacts of development proposals during the planning process. While many aspects of these mechanisms can be criticised, they nevertheless represent an advance over earlier practices. In the third world, where very extensive wetlands are currently subject to development proposals (Maltby 1986), often with funding from aid programmes, evaluation of the consequences is often minimal. However, the pressures for land development to meet the needs of rapidly expanding populations must be recognised. Simply to argue that wetlands should be conserved, without being able to demonstrate the benefits from such a policy or proposing alternative ways of meeting community needs, is unlikely to succeed. The issues are not easily or speedily resolved: would the benefit of reclaiming for rice cultivation, as advocated by Passioura (1986) for upper intertidal wetlands in south-east Asia, which in their natural state may not be directly exploited by humans but which may have considerable indirect value, be greater in the long-term than leaving the areas undeveloped? Answers to such questions are required now; without them, it is probable that the reclamation and development options will be taken.

The problems of saltmarsh conservation in the developing world are part of the wider problems of conservation in these countries. Conservation of saltmarshes in the western world is important in its own right but also serves to provide examples of conservation practice which can be adopted in other situations.

In the western world there has been a long history of saltmarsh conservation. For example, some of the earliest nature reserves in Britain were the saltmarsh, shingle and sand-dune complexes on the North Norfolk coast (Steers 1964). These areas were protected for a variety of reasons: as habitats for rare species, as sites for educational purposes and scientific research, and as areas regarded as being of exceptional beauty. Other saltmarshes were protected by wildfowling interests. Even in the early years of the twentieth century, extensive areas of saltmarsh were recognised as providing one of the few opportunities for 'wilderness' experience in industrialised countries.

Historically, much conservation effort has been concerned with the recognition of sites which could, as judged against various criteria, be regarded as special. Sites which display outstanding development of particular features or provide habitats for rare species were protected by some form of reservation. Ratcliffe (1977) represents one such attempt at site evaluation on a national scale. Although ideally such an approach requires a comprehensive data base on which to conduct an evaluation such information is still unavailable for many areas. Nevertheless, the judgements of those who long ago secured sites for conservation have generally stood the test of time.

The choice of criteria for comparative site evaluation of wetlands remains controversial (see Pressey 1985). Several of the schemes which have been used are weighted in favour of particular attributes (notably habitat value for birds). Unless some special weighting is given to some attributes, saltmarshes tend to score badly as many have relatively low species diversity and the widespread distribution of many of the species means that sites with a concentration of rare species, or species at distribution limits, are uncommon.

Whatever criteria for evaluation are adopted, it is clear that in most countries there is a strong case for inclusion of more saltmarshes in reserves. However, because reservation seeks to protect 'specialness' it is unlikely that more than a small proportion of saltmarshes will be protected in this way.

This raises the prospect at some time in the future of there being a relatively small number of isolated saltmarsh sites as reserves with few other saltmarshes surviving industrialisation and reclamation.

There are strong arguments against such a situation and in favour of protecting the largest possible area of saltmarsh.

The maintenance of saltmarsh areas outside reserves may be necessary for these reserves to fulfill their conservation role. The importance of particular sites as seasonal habitats for birds is often a major justification of their preservation. However, although large flocks of migratory or nomadic species may congregate at particular sites for limited periods, at other times populations may be more dispersed. Exactly why some sites are more favoured than others is not always clear but, nevertheless, en route to and from these major sites there is a need for stop-over or staging sites. The intensity of use of these staging sites may vary from year to year and some degree of substitution between sites may be possible. Nevertheless, for species survival a network of potentially viable sites needs to be maintained linking the major sites. Similarly non-reserved sites may provide reservoirs for the recolonisation of protected sites after major disturbance (for example, major oil pollution or storm damage).

There are also grounds for arguing that wetlands in general may have various attributes which are of direct economic value (Maltby 1986; Adam 1984; Adam *et al.* 1985). In the case of saltmarshes, the normally quoted attributes include shoreline protection, provision of sinks to mop up pollutants (most particularly heavy metals), provision of habitat for economically valuable species, and as sources of detritus to estuarine and offshore food chains. In addition, it can be argued that development of saltmarshes may result in consequent further pollution to the estuary as well as loss of aesthetic value.

The extent to which such claims are justified is unclear. Even if saltmarshes provide all or some of these 'services' the actual importance of them is likely to vary between sites. Attempts have been made to assign economic value to saltmarshes but such exercises are of dubious value given the lack of reliable quantitative data on ecosystem function and the subjectivity which must apply to the assigning of values to some attributes (such as aesthetic values). Even if future research allows such valuations to be improved, who should bear the cost of marsh maintenance? Most of the suggested economic values accrue to society as a whole, not to the owners of individual sites. From the viewpoint of owners, realisation of the development values of a saltmarsh will far outweigh the advantages of maintaining the marshland.

In the United States and elsewhere, conservation of saltmarsh has frequently been justified on the grounds that saltmarshes, through the export of detritus, sustain commercially valuable fisheries. It is now clear that the linkages between saltmarshes and fisheries are complex (see pp.348–355; Nixon 1980) and we are unable to justify any simple generalisation about the importance of saltmarsh to fisheries. Nevertheless, this supposed relationship undoubtedly prevented the destruction of extensive marshes and is still invoked today. Nixon (1980) suggests that this reflects an argument on the lines of 'Yes, perhaps we overstated the case a bit, but it was important to help save the marshes. Now that is done, or at least well along, and we can go back and work on getting our science right'. Such an approach is dangerous and 'it is a bad bargain to trade our credibility for political advantage' (Nixon 1980). As the assessment of environmental issues moves increasingly into the domain of the courts, it is important that the arguments used to justify conservation stand up to scrutiny. If they are proved wrong in one particular instance, then the general case for conservation may fall into disrepute.

Unfortunately although the 'specialness' of outstanding sites may be demonstrable and the evidence withstand cross-examination, the generalisations about the values of wetlands are more difficult to defend (Bardecki 1984). The arguments are all theoretically credible and can be shown to be valid in specific instances (some saltmarsh creeks are nursery grounds for commercially important fish species, some marshes are net exporters of carbon, on the other hand other sites appear to be importers). Faced with such uncertainty, it could be argued that given the relatively limited total area of saltmarsh, the historic losses and degradation and the potential for habitat loss on an even greater scale in the future, it would be prudent to conserve as much saltmarsh as possible. If saltmarshes do have the values ascribed to them, it would be inappropriate to sacrifice them for short-term gain. Unfortunately, developers and politicians are rarely swayed by arguments full of 'ifs' 'buts' and 'maybes' and would be prepared to take the gamble.

There is an urgent need for research on the functioning of saltmarsh ecosystems. Such research will require multidisciplinary teams free from the preconceptions of the recent past. It will also require that a network of sites be studied. Historically there has been an understandable tendency to extrapolate from a few study sites. However, several of these sites which provide the paradigms for our present understanding of saltmarsh are clearly in the 'outstanding' or 'special' category and it is not clear that all the findings are applicable to 'ordinary' sites. This research is both time consuming and relatively expensive. Even if the funding were available, many sites would be threatened before findings were available. To hold the fort, it will be necessary to document the more-easily studied attributes of sites (such as their use by various species such as birds and juvenile fish) and to stress the possibilities (but not certainty) of the importance of other values and urge caution against irrevocable decisions.

The legislative and bureaucratic framework against which decisions affecting saltmarsh are taken varies from country to country. Historically, and to a large extent still at present, decisions have been taken on a local *ad hoc* basis. This has meant that the decision makers have been largely ignorant of the cumulative effect of numerous similar decisions affecting various areas along the coastline with the result that the aggregate extent of habitat modification is large (Odum 1982). Clearly one of the first requirements for rational planning in the coastal zone is an appropriate geographical scale for assessment and evaluation.

In several parts of the world, there is now specific planning legislation which confers blanket protection on at least certain types of wetland. A consequence of this has been that considerable effort has been devoted to defining and delimiting wetland boundaries. Although the boundaries of saltmarsh are sometimes easily detected (as where a seawall provides an artificial limit), in other cases considerable resources may be devoted to chasing a moving target. The upper limit may be defined in terms of a tidal datum – determination of such a boundary in the field requires accurate survey. Alternatively biological definitions can be employed, based either on the distribution limits of particular species or on features such as root distribution or morphological variation in selected species (Seliskar 1983, 1985). Unfortunately, tidal and biological boundaries do not necessarily coincide (Teleky 1978). Although the use of a tidal datum appeals to planners, engineers and lawyers as an absolute measure, free from the qualifications surrounding a biologically defined boundary, it is not constant over time (see pp.2–13). The biological boundary would be expected to vary (possibly with some lag period) in response to movement of the tidal boundary but would also respond to local variation in relative sea level and climatic factors. While the elevation range of this 'zone of uncertainty' might cover only a few centimetres, its horizontal translation may extend across many metres. Legalistic concern over the position of an 'exact' boundary may also divert attention away from the fact that maintenance of a wetland requires consideration of land use in its surroundings. Development right up to an arbitrarily defined boundary may well have deleterious consequences to the wetland. Frequently a buffer zone will be required, but no simple prescription as to its extent can be given; each case will have to be determined individually.

Planning issues involving saltmarsh are often complicated by the involvement of several different arms of government. Saltmarshes occupy the physical boundary between land and sea, a boundary which also marks the limits of jurisdiction of different departments. The intertidal nature of saltmarshes may also lead to uncertainty as to tenure and the rights of traditional users.

Management

Establishment of nature reserves on saltmarshes creates a need for management.

Long-term maintenance of saltmarsh habitats requires that consideration be given to events occurring over a wider area. Pollution does not respect reserve boundaries. Protection of saltmarshes from damage by pollution requires the identification and regulation of sources elsewhere in the estuary catchment. While, at least in theory, regulation of point sources is possible, it is much more difficult to effectively manage diffuse non-point sources of pollution (such as run-off from agriculture), while the risk from accidents (such as shipping accidents) can never be completely eliminated. In most instances the regulation of pollution falls to different government agencies from those charged with management of conservation reserves. Unfortunately, this can mean that pollution regulators are unaware of the susceptibility of saltmarshes to pollutants.

The conservation of migratory birds, for which saltmarshes may be important habitats, cannot be achieved unilaterally but requires international co-operation. [There are already examples of international treaties to this end – for example JAMBA, the Japan–Australia Migratory Birds Agreement.]

The naturally dynamic character of the saltmarsh environment may create difficult management problems. If a reserve is exposed to erosion which might, for some long period of time, result in the loss of much of the saltmarsh should attempts be made to prevent the erosion by means such as training wall construction, channel diversion, or offshore configuration dredging? Frequently the cost of such proposals would rule them out of consideration but there may be occasions when work of this nature is both affordable and feasible from an engineering standpoint. Although cycles of erosion and accretion may occur at many sites, erosion may be aggravated by modifications elsewhere in an estuary. The management response to such circumstances would be the subject of considerable debate but intervention may not necessarily be undesirable.

As has been stressed earlier (pp.188–192) the vegetation of many north European marshes has been determined by past and present grazing regimes. If grazed marshes become nature reserves should grazing be continued? Cessation of grazing will lead to a reduction in both plant species and community diversity in mid- and upper-marsh zones. The presence of taller vegetation may provide more niches for terrestrial invertebrates although the reduction in plant species diversity may mitigate against an increase in total faunal diversity. Tall rank vegetation is generally avoided by wildfowl, whose seasonal presence is often one of the major justifications for protecting saltmarsh sites.

In order to maintain the conservation values of previously grazed sites then it will be necessary to continue or re-introduce grazing. There is scope for experimentation with grazing regimes – stocking rates, periods of the year when grazing occurs, and mix of species could all be varied. Grazing intensities to meet conservation objectives are probably lower than those employed in commercial farming. With the decline in grazing on saltmarshes it may become difficult for reserve managers to find tenant farmers who can incorporate grazing of a saltmarsh in their other operations. Mowing may be viewed as a more practical alternative to maintaining small numbers of livestock. However, Bakker (1985) showed that although species diversity on marshes previously grazed increased quickly initially under both grazing and mowing regimes, this trend was not continued under mowing. The higher species diversity under grazing may reflect the greater spatial diversity created by selectivity in grazing, local nutrient enrichment, and the mosaic of small gaps in the sward generated by grazing and trampling. From Bakker's (1985) results it appears unlikely that, in the long-term, mowing will be a satisfactory substitute for grazing.

In other parts of the world, grazing does not appear to play such a role in community diversification and it may not have a role in conservation management. Further work is required on the effects of other traditional management practices (for example, burning in the southern United States).

While it may be important to maintain grazing regimes in northern

Europe, those few sites which historically have been ungrazed should be protected from the introduction of grazing in order to maintain the diversity of marsh types.

Marsh creation and rehabilitation

In recent years, there has been considerable interest in the establishment of saltmarsh by planting (Lewis 1982). Earlier in the century extensive planting of *S. anglica* at various localities around the world was successful in establishing extensive areas of marsh (Ranwell 1967; Boston 1981). *S. anglica* has more recently been planted on a very large scale in China (Chung 1982, 1985) but elsewhere in the world more attention is given to control rather than further planting. Most of the early plantings of *S. anglica* were for the purpose of stabilising channels or to promote accretion (to be followed by reclamation) rather than to create marsh *per se*. The extensive recent plantings in China have been to create marsh pasture, a source of fodder and green manure, and to provide sites for subsequent reclamation (Chung 1982).

The recent developments in marsh creation has been more concerned with conservation interests – the creation of new marshes to provide new habitat and as a component of mitigation policies in which creation of new habitats is offset against habitat lost during development. Particularly in the United States, saltmarsh creation has filled the dual role of stabilising dredged spoil and providing new habitat. In addition to creation of new habitats, there is an increasing demand for rehabilitation and restoration works of damaged areas. For example, very heavily oil-polluted marshes may need to be excavated and re-established, while on a more local scale, pipeline or powerline easements may require re-vegetation.

The history of *S. anglica* indicates that successful marsh establishment may not be difficult in appropriate environments. However, even now there is considerable uncertainty as to the long-term course of succession in *Spartina* marshes.

If marshes are established by planting low-marsh species then natural processes of accretion may in time lead to the establishment of more diverse communities. On dredge spoil, covering a range of elevations, it is possible to plant a number of species in order to recreate *de novo* the whole zonation pattern.

The ecotypic variation in most saltmarsh plant species (pp.107–131) suggests that care should be taken in choosing propagation material. Simply to regard all individuals of a species as equal in planting programmes is likely to be a recipe for failure. If extensive works are to be undertaken, it will be necessary to establish nursery stocks – collection of material from existing marshes could cause considerable damage (Race 1985).

Mitigation policies may permit marsh re-establishment by removal of sluices or seawalls protecting reclaimed land. Purchase of reclaimed agricultural land for this purpose is likely to be expensive but, if the development for which marsh creation is required in mitigation is a major one, the exercise may be economically viable.

Although many attempts at marsh creation has been undertaken (Lewis 1982) not all have been successful (Knutson, Ford & Inskeep 1981; Race 1985). Race (1985) has pointed out that creation programmes have often been poorly documented and frequently it is very difficult to assess their success. Until existing schemes have been properly assessed, proposals for further mitigation works should be approached with caution.

Prospect

The future of the world's saltmarshes remains uncertain. Perhaps within a few decades a 'Greenhouse effect' induced sea-level rise will have destroyed most saltmarshes. In addition, technology now exists for modifying coastal environments on a vast scale. On the other hand, greater public involvement in the planning process and increased scepticism about the benefits of technological solutions may result in a reduction in the rate of habitat loss. Public pressure may be sufficient to modify major schemes even after work has commenced; the re-appraisal and modification of the Dutch Delta Project is the most spectacular example of this (Saeijis & de Jong 1981).

If decisions on the fate of major projects are to be made on a rational basis both engineers and the public will require information on the environmental consequences of the proposals. Ecologists have not been used to addressing such issues and are often reluctant to commit themselves to firm answers. Existing information is often insufficient for prediction, but retrospective studies of the effects of past developments and better integration of the wealth of empirical observations with developing ecological theory should permit ecologists to enter into debate with greater confidence.

Increased understanding of the ecology of saltmarshes will also aid management. However, there needs to be greater discussion on the aims and objectives of management. Increasing habitat diversity within a saltmarsh will increase its biotic diversity. Diversity is often seen as a desirable attribute of reserved areas but should increasing diversity be an aim of saltmarsh management? Burger, Shisler & Lesser (1982) have suggested that, when considering management for birds, the answer may be no –

> 'The importance of the natural salt marsh ecosystem and its attendant species should not be under-valued because of the higher numbers which impoundments, stop-ditch marshes and other managed areas can support. Bird species using managed areas can use a wider variety of habitats while the typical salt marsh species are limited to large expanses of relatively undisturbed salt marshes.'

Mitigation strategies similarly pose many problems for managers and planners. Mitigation may involve habitat substitution rather than exact replacement; how are the relative merits of different habitats to be assessed? Even if the newly 'created' habitat is similar to that destroyed, should the ability to create habitat be used in partial justification for the destruction of existing wetland areas?

Ecologists have an important role in advising policy makers but ultimately the task of determining the merits of conflicting demands for land use must be a political one. Will politicians heed the words of Gerard Manley Hopkins?

> What would the world be, once bereft
> Of wet and of wildness? Let them be left,
> Oh let them be left, wildness and wet;
> Long live the weeds and wilderness yet.
> (from *Inversnaid* – by
> G. M. Hopkins – Gardner &
> Mackenzie 1967)

References

Abd. Aziz, S. A. & Nedwell, D. B. (1979). Microbial nitrogen transformations in the salt marsh environment. In *Ecological Processes in Coastal Environments*, ed. R.L. Jefferies & A. J. Davy, pp. 385–98. Oxford: Blackwell Scientific Publications.

(1986*a*). The nitrogen cycle of an East Coast, U.K., saltmarsh: I. Nitrogen assimilation during primary production; detrital mineralisation. *Estuarine, Coastal and Shelf Science*, **22**, 559–75.

(1986*b*). The nitrogen cycle of an East Coast, U.K., saltmarsh: II. Nitrogen fixation, nitrification, denitrification, tidal exchange. *Estuarine, Coastal and Shelf Science*, **22**, 689–704.

Abdulrahman, F. S. & Williams, G. J. (1981). Temperature and salinity regulation of growth and gas exchange of *Salicornia fruticosa* (L.) L. *Oecologia (Berlin)*, **48**, 346–52.

Adam, P. (1976*a*). *Plant sociology and habitat factors in British saltmarshes*. Ph.D. thesis, University of Cambridge.

(1976*b*). The occurrence of bryophytes on British saltmarshes. *Journal of Bryology*, **9**, 265–74.

(1977*a*). The ecological significance of 'halophytes' in the Devensian flora. *New Phytologist*, **78**, 237–44.

(1977*b*). On the phytosociological status of *Juncus maritimus* on British saltmarshes. *Vegetatio*, **35**, 81–94.

(1978). Geographical variation in British saltmarsh vegetation. *Journal of Ecology*, **66**, 339–66.

(1981*a*). Vegetation of British saltmarshes. *New Phytologist*, **88**, 143–96.

(1981*b*). Australian saltmarshes. *Wetlands (Australia)*, **1**, 11–19.

(1984). Towards a wetlands conservation strategy. *Wetlands (Australia)*, **4**, 33–48.

Adam, P., Birks, H. J. B. & Huntley, B. (1977). Plant communities of the Island of Arran, Scotland. *New Phytologist*, **79**, 689–712.

Adam, P., Fisher, A. & Anderson, J. M. E. (1987). Pollen collection by honey bees from *Sarcocornia quinqueflora*. *Wetlands (Australia)*, **4**, 25–8.

Adam, P., Urwin, N., Weiner, P. & Sim, I. (1985).

Coastal Wetlands of New South Wales. A Survey and Report Prepared for the Coastal Council of New South Wales. Sydney: Coastal Council of NSW.

Adam, P., Wilson, N. C. & Huntley, B. (1988). The phytosociology of coastal saltmarsh vegetation in New South Wales. *Wetlands (Australia)*, **7**, 35–85.

Adam, P. & Wiecek, B. M. (1983). The salt glands of *Samolus repens*. *Wetlands (Australia)*, **3**, 2–11.

Adams, D. A. (1963). Factors influencing vascular plant zonation in North Carolina salt marshes. *Ecology*, **44**, 445–56.

Ahmad, I., Larher, F., Mann, A. F., McNally, S. F. & Stewart, G. R. (1982). Nitrogen metabolism of halophytes. IV. Characteristics of glutamine synthetase from *Triglochin maritima* L. *New Phytologist*, **91**, 585–95.

Ahmad, I., Larher, F. & Stewart, G. R. (1979). Sorbitol, a compatible osmotic solute in *Plantago maritima. New Phytologist*, **82**, 671–8.

Ahmad, I. & Wainwright, S.J. (1976). Ecotypic differences in leaf surface properties of *Agrostis stolonifera* from salt marsh, spray zone and inland habitats. *New Phytologist*, **76**, 361–6.

(1977). Tolerance to salt, partial anaerobiosis and osmotic stress in *Agrostis stolonifera. New Phytologist*, **79**, 605–12.

Ahmad, I., Wainwright, S. J. & Stewart, G. R. (1981). The solute and water relations of *Agrostis stolonifera* ecotypes differing in their salt tolerance. *New Phytologist*, **87**, 615–29.

Ahmad, N., Wyn Jones, R. G. & Jeschke, W. D. (1987). Effects of exogenous glycinebetaine on Na^+ transport in barley roots. *Journal of Experimental Botany*, **38**, 913–21.

Albert, R. (1975). Salt regulation in halophytes. *Oecologia (Berlin)*, **21**, 57–71.

Albert, R. & Popp, M. (1977). Chemical composition of halophytes from the Neusiedler lake region in Austria. *Oecologia (Berlin)*, **27**, 157–70.

(1978). Zur Rolle der löslichen Kohlenhydrate in Halophyten des Neusiedlersee – Gebietes (Österreich). *Oecologia Plantarum*, **13**, 27–42.

Allaway, W. G. (1982). Mangrove die-back in Botany Bay. *Wetlands (Australia)*, **2**, 2–7.

Allen, E. B. & Cunningham, G. L. (1983). Effects of vesicular–arbuscular mycorrhizae on *Distichlis spicata* under three salinity levels. *New Phytologist*, **93**, 227–36.

Anderson, C.E. (1974). A review of structures of several North Carolina salt marsh plants. In *Ecology of Halophytes*, ed. R. J. Reimold & W. H. Queen, pp. 307–44. New York: Academic Press.

Andersson, B., Critchley, C., Ryrie, I. J., Jansson, C., Larsson, C. & Anderson, J. M. (1984). Modification of the chloride requirement for photosynthetic O_2 evolution. *FEBS Letters*, **168**, 113–17.

Andrews, T. J. & Muller, G. J. (1985). Photosynthetic gas exchange of the mangrove, *Rhizophora stylosa* Griff., in its natural environmental. *Oecologia (Berlin)*, **65**, 449–55.

Anink, P. J., Roberts, D. E., Hunt, D. R. & Jacobs, N. E. (1985). Oil spill in Botany Bay: short term effects and long term implications. *Wetlands (Australia)*, **5**, 32–41.

Anning, P. (1980). Pastures for Cape York Peninsula. *Queensland Agricultural Journal*, (March–April), 148–71.

Antlfinger, A. E. & Dunn, E. L. (1979). Seasonal patterns of CO_2 and water vapour exchange of three salt marsh succulents. *Oecologia (Berlin)*, **43**, 249–60.

Apinis, A. E. & Chesters, C. G. C. (1964). Ascomycetes of some salt marshes and sand dunes. *Transactions of the British Mycological Society*, **47**, 419–35.

ap Rees, T., Jenkin, L. E. T., Smith, A. M. & Wilson, P. M. (1987). The metabolism of flood-tolerant plants. In *Plant Life in Aquatic and Amphibious Habitats*, ed. R. M. M. Crawford, pp. 227–38. Oxford: Blackwell Scientific Publications.

ap Rees, T. & Wilson, P. M. (1984). Effects of reduced supply of oxygen on the metabolism of roots of *Glyceria maxima* and *Pisum sativum. Zeitschrift für Pflanzenphysiologie*, **114**, 493–503.

Armentano, T. V. & Woodwell, G. M. (1975). Sedimentation rates in a Long Island marsh determined by ^{210}Pb dating. *Limnology and Oceanography*, **20**, 452–6.

Armstrong, W. (1978). Root aeration in the wetland condition. In *Plant Life in Anaerobic Environments*, ed. D. D. Hook & R. M. M. Crawford, pp. 269–97. Ann Arbor: Ann Arbor Science Publishers.

(1979). Aeration in higher plants. *Advances in Botanical Research* **7**, 225–32.
(1982). Waterlogged soils. Contributed chapter in *Environmental Plant Ecology*, 2nd edn, ed. J. R. Etherington, pp. 290–330. Chichester: John Wiley.
Armstrong, W. & Beckett, P. M. (1985). Root aeration in unsaturated soil: a multi-shelled mathematical model of oxygen diffusion and distribution with and without sectoral wet-soil blocking of the diffusion path. *New Phytologist*, **100**, 293–311.
Armstrong, W., Wright, E. J., Lythe, S. & Gaynard, T. J. (1985). Plant zonation and the effects of the spring–neap tidal cycle on soil aeration in a Humber salt marsh. *Journal of Ecology*, **73**, 323–39.
Arnold, A.M. (1973). Interactions of light and temperature on the germination of *Plantago maritima* L. *New Phytologist*, **72**, 583–93.
Aspinall, D. & Paleg, L. G. (1981). Proline accumulation: physiological aspects. In *The Physiology and Biochemistry of Drought Resistance in Plants*, ed. L. G. Paleg & D. Aspinall, pp. 205–41. Sydney: Academic Press.
Atkinson, M. R., Findlay, G. P., Hope, A. B., Pitman, M. G., Saddler, H. D. W. & West, K. R. (1967). Salt regulation in the mangroves *Rhizophora mucronata* Lam. and *Aegialitis annulata* R.Br. *Australian Journal of Biological Sciences*, **20**, 589–99.
Ayadi, A. & Hamza, M. (1984). Research on salinity tolerance in Tunisia. In *Salinity Tolerance in Plants – Strategies for Crop Improvement*, ed. R. S. Staples & G. H. Toenniessen, pp. 205–12. New York: John Wiley.
Baker, H. G. (1966). The evolution, functioning and breakdown of heteromorphic incompatibility systems. I. Plumbaginaceae. *Evolution*, **20**, 349–67.
Baker, J. M. (1979). Responses of salt marsh vegetation to oil spills and refinery effluents. In *Ecological Processes in Coastal Environments*, ed. R. L. Jefferies & A. J. Davy, pp. 529–42. Oxford: Blackwell Scientific Publications.
(1983). Impact of oil production on living resources. *The Environmentalist*, **3**, Supplement 4, 1–48.
Baker, S. M. & Bohling, M. H. (1916). On the brown seaweeds of the saltmarsh Pt. II. Their systematic relationships, morphology and ecology. *Journal of the Linnean Society (Botany)*, **43**, 325–80.
Bakker, J. P. (1978). Changes in salt marsh vegetation as a result of grazing and mowing – a five-year study of permanent plots. *Vegetatio*, **38**, 77–87.
(1983). Use and management of salt marshes on sand-dune islands. In *Flora and Vegetation of the Wadden Sea Islands and Coastal Areas*, ed. K. S. Dijkema & W. J. Wolff, pp. 290–302. Rotterdam: Balkema.
(1985). The impact of grazing on plant communities, plant populations and soil conditions on salt marshes. *Vegetatio*, **62**, 391–8.
Bakker, J. P., Dijkstra, M. & Russchen, P. T. (1985). Dispersal, germination and early establishment of halophytes and glycophytes on a grazed and abandoned salt-marsh gradient. *New Phytologist*, **101**, 291–308.
Bakker, J. P. & Ruyter, J. C. (1981). Effects of five years of grazing on a salt marsh vegetation. *Vegetatio*, **44**, 81–100.
Ball, M. C. (1986). Photosynthesis in mangroves. *Wetlands (Australia)*, **6**, 12–22.
Ball, M. C. & Anderson, J. M. (1986). Sensitivity of photosystem II to NaCl in relation to salinity tolerance. Comparative studies with thylakoids of the salt-tolerant mangrove, *Avicennia marina*, and the salt-sensitive pea, *Pisum sativum*. *Australian Journal of Plant Physiology*, **13**, 689–98.
Ball, M. C. & Farquhar, G. D. (1984*a*). Photosynthetic and stomatal responses of two mangrove species, *Aegiceras corniculatum* and *Avicennia marina* to long term salinity and humidity conditions. *Plant Physiology*, **74**, 1–6.
(1984*b*). Photosynthetic and stomatal response of the grey mangrove, *Avicennia marina* to transient salinity conditions. *Plant Physiology*, **74**, 7–11.

Ball, P. W. & Brown, K. G. (1970). A biosystematic and ecological study of *Salicornia* in the Dee estuary. *Watsonia*, **8**, 27–40.

Ball, P. W. & Tutin, T. G. (1959). Notes on annual species of *Salicornia* in Britain. *Watsonia*, **4**, 193–205.

Bardecki, M. J. (1984). What value wetlands? *Journal of Soil and Water Conservation*, **39**, 166–9.

Banwell, A. D. (1951). A new species of *Riella* from Australia. *Transactions of the British Bryological Society*, **1**, 475–8.

Barbour, M. G. (1970). Is any angiosperm an obligate halophyte? *American Midland Naturalist*, **84**, 105–20.

Barbour, M. G. (1978). The effect of competition and salinity on the growth of a salt marsh plant species. *Oecologia (Berlin)*, **37**, 93–9.

Barko, J. W. & Smart, R. M. (1978). The growth and biomass distribution of two emergent freshwater plants, *Cyperus esculentus* and *Scirpus validus*, on different sediments. *Aquatic Botany*, **5**, 109–17.

Barnes, R. D. (1953). The ecological distribution of spiders in non-forest maritime communities at Beaufort, North Carolina. *Ecological Monographs*, **3**, 216–25.

Barnes, R. S. K. (1984). *Estuarine Biology*, 2nd edn. London: Edward Arnold.

Barrett-Lennard, E. G. (1986). Effects of waterlogging on the growth and NaCl uptake by vascular plants under saline conditions. *Reclamation and Revegetation Research*, **5**, 245–61.

Barrett-Lennard, E. G., Malcolm, C. V., Stern, W. R. & Wilkins, S. M., eds. (1986). *Forage and Fuel Production from Salt Affected Wasteland.* (Also published as *Reclamation and Revegetation Research*, **5**). Amsterdam: Elsevier.

Bascand, L. D. (1970). The roles of *Spartina* species in New Zealand. *Proceedings of the New Zealand Ecological Society*, **17**, 33–40.

Bassett, P. A. (1978). The vegetation of a Camargue pasture. *Journal of Ecology*, **66**, 803–27.

(1980). Some effects of defoliation on the vegetation dynamics of two Camargue grasslands. *Acta Oecologia*, **1**, 121–35.

Basson, P. W., Burchard, J. E., Hardy, J. T. & Price, A. R. G. (1977). *Biotopes of the Western Arabian Gulf. Marine Life and Environments of Saudi Arabia.* Dhahran: Aramco Dept. of Loss Prevention and Environmental Affairs, Dhahran.

Bates, J. W. (1976). Cell permeability and regulation of intracellular sodium concentration in a halophytic and glycophytic moss. *New Phytologist*, **77**, 15–23.

Bates, J. W. & Brown, D. H. (1974). The control of cation levels in seashore and inland mosses. *New Phytologist*, **73**, 483–95.

(1975). The effect of seawater on the metabolism of some seashore and inland mosses. *Oecologia (Berlin)*, **21**, 335–44.

Baumeister, W. & Kloos, G. (1974). Über die Salzsekretion bei *Halimione portulacoides* (L.) Aellen. *Flora*, **163**, 310–26.

Bazely, D. R. & Jefferies, R.L. (1985). Goose faeces: a source of nitrogen for plant growth in a grazed salt marsh. *Journal of Applied Ecology*, **22**, 693–703.

(1986). Changes in the composition and standing crop of salt-marsh communities in response to the removal of a grazer. *Journal of Ecology*, **74**, 693–706.

Beadle, C. L., Long, S. P., Inbamba, S. K., Hall, D. O. & Olembo, R. J. (1985). *Photosynthesis in Relation to Plant Production in Terrestrial Environments.* Oxford: UNEP/Tycooly Publishing.

Beard, J. S. (1967). An inland occurrence of mangrove. *West Australian Naturalist*, **10**, 112–15.

Beare, P. A. & Zedler, J. B. (1987). Cattail invasion and persistence in a coastal salt marsh. The role of salinity reduction. *Estuaries*, **10**, 165–70.

Beattie, A. J. (1985). *Evolutionary Ecology of Ant–Plant Mutualisms*. Cambridge: Cambridge University Press.
Beeftink, W. G. (1959). Some notes on Skallingens salt marsh vegetation and its habitat. *Acta Botanica Neerlandica*, **8**, 449–72.
(1962). Conspectus of the phanerogamic salt plant communities in the Netherlands. *Biologisch Jaarboek Dodonaea*, **30**, 325–62.
(1965). De Zoutvegetatie von ZW-Nederland beschouwd in Europees Verband. *Mededelingen van de Landbouwhogeschool te Wageningen*, **65**, 1–167.
(1966). Vegetation and habitat of the salt marshes and beach plains in the south-western part of the Netherlands. *Wentia* **15**, 83–108.
(1968). Die Systematic der Europäischen Salzpflanzengesellschaften. In *Pflanzensociologische Systematik*, ed. R. Tüxen, pp. 239–63. The Hague: Dr Junk.
(1972). Übersicht über die Anzahl der Aufnahmen Europäischer un Nordafrikanischer Salzflanzengesellschaften für das Projekt der Arbeitsgruppe für Datenverarbeitung. In *Grundfragen und Methoden in der Pflanzensociologie*, ed. R. Tüxen, pp. 371–96. The Hague: Dr Junk.
(1975). The ecological significance of embankment and drainage with respect to the vegetation of south-west Netherlands. *Journal of Ecology*, **63**, 423–58.
(1977*a*). The coastal salt marshes of western and northern Europe: an ecological and phytosociological approach. In *Wet Coastal Ecosystems*, ed. V.J. Chapman, pp. 109–55. Amsterdam: Elsevier.
(1977*b*). Salt-marshes. In *The Coastline*, ed. R.S.K. Barnes, pp. 93–121. Chichester: John Wiley.
(1984). Geography of European halophytes. In *Salt Marshes in Europe*, ed. K.S. Dijkema, pp. 15–33. Strasbourg: European Committee for the Conservation of Nature and Natural Resources.
(1985*a*). Population dynamics of annual *Salicornia* species in the tidal salt marshes of the Oosterschelde, The Netherlands. *Vegetatio*, **61**, 127–36.
(1985*b*). Vegetation study as a generator for population biological and physiological research on salt marshes. *Vegetatio*, **62**, 469–86.
Beeftink, W. G., Daane, M.C., de Munck, W. & Nieuwenhuize, J. (1978). Aspects of population dynamics in *Halimione portulacoides* communities. *Vegetatio*, **36**, 31–43.
Beeftink, W. G. & Géhu, J.-M. (1973). *Spartinetea maritimae. Prodrome des Groupements Végétaux d'Europe*, vol. 1. Lehre: Cramer.
Beeftink, W. G., Nieuwenhuize, J., Stoeppler, M. & Mohl, C. (1982). Heavy-metal accumulation in salt marshes from the western and eastern Scheldt. *The Science of the Total Environment*, **25**, 199–223.
Bell, F. G. (1969). The occurrence of southern steppe and halophyte elements in Weichselian (full glacial) floras from southern England. *New Phytologist*, **68**, 913–22.
Bell, S. S., Watzin, M. C. & Coull, B. C. (1978). Biogenic structure and its effects on the spatial heterogeneity of meiofauna in a salt marsh. *Journal of Experimental Marine Biology and Ecology*, **35**, 99–107.
Bellot, F. R. (1966). La vegetación de Galicia. *Anales del institutio botánico A.J. Cavanilles*, **34**, 1–306.
Bertness, M. D. (1984). Ribbed mussels and *Spartina alterniflora* production in a New England salt marsh. *Ecology*, **65**, 1794–807.
(1985). Fiddler crab regulation of *Spartina alterniflora* production on a New England saltmarsh. *Ecology*, **66**, 1042–55.
Bertness, M.D. & Miller, T. (1984). Fiddler crab burrow dynamics across a New England salt marsh. *Journal of Experimental Marine Biology and Ecology*, **83**, 211–37.
Bertness, M.D., Wise, C. & Ellison, A.M. (1987). Consumer pressure and seed set in salt marsh perennial plant community. *Oecologia (Berlin)*, **71**, 190–200.

Bhatti, A. S. & Wieneke, J. (1984). Na^+ and Cl^-: leaf extrusion, retranslocation and root efflux in *Diplachne fusca* (Kallar grass) grown in NaCl. *Journal of Plant Nutrition*, **7**, 1233–50.

Bickenbach, K. (1932). Zur Anatomie und Physiologie einiger Strand und Dunenpflanzen. Beiträge zur Halophytenproblem. *Beiträge zur Biologie der Pflanzen*, **19**, 334–70.

Binet, P. (1965). Action de la température et de la salinité sur la germination des graines de *Glaux maritima* L. *Bulletin de la Société Botanique de France*, **112**, 346–50.

Bird, E. C. F. (1984). *Coasts. An Introduction to Coastal Geomorphology*, 3rd edn. Canberra: A.N.U. Press.

Bird, E. C. F. & Ranwell, D. S. (1964). *Spartina* salt marshes in southern England. IV. The physiography of Poole Harbour, Dorset. *Journal of Ecology*, **52**, 355–66.

Bird, J. F. (1978). The nineteenth century soap industry and the exploitation of intertidal vegetation in eastern Australia. *Australian Geographer*, **14**, 38–41.

(1981). Barilla production in Australia. In *Plants and Man in Australia*, ed. D. J. & S. G. M. Carr, pp. 274–80. Sydney: Academic Press.

Black, R. E. (1958). Effect of sodium chloride on leaf succulence and area of *Atriplex hastata* L. *Australian Journal of Botany*, **6**, 306–21.

Blake, T. J. (1981). Salt tolerance of eucalypt species grown in saline solution culture. *Australian Forest Research*, **11**, 179–83.

Blasco, F. (1977). Outlines of ecology, botany and forestry of the mangals of the Indian subcontinent. In *West Coastal Ecosystems*, ed. V.J. Chapman, pp. 241–60. Amsterdam: Elsevier.

Bleakney, J. S. (1972). Ecological implications of annual variation in tidal extremes. *Ecology*, **53**, 933–8.

Boorman, L. A. (1967). *Limonium vulgare* Mill. and *Limonium humile* Mill. Biological Flora of the British Isles. *Journal of Ecology*, **55**, 221–32.

(1968). Some aspects of the reproductive biology of *Limonium vulgare* Mill. and *Limonium humile* Mill. *Annals of Botany*, **32**, 803–24.

(1971). Studies in salt marsh ecology with special reference to the genus *Limonium*. *Journal of Ecology*, **59**, 103–20.

Boston, K. G. (1981). The introduction of *Spartina townsendii* (s.l.) to Australia. *Occasional Paper – Melbourne State College*, **6**, 1–57.

Boulter, D., Coult, D. A. & Henshaw, G.G. (1963). Some effects of gas concentrations on metabolism of the rhizome of *Iris pseudacorus* (L.). *Physiologia Plantarum*, **16**, 541–8.

Bowden, W. M. (1961). Chromosome numbers and taxonomic notes on northern grasses. IV. Tribe Festuceae: *Poa* and *Puccinellia*. *Canadian Journal of Botany*, **39**, 123–38.

Bradfield, G. F. & Porter, G. L. (1982). Vegetation, structure and diversity components of a Fraser estuary tidal marsh. *Canadian Journal of Botany*, **60**, 440–51.

Bradshaw A. D. & McNeilly, T. (1981). *Evolution and Pollution*. London: Edward Arnold.

Brady, C. J., Gibson, T. S., Barlow, E. W. R., Speirs, J. & Wyn Jones, R.G. (1984). Salt-tolerance in plants. I. Ions, compatible organic solutes and the stability of plant ribosomes. *Plant, Cell and Environment*, **7**, 571–8.

Braun-Blanquet, J. & de Ramm, C. (1957). Contribution à la connaissance de la végétation du littoral méditterané. Les prés salés du Lanquedoc Méditteranéen. *Bulletin de la Muséum Histoire Naturelle Marseille*, **17**, 5–43.

Braun-Blanquet, J., Roussine, N. & Nègre, R. (1952). *Les groupements végétaux de la France méditerranéene*. Montpellier: C.N.R.S.

Brereton, A. J. (1971). The structure of the species populations in the initial stages of salt-marsh succession. *Journal of Ecology*, **59**, 321–38.

Breteler, R. J., Teal, J. M., Giblin, A. E. & Valiela, I. (1981). Trace element enrichments in decomposing litter of *Spartina alterniflora*. *Aquatic Botany*, **11**, 111–20.

Bridgewater, P. B. (1975). Peripheral vegetation of Westernport Bay. *Proceedings of the Royal Society of Victoria*, **87**, 69–78.

(1982). Phytosociology of coastal salt-marshes in the mediterranean climatic region of Australia. *Phytocoenologia*, **10**, 257–96.

Bridgewater, P. B. & Kaeshagen, D. (1979). Changes induced by adventive species in Australian plant communities. In *Werden und Vergehen von Pflanzengesellschaften*, ed. O. Wilmanns & R. Tüxen, pp. 561–79. Lehre: Cramer.

Bridgewater, P. B., Rosser, C. & de Corona, A. (1981). *The Saltmarsh Plants of Southern Australia*. Clayton: Botany Dept. Monash University.

Briens, M. & Larher, F. (1982). Osmoregulation in halophytic higher plants: a comparative study of soluble carbohydrates, polyols, betaines and free proline. *Plant, Cell and Environment*, **5**, 287–92.

(1983). Sorbitol accumulation in Plantaginaceae; further evidence for a function in stress tolerance. *Zeitschrift für Pflanzenphysiologie*, **110**, 447–58.

Brinckhuis, B. H. (1976). The ecology of temperate salt-marsh fucoids I. Occurrence and distribution of *Ascophyllum nodosum* ecads. *Marine Biology*, **34**, 325–38.

Britton, R. H. & Podlejski, V. D. (1981). Inventory and classification of the wetlands of the Camargue (France). *Aquatic Botany*, **10**, 195–228.

Brock, M. A. (1981). Accumulation of proline in a submerged aquatic halophyte, *Ruppia* L. *Oecologia (Berlin)*, **51**, 217–19.

Brown, A. D. & Simpson, J. R. (1972). Water relations of sugar-tolerant yeasts: the role of intracellular polyols. *Journal of General Microbiology*, **72**, 589–91.

Brownell, P. F. (1979). Sodium as an essential micronutrient element for plants and its possible role in metabolism. *Advances in Botanical Research*, **7**, 117–224.

Brownell, P. F. & Wood, J. G. (1957). Sodium as an essential element for *Atriplex vesicaria* Heward. *Nature (London)*, **179**, 635–6.

Broyer, T. C., Carlton, A. B., Johnson, C. M. & Stout, P. R. (1954). Chlorine as a micronutrient. *Plant Physiology*, **29**, 526–32.

Buchsbaum, R., Wilson, J. & Valiela, I. (1986). Digestibility of plant constituents by Canadian geese and Atlantic brant. *Ecology*, **67**, 386–93.

Bülow-Olsen, A. (1983). Germination response to salt in *Festuca rubra* in a population from a salt marsh. *Holarctic Ecology*, **6**, 194–8.

Burbidge, N. T. (1960). The phytogeography of the Australian region. *Australian Journal of Botany*, **8**, 75–209.

Burdick, D. M. & Mendelssohn, I. A. (1987). Waterlogging response in dune, swale and marsh populations of *Spartina patens* under field conditions. *Oecologia (Berlin)*, **74**, 321–9.

Burg, M.E., Tripp, D. R. & Rosenberg, E. S. (1980). Plant associations and primary productivity of the Nisqually salt marsh on Southern Puget Sound, Washington. *Northwest Science*, **54**, 222–36.

Burger, J., Shisler, J. & Lesser, F. H. (1982). Avian utilisation on six salt marshes in New Jersey. *Biological Conservation*, **18**, 187–211.

Buth, G. J. C. & de Wolf (1985). Decomposition of *Spartina anglica*, *Elytrigia pungens* and *Halimione portulacoides* in a Dutch salt marsh in association with faunal and habitat influences. *Vegetatio*, **62**, 337–55.

Cadwalladr, D. A. & Morley, J. V. (1973). Sheep grazing preferences on a saltings at Bridgwater Bay National Nature Reserve, Somerset and their significance for wigeon (*Anas penelope* L.) conservation. *Journal of the British Grassland Society*, **28**, 235–42.

(1974). Further experiments on the management of saltings pasture for wigeon (*Anas penelope* L.) conservation at Bridgwater Bay National Reserve, Somerset. *Journal of Applied Ecology*, **11**, 461–6.

Cadwalladr, D. A., Owen, M., Morley, J. V. & Cook, R. S. (1972). Wigeon (*Anas *penelope* L.) conservation and salting pasture management at Bridgwater Bay National Nature Reserve, Somerset. *Journal of Applied Ecology*, **9**, 417–25.

Caldwell, P. A. (1957). The spatial development of *Spartina* colonies growing without competition. *Annals of Botany*, **21**, 203–14.

Cargill, S. M. & Jefferies, R. L. (1984*a*). Nutrient limitation of primary production in a sub-arctic salt marsh. *Journal of Applied Ecology*, **21**, 657–68.

(1984*b*). The effects of grazing by lesser snow geese on the vegetation of a sub-arctic salt marsh. *Journal of Applied Ecology*, **21**, 669–86.

Carlson, P. R. & Forrest, J. (1982). Uptake of dissolved sulphide by *Spartina alterniflora*: evidence from natural sulphur isotope abundance ratios. *Science*, **216**, 633–5.

Carr, A. P. & Blackley, M. W. L. (1986*a*). The effects and implications of tides and rainfall on the circulation of water within a salt marsh. *Limnology and Oceanography*, **31**, 266–76.

(1986*b*). Implications of sedimentological and hydrological processes on the distribution of radionuclides: the example of a saltmarsh near Ravenglass, Cumbria. *Estuarine, Coastal and Shelf Science*, **22**, 529–43.

Carr, D. J. (1956). Contributions to Australian bryology. I. The structure, development and systematic affinities of *Monocarpus sphaerocarpus* gen. et sp. nov. (Marchantiales). *Australian Journal of Botany*, **4**, 175–91.

Carter, N. (1932). A comparative study of the algal flora of two salt marshes. Part I. *Journal of Ecology*, **20**, 341–70.

(1933*a*). A comparative study of the algal flora of two salt marshes. Part II. *Journal of Ecology*, **21**, 128–208.

(1933*b*). A comparative study of the algal flora of two salt marshes. Part III. *Journal of Ecology*, **21**, 385–403.

Caswell, H., Reed, F., Stephenson, S. N. & Werner, P. A. (1973). Photosynthetic pathways and selective herbivory: a hypothesis. *American Naturalist*, **107**, 465–80.

Catcheside, D. G. (1980). *Mosses of South Australia*. Adelaide: South Australian Government Printer.

Cavalieri, A. J. (1983). Proline and glycinebetaine accumulation by *Spartina alterniflora* Loisel in response to NaCl and nitrogen in a controlled environment. *Oecologia (Berlin)*, **57**, 20–4.

Cavalieri, A. J. & Huang, A. H. C. (1977). Effect of NaCl on the *in vitro* activity of malate dehydrogenase in salt marsh halophytes of the U.S. *Physiologia Plantarum*, **37**, 79–84.

(1979). Evaluation of proline accumulation in the adaptation of diverse species of marsh halophytes to the saline environment. *American Journal of Botany*, **66**, 307–12.

Cavers, P. B. & Harper, J. L. (1967). The comparative biology of closely related species living in the same area IX. *Rumex*: the nature of adaptation to a sea-shore habitat. *Journal of Ecology*, **55**, 73–82.

Chadwick, L. (1982). *In Search of Heathland*. Durham: Dobson.

Chapman, D. M., Geary, M., Roy, P. S. & Thom, B. G. (1982). *Coastal Evolution and Coastal Erosion in New South Wales*. Sydney: Coastal Council of New South Wales.

Chapman, V. J. (1934). The ecology of Scolt Head Island. In *Scolt Head Island*, ed. J. A. Steers, 1st edn, pp. 77–145. Cambridge: Heffer.

(1937). A revision of the marine algae of Norfolk. *Journal of the Linnean Society (Botany)*, **51**, 205–63.

(1938). Studies in salt-marsh ecology. Sections 1 to III. *Journal of Ecology*, **26**, 144–79.

(1939). Studies in salt-marsh ecology. V. The vegetation. *Journal of Ecology*, **27**, 181–201.

(1941). Studies in salt-marsh ecology. VIII. *Journal of Ecology*, **29**, 69–82.

(1942). The new perspective in the halophytes. *Quarterly Review of Biology*, **17**, 291–311.
(1950). *Halimione portulacoides* (L.) Aell. Biological Flora of the British Isles. *Journal of Ecology*, **38**, 214–22.
(1953). Problems in ecological terminology. In *Report of the Twenty-ninth Meeting of ANZAAS*, ed. J.F. Kefford, pp. 252–79. Sydney: NSW Government Printer.
(1959). Saltmarshes and ecological terminology. *Vegetatio*, **8**, 215–34.
(1960*a*). The plant ecology of Scolt Head Island. In *Scolt Head Island*, 2nd edn, ed. J. A. Steers. Cambridge: Heffer.
(1960*b*). *Salt Marshes and Salt Deserts of the World.* London: Leonard Hill.
(1974*a*). *Salt Marshes and Salt Deserts of the World*, 2nd edn. Lehre: Cramer.
(1974*b*). Salt marshes and salt deserts of the world. In *Ecology of Halophytes*, ed. R. J. Reimold & W. H. Queen, pp. 3–19. New York: Academic Press.
(1975). *Mangrove Vegetation*. Lehre: Cramer.
(1976). *Coastal Vegetation*, 2nd edn. Oxford: Pergamon Press.
(1977*a*). Introduction. In *Wet Coastal Ecosystems*, ed. V. J. Chapman, pp. 1–29. Amsterdam: Elsevier.
(1977*b*). Africa B. The remainder of Africa. In *Wet Coastal Ecosystems*, ed. V. J. Chapman, pp. 233–40. Amsterdam: Elsevier.
Chapman, V. J. & Ronaldson, J.W. (1958). The mangrove and salt-marsh flats of the Auckland isthmus, NZ. *DSIR Bulletin*, **125**, 1–79.
Chater, E. M. (1965). Ecological aspects of the dwarf brown form of *Spartina* in the Dovey estuary. *Journal of Ecology*, **53**, 789–97.
(1973). *Spartina* in the Dyfi estuary. In *A Handbook for Ynyslas*, ed. E.E. Watkin, pp. 115–26. Aberystwyth University College of Wales.
Chater, E. M. & Jones, H. (1951). New forms of *Spartina townsendii* Groves. *Nature (London)*, **168**, 126.
Chock, J. S. & Mathieson, A.C. (1976). Ecological studies of the salt-marsh ecad *scorpioides* (Hornemann) Hauck of *Ascophyllum nodosum* (L.) Le Jolis. *Journal of Experimental Marine Biology and Ecology*, **23**, 171–90.
Christiansen, C. & Møller, J. T. (1983). Rate of establishment and seasonal immersion of *Spartina* in Maringer Fjord, Denmark. *Holarctic Ecology*, **6**, 315–19.
Chung, C.-H. (1982). Low marshes, China. In *Creation and Restoration of Coastal Plant Communities*, ed. R. R. Lewis, pp. 131–45. Boca Raton: CRC Press.
(1985). The effects of introduced *Spartina* grass on coastal morphology in China. *Zeitschrift für Geomorphologie, Supplementbände*, **57**, 169–74.
Clapham, A. R., Pearsall, W. H. & Richards, P. W. (1942). *Aster tripolium* L. Biological Flora of the British Isles. *Journal of Ecology*, **30**, 385–95.
Clapham, A. R., Tutin, T. G. & Warburg, E. F. (1962). *Flora of the British Isles*, 2nd edn. London: Cambridge University Press.
Clark, J. S. (1986). Late-Holocene vegetation and coastal processes at a Long Island tidal marsh. *Journal of Ecology*, **74**, 561–78.
Clark, J. S. & Patterson, W. A. (1985). The development of a tidal marsh: upland and oceanic influences. *Ecological Monographs*, **55**, 189–217.
Clarke, L. D. & Hannon, N. J. (1967). The mangrove swamp and salt marsh communities of the Sydney district. I. Vegetation, soils and climate. *Journal of Ecology*, **55**, 753–71.
(1969). The mangrove swamp and salt marsh communities of the Sydney district. II. The holocoenotic complex with particular reference to physiography. *Journal of Ecology*, **57**, 213–34.
(1970). The mangrove swamp and salt marsh communities of the Sydney district III.

Plant growth in relation to salinity and waterlogging. *Journal of Ecology*, **56**, 351–69.
(1971). The mangrove swamp and saltmarsh communities of the Sydney district. IV. The significance of species interaction. *Journal of Ecology*, **59**, 535–53.

Clifford, H. T. & Simon, B. K. (1981). The biogeography of Australian grasses. In *Ecological Biogeography of Australia*, ed. A. Keast, pp. 539–54. The Hague: Dr Junk.

Clipson, N. J. W. (1987). Salt tolerance in the halophyte *Suaeda maritima* L. Dum. Growth, ion and water relations and gas exchange in response to altered salinity. *Journal of Experimental Botany*, **38**, 1996–2004.

Clipson, N. J. W., Tomos, A. D., Flowers, T. J. & Wyn Jones, R. G. (1985). Salt tolerance in the halophyte *Suaeda maritima* (L.) Dum. The maintenance of turgor pressure and water potential gradients in plants growing at different salinities. *Planta*, **165**, 392–6.

Clough, B. F., Andrews, T. J. & Cowan, I. R. (1982). Physiological processes in mangroves. In *Mangrove Ecosystems in Australia – Structure, Function and Management*, ed. B. F. Clough, pp. 193–210. Canberra: AIMS/ANU Press.

Cockayne, L. (1921). *The Vegetation of New Zealand.* Leipzig: Engelmann.
(1967). *New Zealand Plants and Their Story*, 4th. edn (ed. E. J. Godley). Wellington: Government Printer.

Coles, S. M. (1979). Benthic microalgal populations on intertidal sediments and their role as precursors to salt marsh development. In *Ecological Processes in Coastal Environments*, ed. R. L. Jefferies & A. J. Davy, pp. 25–42. Oxford: Blackwell Scientific Publications.

Collander, R. (1941). Selective absorption of cations by plants. *Plant Physiology*, **16**, 691–720.

Colman, J. (1933). The nature of the intertidal zonation of plants and animals. *Journal of the Marine Biological Association of the United Kingdom*, **18**, 435–76.

Conard, H. S. (1924). Second survey of the vegetation of a Long Island salt marsh. *Ecology*, **5**, 379–88.
(1935). The plant associations of Central Long Island. A study in descriptive plant sociology. *American Midland Naturalist*, **16**, 433–516.

Conard, H. S. & Galligar, G. S. (1929). Third survey of a Long Island salt marsh. *Ecology*, **10**, 326–36.

Cook, R. E. (1985). Growth and development in clonal populations. In *Population Biology and Evolution of Clonal Organisms*, ed. J. B. C. Jackson, L. W. Buss & R.E. Cook, pp. 259–96. New Haven: Yale University Press.

Cooper, A. (1982). The effects of salinity and waterlogging on the growth and cation uptake of salt-marsh plants. *New Phytologist*, **90**, 263–75.

Corkhill, P. (1984). *Spartina* at Lindisfarne NNR and details of recent attempts to control its spread. *In* Spartina anglica in *Great Britain*, ed. P. Doody, pp. 60–3. Attingham Park: Nature Conservancy Council.

Corré, J. J. (1979). L'équilibre des biocénoses végétales salées en Basse Camargue. In *Ecological Processes in Coastal Environments*, ed. R.L. Jefferies & A.J. Davy, pp. 65–76. Oxford: Blackwell Scientific Publications.

Cotton, A. D. (1912). Clare Island Survey Pt. 15 Marine Algae. *Proceedings of the Royal Irish Academy B*, **31**, 1–178.

Coughlan, J. (1979). Aspects of reclamation in Southhampton Water. In *Estuarine and Coastal Land Reclamation and Water Storage*, ed. B. Knights & A. J. Phillips, pp. 99–124. Farnborough: Saxon House.

Coughlan, S. J. & Wyn Jones, R. G. (1980). Some responses of *Spinacea oleracea* to salt stress. *Journal of Experimental Botany*, 31, 883–93.

Cowardin, L. M. (1982). Wetlands and deepwater habitats: a new classification. *Journal of Soil and Water Conservation*, **37**, 83–95.

Cowardin, L. M., Carter, V., Golet, F. C. & LaRoe, E. T. (1979). *Classification of*

Wetlands and Deepwater Habitats of the United States. Washington: FWS/OBS–79/31. Fish and Wildlife Service.

Craig, N. J., Turner, R. E. & Day, J. W. (1980). Wetland losses and their consequences in Coastal Louisiana. *Zeitschrift für Geomorphologie Supplementbände*, **34**, 252–41.

Crawford, R. M. M. (1978). Metabolic adaptations to anoxia. In *Plant Life in Anaerobic Environments*, ed. D. D. Hook & R. M. M. Crawford, pp. 119–36. Ann Arbor: Ann Arbor Science Publishers.

(1982). Physiological responses to flooding. In *Encyclopedia of Plant Physiology N.S. Vol. 12B. Water relations and carbon assimilation*, ed. O. L. Lange, P. S. Nobel, C. B. Osmond & H. Ziegler, pp. 452–77. Berlin: Springer-Verlag.

(1983). Root survival in flooded soils. In *Ecosystems of the World. Mires – Swamp, Bog, Fen and Moor*, ed. A. J. P. Gore, pp. 257–83. Amsterdam: Elsevier.

ed. (1987). *Plant Life in Aquatic and Amphibious Habitats.* Oxford: Blackwell Scientific Publications.

Cribb, A. B. & Cribb, J. W. (1975). *Wild Food in Australia.* Sydney: Collins.

Critchley, C. (1982). Stimulation of photosynthetic electron transport in a salt-tolerant plant by high chloride concentrations. *Nature (London)*, **298**, 483–5.

Crow, J. H. (1971). Earthquake-initiated changes in the nesting habitat of the Dusky Canada Goose. In *The Great Alaska Earthquake of 1964: Biology*, pp. 130–6. Washington, D.C: National Academy of Sciences.

(1977). Salt marshes of the Alaska Pacific Coast. In *Terrestrial and Aquatic Ecological Studies of the Northwest*, ed. R. D. Andrews *et al.*, pp. 103–10. Cheney: Eastern Washington State College, Cheney, Washington.

Crow, J. H. & Koppen, J. D. (1977). The salt marsh vegetation of China Poot Bay, Alaska. In *Environmental Studies of Kachemak Bay and Lower Cook Inlet*, Vol. II, ed. L. Trasky, L. Flagg & D. Burbank, pp. 716–44. Juneau: Alaska Department of Fish and Game.

Cruz, A. A. de la & Hackney, C. T. (1977). Energy value, elemental composition, and productivity of belowground biomass of a *Juncus* tidal marsh. *Ecology*, **58**, 1165–70.

Curran, M., Cole, M. & Allaway, W. G. (1986). Root aeration and respiration in young mangrove plants [*Avicennia marina* (Forsk.) Vierh.]. *Journal of Experimental Botany*, **37**, 1225–33.

Dahl, E. & Hadač, E. (1941). Strandgesellschaften der Insel Ostøy in Oslofjord. Eine pflanzensoziologische Studie. *Nytt Magasin for Naturvidenskappene*, **82**, 251–312.

Dahlbeck, N. (1945). Strandwiesen an Südöstlichen Oresund. *Acta Phytogeographica Suecica*, **18**, 1–168.

Daiber, F. C. (1982). *Animals of the Tidal Marsh.* New York: van Nostrand Reinhold.

(1986). *Conservation of Tidal Marshes.* New York: van Nostrand Reinhold.

Daines, R. J. & Gould, A. R. (1985). The cellular basis of salt tolerance studied with tissue cultures of the halophytic grass *Distichlis spicata. Journal of Plant Physiology*, **119**, 269–80.

Dainty, J. (1979). The ionic and water relations of plants which adjust to a fluctuating saline environment. In *Ecological Processes in Coastal Environments*, ed. R. L. Jefferies & A. J. Davy, pp. 201–9. Oxford: Blackwell Scientific Publications.

Dalby, D. H. (1962). Chromosome number, morphology and breeding behaviour in the British *Salicorniae. Watsonia*, **5**, 150–62.

(1987). Salt marshes. In *Biological Surveys of Estuaries and Coasts*, ed. J. M. Baker & W. J. Wolff, pp. 38–80. Cambridge: Cambridge University Press.

Dale, M. R. T. (1984). The contiguity of upslope and downslope boundaries of species in a zoned community. *Oikos*, **42**, 92–6.

Dale, P. E. R., Hulsman, K., Harrison, D. & Congdon, B. (1986). Distribution of the immature stages of *Aedes vigilax* on a coastal salt-marsh in south-east Queensland. *Australian Journal of Ecology*, **11**, 269–78.

Dame, R. F. (1982). The flux of floating macrodetritius in the North Inlet estuarine ecosystem. *Estuarine, Coastal and Shelf Science*, **15**, 337–9.

Daquan, S. (1983). Salt-marsh resources of China. In *Land Resources of the People's Republic of China*, ed. K. Ruddle & W. Chuanjun, pp. 37–44. Tokyo: United Nations University.

Davies, D. D., Nascimento, K. H. & Patil, K. D. (1974). The distribution and properties of NADP malic enzyme in flowering plants. *Phytochemistry*, **13**, 2417–25

Davies, M. S. & Singh, A. K. (1983). Population differentiation in *Festuca rubra* L. and *Agrostis stolonifera* L. in response to soil water logging. *New Phytologist*, **94**, 573–83.

Davies, J. L. (1980). *Geographical Variation in Coastal Development*, 2nd edn. London: Longman.

Davis, L. V. & Gray, I. E. (1966). Zonal and seasonal distribution of insects in North Carolina salt marshes. *Ecological Monographs*, **36**, 275–95.

Davis, P. H. & Heywood, V.H. (1963). *Principles of angiosperm taxonomy*. Edinburgh: Oliver & Boyd.

Davis, P. & Moss, D. (1984). *Spartina* and waders – the Dyfi estuary. In Spartina anglica *in Great Britain*, ed. P. Doody, pp. 37–40. Attingham Park: Nature Conservancy Council.

Davy, A. J. & Jefferies, R. L. (1981). Approaches to the monitoring of rare plant populations. In *The Biological Aspects of Rare Plant Conservation* ed. H. Synge, pp. 219–32. Chichester: Wiley

Davy, A. J. & Smith, H. (1985). Population differentiation in the life-history characteristic of salt-marsh annuals. *Vegetatio*, **61**, 117–25.

Day, J. H. (1981). The estuarine flora. In *Estuarine Ecology with Particular Reference to Southern Africa*, ed. J. H. Day, pp. 77–99. Rotterdam, Balkema.

Dawe, N. K. & White, E. R. (1986). Some aspects of the vegetation ecology of the Nanoose-Bonell estuary, Vancouver Island, British Columbia. *Canadian Journal of Botany*, **64**, 27–34.

DeJong, T. M., Drake, B. G. & Pearcy, R. W (1982). Gas exchange response of Chesapeake Bay tidal marsh species under field and laboratory conditions. *Oecologia (Berlin)*, **52**, 5–11.

de la Cruz, A. A & Hackney, C. T. (1977). Energy value, elemental composition, and productivity of below ground biomass of a *Juncus* tidal marsh. *Ecology*, **58**, 1165–70.

DeLaune, R. D., Buresh, R. J. & Patrick, W. H. (1979). Relationship of soil properties to standing crop biomass of *Spartina alterniflora* in a Louisiana marsh. *Estuarine, Coastal and Marine Science*, **8**, 477–87.

Demmig, B. & Winter, K. (1986). Sodium, potassium, chloride and proline concentrations of chloroplasts isolated from a halophyte, *Mesembryanthemum crystallinum* L. *Planta*, **165**, 421–6.

de Molenaar, J. G. (1974). Vegetation of the Angmagssalik district south east Greenland. I. Littoral vegetation. *Meddelelser om Grønland*, **198** (1), 1–79.

Deschênes, J. & Sérodes, J. B. (1985). The influence of salinity on *Scirpus americanus* tidal marshes in the St. Lawrence River Estuary, Quebec. *Canadian Journal of Botany*, **63**, 920–7.

de Vlaming, V. & Proctor, V. (1968). Dispersal of aquatic organisms: viability of seeds recovered from the droppings of captive kildeer and mallard ducks. *American Journal of Botany*, **55**, 20–6.

Dickinson C. M. (1965). The mycoflora associated with *Halimione portulacoides* III. Fungi on green and moribund leaves. *Transactions of the British Mycological Society*, **48**, 603–10.

Dijkema, K. S. (1983). Use and management of mainland salt marshes and Halligen. In *Flora and Vegetation of the Wadden Sea Islands and Coastal Areas*, ed. K. S. Dijkema & W. J. Wolff, 302–12. Rotterdam: Balkema.

(1984*a*). Development and classification of main salt marsh biotopes in Europe. In *Salt Marshes in Europe*, ed. K. S. Dijkema, pp. 8–15. Strasbourg: European Committee for the Conservation of Nature and Natural Resources.

(1984*b*). Western-European salt marshes. In *Salt Marshes in Europe*, ed. K. S. Dijkema, pp. 82–103. Strasbourg: European Committee for the Conservation of Nature and Natural Resources.

(1984*c*). Wadden Sea, south-west Netherlands and Belgium. In *Salt Marshes in Europe*, ed. K. S. Dijkema, pp. 152–6. Strasbourg: European Committee for the Conservation of Nature and Natural Resources.

(1984*d*). Agriculture. In *Salt Marshes in Europe*, ed. K.S. Dijkema, pp. 152–6. Strasbourg: European Committee for the Conservation of Nature and Natural Resources.

Disraeli, D. J. & Fonda, R. W. (1979). Gradient analysis of the vegetation in a brackish marsh in Bellingham Bay, Washington. *Canadian Journal of Botany*, **57**, 465–75.

Dixon, A. F. G. (1973). *Biology of Aphids*. London: Edward Arnold.

Dobbs, C. G. (1939). The vegetation of Cape Napier, Spitsbergen. *Journal of Ecology*, **27**, 126–48.

Dollenz, O. A. (1977). Estado de la flora vascular en Puerto Espora, Tierra del Fuego, contaminada por el petroleo del B/T "Metula". I. Reconocimiento de la entrada de mar noreste. *Anales del Instituto de la Patagonia*, **8**, 251–61.

Donovan, L. A. & Gallahger, J. L. (1984). Anaerobic substrate tolerance in *Sporobolus virginicus* (L.) Kunth. *American Journal of Botany*, **71**, 1424–31.

Doody, J. P. (ed.) (1984*a*) Spartina anglica *in Great Britain*. Attingham Park: Nature Conservancy Council.

(1984*b*). Threats and decline – Great Britain and Ireland. In *Salt Marshes in Europe*, ed. K. S. Dijkema, pp. 162–6. Strasbourg: European Committee for the Conservation of Nature and Natural Resources.

Doody, J. P., Langslow, D. A. & Stubbs, A. E. (1984). Great Britain and Ireland. In *Salt Marshes in Europe*, ed. K. S. Dijkema, pp. 103–19. Strasbourg: European Committee for the Conservation of Nature and Natural Resources.

Doornbos, G., Groenendijk, A. M. & Jo, Y. W. (1986). Nakdong estuarine barrage and reclamation project: preliminary results of the botanical, macrozoobenthic and ornithological studies. *Biological Conservation*, **38**, 115–42.

Doty, M. S. (1946). Critical tide factors that are correlated with the vertical distribution of marine algae and other organisms along the Pacific Coast. *Ecology*, **27**, 315–28.

Dowling, R. M. & McDonald, T. J. (1982). Mangrove communities of Queensland. In *Mangrove Ecosystems in Australia. Structure, Function and Management*, ed. B. F. Clough, pp. 79–93. Canberra: ANU Press/AIMS.

Drifmeyer, J. E. & Rublee P. A. (1981). Mn, Fe, Cu and Zn in *Spartina alterniflora* detritus and microorganisms. *Botanica Marina*, **24**, 251–6.

Dring, M. J. (1982). *The Biology of Marine Plants*. London: Edward Arnold.

Duncan, W. H. (1974). Vascular halophytes of the Atlantic and Gulf coasts of North America north of Mexico. In *Ecology of Halophytes*, ed. R. J. Reimold & W. H. Queen, pp. 23–50. New York: Academic Press.

Dunn, R., Thomas, S. M., Keys, A. J. & Long, S. P. (1987). A comparison of the growth of the C_4 grass *Spartina anglica* with the C_3 grass *Lolium perenne* at different temperatures. *Journal of Experimental Botany*, **38**, 433–41.

Duvigneaud, P. & Jacobs, M. (1971). Ecologie génétique de population halophytiques d'*Aster tripolium* L. dans l'estuaries de L'Yser (Belgique). *Bulletin du Jardin botanique de l'État à Bruxelles*, **41**, 81–91.

Dyer, K. R. (ed.). (1979). *Estuarine Hydrography and Sedimentation. A Handbook*. London: Cambridge University Press.

Ebbinge, B., Canters, K. & Drent, R. (1975). Foraging routines and estimated daily food

intake in Barnacle Geese wintering in the northern Netherlands. *Wildfowl*, **26**, 5–19.
Edwards, A. C. & Davis, D. E. (1974). Effects of herbicides on the *Spartina* salt marsh. In *Ecology of Halophytes*, ed. R. J. Reimold & W. H. Queen, pp. 531–45. New York: Academic Press.
Eilers, H. P. (1979). Production ecology in an Oregon coastal salt marsh. *Estuarine and Coastal Marine Science*, **8**, 399–410.
Eisikowitch, D. & Woodell, S. R. J. (1975). Some aspects of pollination ecology of *Armeria maritima* (Mill.) Willd. in Britain. *New Phytologist*, **74**, 303–22.
Eleuterius, L. N. (1972). The marshes of Mississippi. *Castanea*, **37**, 153–68.
(1974). Flower morphology and plant types within *Juncus roemerianus*. *Bulletin of Marine Science*, **24**, 493–7.
(1976). The distribution of *Juncus roemerianus*, in the saltmarshes of North America. *Chesapeake Science*, **17**, 289–92.
(1978). A revised description of the salt marsh rush, *Juncus roemerianus*. *Sida*, **7**, 355–60.
(1984). Sex distribution in the progeny of the salt marsh rush, *Juncus roemerianus* in Mississippi. *Castanea*, **49**, 35–8.
Eleuterius, L. N. & Caldwell, J. D. (1981). Growth kinetics and longevity of the salt marsh rush *Juncus roemerianus*. *Gulf Research Reports*, **7**, 27–34.
Eleuterius, L. N. & Eleuterius, C. K. (1979). Tide levels and salt marsh zonation. *Bulletin of Marine Science*, **29**, 394–400.
Elliott, J. S. B. (1930). The soil fungi of the Dovey salt marshes. *Annals of Applied Biology*, **17**, 284–305.
Ellis, E. A. (1960*a*). The lichens. In *Scolt Head Island*, 2nd edn, ed. J. A. Steers, pp. 177–8. Cambridge: Heffer.
(1960*b*). An annotated list of fungi. In *Scolt Head Island*, 2nd edn, ed. J. A. Steers, pp. 179–82. Cambridge: Heffer.
Ellison, A. M. (1987*a*). Effects of competition, disturbance and herbivory on *Salicornia europaea*. *Ecology*, **68**, 576–86.
(1987*b*). Density-dependent dynamics of *Salicornia europaea* monocultures. *Ecology*, **68**, 737–41.
Elsol, J. A. (1985). Illustrations of the use of higher plant taxa in biogeography. *Journal of Biogeography*, **12**, 433–44.
Engel, J. J. & Schuster, R. M. (1973). On some tidal zone Hepaticae from South Chile, with comments on marine dispersal. *Bulletin of the Torrey Botanical Club*, **100**, 29–35.
Epstein, E. (1980). Responses of plants to saline environments. In *Genetic Engineering of Osmoregulation*, ed. D. W. Rains, R. C. Valentine & A. Hollaender, pp. 7–21. New York: Plenum Press.
(1985). Salt-tolerant crops: origins development, and prospects of the concept. *Plant and Soil*, **89**, 187–98.
Epstein, E. & Rains, D. W. (1987). Advances in salt tolerance. *Plant and Soil*, **99**, 17–29.
Erdei, L. & Kuiper, P. J. C. (1979). The effect of salinity on growth; cation content, Na^+-uptake and translocation in salt-sensitive and salt-tolerant *Plantago* species. *Physiologia Plantarum*, **47**, 95–9.
Erdei, L., Stuiver, C. E. E. & Kuiper, P. J. C. (1980). The effect of salinity on lipid composition and an activity of Ca^{2+} and Mg^{2+} stimulated ATPases in salt-sensitive and salt-tolerant *Plantago* species. *Physiologia Plantarum*, **49**, 315–19.
Ericson, L. & Wallentinus, H. G. (1979). Seashore vegetation around the Gulf of Bothnia. *Wahlenbergia*, **5**, 1–142.
Ernst, W. H. O. (1985). Some considerations of and perspectives in coastal ecology. *Vegetatio*, **62**, 533–45.

Eshel, A. (1985). Response of *Suaeda aegyptiaca* to KCl, NaCl and Na_2SO_4 treatments. *Physiologia Plantarum*, **64**, 308–15.

Evenari, M., Gutterman, Y. & Gavish, E. (1985). Botanical studies on coastal salinas and sabkhas of the Sinai. In *Hypersaline Ecosystems. The Gavish Sabkha*, ed. G. M. Friedman & W. E. Krumbein, pp. 145–82. Berlin: Springer-Verlag.

Ewing, K. & Kershaw, K. A. (1986). Vegetation patterns in James Bay coastal marshes. I. Environmental factors on the south coast. *Canadian Journal of Botany*, **64**, 217–26.

Faegri, K. & van der Pijl, L. (1979). *The Principles of Pollination Ecology*. Oxford: Pergamon.

Fahn, A. (1979). *Secretory Tissues in Plants*. London: Academic Press.

(1988). Tansley review no.14. Secretory tissue in vascular plants. *New Phytologist*, **108**, 229–57.

Farquahar, G. D., Ball, M. C., von Caemmerer, S. & Roksandic, Z. (1982). Effect of salinity and humidity on $\delta^{13}C$ value of halophytes – evidence for diffusional isotope fractionation determined by the ratio of intercellular/atmospheric partial pressure of CO_2 under different environmental conditions. *Oecologia (Berlin)*, **52**, 121–4.

Farquahar, G. D. & Richards, R. A. (1984). Isotopic composition of plant carbon correlates with water-use efficiency of wheat genotypes. *Australian Journal of Plant Physiology*, **11**, 539–52.

Fenner, M. (1985). *Seed Ecology*. London: Chapman and Hall.

Ferrell, R. E., Seneca, E. D. & Linthurst, R. A. (1984). The effects of crude oil on the growth of *Spartina alterniflora* Loisel, and *Spartina cynosuroides* (L.) Roth. *Journal of Experimental Marine Biology and Ecology*, **83**, 27–39.

Fischer, R. A. & Turner, N. C. (1978). Plant productivity in the arid and semiarid zones. *Annual Reviews of Plant Physiology*, **29**, 277–317.

Flowers, T. J. (1975). Halophytes. In *Ion Transport in Plant Cells and Tissues*, ed. D. A. Baker & J. L. Hall, pp. 309–34. North Holland.

(1985). Physiology of halophytes. *Plant and Soil*, **89**, 41–56.

Flowers, T. J., Hajibagheri, M. A. & Clipson, N. J. W. (1986). Halophytes. *Quarterly Review of Biology*, **61**, 313–37.

Flowers, T. J. & Hall, J. L. (1978). Salt tolerance in the halophyte, *Suaeda maritima* (L.) Dum. The influence of the salinity of the culture solution on the content of various organic compounds. *Annals of Botany*, **42**, 1057–63.

Flowers, T. J., Hall, J. L. & Ward, M. E. (1978). Salt tolerance in the halophyte *Suaeda maritima* (L.) Dum. Properties of malic enzyme and PEP carboxylase. *Annals of Botany*, **42**, 1065–74.

Flowers, T. J. & Läuchli, A. (1983). Sodium versus potassium: substitution and compartmentation. In *Inorganic Plant Nutrition. Encyclopedia of Plant Physiology*, new series 15B, ed. A. Läuchli & R.L. Bieleski, pp. 651–81. Berlin: Springer-Verlag.

Flowers, T. J., Troke, P. F. & Yeo, A. R. (1977). The mechanism of salt tolerance in halophytes. *Annual Reviews of Plant Physiology*, **28**, 89–121.

Flowers, T. J. & Yeo, A. R. (1986). Ion relation of plants under drought and salinity. *Australian Journal of Plant Physiology*, **13**, 75–91.

Foster, W. A. (1983). Activity rhythms and the tide in a saltmarsh beetle *Dicheirotrichus gustavi*. *Oecologia (Berlin)*, **60**, 111–13.

(1984). The distribution of the sea-lavender aphid *Staticobium staticis* on a marine saltmarsh and its effect on host plant fitness. *Oikos*, **42**, 97–104.

Foster, W. A. & Moreton, R. B. (1981). Synchronization of activity rhythms with the tide in a saltmarsh collembolan *Anurida maritima*. *Oecologia (Berlin)*, **50**, 265–70.

Foster, W. A. & Treherne, J. E. (1975). The distribution of an intertidal aphid. *Pemphigus trehernei* Foster on marine saltmarsh. *Oecologia (Berlin)*, **21**, 141–55.

(1976). Insects of marine saltmarshes: problems and adaptations. In *Marine Insects*, ed.

L. Cheng, pp. 5–42. Amsterdam: North-Holland Publishing Company.
Freeman, D. C., Klikoff, L. G. & Harper, K. T. (1976). Differential resource utilization by the sexes of dioecious plants. *Science*, **193**, 597–9.
Frey, R. W. & Basan, P. B. (1985). Coastal salt marshes. In *Coastal Sedimentary Environments*, 2nd edn, ed. R. A. Davis, pp. 225–301. New York: Springer-Verlag.
Fries, N. (1944). Beobachtungen über die thamniscophage mykorrhiza einiger Halophyten. *Botaniska Notiser*, **2**, 255–64.
Fritsch, F. E. (1935). *Structure and Reproduction of the Algae*, vol. 2. Cambridge: Cambridge University Press.
Froomer, N. (1980). Morphologic changes in some Chesapeake Bay tidal marshes resulting from accelerated soil erosion. *Zeitschrift für Geomorphologie Supplementbände*, **34**, 242–54.
Fuller, R. J. (1982). *Bird Habitats in Britain*. Calton: T. & A. D. Poyser.
Gale, J., Naaman, R., Poljakoff-Mayber, A. (1970). Growth of *Atriplex halimus* L. in sodium chloride salinated culture solutions as affected by the relative humidity of the air. *Australian Journal of Biological Science*, **23**, 947–52.
Gale, J. & Zeroni, M. (1985). The cost to plants of different strategies of adaptation to stress and the alleviation of stress by increasing assimilation. *Plant and Soil*, **89**, 57–67.
Gallagher, J. L. (1979). Growth and element compositional responses of *Sporobolus virginicus* (L.) Kunth. to substrate salinity and nitrogen. *American Midland Naturalist*, **102**, 68–75.
Gallagher, J. L. & Kibby, H. V. (1980). Marsh plants as vectors in trace metal transport in Oregon tidal marshes. *American Journal of Botany*, **67**, 1079–84.
(1981). The streamside effect in a *Carex lyngbei* estuarine marsh; the possible role of recoverable underground reserves. *Estuarine and Coastal Shelf Science*, **12**, 451–60.
Gallagher, J. L., Reimold, R. J., Linthurst, R. A. & Pfeiffer, W.J. (1980). Aerial production, mortality, and mineral accumulation – export dynamics in *Spartina alterniflora* and *Juncus roemerianus* plant stands. *Ecology*, **61**, 303–12.
Gambrell, R. P. & Patrick, W. H. (1978). Chemical and microbiological properties of anaerobic soils and sediments. In *Plant Life in Anaerobic Environments*, ed. D. D. Hook & R. M. M. Crawford, pp. 119–36. Ann Arbor: Ann Arbor Press.
Ganong, W. F. (1903). The vegetation of the Bay of Fundy salt and diked marshes: an ecological study. Contributions to the ecological plant-geography of the Province of New Brunswick No. 3. *Botanical Gazette*, **36**, 161–86.
Ganzmann, R. J. & von Willert, D.J. (1972). Nachweis eines diurnalen Säurerhythmus beim Halophyten *Aster tripolium*. *Naturwissenschaften*, **59**, 422–3.
Gardner, W. H. & Mackenzie, N. D. (eds.) (1967). *The Poems of Gerard Manley Hopkins*, 4th edn. London: Oxford University Press.
Garofalo, D. (1980). The influence of wetland vegetation on tidal stream channel migration and morphology. *Estuaries*, **3**, 258–70.
Géhu, J.-M. (1976). Approache phytosociologique synthétique de la végétation des vases salées du littoral Atlantique Français (Sysystématique & Synchorologie). *Colloques Phytosociologiques*, **4**, 395–462.
(1984*a*). France and northwest Iberian Peninsula. In *Salt Marshes in Europe*, ed. K. S. Dijkema, pp. 119–28. Strasbourg: European Committee for the Conservation of Nature and Natural Resources.
(1984*b*). Mediterranean salt marshes and salt steppes. In *Salt Marshes in Europe*, ed. K. S. Dijkema, pp. 129–42. Strasbourg: European Committee for the Conservation of Nature and Natural Resources.
(1984*c*). Threats and decline – Southern Europe. In *Salt Marshes in Europe*, ed. K.S. Dijkema, pp. 166–7. Strasbourg: European Committee for the Conservation of Nature and Natural Resources.

Géhu, J.-M. & Rivas-Martinez (1984). Classification of European salt plant communities. In *Salt Marshes in Europe*, ed. K. S. Dijkema, pp. 34–40. Strasbourg: European Committee for the Conservation of Nature and Natural Resources.

Gessner, R. V. (1977). Seasonal occurrence in distribution of fungi associated with *Spartina alterniflora* from a Rhode Island estuary. *Mycologia*, **69**, 477–91.

Gessner, R. V. & Goos, R. D. (1973). Fungi from decomposing *Spartina alterniflora. Canadian Journal of Botany*, **51**, 51–5.

Gessner, R. V., Goos, R. D. & Sieburth, J.McN. (1972). The fungi microcosm of the internodes of *Spartina alterniflora. Marine Biology*, **16**, 269–73.

Gessner, R. V. & Kohlmeyer, J. (1976). Geographical distribution and taxonomy of fungi from salt marsh *Spartina. Canadian Journal of Botany*, **54**, 2023–37.

Gettys, K. L., Hancock, J. F. & Cavalieri, A. J. (1980). Salt tolerance of *in vitro* activity of leucine aminopeptidase, peroxidase, and malate dehydrogenase in the halophytes *Spartina alterniflora* and *S. patens. Botanical Gazette*, **141**, 453–7.

Gibb, D. C. (1957). The free-living forms of *Ascophyllum nodosum* (L.) Le Jol. *Journal of Ecology*, **45**, 49–84.

Giblin, A. E. Bourg, A., Valiela, I. & Teal, J. M. (1980). Uptake and losses of heavy metals in sewage sludge by a New England salt marsh. *American Journal of Botany*, **67**, 1059–68.

Gibson, T. S., Speirs, J. & Brady, C. J. (1984). Salt-tolerance in plants. II. *In vitro* translation of m-RNAs from salt-tolerant and salt-sensitive plants on wheat germ ribosomes. Response to ions and compatible organic solutes. *Plant, Cell and Environment*, **7**, 579–87.

Gilbertson, D. D., Kent, M. & Pyatt, F. B. (1985). *Practical Ecology for Geography and Biology. Survey, Mapping and Data Analysis.* London: Hutchinson.

Gillham, M. E. (1957*a*). Vegetation of the Exe estuary in relation to water salinity. *Journal of Ecology*, **45**, 735–56.

(1957*b*). Coastal vegetation of Mull and Iona in relation to salinity and soil reaction. *Journal of Ecology*, **45**, 757–78.

Gillner, V. (1955). Strandängsvegetation i Nord-Norge. *Svensk botanisk tidskrift*, **49**, 217–28.

(1960). Vegetations- und Standortsuntersuchungen in den Strandwiesen der Schwedischen Westkuste. *Acta phytogeographica Suecica*, **43**, 1–198.

(1965). Salt marsh vegetation in southern Sweden. *Acta phytogeographica Suecica*, **50**, 97–104.

Gimingham, C. H. (1964). Maritime and sub-maritime communities. In *The Vegetation of Scotland*, ed. J. H. Burnett, pp. 67–142. Edinburgh: Oliver & Boyd.

Gleason, M. L. & Zieman, J. C. (1981). Influence of tidal inundation of internal oxygen supply of *Spartina alterniflora* and *Spartina patens. Estuarine, Coastal and Shelf Science*, **13**, 47–57.

Glenn, E. P. (1987). Relationship between cation accumulation and water content of salt-tolerant grasses and a sedge. *Plant, Cell and Environment*, **10**, 205–12.

Glenn, E. P. & O'Leary, J. W. (1984). Relationship between salt accumulation and water content of dicotyledonous halophytes. *Plant, Cell and Environment*, **7**, 253–61.

Glooschenko, W. A. (1978). Above-ground biomass of vascular plants in a subarctic James Bay salt marsh. *Canadian Field Naturalist*, **92**, 30–2.

(1980*a*). Coastal ecosystems of the James/Hudson Bay area of Ontario, Canada. *Zeitschrift für Geomorphologie Supplementbände*, **34**, 214–24.

(1980*b*). Coastal salt marshes in Canada. In *Environment Canada Lands Directorate. Ecological Land Classification Series*, vol. 2, ed. C. P. A. Rubec & F. C. Pollet, pp. 39–47.

Glooschenko, W. A. & Harper, N. S. (1982). Net aerial primary production of a James Bay, Ontario, salt marsh. *Canadian Journal of Botany*, **60**, 1060–7.

Glooschenko, W. A. & Martini, I. P. (1978). Hudson Bay lowlands baseline study. In *Coastal Zone '78. Symposium on Technical Environmental, Socioeconomic and Regulatory Aspects of Coastal Zone Management*, Vol. II, pp. 663–79. New York: American Society of Civil Engineers.

(1981). Salt marshes of the Ontario coast of Hudson Bay. Canada. *Wetlands* (U.S.A.), **1**, 9–18.

Godfrey, P. J., Leatherman, S. P. & Buckley, P. A. (1978). Impact of off-road vehicles on coastal ecosystems. In *Coastal Zone '78. Symposium on Technical, Environmental, Socioeconomic and Regulatory Aspects of Coastal Zone Management*, vol. 2, pp. 581–600. New York: American Society of Civil Engineers.

Godwin, H. (1975). *The History of the British Flora*, 2nd edn. London: Cambridge University Press.

Good, R. E. (1965). Salt marsh vegetation, Cape May, New Jersey. *Bulletin New Jersey Academy of Science*, **10**, 1–11.

Good, R. E., Whigham, D. F. & Simpson, R. L. (eds.) (1978). *Freshwater Wetlands: Ecological Processes and Management Potential.* New York: Academic Press.

Goodin, J. R. (1977). Salinity effects on range plants. In *Rangeland Plant Physiology*, ed. R. E. Sosobee, 141–53. Denver: Society for Range Management.

Goodman, P. J. (1959). The possible role of pathogenic fungi in die-back of *Spartina townsendii* agg. *Transactions of the British Mycological Society*, **42**, 409–15.

Goodman, P. J., Braybrooks, E. M. & Lambert, J. M. (1959). Investigations into 'Die-Back' in *Spartina townsendii* agg. I. The present state of *Spartina townsendii* in Britain. *Journal of Ecology*, **47**, 651–77.

Goodman, P. J., Braybrooks, E. M., Marchant, C. J. & Lambert, J. M. (1969). *Spartina* × *townsendii*, H. & W. Groves *sensu lato*. Biological Flora of the British Isles. *Journal of Ecology*, **57**, 298–313.

Goodman, P. J. & Williams, W. T. (1961). Investigations into 'die-back' in *Spartina townsendii* agg. III: Physiological correlates of 'die-back'. *Journal of Ecology*, **49**, 391–8.

Gorham, A. V. & Gorham, E. V. (1955). Iron, manganese, ash and nitrogen in some plants from salt marsh and shingle habitats. *Annals of Botany*, **19**, 571–7.

Gorham, J. (1987). Photosynthesis, transpiration and salt fluxes through leaves of *Leptochloa fusca* L. Kunth. *Plant, Cell and Environment*, **10**, 191–6.

Gorham, J., Budrewicz, E., McDonnell, E. & Wyn Jones, R. G. (1986). Salt tolerance in the Triticeae: salinity-induced changes in the leaf solute composition of some perennial Triticeae. *Journal of Experimental Botany*, **37**, 1114–28.

Gorham, J., Hughes, LL. & Wyn Jones, R. G. (1980). Chemical composition of salt-marsh plants from Ynys Môn (Anglesey): the concept of physiotypes. *Plant, Cell and Environment*, **3**, 309–18.

(1981). Low-molecular-weight carbohydrates in some salt-stressed plants. *Physiologia Plantarum*, **53**, 27–33.

Gorham, J., McDonnell, E. & Wyn Jones, R. G. (1984*a*). Pinitol and other solutes in salt stressed *Sesbania aculeata. Zeitschrift für Pflanzenphysiologie*, **114**, 173–8.

(1984*b*). Salt tolerance in the Triticeae: *Leymus sabulosus. Journal of Experimental Botany*, **35**, 1200–9.

(1985). Some mechanisms of salt tolerance in crop plants. *Plant and Soil*, **89**, 15–40.

Gorham, J., McDonnell, E., Budrewicz, E. & Wyn Jones, R. G. (1985). Salt tolerance in the Triticeae: growth and solute accumulation in leaves of *Thinopyrum bessarabicum. Journal of Experimental Botany*, **36**, 1021–31.

Gorham, J. & Wyn Jones, R. G. (1983). Solute distribution in *Suaeda maritima. Planta*, **157**, 344–9.

Gosselink, J. G. & Baumann, R. H. (1980). Wetland inventories: wetland loss along the United States coast. *Zeitschrift für Geomorphologie Supplementbände* **34**, 173–87.
Gradstein, S. R. & Smittenberg, J. H. (1977). The hydrophilous vegetation of Western Crete. *Vegetatio*, **34**, 65–86.
Gravesen, P. (1972). Plant communities of salt-marsh origin at Tipperne, Western Jutland. *Botanisk Tidsskrift*, **67**, 1–32.
Gravesen, P. & Vestergaard, P. (1969). Vegetation of a Danish off-shore barrier island. *Botanisk Tidsskrift*, **65**, 44–99.
Gray, A. J. (1970). The colonisation of estuaries following barrage building. In *The Flora of a Changing Britain*, ed. F. Perring, pp. 63–72. Hampton: E.W. Classey.
(1971). *Variation in* Aster tripolium *L. with particular reference to some British Populations*. Ph.D. thesis, University of Keele.
(1972). The ecology of Morecambe Bay V. The salt marshes of Morecambe Bay. *Journal of Applied Ecology*, **9**, 207–20.
(1974). The genecology of salt marsh plants. *Hydrobiological Bulletin (Amsterdam)*, **8**, 152–65.
(1976). The Ouse washes and the Wash. In *Nature in Norfolk – a Heritage in Trust*, pp. 123–9. Norwich: Jarrold.
(1977). Reclaimed land. In *The Coastline*, ed. R. S. K. Barnes, pp. 253–70. Chichester: John Wiley.
(1979*a*). The ecological implications of estuarine and coastal land reclamation. In *Estuarine and Coastal Land Reclamation and Water Storage*, ed. B. Knights & A.J. Phillips, pp. 177–94. Farnborough: Saxon House.
(1979*b*). The banks of estuaries and their management. In *Tidal Power and Estuary Management*, ed. R. T. Severn, D. Dinoley & L. E. Hawker, pp. 235–44. Bristol: Scientechnica.
(1985*a*). Adaptation in perennial coastal plants – with particular reference to heritable variation in *Puccinellia maritima* and *Ammophila arenaria*. *Vegetatio*, **61**, 179–88.
(1985*b*). *Poole Harbour. Ecological Sensitivity Analysis of the Shoreline*. Abbots Ripton: NERC.
Gray, A. J. & Adam, P. (1974). The reclamation history of Morecambe Bay. *Nature in Lancs*, **4**, 13–20.
Gray, A. J. & Bunce, R. G. M. (1972). The ecology of Morecambe Bay VI. Soils and vegetation of the salt marshes: a multivariate approach. *Journal of Applied Ecology*, **9**, 221–34.
Gray, A. J., Parsell, R. J. & Scott, R. (1979). The genetic structure of plant populations in relation to the development of salt marshes. In *Ecological Processes in Coastal Environments*, ed. R. L. Jefferies & A. J. Davy, pp. 43–64. Oxford: Blackwell Scientific Publications.
Gray, A. J. & Pearson, J. M. (1984). *Spartina* marshes in Poole Harbour, Dorset. In Spartina anglica *in Great Britain*, ed. P. Doody, pp. 11–14. Attingham Park: Nature Conservancy Council.
Gray, A. J. & Scott, R. (1977*a*). *Puccinellia maritima* (Huds.) Parl. Biological Flora of the British Isles. *Journal of Ecology*, **65**, 699–716.
(1977*b*). The ecology of Morecambe Bay. VII. The distribution of *Puccinellia maritima*, *Festuca rubra* and *Agrostis stolonifera* in the salt marshes. *Journal of Applied Ecology*, **14**, 229–41.
(1980). A genecological study of *Puccinellia maritima* Huds. (Parl.) I. Variation estimated from single plant samples from British populations *New Phytologist*, **85**, 89–107.
(1987). Salt marshes. In *Morecambe Bay. An Assessment of Present Ecological*

Knowledge, ed. N. A. Robinson & A. W. Pringle, pp. 97–117. Lancaster: Morecambe Bay Study Group.

Green, J. (1968). *The Biology of Estuarine Animals*. London: Sidgwick Jackson.

Greensmith, J. T. & Tucker, E. V. (1966). Morphology and evolution of inshore shell ridges and mud-mounds on modern intertidal flats at Bradwell, Essex. *Proceedings of the Geological Association*, **77**, 329–46.

Greenway, H. (1973). Salinity, plant growth and metabolism. *Journal of the Australian Institute of Agricultural Science*, **39**, 24–34.

Greenway, H. & Munns, R. (1980). Mechanisms of salt tolerance in nonhalophytes. *Annual Reviews of Plant Physiology*, **31**, 149–90.

(1983). Interactions between growth, uptake of Cl^- and Na^+, and water relations of plants in saline environments. II. Highly vacuolate cells. *Plant, Cell and Environment*, **6**, 575–89.

Greenway, H., Munns, R. & Wolfe, J. (1983). Interaction between growth, Cl^- and Na^+ uptake, and water relations of plants in saline environments. I. Slightly vacuolated cells. *Plant, Cell and Environment*, **6**, 567–74.

Greenway, H. & Sims, A. P. (1974). Effects of high concentrations of KCl and NaCl on responses of malate dehydrogenase (decarboxylating) to malate and various inhibitors. *Australian Journal of Plant Physiology*, **1**, 15–29.

Gregor, J. W. (1938). Experimental taxonomy II. Initial population differentiation in *Plantago maritima* L. of Britain. *New Phytologist*, **37**, 15–49.

(1939). Experimental taxonomy IV. Population differentiation in North American and European sea plantains allied to *Plantago maritima* L. *New Phytologist*, **38**, 393–422.

(1944). The ecotype. *Biological Reviews*, **14**, 20–30.

(1946). Ecotypic differentiation. *New Phytologist*, **45**, 254–70.

(1956). Adaptation and ecotypic components. *Proceedings of the Royal Society B*, **45**, 333–7.

Grime, J. P. (1979). *Plant Strategies and Vegetation Processes*. Chichester: John Wiley.

Grimshaw, J. F. (1982). A checklist of spiders known from the mangrove forests and associated tidal marshes of northern and eastern Australia. *Operculum*, **5**, 158–61.

Groenendijk, A. M. & Vink-Lievaart, M.A. (1987). Primary production and biomass on a Dutch salt marsh: emphasis on the below-ground component. *Vegetatio*, **70**, 21–7.

Grof, C. P. L., Johnston, M. & Brownell, P. F. (1986). *In vivo* chlorophyll *a* fluorescence in sodium-deficient C_4 plants. *Australian Journal of Plant Physiology*, **13**, 589–95.

Grondin, P. & Melançon, M. (1980). La végétation du marais sale de l'ile à Samuel, archipel de Mingan, Québec. *Phytocoenologia*, **7**, 336–55.

Grubb, P. J. (1977). The maintenance of species-richness in plant communities: the importance of the regeneration niche. *Biological Reviews*, **52**, 107–45.

(1986). Problems posed by sparse and patchily distributed species in species-rich plant communities. In *Community Ecology*, ed. J. Diamond & T. J. Case, pp. 207–25. New York: Harper & Row.

Guy, R. D. & Reid, D. M. (1986). Photosynthesis and the influence of CO_2-enrichment on $\delta^{-13}C$ values in a C_3 halophyte. *Plant, Cell and Environment*, **9**, 65–72.

Guy, R. D., Reid, D. M. & Krouse, H. R. (1980). Shifts in carbon isotope ratios of two C_3 halophytes under natural and artificial conditions. *Oecologia (Berlin)*, **44**, 241–7.

Guy, R. D., Warne, P. G. & Reid, D. M. (1984). Glycine content of halophytes: improved analysis by liquid chromatography and interpretation of results. *Physiologia Plantarum*, **61**, 195–202.

Hackney, C. T. & Bishop, T. D. (1981). A note on the relocation of marsh debris during a storm surge. *Estuarine, Coastal and Shelf Science*, **12**, 621–4.

Hackney, C. T. & de la Cruz, A. A. (1981). Effects of fire on brackish marsh communities: management implications. *Wetlands (USA)*, **1**, 75–86.

Hadač, E. (1946). The plant-communities of Sassen Quarter, Vestspitsbergen. *Studia Botanica Cechoslovaca*, **7**, 127–64.

(1970). Sea-shore communities of Reykjanes Peninsula, SW. Iceland (plant communities of Reykjanes Peninsula, Part 2). *Folia geobotanica phytotaxonomica*, **5**, 133–44.

Haines, B. L. & Dunn, E. L. (1976). Growth and resource allocation responses of *Spartina alterniflora* Loisel to three levels of NH_4 –N, Fe, and NaCl in solution culture. *Botanical Gazette*, **137**, 224–30.

Haines, E. B. (1976). Stable carbon isotope ratios in the biota, soils and tidal water of a Georgia salt marsh. *Estuarine and Coastal Marine Science*, **4**, 609–16.

(1977). The origins of detritus in Georgia salt marsh estuaries. *Oikos*, **29**, 254–60.

(1979). Interactions between Georgia salt marshes and coastal waters: a changing paradigm. In *Ecological Processes in Coastal and Marine Systems*, ed. R.J. Livingstone, pp. 35–46. New York: Plenum Press.

Haines, E. B. & Montague, C. L. (1979). Food sources of estuarine invertebrates analysed using $^{13}C/^{12}C$ ratios. *Ecology*, 60, 48–56.

Hajibagheri, M. A. & Flowers, T. J. (1985). Salt tolerance in the halophyte *Suaeda maritima* (L.) Dum. The influence of the salinity of the culture solution of leaf starch and phosphate content. *Plant, Cell and Environment*, **8**, 261–7.

Hajibagheri, M. A., Hall, J. A. & Flowers, T. J. (1983). The structure of the cuticle in relation to cuticular transpiration in leaves of the halophyte *Suaeda maritima* (L.) Dum. *New Phytologist*, **94**, 125–31.

(1984). Stereological analysis of leaf cells of the halophyte *Suaeda maritima* (L.) Dum. *Journal of Experimental Botany*, **35**, 1547–57.

Hajibagheri, M. A., Yeo, A. R. & Flowers, T. J. (1985). Salt tolerance in *Suaeda maritima* (L.). Dum. Fine structure and ion concentrations in the apical region of roots. *New Phytologist*, **99**, 331–43.

Hale, W. C. (1980). *Waders*. London: Collins.

Hall, J. L., Harvey, D. M. R. & Flowers, T. J. (1978). Evidence for the cytoplasmic localisation of betaine in leaf cells. *Planta*, **140**, 59–62.

Handley, J. F. & Jennings, D. H. (1977). The effect of ions on growth and leaf succulence of *Atriplex hortensis* var. *cupreata*. *Annals of Botany*, **41**, 1109–12.

Hannon, N. J. & Barber, H. N. (1972). The mechanism of salt tolerance in naturally selected populations of grasses. *Search*, **3**, 259–60.

Hannon, N. & Bradshaw, A. D. (1968). Evolution of salt tolerance in two coexisting species of grass. *Nature (London)*, **220**, 1342–3.

Hansen, A. D. & Nelson, C. E. (1978). Betaine accumulation and ^{14}C-formate metabolism in water stressed barley leaves. *Plant Physiology*, **62**, 305–12.

Hansen, A. D., May, A. M., Grumet, R., Bode, J., Jamieson, G. C. & Rhodes, D. (1985). Betaine synthesis in chenopods: localisation in chloroplasts. *Proceedings of the National Academy of Science, U.S.A.*, **82**, 3678–82.

Hanson, H. C. (1951). Characteristics of some grassland, marsh and other plant communities in western Alaska. *Ecological Monographs*, **21**, 317–78.

(1953). Vegetation types in northwestern Alaska and comparisons with communities in other Arctic regions. *Ecology*, **34**, 111–40.

Hardisky, M. A., Gross, M. F. & Klemas, V. (1986). Remote sensing of coastal wetlands. *Bioscience*, **36**, 453–60.

Harmsworth, G. C. & Long, S. P. (1986). An assessment of saltmarsh erosion in Essex, England, with reference to the Dengie Peninsula. *Biological Conservation*, **35**, 377–87.

Harper, J. L. (1985). Modules, branches, and the capture of resources. In *Population Biology and Evolution of Clonal Organisms*, ed. J. B. C. Jackson, L. W. Buss & R. E. Cook, pp. 1–33. New Haven: Yale University Press.

Harrison, E. Z. & Bloom, A. L. (1977). Sedimentation rates on tidal salt marshes in Connecticut. *Journal of Sedimentary Petrology*, **47**, 1484–90.

Hartman, J., Caswell, H. & Valiela, I. (1983). Effects of wrack accumulation on salt marsh vegetation. *Oceanologica Acta 1983*, pp. 99–102.

Hartnoll, R. G. & Hawkins, S. J. (1982). The emersion curve in semidiurnal tidal regimes. *Estuarine, Coastal and Shelf Science*, **15**, 365–71.

Harvey, D. M. R. & Flowers, T. J. (1978). Determination of the sodium, potassium and chloride ion concentrations in the chloroplasts of the halophyte *Suaeda maritima*, by non-aqueous cell fractionation. *Protoplasma*, **97**, 337–49.

Harvey, D. M. R., Hall, J. L., Flower, T. J. & Kent, B. (1981). Quantitative ion localization within *Suaeda maritima* leaf mesophyll cells. *Planta*, **151**, 555–60.

Harvey, D. M. R., Stelzer, R., Brandtner, R. & Kramer, D. (1985). Effects of salinity on ultrastructure and ion distribution in roots of *Plantago coronopus*. *Physiologia Plantarum*, **66**, 328–38.

Harvey, H. W. (1957). *The Chemistry and Fertility of Sea Waters*, 2nd edn. Cambridge: Cambridge University Press.

Hauman, L. (1926). Étude phytogéographique de la Patagonie. *Bulletin de la Société royale de botanique de Belgique*, **58**, 105–79.

Havill, D. C., Ingold, A. & Pearson, J. (1985). Sulphide tolerance in coastal halophytes. *Vegetatio*, **62**, 279–85.

Haynes, F. N. (1984). *Spartina* in Langstone Harbour, Hampshire. In Spartina anglica *in Great Britain*, ed. P. Doody, pp. 5–10. Attingham Park: Nature Conservancy Council.

Haynes, F. N. & Coulson, M. G. (1982). The decline of *Spartina* in Langstone Harbour, Hampshire. *Proceedings of the Hampshire Field Club and Archaeological Society*, **38**, 5–18.

Haynes, J. & Dobson, M. (1969). Physiography, Foraminifera and sedimentation in the Dovey estuary (Wales). *Geological Journal*, **6**, 217–56.

Heckman, C. W. (1986). The role of marsh plants in the transport of nutrients as shown by a quantitative model for the freshwater section of the Elbe estuary. *Aquatic Botany*, **25**, 139–51.

Hellebust, J. A. (1976). Osmoregulation. *Annual Review of Plant Physiology*, **27**, 485–505.

Hendrarto, I. B. & Dickinson, C. H. (1984). Soil and root micro-organisms in four salt marsh communities. *Transactions of the British Mycological Society*, **83**, 615–20.

Henriksen, K. & Jensen, A. (1979). Nitrogen mineralization in a salt marsh ecosystem dominated by *Halimione portulacoides*. In *Ecological Process in Coastal Environments*, ed. R. L. Jefferies & A. J. Davy, pp. 373–84. Oxford: Blackwell Scientific Publications.

Hepburn, I. (1952). *Flowers of the Coast*. London: Collins.

Heslop-Harrison, J. (1964). Forty years of genecology. *Advances in Ecological Research*, **2**, 159–247.

Heslop-Harrison, Y. & Heslop-Harrison, J. (1957). Ring formation by *Triglochin maritima* in eastern Irish salt marshes. *Irish Naturalists Journal*, **12**, 223–9.

Heydemann, B. (1979). Responses of animals to spatial and temporal environmental heterogeneity within salt marshes. In *Ecological Processes in Coastal Environments*, ed. R. L. Jefferies & A. J. Davy, pp. 145–63. Oxford: Blackwell Scientific Publications.

(1981). Ecology of the arthropods of the lower salt marsh. In *Terrestrial and Freshwater Fauna of the Wadden Sea Area*, ed. C. J. Smit, J. den Hollander, W.K.R.E. Wingerden & W. J. Wolff, pp. 35–57. Rotterdam: Balkema.

Heyligers, P. (1984). Beach invaders. *Australian Natural History*, **21**, 212–14.

Hill, A. E. & Hill, B. S. (1976). Mineral ions. In *Transport in Plants II. Part B. Tissues*

and Organs, Encyclopedia of Plant Physiology N.S., vol. 1, ed. U. Luttge & M. C. Pitman, pp. 225–243, Berlin: Springer.
Hill, A. W. (1927). The genus *Lilaeopsis*: a study in geographical distribution. *Linnean Society Journal (Botany)*, **47**, 525–31.
Hill, T. G. (1909). The Bouche D'Erquy in 1908. *New Phytologist*, **8**, 97–103.
Hochachka, P. W. & Somero, G. N. (1984). *Biochemical Adaptation*, 2nd edn. Princeton: Princeton University Press.
Hodgkin, E. P. (1978). Blackwood River estuary. An environmental study of the Blackwood River estuary Western Australia 1974–75. *Report No. 1 Department of Conservation and Environment*, pp. 1–78.
Hodgkin, E. P., Black, R. E., Birch, P. B. & Hillman, K. (1985). *The Peel-Harvey Estuarine System. Proposals for Management*. Report 14. Perth: Department of Conservation and Environment. Western Australia.
Hodson, M. J., Öpik, H. & Wainwright, S. J. (1985). Changes in ion and water content of individual shoot organs in a salt-tolerant and a salt-sensitive clone of *Agrostis stolonifera* L. during and subsequent to treatment with sodium chloride. *Plant, Cell and Environment*, **8**, 677–88.
Hodson, M. J., Smith, M. M., Wainwright, S. J. & Öpik, H. (1982). Cation cotolerance in a salt-tolerant clone of *Agrostis stolonifera* L. *New Phytologist*, **90**, 253–61.
(1985). The effects of the interaction between salinity and nitrogen limitation in *Agrostis stolonifera* L. *Vegetatio*, **61**, 255–63.
Hofmann, W. (1969). Das Puccinellietum phryganodis in Südost-Spitzbergen. *Mitteilungen der floristisch-soziologischen Arbeitsgemeinschaft N.F.*, **14**, 224–30.
Hollaender, A., Aller, J. C., Epstein, E., San Pietro, A. A. & Zaborsky, O. R. (eds.) (1979). *The Biosaline Concept*. New York: Plenum Press.
Hook, D. D. & Crawford, R. M. M. (eds). (1978). *Plant Life in Anaerobic Environments*. Ann Arbor: Ann Arbor Science Publishers.
Hopkins, D. R. & Parker, V. T. (1984). A study of the seed bank of a salt marsh in northern San Francisco Bay. *American Journal of Botany*, **71**, 348–55.
Hopmans, P., Douglas, L. A. & Chalk, P. M. (1984). Effects of soil salinity and mineral nitrogen on the acetylene reduction activity of *Trifolium subterraneum* L. *Australian Journal of Agricultural Research*, **35**, 9–15.
Howarth, R. W. (1979). Pyrite: its rapid formation in a salt marsh and its importance in ecosystem metabolism. *Science*, **203**, 49–51.
Howarth, R. W. & Teal, J. M. (1979). Sulfate reduction in a New England salt marsh. *Limnology and Oceanography*, **24**, 999–1013.
(1980). Energy flow in a salt marsh ecosystem: the role of reduced inorganic sulfur compounds. *American Naturalist*, **116**, 862–72.
Howes, B. L., Dacey, J. W. H. & Goehringer, D. D. (1986). Factors controlling the growth form of *Spartina alterniflora*: feedbacks between above-ground production, sediment oxidation, nitrogen and salinity. *Journal of Ecology*, **74**, 881–98.
Hsiao, T. C. (1973). Plant responses to water stress. *Annual Reviews of Plant Physiology*, **24**, 519–70.
Hubbard, J. C. E. (1965). *Spartina* marshes in Southern England VI. Pattern of invasion in Poole Harbour. *Journal of Ecology*, **53**, 799–813.
(1969). Light in relation to tidal immersion and the growth of *Spartina townsendii* (s.l.). *Journal of Ecology*, **57**, 795–804.
(1970). Effects of cutting and seed production in *Spartina anglica*. *Journal of Ecology*, **58**, 329–34.
Hubbard, J. C. E. & Ranwell, D. S. (1966). Cropping *Spartina* salt marsh for silage. *Journal of the British Grassland Society*, **21**, 214–17.
Hubbard, J. C. E. & Stebbings, R. E. (1967). Distribution, dates of origin and acreage of

Spartina townsendii (s.l.) marshes in Great Britain. *Proceedings of the Botanical Society of the British Isles*, **7**, 1–7.

(1968). *Spartina* marshes in southern England. VII. Stratigraphy of the Keysworth Marsh, Poole Harbour. *Journal of Ecology*, **56**, 707–22.

Huiskes, A. H. L., van Soelen, J. & Markusse, M. M. (1985). Field studies on the variability of populations of *Aster tripolium* L. in relation to salt-marsh zonation. *Vegetatio*, **61**, 163–9.

Humphreys, M. O. (1982). The genetic basis of tolerance to salt spray in populations of *Festuca rubra* L. *New Phytologist*, **91**, 287–96.

Hunter, G. T. (1970). Post-glacial uplift at Fort Albany, James Bay. *Canadian Journal of Earth Sciences*, **7**, 547–8.

Huntley, B. & Birks, H. J. B. (1983). *An Atlas of Past and Present Pollen Maps for Europe: 0–13,000 Years Ago*. Cambridge: Cambridge University Press.

Hussey, A. & Long, S. P. (1982). Seasonal changes in weight of above- and below-ground material in a salt marsh at Colne Point, Essex. *Journal of Ecology*, **70**, 757–71.

Hutchings, P. & Saenger, P. (1987). *Ecology of Mangroves*. St Lucia: University of Queensland Press.

Hutchinson, I. (1982). Vegetation–environment relations in a brackish marsh, Lulu Island, Richmond, B. C. *Canadian Journal of Botany*, **60**, 452–62.

Inglis, C. C. & Kestner, F. J. T. (1958). The long-term effects of training walls, reclamation and dredging on estuaries. *Proceedings of the Institution of Civil Engineers*, **9**, 193–216.

Ingold, A. & Havill, D. C. (1984). The influence of sulphide on the distribution of higher plants in salt marshes. *Journal of Ecology*, **72**, 1043–54.

Ingram, H. A. P., Barclay, A. M., Coupar, A. M., Glover, J. G., Lynch, B. M. & Sprent, J. I. (1980). *Phragmites* performance in reed beds in the Tay estuary. *Proceedings of the Royal Society of Edinburgh*, **78B**, 89–107.

Ingrouille, M. J. & Pearson, J. (1987). The pattern of morphological variation in the *Salicornia europaea* L. aggregate (Chenopodiaceae). *Watsonia*, **16**, 269–81.

Ishizuka, K. (1974). Maritime vegetation. In *The Flora and Vegetation of Japan*, ed. M. Numata, pp. 151–72. Amsterdam: Elsevier.

Iversen, J. (1936). Biologische Pflanzentypen als Hilfsmittel in der Vegetationsforschung. Ein Beitrag zur Ökologischen Characterisierung und Anordnung der Pflanzengesellschaften. *Meddelelser fra Skalling-Laboratoriet*, **5**, 1–224.

Jackson, D., Harkness, D. D., Mason, C. F. & Long, S. P. (1986). *Spartina anglica* as a carbon source for salt-marsh invertebrates: a study using $\delta^{13}C$ values. *Oikos*, **46**, 163–70.

Jackson, D., Long, S.P. & Mason, C. F. (1986). Net primary production, decomposition and export of *Spartina anglica* on a Suffolk salt-marsh. *Journal of Ecology*, **74**, 647–62.

Jackson, D., Mason, C. F. & Long, S. P. (1985). Macro-invertebrate populations and production on a salt-marsh in east England dominated by *Spartina anglica*. *Oecologia (Berlin)*, **65**, 406–411.

Jackson, M. M., Herman, B. & Goodenough, A. (1982). An examination of the importance of ethanol in causing injury to flooded plants. *Plant, Cell and Environment*, **5**, 163–72.

Jacobson, H. A., Jacobson, G. L. & Kelley, J. T. (1987). Distribution and abundance of tidal marshes along the coast of Maine. *Estuaries*, **10**, 126–31.

Jakobsen, B. (1964). Vadehavets morfologi. En geografisk analyse af vadelandskabets formuduikling med saerlig hensyntagen til Juvre Dybs Tidevandsområde. *Folia Geografiska Danica*, **11**, 1–176.

Jefferies, R. L. (1972). Aspects of saltmarsh ecology with particular reference to inorganic plant nutrition. In *The Estuarine Environment*, ed. R. S. K. Barnes, pp. 61–85. Barking: Applied Science Publishers.

(1973). The ionic relations of the halophyte *Triglochin maritima* L. In *Ion Transport in Plants*, ed. W. P. Anderson, pp. 297–321. London: Academic Press.

(1976). The North Norfolk coast. In *Nature in Norfolk – a Heritage in Trust*, pp. 130–8. Norwich: Jarrold.

(1977*a*). The vegetation of salt marshes at some coastal sites in arctic North America. *Journal of Ecology*, **65**, 661–72.

(1977*b*). Growth responses of coastal halophytes to inorganic nitrogen. *Journal of Ecology*, **65**, 847–65.

(1980). The role of organic solutes in osmoregulation in halophytic higher plants. In *Genetic Engineering of Osmoregulation*, ed. D. W. Rains, R. C. Valentine & A. Hollaender, pp. 135–54. New York: Plenum Press.

(1981). Osmotic adjustment and the response of halophytic plants to salinity. *BioScience*, **31**, 42–6.

Jefferies, R. L., Davy, A. J. & Rudmik, T. (1979). The growth strategies of coastal halophytes. In *Ecological Processes in Coastal Environments*, ed. R. L. Jefferies & A. L. Davy, pp. 243–68. Oxford: Blackwell Scientific Publications.

(1981). Population biology of the salt marsh annual *Salicornia europaea* agg. *Journal of Ecology*, **69**, 17–31.

Jefferies, R. L. & Gottlieb, L. D. (1982). Genetic differentiation of the microspecies *Salicornia europaea* L. (*sensu stricto*) and *S. ramosissima* J. Woods. *New Phytologist*, **92**, 123–9.

(1983). Genetic variation within and between populations of the asexual plant *Puccinellia phryganodes*. *Canadian Journal of Botany*, **61**, 774–9.

Jefferies, R. L., Jensen, A. & Abraham, K. F. (1979). Vegetational development and the effect of geese on vegetation at La Perouse Bay, Manitoba. *Canadian Journal of Botany*, **57**, 1439–50.

Jefferies, R. L., Jensen, A. & Bazely, D. (1983). The biology of the annual *Salicornia europaea* agg. at the limits of its range in Hudson Bay. *Canadian Journal of Botany*, **61**, 762–73.

Jefferies, R. L. & Perkins, N. (1977). The effects on the vegetation of the additions of inorganic nutrients to salt marsh soils at Stiffkey, Norfolk. *Journal of Ecology*, **65**, 867–82.

Jefferies, R. L. & Rudmik, T. (1984). The responses of halophytes to salinity: an ecological perspective. In *Salinity Tolerance in Plants – Strategies for Crop Improvement*, ed. R. C. Staples & G. H. Toenniessen, pp. 213–27. New York: John Wiley.

Jefferies, R. L., Rudmik, T. & Dillon, E. M. (1979). Responses of halophytes to high salinities and low water potentials. *Plant Physiology*, **64**, 889–994.

Jeffrey, D. W. (1984). Case history: North Bull Island. An assessment of a nature conservation resource in Ireland. In *Nature Conservation Progress and Problems*, ed. D. W. Jeffrey, pp. 67–80. Dublin: Royal Irish Academy.

(1987). *Soil–Plant Relationships. An Ecological Approach.* Beckenham: Croom Helm.

Jennings, D. H. (1968). Halophytes, succulence and sodium in plants – a unified theory. *New Phytologist*, **67**, 899–911.

(1976). The effect of sodium chloride on higher plants. *Biological Reviews*, **51**, 453–86.

Jensen, A. (1985*a*). The effect of cattle and sheep grazing on salt marsh vegetation at Skallingen, Denmark. *Vegetatio*, **60**, 37–48.

(1985*b*). On the ecophysiology of *Halimione portulacoides*. *Vegetatio*, **61**, 231–40.

Jensen, A., Henriksen, K. & Rasmussen, M. B. (1985). The distribution and interconversion of ammonium and nitrate in the Skallingen salt marsh (Denmark) and their exchange with the adjacent coastal water. *Vegetatio*, **62**, 357–66.

Jensen, A. & Jefferies, R. L. (1984). Fecundity and mortality in populations of *Salicornia europaea* agg. at Skallingen, Denmark. *Holarctic Ecology*, **7**, 399–412.

Jerling, L. (1981). Effects of microtopography on the summer survival of *Plantago maritima* seedlings. *Holarctic Ecology*, **4**, 120–6.

(1983). Composition and viability of the seed bank along a successional gradient on a Baltic sea shore meadow. *Holarctic Ecology*, **6**, 150–6.

(1984). The impact of some environmental factors on the establishment of *Plantago maritima* seedlings and juveniles along a distributional gradient. *Holarctic Ecology*, **7**, 271–9.

(1985). Population dynamics of *Plantago maritima* along a distributional gradient on a Baltic seashore meadow. *Vegetatio*, **61**, 155–61.

Jerling, L. & Andersson, M. (1982). Effects of selective grazing by cattle on the reproduction of *Plantago maritima*. *Holarctic Ecology*, **5**, 405–11.

Jerling, L. & Liljelund, L-E. (1984). Dynamics of *Plantago maritima* along a distributional gradient: a demographic study. *Holarctic Ecology*, **7**, 280–8.

Jeschke, W. D. (1984). K^+–Na^+ exchange at cellular membranes, intracellular compartmentation of cations and salt tolerance. In *Salinity Tolerance in Plants – Strategies for Crop Improvement*, ed. R. C. Staples & G. H. Toenniessen, pp. 37–66. New York: John Wiley.

Joenje, W. (1985). The significance of waterfowl grazing in the primary vegetation succession on embanked sandflats. *Vegetatio*, **62**, 399–406.

Johnson, D. S. & York, H. H. (1915). *The relation of plants to tide-levels. A study of factors affecting the distribution of marine plants.* Washington: Carnegie Institution of Washington.

Jolivet, Y., Hamelin, J. & Larher, F. (1983). Osmoregulation in halophytic higher plants: the protective effects of glycine betaine and other related solutes against the oxalate destabilization of membranes in beet root cells. *Zeitschrift für Pflanzenphysiologie*, **109**, 171–80.

Jørgensen, B. B. (1977). Bacterial sulfate reduction within reduced microniches of oxidized marine sediments. *Marine Biology*, **41**, 7–17.

Jones, K. (1974). Nitrogen fixation in a salt marsh. *Journal of Ecology*, **62**, 553–65.

Kaiser, W. M., Weber, H. & Sauer, M. (1983). Photosynthetic capacity, osmotic responses and solute content of leaves and chloroplasts from *Spinacia oleracea* under salt stress. *Zeitschrift für Pflanzenphysiologie*, **113**, 15–27.

Kalela, A. (1939). Über Wiesen und wiesenartige Pflanzengesellschaften auf der Fischer-Halbinsel in Petsamo-Lappland. *Acta Forestalia Fennica*, **48(2)**, 1–523.

Kaplan, A. & Gale, J. (1972). Effect of sodium chloride salinity on the water balance of *Atriplex halimus*. *Australian Journal of Biological Science*, **25**, 895–903.

Karimi, S. H. & Ungar, I. A. (1986). Oxalate and inorganic ion concentrations in *Atriplex triangularis* Willd. organs in response to salinity, light level, and aeration. *Botanical Gazette*, **147**, 65–70.

Kassas, M. & Zahran, M. A. (1962). Studies on the ecology of the Red Sea coastal land. I. The district of Gebel Ataga and El-Galala, El-Bahariya. *Bulletin de la Société de Geographie d'Egypte*, **35**, 129–75.

(1965). Studies on the ecology of the Red Sea coastal land. II. The district from El-Galala, El-Qibliya to Hurghada. *Bulletin de la Société de Géographie d'Egypte*, **38**, 155–93.

Kay, B. H., Sinclair, P. & Marks, E. N. (1981). Mosquitoes: their interrelationships with man. In *The Ecology of Pests*, ed. R. L. Kitching & R. E. Jones, pp. 157–74. Melbourne: CSIRO.

Kay, Q. O. N. & Rojanavipart, P. (1977). Saltmarsh ecology and trace-metal studies. In *Problems of a Small Estuary*, ed. A. Nelson-Smith & E. M. Bridges, pp. 2:2/1–2:2/16. Swansea: Institute of Marine Studies.

Kay, Q. O. N. & Woodell, S. R. J. (1976). The vegetation of anthills in West Glamorgan saltmarshes. *Nature in Wales*, **15**, 81–7.

Keeley, J. E. (1978). Malic acid accumulation in roots in response to flooding: evidence contrary to its role as an alternative to ethanol. *Journal of Experimental Botany*, **29**, 1345–9.

Keighery, G. J. (1979). Insect pollination of *Suaeda australis* (Chenopodiaceae). *West Australian Naturalist*, **14**, 154.

Kellerhals, P. & Murray, J. W. (1969). Tidal flats at Boundary Bay. Fraser River delta, British Columbia. *Bulletin of Canadian Petroleum Geology*, **17**, 67–91.

Kemp, P. R. & Cunningham, G. L. (1981). Light, temperature and salinity effects of growth, leaf anatomy and photosynthesis of *Distichlis spicata* (L.) Greene. *American Journal of Botany*, **68**, 507–16.

Kennish, M. J. (1986). *Ecology of estuaries, vol. I. Physical and Chemical Aspects*. Boca Raton: CRC Press.

Kershaw, K. A. (1976). The vegetational zonation of the East Pen Island salt marshes, Hudson Bay. *Canadian Journal of Botany*, **54**, 5–13.

Kesel, R. H. & Smith, J.S. (1978). Pan and creek formation in intertidal salt-marshes. *Scottish Geographical Magazine*, **94**, 159–68.

Kestner, F. J. T. (1962). The old coastline of the Wash. A contribution to the understanding of loose boundary processes. *Geographical Journal*, **128**, 457–78.

(1979). Loose boundary hydraulics and land reclamation. In *Estuarine and Coastal Land Reclamation and Water Storage*, ed. B. Knights & A. J. Phillips, pp. 23–47. Farnborough: Saxon House.

Kestner, F. J. T. & Inglis, C. C. (1956). A study of erosion and accretion during cyclic changes in an estuary and their effect on reclamation of marginal land. *Journal of Agricultural Engineering Research*, **1**, 63–7.

King, C. A. (1982). Classification. In *Encyclopedia of Beaches and Coastal Environments*, ed. M. L. Schwartz, pp. 210–22. Stroudsburg: Hutchinson Ross.

King, G. M., Klug, M. J., Weigert, R. G. & Chalmers, A. G. (1982). Relation of soil water movement and sulphide concentrations to *Spartina alterniflora* production in a Georgia, U.S.A. salt-marsh. *Science*, **218**, 61–3.

King, R. J. (1981*a*). Mangrove and saltmarsh plants. In *Marine Botany – An Australian Perspective*, ed. M. N. Clayton & R. J. King, pp. 308–28. Melbourne: Longman.

(1981*b*). The free-living *Hormosira banksii* (Turner) Decaisne associated with mangroves in temperate Eastern Australia. *Botanica Marina*, **24**, 569–76.

Kingsbury, R. W., Radlow, A., Mudie, P. J., Rutherford, J. & Radlow, R. (1976). Salt stress responses in *Lasthenia glabrata*, a winter annual composite endemic to saline soils. *Canadian Journal of Botany*, **54**, 1377–85.

Kirkby, R. (1984). The recent history of the Lower Medway salt marshes. In Spartina anglica *in Great Britain*, ed. P. Doody, pp. 18–20. Attingham Park: Nature Conservancy Council.

Kirkpatrick, J. B. (1981). Coastal, heath and wetland vegetation. In *The Vegetation of Tasmania*, ed. W. D. Jackson, pp. 36–54. Canberra: Australian Academy of Science.

Kirkpatrick, J. B. & Glasby, J. (1981). Salt marshes in Tasmania – distribution, community composition and conservation. *Occasional paper 8, Department of Geography University of Tasmania*, pp. 1–62.

Klecka, A. & Vukolov, V. (1937). Comparative studies of the mycorrhiza of meadow halophytes (translated abstract). *Review of Applied Mycology*, **16**, 768.

Kloot, P. M. (1983). The role of common iceplant (*Mesembryanthemum crystallinum*) in the deterioration of medic pastures. *Australian Journal of Ecology*, **8**, 301–6.

Knox, G. A. (1986). *Estuarine Ecosystems: A Systems Approach.* 2 vols. Boca Raton: CRC Press.

Knutson, P. L., Ford, J. C. & Inskeep, M. R. (1981). National survey of planted salt marshes (vegetative stabilization and wave stress). *Wetlands (USA)*, **1**, 129–57.

Kohlmeyer, J. & Kohlmeyer, E. (1979). *Marine Mycology – The Higher Fungi.* New York: Academic Press.

Kozlowski, T. T. (1984*a*). Plant responses to flooding of soil. *BioScience*, **34**, 163–7.

Kozlowski, T. T., ed. (1984*b*). *Flooding and Plant Growth.* New York: Academic Press.

Kramer, D. (1983). Genetically determined adaptations in roots to nutritional stress. Correlation of structure and function. *Plant and Soil*, **72**, 167–73.

(1984). Cytological aspects of salt tolerance in higher plants. In *Salinity Tolerance in Plants – Strategies for Crop Improvement*, ed. R. C. Staples & G. H. Toenniessen, pp. 3–15. New York: John Wiley.

Krasilov, V. A. (1975). *Paleoecology of Terrestrial Plants. Basic Principles and Techniques* (translated by H. Hardin). Chichester: John Wiley.

Kraeuter, J. N. & Wolf, P. L. (1974). The relationship of marine macroinvertebrates to salt marsh plants. In *Ecology of Halophytes*, ed. R. J. Reimold & W. H. Queen, pp. 449–62. New York: Academic Press.

Kristiansen, J. N. (1977). A phytosociological and synchorological contribution to the Caricetum subspathaceae and the Festuco–Caricetum glareosae on salt marshes in northern Norway. *Astarte*, **10**, 107–21.

Kristjansson, J. K. & Schönheit, P. (1983). Why do sulfate-reducing bacteria outcompete methanogenic bacteria for substrates? *Oecologia (Berlin)*, **60**, 264–6.

Kuenzler, E. J. (1961*a*). Structure and energy flow of a mussel population in a Georgia salt marsh. *Limnology & Oceanography*, **6**, 191–204.

(1961*b*). Phosphorus budget of a mussel population. *Limnology & Oceanography*, **6**, 400–15.

Kuiper, P. J. C. (1984). Functioning of plant cell membranes under saline conditions: membrane lipid composition and ATPases. In *Salinity Tolerance in Plants – Strategies for Crop Improvement*, ed. R. C. Staples & G. H. Toenniessen, pp. 77–91. New York: Wiley.

Kuraishi, S., Kiyohide, K., Miyauchi, H., Sakurai, N., Tsobota, H., Ninaki, M., Goto, I. & Sugo, J. (1985). Brackish water and soil components of mangrove forests on Iriomote Island, Japan. *Biotropica*, **17**, 277–86.

Kuramato, R. T. & Brest, D. E. (1979). Physiological response to salinity by four salt marsh plants. *Botanical Gazette*, **140**, 295–8.

LaHaye, P. A. & Epstein, E. (1969). Salt toleration by plants: enchancement with calcium. *Science*, **166**, 395–6.

Larher, F. & Hamelin, J. (1975). L'acide trimethylaminoproprionique des rameaux de *Limonium vulgare*. *Phytochemistry*, **14**, 1798–800.

Larher, F., Hamelin, J. & Stewart, G. R. (1977). L'acide dimethylsulfonium-3 propanoique de *Spartina anglica*. *Phytochemistry*, **16**, 2019–20.

Larher, F., Jolivet, Y., Briens, M. & Goas, M. (1982). Osmoregulation in higher plant halophytes: organic nitrogen accumulation in glycine betaine and proline during the growth of *Aster tripolium* and *Suaeda macrocarpa* under saline conditions. *Plant Science Letters*, **24**, 201–10.

Larkum, A. W. D. & Wyn Jones, R. G. (1979). Carbon dioxide fixation by chloroplasts isolated in glycinebetaine. A putative cytoplasmic osmoticum. *Planta*, **145**, 393–4.

Läuchli, A. (1984). Salt exclusion: an adaptation of legumes for crops and pastures under saline conditions. In *Salinity Tolerance in Plants – Strategies for Crop Improvement*, ed. R. C. Staples & G. H. Toenniessen, pp. 171–87. New York: John Wiley.

Lawton, J. R., Todd, A. & Naidoo D. K. (1981). Preliminary investigations into the structure of the roots of the mangroves, *Avicennia marina* and *Bruguiera gymnorrhiza*, in relation to ion uptake. *New Phytologist*, **88**, 713–22.

Leach, S. J. (1988). Rediscovery of *Halimione pedunculata* (L.) Aellen in Britain. *Watsonia*, **17**, 170–1.

Leach, S. J. & Phillipson P. H. (1985). The saltmarsh and brackish swamp vegetation of the Fife peninsula. *Transactions of the Botanical Society of Edinburgh*, **44**, 357–73.

Lee, J. A. (1977). The vegetation of British inland salt marshes. *Journal of Ecology*, **65**, 673–98.

Leereveld, H., Meeuse, A. D. J. & Stelleman, P. (1981). Anthecological relations between reputedly anemophilous flowers and syrphid flies IV. A note on the anthecology of *Scirpus maritimus* L. *Acta botanica Neerlandica*, **30**, 465–73.

Leigh, R. A., Ahmad, N. & Wyn Jones, R. G. (1981). Assessment of glycinebetaine and proline compartmentation by analysis of isolated beet vacuoles. *Planta*, **153**, 34–41.

Leopold, A. C. & Willing, R. P. (1984). Evidence for toxicity effects of salt on membranes. In *Salinity Tolerance in Plants – Strategies for Crop Improvement*, ed. R. C. Staples & G. H. Toenniessen, pp. 67–76. New York: John Wiley.

Lerner, H. R. (1985). Adaptation to salinity at the plant cell level. *Plant and Soil*, **89**, 3–14.

Lessani, H. & Marschner, H. (1978). Relation between salt tolerance and long-distance transport of sodium and chloride in various crop species. *Australian Journal of Plant Physiology*, **5**, 27–37.

Levins, R. (1962). Theory of fitness in a heterogeneous environment I. The fitness set and its adaptive function. *American Naturalist*, **96**, 361–78.

(1963). Theory of fitness in a heterogeneous environment II. Developmental flexibility and niche selection. *American Naturalist*, **97**, 75–90.

Lewinton, R. C. (1974). *The Genetic Basis of Evolutionary Change*. New York: Columbia University Press.

Lewis, J. R. (1964). *Ecology of Rocky Shores*. London: English Universities Press.

Lewis, R. R. (ed.) (1982). *Creation and Restoration of Coastal Plant Communities*. Boca Raton: CRC Press.

Linhart, Y. B. (1976). Density-dependent seed germination strategies in colonising versus non-colonising plant species. *Journal of Ecology*, **64**, 375–80.

Linthurst, R. A. & Seneca, E. D. (1981). Aeration, nitrogen and salinity as determinants of *Spartina alterniflora* Loisel. growth response. *Estuaries*, 4, 59–63.

Liphschitz, N. & Waisel, Y. (1974). Existence of salt glands in various genera of the Gramineae. *New Phytologist*, **73**, 507–13.

(1982). Adaptations of plants to saline environments: salt excretion and glandular structure. In *Contributions to the Ecology of Halophytes*, ed. P. S. Sen & K. S. Rajpurohit, pp. 197–214. The Hague: Dr Junk.

Lockwood, A. P. M. & Inman, C. B. E. (1979). Ecophysiological responses of *Gammarus duebeni* to salinity fluctuations. In *Ecological Processes in Coastal Environments*, ed. R. L. Jefferies & A. J. Davy, pp. 269–84. Oxford: Blackwell Scientific Publications.

Long, S. P. (1983). C_4 photosynthesis at low temperatures. *Plant, Cell and Environment*, **6**, 345–63.

Long, S. P. & Mason, C. F. (1983). *Saltmarsh Ecology*. Glasgow, Blackie.

Long, S. P. & Woolhouse, H. W. (1979). Primary production in *Spartina* marshes. In *Ecological Processes in Coastal Environments*, ed. R. L. Jefferies & A. J. Davy, pp. 333–52. Oxford: Blackwell Scientific Publications.

Longstreth, D. J. & Nobel, P. S. (1979). Salinity effects on leaf anatomy: consequences for photosynthesis. *Plant Physiology*, **63**, 700–3.

Longstreth, D. J. & Strain, B. R. (1977). Effects of salinity and illumination on photosynthesis and water balance of *Spartina alterniflora* Loisel. *Oecologia (Berlin)*, **38**, 303–16.

Lopez, G. R., Levinton, J. S. & Slobodkin, L. G. (1977). The effect of grazing by the

detritivore *Orchestia grillus* on *Spartina* litter and its associated microbial community. *Oecologia (Berlin)*, **30**, 111–27.
Lugo, A. F. (1980). Mangrove ecosystems: successional or steady state? *Biotropica Special Issue – Tropical Succession*, pp. 65–72.
Lüttge, U. (1975). Salt glands. In *Ion Transport in Plant Cells and Tissues*, ed. D. A. Baker & J. L. Hall, pp. 335–76. Amsterdam: North-Holland Publishing Company.
Lüttge, U. & Smith, J. A. C. (1984). Structural, biophysical and biochemical aspects of the role of leaves in plant adaptation to salinity and water stress. In *Salinity Tolerance in Plants – Strategies for Crop Improvement*, ed. R. C. Staples & G. H. Toenniessen, pp. 125–50. New York: John Wiley.
Lynch, J. & Läuchli, A. (1985). Salt stress disturbs the calcium nutrition of barley (*Hordeum vulgare* L.). *New Phytologist*, **99**, 345–54.
Lynch, J. J., O'Neil, T. & Lay, D. W. (1947). Management significance of damage by geese and muskrats to Gulf Coast marshes. *Journal of Wildlife Management*, **11**, 50–76.
Lytle, R. W. & Hull, R. J. (1980*a*). Photoassimilate distribution in *Spartina alterniflora* Loisel. I. Vegetative and floral development. *Agronomy Journal*, **72**, 933–8.
(1980*b*). Photoassimilate distribution in *Spartina alterniflora* Loisel. II. Autumn and winter storage and spring regrowth. *Agronomy Journal*, **72**, 938–42.
Maas, E. V. (1985). Crop tolerance to saline sprinkling water. *Plant and Soil*, **89**, 273–84.
MacArthur, R. H. & Wilson, E. O. (1967). *The Theory of Island Biogeography*. Princeton: University of Princeton Press.
McCree, K. J. (1986). Whole-plant carbon balance during osmotic adjustment to drought and salinity stress. *Australian Journal of Plant Physiology*, **13**, 33–43.
Macdonald, K. B. (1977*a*). Plant and animal communities of Pacific North American salt marshes. In *Wet Coastal Ecosystems*, ed. V. J. Chapman, pp. 167–91. Amsterdam: Elsevier.
(1977*b*). Coastal salt marsh. In *Terrestrial Vegetation of California*, ed. M. G. Barbour & J. Major, pp. 263–94. New York: Wiley.
Macdonald, K. B. & Barbour, M. C. (1974). Beach and salt marsh vegetation of the North American Pacific Coast. In *Ecology of Halophytes*, ed. R. J. Reimold & W. H. Queen, pp. 175–233. New York: Academic Press.
McGrail, S. (1981). The environment. In *The Brigg Raft and Her Prehistoric Environments*, ed. S. McGrail, pp. 271–4. Oxford: British Archaeological Reports.
McLusky, D. S. (1981). *The Estuarine Ecosystem*. Glasgow: Blackie.
McMannon, M. & Crawford, R. M. M. (1971). A metabolic theory of flooding tolerance: the significance of enzyme distribution and behaviour. *New Phytologist*, **70**, 299–306.
McNeilly, T., Ashraf, M. & Veltkamp, C. (1987). Leaf micromorphology of sea cliff and inland plants of *Agrostis stolonifera* L., *Dactylis glomerata* L. and *Holcus lanatus* L. *New Phytologist*, **106**, 261–9.
Mahall, B. E. & Park, R. B. (1976*a*). The ecotone between *Spartina foliosa* Trin. and *Salicornia virginica* L. in salt marshes of Northern San Francisco Bay. *Journal of Ecology*, **64**, 421–33.
(1976*b*). The ecotone between *Spartina foliosa* Trin. and *Salicornia virginica* L. in salt marshes of Northern San Francisco Bay II. Soil water and salinity. *Journal of Ecology*, **64**, 793–809.
(1976*c*). The ecotone between *Spartina foliosa* Trin. and *Salicornia virginica* L. in salt marshes of Northern San Francisco Bay III. Soil aeration and tidal immersion. *Journal of Ecology*, **64**, 811–19.
Mäkirinta, A. M. (1970). Über das Vorkommen und die Ökologie von *Juncus gerardii* Lois. an der Nordküste des Bottanischen Meerbusens. *Aquilo Seria Botanica*, **9**, 110–26.

Maltby, E. (1986). *Waterlogged Wealth. Why Waste the World's Wet Places?* London: International Institute for Environment and Development.

Manetas, Y., Petropoulou, Y. & Karabourniotis (1986). Compatible solutes and their effects on phosphoenolpyruvate carboxylase of C_4 – halophytes. *Plant, Cell and Environment*, **9**, 145–51.

Mann, K. M. (1979). Nitrogen limitations on the productivity of *Spartina* marshes, *Laminaria* kelp beds and higher trophic levels. In *Ecological Processes in Coastal Environments*, ed. R. L. Jefferies & A. J. Davy, pp. 363–70. Oxford: Blackwell.

Mann, K. H. (1982). *Ecology of Coastal Waters. A Systems Approach.* Oxford: Blackwell.

Marchant, C. J. (1967). Evolution in *Spartina* (Gramineae) I. The history and morphology of the genus in Britain. *Journal of the Linnean Society of London (Botany)*, **60**, 1–24.

(1968). Evolution in *Spartina* (Gramineae) II. Chromosomes, basic relationships and the problem of the *S.* × *townsendii* agg. *Journal of the Linnean Society (Botany)*, **60**, 381–409.

(1975). *Spartina* Schreb. In *Hybridization and the Flora of the British Isles*, ed. C. A. Stace, pp. 586–7. London: Academic Press.

Marchant, C. J. & Goodman, P. J. (1969*a*). *Spartina maritima* (Curtis) Fernald. Biological Flora of the British Isles. *Journal of Ecology*, **57**, 287–91.

(1969*b*). *Spartina alterniflora* Loisel. Biological Flora of the British Isles. *Journal of Ecology*, **57**, 291–7.

Marker, M. E. (1967). The Dee estuary: its progressive silting and saltmarsh development. *Transactions of the Institute of British Geographers*, **41**, 65–71.

Marks, T. C. & Truscott, A. J. (1985). Variation in seed production and germination of *Spartina anglica* within a zoned saltmarsh. *Journal of Ecology*, **73**, 695–705.

Marschner, H., Kuiper, P. J. C. & Kylin, A. (1981). Genotypic differences in the response of sugar beet plants to replacement of potassium by sodium. *Physiologia Plantarum*, **51**, 239–44.

Marschner, H., Kylin, A. & Kuiper, P. J. C. (1981). Differences in salt tolerance of three sugar beet genotypes. *Physiologia Plantarum*, **51**, 234–8.

Marsh, A. S. (1915). The maritime ecology of Holme-next-the-Sea, Norfolk. *Journal of Ecology*, **3**, 65–93.

Mason, E. (1928). Notes on the presence of mycorrhiza on the roots of salt marsh plants. *New Phytologist*, **27**, 193–5.

Mathieson, A. C., Penniman, C. A., Busse, P. K. & Tveter-Gallagher, E. (1982). The effects of ice on *Ascophyllum nodosum* within the Great Bay Estuary System of New Hampshire-Maine. *Journal of Phycology*, **18**, 331–6.

Matthews, J. R. (1955). *Origin and Distribution of the British Flora.* London: Hutchinson.

Meiri, A. & Plaut, Z. (1985). Crop production and management under saline conditions. *Plant and Soil*, **89**, 253–71.

Mendelssohn, I. A. (1979*a*). Nitrogen metabolism in the height forms of *Spartina alterniflora* in North Carolina. *Ecology*, **60**, 574–84.

(1979*b*). The influence of nitrogen level, form and application method on the growth response of *Spartina alterniflora* in North Carolina. *Estuaries*, **2**, 106–12.

Mendelssohn, I. A., McKee, K. L. & Patrick, W. H. (1981). Oxygen deficiency in *Spartina alterniflora* roots: metabolic adaptation to anoxia. *Science*, **214**, 439–41.

Mendelssohn, I. A. & Postek, M. T. (1982). Elemental analysis of deposits on the roots of *Spartina alterniflora* Loisel. *American Journal of Botany*, **69**, 904–12.

Metcalfe, W. S., Ellison, A. M. & Bertness, M. C. (1986). Survivorship and spatial development of *Spartina alterniflora* Loisel. (Gramineae) seedlings in a New England salt marsh. *Annals of Botany*, **58**, 249–58.

Mikkelsen, V. M. (1949). Ecological studies of the salt marsh vegetation in Isefjord. *Dansk Botanisk Arkiv*, **13(2)**, 1–48.

Millard, A. V. & Evans, P. R. (1984). Colonization of mudflats by *Spartina anglica*: some effects on invertebrate and shore bird populations at Lindisfarne. In Spartina anglica *in Great Britain*, ed. P. Doody, pp. 41–8. Attingham Park: Nature Conservancy Council.

Milledge, D. (ed.) (1980). *A Survey of the Estuarine Fauna of Eurobodalla Shire*. Sydney: The Australian Museum.

Miller, W. R. & Egler, F. E. (1950). Vegetation of the Wequetequock-Pawcatuck tidal marshes, Connecticut. *Ecological Monographs*, **20**, 143–72.

Milton, W. E. J. (1939). The occurrence of buried viable seeds in soils at different elevations and on a salt marsh. *Journal of Ecology*, **27**, 149–59.

Miyawaki, A. & Ohba, T. (1965). Studien über Strands-salzwiesengesellschaften auf Ost-Hokkaido (Japan). *Science Report Yokahama National University*. Sect. II, **12**, 1–25.

(1969). Studien über die Strandsalzwiesengesellschaften auf Honshu, Shikoku und Kyushu (Japan). *Science Report Yokohama National University*. Sect. II. **15**, 1–23.

Mobberley, D. G. (1956). Taxonomy and distribution of the genus *Spartina*. *Iowa State College Journal of Science*, **30**, 471–574.

Molinier, R. & Tallon, G. (1970). Prodrome des unités phytosociologiques observées en Camargue. *Bulletin du Muséum d'Histoire Naturelle de Marseille*, **30**, 5–110.

(1974). Documents pour un inventaire des plantes vasculaires de la Camargue. *Bulletin du Muséum d'Histoire Naturelle de Marseille*, **34**, 7–165.

Monk, L. S., Crawford, R. M. M. & Brändle, R. (1984). Fermentation rates and ethanol accumulation in relation to flooding tolerance in rhizomes of monocotyledonous species. *Journal of Experimental Botany*, **35**, 738–45.

Montague, C. L. (1982). The influence of fiddler crab burrows and burrowing on metabolic processes in salt marsh sediments. In *Estuarine Comparisons*, ed. V. S. Kennedy, pp. 283–301. New York: Academic Press.

Montague, C. L., Bunker, S. M., Haines, E. B. & Pace, M. L. (1981). Aquatic macroconsumers. In *The Ecology of a Salt Marsh*, ed. L.R. Pomeroy & R.G. Weigert, pp. 69–85. New York: Springer-Verlag.

Montfort, C. & Brandrup, W. (1927). Physiologische und Pflanzengeographische Seesalzwirkungen. I. Ökologische Studien über Keimung und erste Entwicklung bei Halophyten. *Jahrbücher für wissenschaftliche Botanik*, **66**, 902–46.

(1928). Physiologische und Pflanzengeographische Seesalzwirkungen. III. Vergleichende Untersuchungen der Salzwachstumsreaktion von Würzeln. *Jahrbücher für wissenschaftliche Botanik*, **67**, 105–73.

Moon, G. J., Clough, B. F., Peterson, C. A. & Allaway, G. W. (1986). Apopolastic and symplastic pathways in *Avicennia marina* (Forsk.) Vierh. roots revealed by fluorescent tracer dyes. *Australian Journal of Plant Physiology*, **13**, 637–48.

Mooring, M. T., Cooper, A. W. & Seneca, E. D. (1971). Seed germination response and evidence for height ecophenes in *Spartina alterniflora* from North Carolina. *American Journal of Botany*, **58**, 48–55.

Morley, J. V. (1973). Tidal immersion of *Spartina* marsh at Bridgwater Bay, Somerset. *Journal of Ecology*, **61**, 383–6.

Morris, A. W., Mantoura, R. F. C., Bale, A. J. & Howland, R. J. M. (1978). Very low salinity regions of estuaries: important sites for chemical and biological reactions. *Nature (London)*, **274**, 678–80.

Morris, J. T. (1980). The nitrogen uptake kinetics of *Spartina alterniflora* in culture. *Ecology*, **61**, 1114–21.

(1982). A model of growth responses by *Spartina alterniflora* to nitrogen limitation. *Journal of Ecology*, **70**, 25–42.

(1984). Effects of oxygen and salinity on ammonium uptake by *Spartina alterniflora* Loisel. and *Spartina patens* (Aiton) Muhl. *Journal of Experimental Marine Biology and Ecology*, **78**, 87–98.

Morris, J. T. & Dacey, J. W. H. (1984). Effects of O_2 on ammonium uptake and root respiration by *Spartina alterniflora. American Journal of Botany*, **71**, 979–85.

Morton, R. M., Pollock B. R. & Beumer, J. P. (1987). The occurrence and diet of fishes in a tidal inlet to a saltmarsh in southern Moreton Bay, Queensland. *Australian Journal of Ecology*, **12**, 217–37.

Mueller-Dombois, D. & Ellenberg, H. (1974). *Aims and Methods of Vegetation Ecology*. New York: John Wiley.

Muller, J. (1970). Palynological evidence on early differentiation of angiosperms. *Biological Reviews*, **45**, 417–50.

Munns, R., Greenway, H. & Kirst, G. O. (1983). Halotolerant eukaryotes. In *Physiological Plant Ecology III. Responses to the Chemical and Biological Environment*, ed. O. L. Lange, P. S. Nobel, C. B. Osmond & H. Ziegler, pp. 59–135. Berlin: Springer-Verlag.

Munns, R. & Termaat, A. (1986). Whole-plant responses to salinity. *Australian Journal of Plant Physiology*, **13**, 143–60.

Neales, T. F. & Sharkey, P. J. (1981). Effect of salinity on growth and on mineral and organic constituents of the halophyte *Disphyma australe* (Soland.), J. M. Black. *Australian Journal of Plant Physiology*, **8**, 165–79.

Nelson, E. C. (1978). Tropical drift fruits and seeds on coasts in the British Isles and western Europe, I. Irish beaches. *Watsonia*, **12**, 103–12.

Neuenschwander, L. F., Thorsted, T. H. Jr. & Vogl, R. J. (1979). The salt marsh and transitional vegetation of Bahia de San Quintin. *Bulletin of the Southern California Academy of Sciences*, **78**, 163–82.

Nicol, E. A. T. (1935). The ecology of a salt-marsh. *Journal of the Marine Biology Association of the United Kingdom*, **20**, 203–61.

Nichols, F. H., Cloern, J. E., Luoma, S. N. & Peterson, D. H. (1986). The modification of an estuary. *Science*, **231**, 567–73.

Nichols, G. E. (1918). The vegetation of northern Cape Breton Island, Nova Scotia. *Transactions of the Connecticut Academy*, **22**, 249–467.

Nichols, M. M. & Biggs, R. B. (1985). Estuaries. In *Coastal Sedimentary Environments*, 2nd edn, ed. R. A. Davis, pp. 77–186. New York: Springer-Verlag.

Nielsen, M. G. (1981). The ant fauna of the high salt marsh. In *Terrestrial and Freshwater Fauna of the Wadden Sea Area*, ed. C. J. Smit, J. Den Hollander, W.K.R.E. van Wingerden & W. J. Wolff, pp. 68–70. Rotterdam: Amsterdam.

Nienhuis, P. H. (1970). The benthic algal communities of flats and salt marshes in the Grevelingen, a sea-arm in the south-western Netherlands. *Netherlands Journal of Sea Research*, **5**, 20–49.

Niering, W. A. & Warren, R. S. (1980). Vegetation patterns and processes in New England salt marshes. *BioScience*, **30**, 301–7.

Ní Lamhna, E. (1982). The vegetation of saltmarshes and sand-dunes of Malahide Island, County Dublin. *Journal of Life Sciences of the Royal Dublin Society*, **3**, 111–29.

Nixon, S. W. (1980). Between coastal marshes and coastal waters – a review of twenty years of speculation and research on the role of salt marshes in estuarine productivity and water chemistry. In *Estuarine and Wetlands Processes with Emphasis on Modeling*, ed. P. Hamilton & K. B. Macdonald, pp. 437–525. New York: Plenum Press.

Nobel, P. S. (1974). *Introduction to Biophysical Plant Physiology*. San Francisco: W. H. Freeman.

Noble, C. L., Halloran, G. M. & West, D. W. (1984). Identification and selection for salt tolerance in lucerne (*Medicago sativa* L.). *Australian Journal of Agricultural Research*, **35**, 239–52.

Nordhagen, R. (1940). Studien über die maritime Vegetation Norwegens. I. Die Pflanzengesellschaften der Tangwälle. *Bergens Museums Årbok 1939–40. Naturvitenskapelig rekke* **2**, 1–123.

(1954). Studies on the vegetation of salt and brackish marshes in Finmark (Norway). *Vegetatio*, **5/6**, 381–94.

Norton, T. A. & Mathieson, A. C. (1983). The biology of unattached seaweeds. *Progress in Phycological Research*, **2**, 333–86.

Numata, M. (1984). Analysis of seeds in the soil. In *Sampling Methods and Taxon Analysis in Vegetation Science*, ed. R. Knapp, pp. 161–9. The Hague: Dr Junk.

Odum, E. P. & de la Cruz, A. A. (1967). Particulate organic detritus in a Georgia salt marsh-estuarine system. In *Estuaries*, ed. G. H. Lauff, pp. 383–5. Washington: AAAS Publication 83.

Odum, W. E. (1978). The importance of tidal freshwater wetlands in coastal zone management. In *Coastal Zone '78. Symposium on Technical, Environmental, Socioeconomic and Regulatory Aspects of Coastal Zone Management*, vol. II, pp. 1196–203. New York: American Society of Civil Engineers.

(1982). Environmental degradation and the tyranny of small decisions. *BioScience*, **32**, 728–9.

(1988). Comparative ecology of tidal freshwater and salt marshes. *Annual Reviews of Ecology and Systematics*, **19**, 147–76.

Odum, W. E., Fisher, J. S. & Pickral, J. C. (1979). Factors controlling the flux of particulate organic carbon from estuarine wetlands. In *Ecological Processes in Coastal and Marine Systems*, ed. R. C. Livingston, pp. 69–80. New York: Plenum Press.

Odum, W. E., McIvor, C. C. & Smith, T. J. (1982). *The Ecology of the Mangroves of South Florida: A Community Profile*. Washington: U.S. Fish and Wildlife Service. FWS/OBS-81/24.

Oertli, J. J. (1968). Extracellular salt accumulation, a possible mechanism of salt injury in plants. *Agrochimica*, **12**, 461–9.

Oertli, J. J. & Richardson, W.F. (1968). Effect of external salt concentration on water relations in plants. IV. The compensation of osmotic and hydrostatic water potential differences between root xylem and external medium. *Soil Science*, **105**, 177–83.

Officer, C. B. (1976). *Physical Oceanography of Estuaries (and Associated Coastal Waters)*. New York: John Wiley.

Ogden, J. G. (1981). Geology and hydrology of salt marshes in Nova Scotia. In *Salt marshes in Nova Scotia. A Status Report of the Salt Marsh Working Group*, ed. A. Hatcher & D. G. Patriquin, pp. 28–43. Halifax: Dalhousie University.

Ogilvie, M. A. (1978). *Wild Geese*. Berkhamsted: T. & A.D. Poyser.

Okusanya, O. T. & Fawole, T. (1985). The possible role of phosphate in the salinity tolerance of *Lavatera arborea*. *Journal of Ecology*, **73**, 317–22.

Okusanya, O. T. & Ungar, I. A. (1983). The effects of time of seed production on the germination response of *Spergularia marina*. *Physiologia Plantarum*, **59**, 335–42.

Okusanya, O. T. & Ungar, I. A. (1984). The growth and mineral composition of three species of *Spergularia* as affected by salinity and nutrients at high salinity. *American Journal of Botany*, **71**, 439–47.

O'Leary, J. W. (1969). The effect of salinity on permeability of roots to water. *Israel Journal of Botany*, **18**, 1–9.

(1984). The role of halophytes in irrigated agriculture. In *Salinity Tolerance in Plants – Strategies for Crop Improvement*, ed. R. C. Staples & G. H. Toenniessen, pp. 285–300. New York: John Wiley.

O'Leary, J. W., Glenn, E. P. & Watson, M. C. (1985). Agricultural production in halophytes irrigated with seawater. *Plant and Soil*, **89**, 311–21.
Oliveira, F. E. C. de & Fletcher, A. (1880). Taxonomic and ecological relationships between rocky-shore and saltmarsh populations of *Pelvetia canaliculata* (Phaeophyta) at Four Mile Bridge, Anglesey, U.K. *Botanica Marina*, **23**, 409–17.
Oliver, F. W. (1906). The Bouche d'Erquy in 1906. *New Phytologist*, **5**, 189–95.
(1907). The Bouche d'Erquy in 1908. *New Phytologist*, **6**, 244–52.
(1925). *Spartina townsendii*, its mode of establishment, economic uses and taxonomic status. *Journal of Ecology*, **13**, 74–91.
Oliver, J. (1982). The geographic and environmental aspects of mangrove communities: climate. In *Mangrove Ecosystems in Australia – Structure, Function and Management*, ed. B. F. Clough, pp. 19–30. Canberra: AIMS/ANU Press.
Olney, P. J. S. (1963). The food and feeding habits of teal *Anas crecca crecca* L. *Proceedings of the Zoological Society of London*, **140**, 169–210.
Oross, J. W., Leonard, R. T. & Thomson, W. W. (1985). Flux rate and a secretion model for salt glands of grasses. *Israel Journal of Botany*, **34**, 69–77.
Osenga, G. A. & Coull, B. C. (1983). *Spartina alterniflora* Loisel root structure and meiofaunal abundance. *Journal of Experimental Marine Biology and Ecology*, **67**, 221–5.
Osmond, C. B. (1963). Oxalates and ionic equilibrium in Australian saltbushes (*Atriplex*). *Nature (London)*, **198**, 503–4.
(1980). Integration of photosynthetic carbon metabolism during stress. In *Genetic Engineering of Osmoregulation*, ed. D. W. Rains, R. C. Valentine & A. Hollaender, pp. 171–85. New York: Plenum Press.
Osmond, C. B., Björkman, O. & Anderson, D. J. (1980). *Physiological Processes in Plant Ecology. Toward a Synthesis with* Atriplex. New York: Springer-Verlag.
Ovenshine, A. T. & Bartsch-Winkler, S. (1978). Portage, Alaska: case history of an earthquake's impact on an estuarine system. In *Estuarine Interactions*, ed. M. L. Wiley, pp. 275–84. New York: Academic Press.
Owen, D. F. & Wiegert, R. F. (1976). Do consumers maximise plant fitness? *Oikos*, **27**, 448–92.
Ownbey, R. S. & Mahall, B. E. (1983). Salinity and root conductivity: differential responses of a coastal succulent halophyte, *Salicornia virginica*, and a weedy glycophyte, *Raphanus sativus*. *Physiologia Plantarum*, **57**, 189–95.
Packham, J. R. & Liddle, M. J. (1970). The Cefni saltmarsh and its recent development. *Field Studies*, **3**, 331–56.
Paleg, L. G., Douglas, T. J., van Daad, A. & Keech, D. B. (1981). Proline, betaine and other organic solutes protect enzymes against heat activation. *Australian Journal of Plant Physiology*, **8**, 107–14.
Paleg, L. G., Stewart, G. R. & Starr, R. (1985). The effect of compatible solutes on proteins. *Plant and Soil*, **89**, 83–94.
Papp, J. C., Ball, M. C. & Terry, N. (1983). A comparative study of the effects of NaCl salinity on respiration, photosynthesis and leaf extension growth in *Beta vulgaris* L. (sugar beet). *Plant, Cell and Environment*, **6**, 675–7.
Parker, V. T. & Leck, M. A. (1985). Relationships of seed banks to plant distribution patterns in a freshwater tidal wetland. *American Journal of Botany*, **72**, 161–74.
Parkinson, M. (1980). Salt marshes of the Exe estuary. *Report and Transactions of the Devonshire Association for the Advancement of Science*, **112**, 17–41.
(1985). The Axe estuary and its marshes. *Report and Transactions of the Devonshire Association for the Advancement of Science*, **117**, 19–62.
Parrondo, R. T., Gosselink, J. G. & Hopkinson, C. S. (1978). Effects of salinity and drainage on the growth of three salt marsh grasses. *Botanical Gazette*, **139**, 102–7.

Parsons, K. A. & de la Cruz, A. (1980). Energy flow and grazing behaviour of concephaline grasshoppers in a *Juncus roemerianus* marsh. *Ecology*, **61**, 1045–50.
Passioura, J. B. (1986). Resistance to drought and salinity: avenues for improvement. *Australian Journal of Plant Physiology*, **13**, 191–201.
Pasternak, D. & San Pietro, A. (ed.) (1985). *Biosalinity in Action: Bioproduction with Saline Water*. (Also published as *Plant and Soil*, **89**). Dordrecht: Martinus Nijhoff.
Patriquin, D. G. (1978). Factors affecting nitrogenase activity (acetylene reducing activity) associated with excised roots of the emergent halophyte *Spartina alterniflora* Loisel. *Aquatic Botany*, **4**, 193–210.
Patriquin, D. G. & Keddy, C. (1978). Nitrogenase activity (acetylene reduction) in a Nova Scotia salt water marsh: its association with angiosperms and the influence of some edaphic factors. *Aquatic Botany*, **4**, 227–44.
Patriquin, D. G. & McClung, C. R. (1978). Nitrogen accretion and the nature and possible significance of N_2 fixation (acetylene reduction) in a Nova Scotian *Spartina alterniflora* stand. *Marine Biology*, **47**, 227–42.
Paviour-Smith, K. (1956). The biotic community of a salt meadow in New Zealand. *Transactions of the Royal Society of New Zealand*, **83**, 525–54.
Pearcy, R. W. & Ustin, S. L. (1984). Effects of salinity on growth and photosynthesis of three California tidal marsh species. *Oecologia (Berlin)*, **62**, 68–73.
Pegg, K. G. & Foresberg, L. I. (1981). *Phytophthora* in Queensland Mangroves. *Wetlands (Australia)*, **1**, 2–3.
Pemadasa, M. A., Balasubramanian, S., Wijewansa, H. G. & Amarasinghe, L. (1979). The ecology of a saltmarsh in Sri Lanka. *Journal of Ecology*, **67**, 41–63.
Pen, L. (1983). *Peripheral Vegetation of the Swan and Canning Estuaries 1981*. Perth: Department of Conservation and Environment Western Australia. Bulletin 113.
(1987). Peripheral vegetation of the Swan-Canning estuary. Past, present and future. In *Swan River Estuary, Ecology and Management*, ed. J. John, pp. 221–31. Bentley: Curtin University Environmental Studies Group.
Penfound, W. T. (1952). Southern swamps and marshes. *Botanical Review*, **18**, 413–46.
Penfound, W. T. & Hathaway, E. S. (1938). Plant communities in the marshlands of south eastern Louisiana. *Ecological Monographs*, **8**, 3–56.
Perring, F. H. & Farrell, L. (1983). *British Red Data Books: I. Vascular Plants*, 2nd edn. Nettlefold: SPNC.
Perring, F. H. & Walters, S. M. (1962). *Atlas of the British Flora*. London: Nelson.
Peterson, B. J., Howarth, R. W. & Garritt, R. H. (1985). Multiple stable isotopes used to trace the flow of organic matter in estuarine food webs. *Science*, **227**, 1361–3.
(1986). Sulfur and carbon isotopes as tracers of salt-marsh organic matter flow. *Ecology*, **67**, 865–74.
Pethick, J. S. (1974). The distribution of salt pans on tidal salt marshes. *Journal of Biogeography*, **1**, 57–62.
(1984). *An Introduction to Coastal Geomorphology*. London: Edward Arnold.
Pezeshki, S. R., De Laune, R. D. & Patrick, W. H. (1987). Response of *Spartina patens* to increasing levels of salinity in rapidly subsidising marshes of the Mississippi River deltaic plain. *Estuarine, Coastal and Shelf Science*, **24**, 389–99.
Pfeiffer, W. J. & Wiegert, R. G. (1981). Grazers on *Spartina* and their predators. In *The Ecology of a Salt Marsh*, ed. L. R. Pomeroy & R. G. Weigert, pp. 88–112. New York: Springer-Verlag.
Pielou, E. C. & Routledge, R. D. (1976). Salt marsh vegetation: latitudinal gradients in the zonation patterns. *Oecologia (Berlin)*, **24**, 311–21.
Pierce, S. M. (1982). What is *Spartina* doing in our estuaries? *South African Journal of Science*, **78**, 229–30.

Pigott, C. D. (1969). Influence of mineral nutrition on the zonation of flowering plants in coastal saltmarshes. In *Ecological Aspects of the Mineral Nutrition of Plants*, ed. I. H. Rorison, pp. 25–35. Oxford: Blackwell.

Pigott, C. D. & Walters, S. M. (1954). On the interpretation of the discontinuous distributions shown by certain British species of open habitats. *Journal of Ecology*, **42**, 95–116.

Pinto da Silva, A. R. (1972). *Armeria*. in *Flora Europaea*, vol. **3**, ed. T. G. Tutin *et al.*, pp. 30–8. Cambridge: Cambridge University Press.

Pitelka, L. F. & Ashmun, J. W. (1985). Physiology and integration of ramets in clonal plants. In *Population Biology and Evolution of Clonal Organisms*, ed. J. B. C. Jackson, L. W. Buss & R. E. Cook, pp. 399–435. New Haven: Yale University Press.

Pitman, M. G. (1965). Transpiration and selective uptake of potassium by barley seedlings (*Hordeum vulgare*). *Australian Journal of Biological Science*, **18**, 987–98.

(1977). Ion transport into the xylem. *Annual Review of Plant Physiology*, **28**, 71–88.

(1984). Transport across the root and shoot/root interactions. In *Salinity Tolerance in Plants – Strategies for Crop Improvement*, ed. R. C. Staples & G. H. Toenniessen, pp. 93–124. New York: Wiley.

Pojar, J. (1973). Pollination of typically anemophilous salt marsh plants by bumble bees, *Bombus terricola occidentalis* Grne. *American Midland Naturalist*, **89**, 448–51.

Polderman, P. J. G. (1975). The algal communities of the northeastern part of the saltmarsh "De Mok" on Texel (the Netherlands). *Acta Botanica Neerlandica*, **24**, 361–78.

(1976). *Wittrockiella paradoxa* Wille (Cladophoraceae) in N.W. European saltmarshes. *Hydrobiological Bulletin (Amsterdam)*, **10**, 98–103.

(1978). Algae of saltmarshes on the south and southwest coasts of England. *British Phycological Journal*, **13**, 235–40.

(1979*a*). The saltmarsh algal communities in the Wadden area, with reference to their distribution and ecology in N.W. Europe. I. The distribution and ecology of the algal communities. *Journal of Biogeography*, **6**, 225–66.

(1979*b*). The saltmarsh algae of the Wadden area. In *Flora and Vegetation of the Wadden Sea*, ed. W. J. Wolff, pp. 124–60. Rotterdam: Balkerma.

(1980*a*). The saltmarsh algal communities in the Wadden area, with reference to their distribution and ecology in N.W. Europe. II. The zonation of algal communities in the Wadden area. *Journal of Biogeography*, **7**, 85–95.

(1980*b*). The saltmarsh algal communities in the Wadden area with reference to their distribution and ecology in N.W. Europe. III. The classificatory and semantic problems of saltmarsh algal communities. *Journal of Biogeography*, **7**, 115–26.

Polderman, P. J. G. & Polderman-Hall, R. A. (1980). Algal communities in Scottish saltmarshes. *British Phycological Journal*, **15**, 59–71.

Pollard, A. & Wyn Jones, G. (1979). Enzyme activities in concentrated solutions of glycinebetaine and other solutes. *Planta*, **144**, 291–8.

Polunin, N. (1948). Botany of the Canadian Eastern Arctic, III: Vegetation and ecology. *Bulletin of the National Museum of Canada*, **104**, 1–304.

Polunin, O. & Walters, M. (1985). *A Guide to the Vegetation of Britain and Europe*. Oxford: Oxford University Press.

Pomeroy, L. R., Darley, W. M., Dunn, E. L., Gallagher, J. L., Haines, E. B. & Whitney, D. M. (1981). Primary production. In *The Ecology of Salt Marsh*, ed. L. R. Pomeroy & R. G. Wiegert, pp. 39–67. New York: Springer-Verlag.

Pomeroy, L. R. & Imberger, J. (1981). The physical and chemical environment. In *The Ecology of a Salt Marsh*, ed. L. R. Pomeroy & R. G. Weigert, pp. 21–36. New York: Springer-Verlag.

Pomeroy, L. R. & Wiegert, R. G. (eds.) (1981). *The Ecology of a Salt Marsh.* New York: Springer-Verlag.

Ponnamperuma, F. N. (1972). The chemistry of submerged soils. *Advances in Agronomy*, **24**, 29–96.

Popp, M. (1984*a*). Chemical composition of Australian mangroves. I. Inorganic ions and organic acids. *Zeitschrift für Pflanzenphysiologie*, **113**, 395–409.

(1984*b*). Chemical composition of Australian mangroves. II. Low molecular weight carbohydrates. *Zeitschrift für Pflanzenphysiologie*, **113**, 411–21.

Popp, M., Larher, F. & Weigel, P. (1984). Chemical composition of Australian mangroves. III. Free amino acids, total methylated onium compounds and total nitrogen. *Zeitschrift für Pflanzenphysiologie*, **114**, 15–25.

(1985). Osmotic adaptation in Australian mangroves. *Vegetatio*, **61**, 247–53.

Praeger, R. L. (1913). On the buoyancy of the seeds of some Britannic plants. *Scientific Proceedings of the Royal Dublin Society*, **14(3)**, 13–62.

Pressey, R. L. (1985). Some problems with wetland evaluation. *Wetlands (Australia)*, **5**, 42–51.

Preston, C. & Critchley, C. (1986). Differential effects of K^+ and Na^+ on oxygen evolution activity of photosynthetic membranes from two halophytes and spinach. *Australian Journal of Plant Physiology*, **13**, 491–8.

Prince, S. D. & Hare, A. D. R. (1981). *Lactuca saligna* and *Pulicaria vulgaris* in Britain. In *The Biological Aspects of Rare Plant Conservation*, ed. H. Synge, pp. 379–88. Chichester: John Wiley.

Proctor, J., Fraser, M. W. & Thompson, J. (1983). Saltmarshes of the Upper Forth estuary. *Transactions of the Botanical Society at Edinburgh*, **44**, 95–102.

Proctor, V. W. (1968). Long-distance dispersal of seeds by retention in the digestive tract of birds. *Science*, **160**, 321–2.

Pugh, G. J. F. (1960). The fungal flora of tidal mud flats. In *The Ecology of Soil Fungi*, ed. D. Parkinson & J. S. Waid, pp. 202–8. Liverpool: Liverpool University Press.

(1962). Studies on fungi in coastal soils II. *Transactions of the British Mycological Society*, **45**, 560–6.

(1974). Fungi in intertidal regions. *Veröffenthlichungen des Instituts für Meeresforschung in Bremerhaven Supplementbände*, **5**, 403–18.

(1979). The distribution of fungi in coastal regions. In *Ecological Processes in Coastal Environments*, ed. R. L. Jefferies & A. J. Davy, pp. 415–27. Oxford: Blackwell Scientific Publications.

Pugh, G. J. F. & Beeftink, W. G. (1980). Fungi in coastal and inland salt marshes. *Botanica Marina*, **13**, 651–6.

Pugh, G. J. F. & Lindsey, B. I. (1975). Studies of *Sporobolomyces* in a maritime habitat. *Transactions of the British Mycological Society*, **65**, 201–9.

Purer, E. A. (1942). Plant ecology of the coastal salt marshland of San Diego county, California. *Ecological Monographs*, **12**, 82–111.

Queen, W. H. (1975). Halophytes: adaptive mechanisms. In *Physiological Ecology of Estuary Organisms*, ed. V. J. Vernberg, pp. 205–11. Columbia: University of South Carolina Press.

Rabinowitz, D. (1978). Early growth of mangrove seedlings in Panama, and an hypothesis concerning the relationship of dispersal and zonation. *Journal of Biogeography*, **5**, 113–33.

Race, M. S. (1985). Critique of present wetlands mitigation policies in the United States based on an analysis of past restoration projects in San Francisco Bay. *Environmental Management*, **9**, 71–82.

Ragsdale, H. L. & Thorhaug, A. (1980). Trace metal cycling in the U.S. coastal zone: a synthesis. *American Journal of Botany*, **67**, 1102–12.

Rains, D. W. (1979). Salt tolerance of plants: strategies of biological systems. In *The*

Biosaline Concept, ed. A. Hollaender, J. C. Aller, E. Epstein, A. San Pietro & O. R. Zaborsky, pp. 47–67. New York: Plenum Press.

Rammert, H. (1981). The wrackbeds and their fauna. In *Terrestrial and Freshwater Fauna of the Wadden Sea Area*, ed. C. J. Smit, J. den Hollander, W.K.R.E. van Wingerden & W. J. Wolff, pp. 70–84. Rotterdam: Balkema.

Randerson, P. F. (1979). A simulation model of salt-marsh development and plant ecology. In *Estuarine and Coastal Land Reclamation and Water Storage*, ed. B. Knights & A. J. Phillips, pp. 48–67. Farnborough: Saxon House.

Ranwell, D. S. (1961). *Spartina* salt marshes in southern England. I. The effects of sheep grazing at the upper limits of *Spartina* marsh in Bridgwater Bay. *Journal of Ecology*, **49**, 325–40.

(1964*a*). *Spartina* salt marshes in southern England. II. Rate and seasonal pattern of sediment accretion. *Journal of Ecology*, **52**, 79–94.

(1964*b*). *Spartina* salt marshes in southern England. III. Rates of establishment, succession and nutrient supply at Bridgwater Bay, Somerset. *Journal of Ecology*, **52**, 95–105.

(1967). World resources of *Spartina townsendii* (*sensu lato*) and economic use of *Spartina* marshland. *Journal of Applied Ecology*, **4**, 239–56.

(1968). Coastal marshes in perspective. *Regional Studies Group Bulletin, University of Strathclyde*, **9**, 1–26.

(1972). *Ecology of Salt Marshes and Sand Dunes*. London: Chapman and Hall.

(1974). The salt marsh to tidal woodland transition. *Hydrobiological Bulletin* (*Amsterdam*), **8**, 139–51.

(1981*a*). Introduced coastal plants and rare species in Britain. In *The Biological Aspects of Rare Plant Conservation*, ed. H. Synge, pp. 413–19. Chichester: John Wiley.

(1981*b*). Saltmarsh – uses and restoration. In *Solent Saltmarsh Symposium*, ed. F. Stranack & J. Coughlan, pp. 14–21. Winchester: Solent Protection Society.

(1981*c*). *Spartina*. In *The Brigg Raft and Her Prehistoric Environment*, ed. S. McGrail, p. 275. Oxford: British Archaeological Reports.

Ranwell, D. S., Bird, E. C. F., Hubbard, J. C. E. & Stebbings, R. E. (1964). *Spartina* salt marshes in southern England. V. Tidal submergence and chlorinity in Poole Harbour. *Journal of Ecology*, **52**, 627–41.

Ratcliffe, D. A. (1977). *A Nature Conservation Review*. 2 vols. Cambridge: Cambridge University Press.

Raunkiaer, C. (1934). *The Life-Forms of Plants and Statistical Plant Geography*. Oxford: Clarendon Press.

Raven, J. A. (1977). H^+ and Ca^{2+} in phloem and symplast: relation of relative immobility of the ions to the cytoplasmic nature of the transport paths. *New Phytologist*, **79**, 465–80.

(1985). Regulation of pH and generation of osmolarity in vascular plants: a cost-benefit analysis in relation to efficiency of use of energy, nitrogen and water. *New Phytologist*, **101**, 25–77.

Reddell, P., Foster, R. C. & Bowen, G. D. (1986). The effects of sodium chloride on growth and nitrogen fixation in *Casuarina obesa* Miq. *New Phytologist*, **102**, 397–408.

Redfield, A. C. (1972). Development of a New England salt marsh. *Ecological Monographs*, **42**, 201–37.

Reed, A. & Moisan, G. (1971). The *Spartina* tidal marshes of the St. Lawrence estuary and their importance to aquatic birds. *Naturaliste Canadien*, **98**, 905–22.

Reed, D. J., Stoddart, D. R. & Bayliss-Smith, T. P. (1985). Tidal flows and sediment budgets for a salt-marsh system. Essex, England. *Vegetatio*, **62**, 375–80.

Reed, R. H. (1984). Use and abuse of osmo-terminology. *Plant, Cell and Environment*, **7**, 165–70.

Rees, T. K. (1935). The marine algae of Lough Ine. *Journal of Ecology*, **23**, 69–133.

Reidenbaugh, T. G. (1983). Tillering and mortality of the salt marsh cordgrass, *Spartina alterniflora*. *American Journal of Botany*, **70**, 47–52.
Reimold, R. J. (1977). Mangals and salt marshes of the eastern United States. In *Wet Coastal Ecosystems*, ed. V. J. Chapman, pp. 157–66. Amsterdam: Elsevier.
Reimhold, R. J., Linthurst, R. A. & Wolf, P. L. (1975). Effects of grazing on a salt marsh. *Biological Conservation*, **8**, 105–25.
Rennenberg, H. (1984). The fate of excess in sulphur higher plants. *Annual Review of Plant Physiology*, **35**, 121–53.
Retallack, G. (1975). The life and times of a Triassic lycopod. *Alcheringa*, **1**, 3–29.
Rey, J. R. (1984). Experimental tests of island biogeographic theory. In *Ecological Communities. Conceptual Issues and the Evidence*, ed. D. R. Strong, D. Simberloff, L. G. Abele & A. B. Thistle, pp. 101–12. Princeton: Princeton University Press.
Rhebergen, L. J. & Nelissen, H. J. M. (1985). Ecotypic differentiation within *Festuca rubra* L. occurring in a heterogeneous coastal environment. *Vegetatio*, **61**, 197–202.
Rice, B. & Westoby, M. (1983). Plant species richness at the 0.1 hectare scale in Australian vegetation compared to other continents. *Vegetatio*, **52**, 129–40.
Richard, G. A. (1978). Seasonal and environmental variations in sediment accretion in a Long Island salt marsh. *Estuaries*, **1**, 29–35.
Richards, F. J. (1934). The salt marshes of the Dovey estuary. IV. The rates of vertical accretion, horizontal extension and scarp erosion. *Annals of Botany*, **48**, 225–59.
Richards, R. A. (1983). Should selection for yield in saline regions by made on saline or non-saline soils? *Euphytica*, **32**, 431–8.
Riley, J. L. & McKay, S. M. (1980). The vegetation and phytogeography of coastal southwestern James Bay. *Life Science Contributions, Royal Ontario Museum*, **124**, 1–81.
Ringius, G. S. (1980). Vegetation survey of a James Bay coastal marsh. *Canadian Field Naturalist*, **94**, 110–20.
Roberts, R. H. (1975). *Frankenia laevis* L. in Anglesey. *Watsonia*, **10**, 291–2.
Roberts, B. A. & Robertson, A. (1986). Salt marshes of Atlantic Canada: their ecology and distribution. *Canadian Journal of Botany*, **64**, 455–67.
Robertson, A. I. (1986). Leaf-burying crabs: their influence on energy flow and export from mixed mangrove forests (*Rhizophora* spp.) in northeastern Australia. *Journal of Experimental Marine Biology and Ecology* **102**, 237–48.
Robertson, K. P. & Wainwright, S. J. (1987). Photosynthetic responses to salinity in two clones of *Agrostis stolonifera*. *Plant, Cell and Environment*, **10**, 45–52.
Robinson, K. I. M., Gibbs, P. J., van der Velde, J. & Barclay, J. B. (1983). Temporal changes in the estuarine benthic fauna of Towra Point, Botany. *Wetlands (Australia)*, **3**, 22–33.
Robinson, S. P., Downton, W. J. S. & Millhouse, J. A. (1983). Photosynthesis and ion content of leaves and isolated chloroplasts of salt-stressed spinach. *Plant Physiology*, **73**, 238–42.
Robinson, S. P. & Jones, G. P. (1986). Accumulation of glycinebetaine in chloroplasts provides osmotic adjustment during salt stress. *Australian Journal of Plant Physiology*, **13**, 659–68.
Roozen, A. J. M. & Westhoff, V. (1985). A study on long-term salt-marsh succession using permanent plots. *Vegetatio*, **61**, 23–32.
Rounick, J. S. & Winterbourn, M. J. (1986). Stable carbon isotopes and carbon flow in ecosystems. *BioScience*, **36**, 171–7.
Rowan, W. (1913). Note on the food plants of rabbits on Blakeney Point, Norfolk. *Journal of Ecology*, **1**, 273–4.
Roy, P. S. (1984). New South Wales estuaries: their origin and evolution. In *Coastal Geomorphology in Australia*, ed. B. G. Thom, pp. 99–121. Sydney: Academic Press.

Rozema, J. (1975). The influence of salinity, inundation and temperature on the germination of some halophytes and non-halophytes. *Oecologia Plantarum*, **10**, 317–29.

Rozema, J., Arp, W., van Diggelen, J., van Esbroek, M., Broekman, R. & Punte, H. (1986). Occurrence and ecological significance of vesicular arbuscular mycorrhiza in the salt marsh environment. *Acta Botanica Neerlandica*, **35**, 457–67.

Rozema, J., Bijwaard, P., Prast, G. & Broekman, R. (1985). Ecophysiological adaptions of coastal halophytes from foredunes and salt-marshes. *Vegetatio*, **62**, 499–521.

Rozema, J. & Blum, B. (1977). Effects of salinity and inundation on the growth of *Agrostis stolonifera* and *Juncus gerardii*. *Journal of Ecology*, **65**, 213–22.

Rozema, J., Broekman, R., Arp, W., Letschert, J., van Esbroek, M. & Punte, H. (1986). A comparison of the mineral relations of a halophytic hemiparasite and holoparasite. *Acta Botanica Neerlandica*, **35**, 105–9.

Rozema, J., Buizer, D. A. G. & Fabritius, H. F. (1978). Population dynamics of *Glaux maritima* and ecophysiological adaptations to salinity and inundation. *Oikos*, **30**, 539–48.

Rozema, J., Gude, H., Bijl, F. & Wesselman, H. (1981). Sodium concentration in xylem sap in relation to ion exclusion, accumulation and secretion in halophytes. *Acta Botanica Neerlandica*, **30**, 309–11.

Rozema, J., Gude, H. & Pollack, G. (1981). An ecophysiological study of the salt secretion of four halophytes. *New Phytologist*, **81**, 201–17.

Rozema, J., Luppes, E. & Broekman, R. (1985). Differential responses of saltmarsh species to variation of iron and manganese. *Vegetatio*, **62**, 293–301.

Rozema, J., Riphagen, I. & Sminia, T. (1977). A light and electron-microscopical study on the structure and function of the salt gland of *Glaux maritima* L. *New Phytologist*, **79**, 665–71.

Rozema, J., Rozema-Dijst, E., Freijse, A. H. J. & Huber, J. J. C. (1978). Population differentiation within *Festuca rubra* L. with regard to soil salinity and soil water. *Oecologia (Berlin)*, **34**, 329–41.

Rozema, J., van Manen, Y., Vugts, H. F. & Leusink, A. (1983). Airborne and soilborne salinity and the distribution of coastal and inland species of the genus *Elytrigia*. *Acta Botanica Neerlandica*, **32**, 447–56.

Russell, R. S. & Wellington, P. S. (1940). Physiological and ecological studies on an Arctic vegetation. I. The vegetation of Jan Mayen Island. *Journal of Ecology*, **28**, 153–79.

Rymer, L. (1974). The Scottish kelp industry. *Scottish Geographical Magazine*, **90**, 142–52.

Sacher, R. F. & Staples, R. C. (1984). Chemical microscopy for study of plants in saline environments. In *Salinity Tolerance in Plants – Strategies for Crop Improvement*, ed. R. C. Staples & G. H. Toenniessen, pp. 17–35. New York: John Wiley.

Saeijis, H. L. F. & de Jong, A. (1981). The Oosterschelde and the protection of the environment. A policy-plan for a changing estuary. *Land & Water International*, **46**, 15–34.

Saenger, P. (1982). Morphological, anatomical and reproductive adaptations of Australian mangroves. In *Mangrove Ecosystems in Australia – Structure, Function and Management*, ed. B. F. Clough, pp. 153–91. Canberra: AIMS/ANU Press.

Saenger, P., Specht, M. M., Specht, R. L. & Chapman, V. J. (1977). Mangal and coastal salt-marsh communities in Australasia. In *Wet Coastal Ecosystems*, ed. V. J. Chapman, pp. 293–345. Amsterdam: Elsevier.

Salzman, A. G. & Parker, M. A. (1985). Neighbors ameliorate local salinity stress for a rhizomatous plant in a heterogeneous environment. *Oecologia (Berlin)*, **65**, 273–7.

Sandhu, G. R., Aslam, Z., Salim, M., Sattar, A., Quereshi, R. H., Ahmad, N. & Wyn

Jones, R. G. (1981). The effect of salinity on the yield and composition of *Diplachne fusca* (Kallar grass). *Plant, Cell and Environment*, **4**, 177–81.
Savile, D. B. O. (1972). Arctic adaptations in plants. *Canada Department of Agriculture Monograph*, **6**, 1–81.
Schat, H. & Scholten, M. (1985). Comparative population ecology of dune slack species: the relation between population stability and germination behaviour in brackish environments. *Vegetatio*, **61**, 189–95.
(1986). Effects of salinity on growth, survival and life history of four short-lived pioneers from brackish dune slacks. *Acta Oecologia*, **7**, 221–31.
Schaefer, M. (1981). The arthropod fauna of high salt marshes. In *Terrestrial and Freshwater Fauna of the Wadden Sea Area*, ed. C. J. Smit, J. den Hollander, W.K.R.E. van Wingerden & W. J. Wolff, pp. 57–67. Rotterdam: Balkema.
Schimper, A. F. W. (1903). *Plant Geography upon a Physiological Basis*. Translated by W. R. Fisher, revised and edited by P. Groom & I. B. Balfour. Oxford: Clarendon Press.
Schirmer, U. & Breckle, S. W. (1982). The role of bladders for salt removal in some Chenopodiaceae (mainly *Atriplex* species). In *Contributions to the Ecology of Halophytes*, ed. D. N. Sen & K. S. Rajpurohit, pp. 215–31. The Hague: Dr Junk.
Scholander, P. F. (1968). How mangroves desalinate seawater. *Physiologia Plantarum*, **21**, 251–61.
Scholander, P. F., Bradstreet, E. D., Hammel, H. T. & Hemmingsen, E. A. (1966). Sap concentrations in halophytes and some other plants. *Plant Physiology*, **41**, 529–32.
Scholander, P. F., Hammel, H. T., Hemmingsen, E. A. & Garey, W. (1962). Salt balance in mangroves. *Plant Physiology*, **37**, 722–9.
Schubel, J. R. & Hirschberg, D. J. (1978). Estuarine graveyards, climatic change, and the importance of the estuarine environment. In *Estuarine Interactions*, ed. M. L. Wiley, pp. 285–303. New York: Academic Press.
(1982). The Chang Jiang (Yangtze) estuary: establishing its place in the community of estuaries. In *Estuarine Comparisons*, ed. V. S. Kennedy, pp. 649–54. New York: Academic Press.
Scott, A. J. (1977). Reinstatement and revision of Salicorniaceae J. Agardh (Caryophyllales). *Botanical Journal of the Linnean Society*, **75**, 357–74.
Scott, R. E. & Gray, A. J. (1976). Chromosome number of *Puccinellia maritima* (Huds.) Parl. in the British Isles. *Watsonia*, **11**, 53–7.
Seliskar, D. M. (1983). Root and rhizome distribution as an indicator of upper salt marsh wetland limits. *Hydrobiologia*, **107**, 231–6.
(1985). Morphometric variations of five tidal marsh halophytes along environmental gradients. *American Journal of Botany*, **72**, 1340–52.
Seliskar, D. M. & Gallagher, J. L. (1983). *The Ecology of Tidal Marshes of the Pacific Northwest Coast: a Community Profile*. Washington: U.S. Fish and Wildlife Service, Division of Biological Services. FWS/OBS-82/32.
Seneca, E. O. (1974). Germination and seedling response of Atlantic and Gulf Coast populations of *Spartina alterniflora*. *American Journal of Botany*, **61**, 947–56.
Seward, A. C. (1910). *Fossil Plants*, vol. 2. London: Cambridge University Press.
Shah, S. H., Gorham, J., Forster, B. P. & Wyn Jones, R. G. (1987). Salt tolerance in the Triticeae: the contribution of the D genome to cation selectivity in hexaploid wheat. *Journal of Experimental Botany*, **38**, 254–69.
Shannon, M. C. (1985). Principles and strategies in breeding for higher salt tolerance. *Plant and Soil*, **89**, 227–41.
Sharrock, J. T. R. (1967). *A Study of Morphological Variation in* Halimione portulacoides *(L.) Aell. in Relation to Variations in the Habitat*. Ph.D thesis University of Southampton.

Shea, M. L., Warren, R. S. & Niering, W. A. (1975). Biochemical and transplantational studies of the growth form of *Spartina alterniflora* on Connecticut salt marshes. *Ecology*, **56**, 461–6.

Sheehy Skeffington, M. J. & Jeffrey, D. W. (1985). Growth performance of an inland population of *Plantago maritima* in response to nitrogen and salinity. *Vegetatio*, **61**, 265–72.

Shennan, C. (1987). Salt tolerance in *Aster tripolium* L. III. Na and K fluxes in intact seedlings. *Plant, Cell and Environment*, **10**, 75–81.

Shennan, C., Hunt, R. & MacRobbie, E. A. C. (1987*a*). Salt tolerance in *Aster tripolium* L. I. The effect of salinity on growth. *Plant, Cell and Environment*, **10**, 59–65.

(1987*b*). Salt tolerance in *Aster tripolium* L. II. Ionic regulation. *Plant, Cell and Environment*, **10**, 67–74.

Shepard, F. P. (1977). *Geological Oceanography – Evolution of Coasts, Continental Margins, and the Deep-sea Floor*. St. Lucia: University of Queensland Press.

Shepherd, U. H. & Bowling, D. J. F. (1979). Sodium fluxes in roots of *Eleocharis uniglumis*, a brackish water species. *Plant, Cell and Environment*, **2**, 127–30.

Shimwell, D. W. (1971). *The Description and Classification of Vegetation*. London: Sidgwick & Jackson.

Shomer-Ilan, A., Moualem-Beno, D. & Waisel, Y. (1985). Effects of NaCl on the properties of phosphoenolpyruvate carboxylase from *Suaeda monoica* and *Chloris gayana*. *Physiologia Plantarum*, **65**, 72–8.

Shomer-Ilan, A. & Waisel, Y. (1986). Effects of stabilizing solutes on salt activation of phosphoenolpyruvate carboxylase from various plant sources. *Physiologia Plantarum*, **67**, 408–14.

Sibly, R. M. (1981). Strategies of digestion and defecation. In *Physiological Ecology. An Evolutionary Approach to Resource Use*, ed. C. R. Townsend & P. Calow, pp. 109–39. Oxford: Blackwell Scientific Publications.

Siira, J. (1970). Studies in the ecology of the sea-shore meadows of the Bothnian Bay with special references to the Liminka area. *Aquilo series Botanica*, **9**, 1–109.

Silander, J. A. (1985). Microevolution in clonal plants. In *Population Biology and Evolution of Clonal Organisms*, ed. J. B. C. Jackson, L. W. Buss & R. E. Cook, pp. 107–52. New Haven: Yale University Press.

Singer, C. E. & Havill, D. C. (1985). Manganese as an ecological factor in salt marshes. *Vegetatio*, **62**, 287–92.

Singleton, P. W. & Bohlool, B. B. (1984). Effect of salinity on nodule formation by soybean. *Plant Physiology*, **74**, 72–6.

Sivanesan, A. & Manners, J. G. (1970). Fungi associated with *Spartina townsendii* in healthy and 'die-back' sites. *Transactions of the British Mycological Society*, **55**, 191–204.

Skyring, G. W., Oshrain, R. L. & Wiebe, W. J. (1979). Assessment of sulfate reduction rates in Georgia marshland soils. *Geomicrobiology Journal*, **1**, 389–400.

Smart, R. M. (1982). Distribution and environmental control of productivity and growth form of *Spartina alterniflora* (Loisel.). In *Contributions to the Ecology of Halophytes*, ed. D. N. Sen & K. S. Rajpurohit, pp. 127–42. The Hague: Dr Junk.

Smart, R. M. & Barko, J. W. (1980). Nitrogen nutrition and salinity tolerance of *Distichlis spicata* and *Spartina alterniflora*. *Ecology*, **61**, 630–8.

Smirnoff, N. & Stewart, R. (1985*a*). Stress metabolites and their role in coastal plants. *Vegetatio*, **62**, 273–8.

(1985*b*). Nitrate assimilation and translocation by higher plants: comparative physiology and ecological consequences. *Physiologia Plantarum*, **64**, 133–40.

Smit, C. J. (1981). Distribution, ecology and zoogeography of breeding birds on the

Wadden Sea Islands. In *Terrestrial and Freshwater Fauna of the Wadden Sea Area*, ed. C. J. Smit, J. den Hollander, W.K.R.E. van Wingerden & W. J. Wolff, pp. 169–231. Rotterdam: Balkema.

Smit, C. J., den Hollander, J., van Wingerden, W.K.R.E. & Wolf, W. J. (eds.) (1981). *Terrestrial and Freshwater Fauna of the Wadden Sea Area. Report 10 of the Wadden Sea Working Group*. Rotterdam: Balkema.

Smith, A. M. & ap Rees, T. (1979). Pathways of carbohydrate fermentation in the roots of marsh plants. *Planta*, **146**, 327–34.

Smith, C. J. (1980). *Ecology of the English Chalk*. London: Academic Press.

Smith, G. S., Clark, C. J. & Holland, P. T. (1987). Chlorine requirement of kiwifruit (*Actinidia deliciosa*). *New Phytologist*, **106**, 71–80.

Smith, J. M. & Frey, R. W. (1985). Biodeposition by the ribbed mussel *Geukensia demissa* in a salt marsh, Sapelo Island, Georgia. *Journal of Sedimentary Petrology*, **55**, 817–28.

Smith, J. S. (1979). A new method of salt pan formation. *Transactions of the Botanical Society of Edinburgh*, **43**, 127-30.

(1982). The *Spartina* communities of the Cromarty Firth. *Transactions of the Botanical Society of Edinburgh*, **44**, 27–30.

Smith, M. K. & McComb, J. A. (1981). Effect of NaCl on the growth of whole plants and their corresponding callus cultures. *Australian Journal of Plant Physiology*, **8**, 267–325.

Smith, T. J. (1983). Alteration of salt marsh plant community composition by grazing snow geese. *Holarctic Ecology*, **6**, 204–10.

Smith, T. J. & Duke, N. C. (1987). Physical determinants of inter-estuary variation in mangrove species richness around the tropical coastline of Australia. *Journal of Biogeography*, **14**, 9–19.

Smith, T. J. & Odum, W. E. (1981). The effects of grazing by snow geese on coastal salt marshes. *Ecology*, **62**, 98–106.

Smith-White, A. R. (1979). Polyploidy in *Sporobolus virginicus* (L.) Kunth. *Australian Journal of Botany*, **27**, 429–37.

(1981). Physiological differentiation in a salt-marsh grass. *Wetlands (Australia)*, **1**, 20.

(1984). *A Genecological Study of* Sporobolus virginicus *(L.) Kunth*. Ph.D thesis, University of New South Wales.

(1988). *Sporobolus virginicus* (L.) Kunth in coastal Australia: the reproductive behaviour and the distribution of morphological types and chromosome races. *Australian Journal of Botany*, **36**, 23–39.

Snedaker, S. C. (1982). Mangrove species zonation: why? In *Contributions to the Ecology of Halophytes*, ed. D. N. Sen & K. Rajpurohit, pp. 111–25. The Hague: Dr Junk.

Snow, A. A. & Vince, S. W. (1984). Plant zonation in an Alaskan salt marsh. II. An experimental study of the role of edaphic conditions. *Journal of Ecology*, **72**, 669–84.

Sørensen, T. (1953). A revision of the Greenland species of *Puccinellia* Parl., with contributions to our knowledge of the Arctic *Puccinellia* flora in general. *Meddelelser om Grønland*, **136(3)**, 1–179.

Specht, R. L. (1981*a*). Major vegetation formations in Australia. In *Ecological Biogeography of Australia*, ed. A. Keast, pp. 165–297. The Hague: Dr Junk.

(1981*b*). Biogeography of halophytic angiosperms (salt-marsh, mangrove and sea-grass). In *Ecological Biogeography of Australia*, ed. A. Keast, pp. 577–89. The Hague: Dr Junk.

Spiegel-Roy, P. & Ben-Hayyim, G. (1985). Selection and breeding for salinity tolerance *in vitro*. *Plant and Soil*, **89**, 243–52.

Sporne, K. R. (1974). *The Morphology of Angiosperms*. London: Hutchinson.

Stalter, R. & Batson, W. T. (1969). Transplantation of saltmarsh vegetation, Georgetown South Carolina. *Ecology*, **50**, 1087–9.
State Pollution Control Commission (1983). *Water Quality in the Hawkesbury-Nepean River. A Study and Recommendations.* Sydney: SPCC.
Staples, R. C. & Toenniessen, G. H. (eds.) (1984). *Salinity Tolerance in Plants–Strategies for Crop Improvement.* New York: John Wiley.
Stavarek, S. J. & Rains, D. W. (1984). Cell culture techniques: selection and physiological studies of salt tolerance. In *Salinity Tolerance in Plant – Strategies for Crop Improvement*, ed. R. C. Staples & G. H. Toenniessen, pp. 321–34. New York: John Wiley.
Steers, J. A. (1938). The rate of sedimentation on salt marshes on Scolt Head Island. Norfolk. *Geological Magazine*, **75**, 26–39.
(1948). Twelve years' measurement of accretion on Norfolk salt marshes. *Geological Magazine*, **85**, 163–6.
(1964). *The Coastline of England and Wales.* Cambridge: Cambridge University Press.
(1977). Physiography. In *Wet Coastal Ecosystems*, ed. V. J. Chapman, pp. 31–60. Amsterdam: Elsevier.
Steindorsson, S. (1954). The coastline vegetation at Gasar in Eyjafjordor in the north of Iceland. *Norwegian Journal of Botany*, **3**, 203–12.
(1974). A list of Icelandic plantsociations: *Research Institute Nedri As, Hueragerdi*, Publication **17**, pp. 1–23.
(1976). Some notes on the shore vegetation of Iceland. *Acta Botanica Islandica*, **4**, 19–35.
Steiner, M. (1935). Zur Öekologie der Salzmarschen der nordöstlichen Vereinigten Staaten von Nordamerika. *Jahrbücher für wissenschaftliche Botanik*, **81**, 94–202.
(1939). Die Zusammensetzung des Zellsaftes bei höheren Pflanzen in ihrer ökologischen Bedeutung. *Ergebnisse der Biologie*, **17**, 151–254.
Stephenson, T. A. & Stephenson, A. (1972). *Life Between Tidemarks on Rocky Shores.* San Francisco: W.H. Freeman.
Steponkus, P. L. (1980). A unified concept of stress in plants? in *Genetic Engineering of Osmoregulation*, ed. R. W. Rains, R. C. Valentine & A. Hollaender, pp. 235–55, New York: Plenum Press.
Sterk, A. A. & Wijnands, D. O. (1970). On the variation in the flower heads of *Aster tripolium* L. in the Netherlands. *Acta Botanica Neerlandica*, **19**, 436–44.
Stewart, G. R. & Ahmad, I. (1983). AdPptation to salinity in angiosperm halophytes. In *Metals and Micronutrients: Uptake and Utilization by Plants*, ed. D. A. Robb & W. S. Pierpoint pp. 33–50. London: Academic Press.
Stewart, G. R., Larher, F., Ahmad, I. & Lee, J. A. (1979). Nitrogen metabolism and salt-tolerance in higher plant halophytes. In *Ecological Processes in Coastal Environments*, ed. R. L. Jefferies & A. J. Davy, pp. 211–27. Oxford: Blackwell Scientific Publications.
Stewart, G. R. & Lee, J. A. (1974). The role of proline accumulation in halophytes. *Planta*, **120**, 279–89.
Stewart, G. R., Lee, J. A. & Orebamjo, T. O. (1972). Nitrogen metabolism of halophytes. I. Nitrate reductase activity in *Suaeda maritima. New Phytologist*, **71**, 263–7.
(1973). Nitrogen metabolism of halophytes II. Nitrate availability and utilization. *New Phytologist*, **72**, 539–46.
Stewart, G. R. & Popp, M. (1987). The ecophysiology of mangroves. In *Plant Life in Aquatic and Amphibious Habitats*, ed. R. M. M. Crawford, pp. 333–45. Oxford: Blackwell Scientific Publications.
Stienstra, A. W. (1986). Nitrate accumulation and growth of *Aster tripolium* L. with a

continuous and intermittent nitrogen supply. *Plant, Cell and Environment*, **9**, 307–13.
Stocker, O. (1928). Das Halophytenproblem. *Ergebnisse der Biologie*, **3**, 265–354.
Storey, R., Ahmad, N. & Wyn Jones, R. G. (1977). Taxonomic and ecological aspects of the distribution of glycinebetaine and related compounds in plants. *Oecologia (Berlin)*, **27**, 319–32.
Storey, R. & Wyn Jones, R. G. (1975). Betaine and choline levels in plants and their relationship to NaCl stress. *Plant Science Letters*, **4**, 161–8.
(1977). Quaternary ammonium compounds in plants in relation to salt resistance. *Phytochemistry*, **16**, 447–53.
(1978). Salt stress and comparative physiology in the Gramineae. III. Effect of salinity upon ion relations and glycinebetaine and proline levels in *Spartina townsendii*. *Australian Journal of Plant Physiology*, **5**, 831–8.
(1979). Responses of *Atriplex spongiosa* and *Suaeda monoica* to salinity. *Plant Physiology*, **63**, 156–62.
Strogonov, B. P. (1964). *Physiological Basis of Salt Tolerance of Plants. (As Affected by Various Types of Salinity.)* Translation by A. Poljakoff-Mayber & A.M. Mayer. Jerusalem: Israel Program for Scientific Translations.
Stumpf, R. P. (1983). The process of sedimentation on the surface of a salt marsh. *Estuarine and Coastal Shelf Science*, **17**, 495–508.
Stumpf, D. K. & O'Leary, W. (1985). The distribution of Na^+, K^+ and glycinebetaine in *Salicornia bigelovii*. *Journal of Experimental Botany*, **36**, 550–5.
Stumpf, D. K., Prisco, J. T., Weeks, J. R., Lindley, V. A. & O'Leary, J. W. (1986). Salinity and *Salicornia bigelovii* Torr. seedling establishment. Water relations. *Journal of Experimental Botany*, **37**, 160–9.
Swinbanks, D. D. (1982). Intertidal exposure zones: a way to subdivide the shore. *Journal of Experimental Marine Biology and Ecology*, **62**, 69–86.
Synge, H. (ed.) (1981). *The Biological Aspects of Rare Plant Conservation*. Chichester: John Wiley.
Szwarcbaum, I. & Waisel, Y. (1973). Inter-relationship between halophytes and glycophytes grown on saline and non-saline media. *Journal of Ecology*, **61**, 775–86.
Tadmor, N. H., Koller, D. & Rawitz, E. (1958). Experiments in the propagation of *Juncus maritimus* Lam. II. Germination and field trials in the Arava. *Ktavim*, **9**, 177–205.
Tal, M. (1984). Physiological genetics of salt resistance in higher plants: studies on the level of the whole plant and isolated organs, tissues and cells. In *Salinity Tolerance in Plants – Strategies for Crop Improvement*, ed. R. C. Staples & G. H. Toeenniessen, pp. 301–20. New York: John Wiley.
(1985). Genetics of salt tolerance in higher plants: theoretical and practical considerations. *Plant and Soil*, **89**, 199–226.
Tansley, A. G. (ed.) (1911). *Types of British Vegetation*. Cambridge: Cambridge University Press.
(1939). *The British Islands and Their Vegetation*. London: Cambridge University Press.
Taylor, L. E. (1979). The concept of freshwater storage in estuaries. In *Estuarine and Coastal Land Reclamation and Water Storage*, ed. B. Knights & A. J. Phillips, pp. 135–51. Farnborough: Saxon House.
Taylor, M. C. & Burrows, E. M. (1968). Studies in the biology of *Spartina* in the Dee estuary, Cheshire. *Journal of Ecology*, **56**, 795–809.
Teal, J. M. (1962). Energy flow in the salt marsh ecosystem of Georgia. *Ecology*, **43**, 614–24.
Teal, J. & Teal, M. (1969). *Life and Death of the Salt Marsh*. Boston: Little, Brown and Co.

Teal, J. M. & Kanwisher, J. W. (1966). Gas transport in a marsh grass *Spartina alterniflora*. *Journal of Experimental Botany*, **17**, 355–61.
Teal, J. M. & Wieser, J. W. (1966). The distribution of ecology of nematodes in a Georgia salt marsh. *Limnology and Oceanography*, **11**, 217–22.
Teleky, L. S. (1978). Are the cries of Hermes being heard in the wetlands? In *Coastal Zone '78. Symposium on Technical, Environmental, Socioeconomic and Regulatory Aspects of Coastal Zone Management*, vol, **3**, pp. 2026–935. New York: American Society of Civil Engineers.
Terchunian, A., Klemas, V., Segovia, N., Alvarez, A., Vasconez, B. & Guerrero, L. (1986). Mangrove mapping in Ecuador: the impact of shrimp pond construction. *Environmental Management*, **10**, 345–50.
Termaat, A. & Munns, R. (1986). Use of concentrated macronutrients solutions to separate osmotic from NaCl – specific effects on plant growth. *Australian Journal of Plant Physiology*, **13**, 509–22.
Thannheiser, D. (1984). The coastal vegetation of eastern Canada. *Memorial University of Newfoundland Occasional Papers in Biology*, **8**, 1–212.
(1987). Die Pflanzengesellschaften der isländischen Salzwiesen. *Acta Botanica Islandica*, **9**, 35–60.
Thom, B. G. (1967). Mangrove ecology and deltaic geomorphology: Tabasco, Mexico. *Journal of Ecology*, **55**, 301–43.
Thom, B. G., Wright, L.D. & Coleman, J.M. (1975). Mangrove ecology and deltaic-estuarine geomorphology: Cambridge Gulf–Ord River, Western Australia. *Journal of Ecology*, **63**, 203–32.
Thompson, K. (1986). Small-scale heterogeneity in the seed bank of an acidic grassland. *Journal of Ecology*, **74**, 733–8.
Thomson, W. W. (1975). The structure and function of salt glands. In *Plants in Saline Environments*, ed. A. Poljakoff-Mayber & J. Gale, pp. 118–48. Berlin: Springer-Verlag.
Thorogood, C. A. (1985). Changes in the distribution of mangroves in the Port Jackson–Parramatta River estuary from 1930–1985. *Wetlands (Australia)*, **5**, 91–6.
Tiku, B. L. & Snaydon, R.W. (1971). Salinity tolerance within the grass species *Agrostis stolonifera* L. *Plant and Soil*, **35**, 421–31.
Tomlinson, P. B. (1986). *The Botany of Mangroves*. Cambridge: Cambridge University Press.
Tooley, M. J. (1978). *Sea-level Changes. North-West England during the Flandrian Stage*. Oxford: Clarendon Press.
Torzilli, A. P. & Andrykovitch, G. (1986). Degradation of *Spartina* lignocellulose by individual and mixed cultures of salt-marsh fungi. *Canadian Journal of Botany*, **64**, 2211–15.
Treherne, J. E. & Foster, W. A. (1977). Diel activity of an intertidal beetle, *Dicheirotrichus gustavi* Crotch. *Journal of Animal Ecology*, **46**, 127–38.
(1979). Adaptive strategies of air-breathing arthropods from marine salt marshes. In *Ecological Processes in Coastal Environments*, ed. R. L. Jefferies & A. J. Davy, pp. 165–75. Oxford: Blackwell Scientific Publications.
Truscott, A. (1984). Control of *Spartina anglica* on the amenity beaches of Southport. In Spartina anglica *in Great Britain*, ed. P. Doody, pp. 64–9. Attingham Park: Nature Conservancy Council.
Tsopa, E. (1939). La végétation des halophytes du nord de la Roumanie en connexion avec celle du reste du pays. *S.I.G.M.A.*, **70**, 1–22.
Tukey, H. G. (1970). The leaching of substances from plants. *Annual Reviews of Plant Physiology*, **21**, 305–24.

Turesson, G. (1922). The genotypical response of the plant species to the habitat. *Hereditas*, **3**, 211–350.

Turner, R. E. (1976). Geographic variations in salt marsh macrophyte production: a review. *Contributions in Marine Science*, **20**, 47–68.

Tutin, T. G. (1942). *Zostera* L. Biological Flora of the British Isles. *Journal of Ecology*, **30**, 217–26.

Tyler, G. (1967). On the effect of phosphorus and nitrogen, supplied to Baltic shore-meadow vegetation. *Botaniska Notiser*, **120**, 433–47.

(1968). Studies in the ecology of Baltic sea-shore meadows. I. Some chemical properties of Baltic shore-meadow clays. *Botaniska Notiser*, **12**, 89–113.

(1969*a*). Regional aspects of Baltic shore-meadow vegetation. *Vegetatio*, **29**, 60–86.

(1969*b*). Studies in the ecology of Baltic sea-shore meadows. II. Flora and vegetation. *Opera Botanica*, **25**, 1–10.

Underwood, A. J. (1978). A refutation of critical tidal levels as determinants of the structure of intertidal communities on British shores. *Journal of Experimenal Marine Biology and Ecology*, **33**, 261–76.

Ungar, I.A. (1974). Inland halophytes of the United States. In *Ecology of Halophytes*, ed. R. J. Reimold & W. H. Queen, pp. 235–305. New York: Academic Press.

(1978). Halophyte seed germination. *Botanical Review*, **44**, 233–64.

(1979). Seed dimorphism in *Salicornia europaea* L. *Botanical Gazette*, **140**, 102–8.

(1982). Germination ecology of halophytes. In *Contributions to the Ecology of Halophytes*, ed. D. N. Sen & K. Rajpurohit, pp. 143–54. The Hague: Dr Junk.

(1987*a*). Population ecology of halophyte seeds. *Botanical Review*, **53**, 301–34.

(1987*b*). Population characteristics, growth, and survival of the halophyte *Salicornia europaea. Ecology*, **68**, 569–75.

Ursin, M. J. (1972). *Life in and around the Salt Marshes*. New York: Thomas Y. Cromwell.

Van der Valk, A. G. & Davis, C. B. (1978). The role of seed banks in the vegetation dynamics of prairie glacial marshes. *Ecology*, **59**, 322–35.

Valiela, I. & Rietsma, C. S. (1984). Nitrogen, phenolic acids, and other feeding cues for salt marsh detritivores. *Oecologia (Berlin)*, **63**, 350–6.

Valiela, I. & Teal, J. M. (1974). Nutrient limitation in salt marsh vegetation. In *Ecology of Halophytes*, ed. R. J Reimold & W. H. Queen, pp. 547–63. New York: Academic Press.

(1979*a*). The nitrogen budget of a salt marsh ecosystem. *Nature (London)*, **280**, 652–6.

(1979*b*). Inputs, outputs and interconversions of nitrogen in a salt marsh ecosystem. In *Ecological Processes in Coastal Environments*, ed. R. L. Jefferies & A. J. Davy, pp. 399–419. Oxford: Blackwell Scientific Publications.

Valiela, I., Teal, J. M. & Deuser, W. G. (1978). The nature of growth forms in the salt marsh grass *Spartina alterniflora. American Naturalist*, **112**, 462–70.

Vanden Berghen, C. (1965*a*). Notes sur la végétation de sud-ouest de la France. III. La végétation de quelques près salés d'Oléron (Charente-maritime). *Bulletin du Jardin botanique de L'État à Bruxelles*, **35**, 363–9.

(1965*b*). La végétation de l'ile Hoëdoc (Morbihan, France). *Bulletin de la société royale de botanique de Belgique*, **98**, 275–94.

van der Valk, A. G. & Davis, C. B. (1978). The role of seed banks in the vegetation dynamics of prairie glacial marshes, *Ecology*, **59**, 322–35.

Van Diggelen, J., Rozema, J., Dickson, D. M. J. & Broekman, R. (1986). β-3-Dimethylsulphoniopropionate, proline and quaternary ammonium compounds in *Spartina anglica* in relation to sodium chloride, nitrogen and sulphur. *New Phytologist*, **103**, 573–86.

van Eerdt, M. M. J. (1985). The influence of vegetation on erosion and accretion in salt marshes of the Oosterschelde, The Netherlands. *Vegetatio*, **62**, 367–73.

van Eijk, M. (1939). Analyse de Wirkung des NaCl auf die Entwicklung Sukkulenze und Transpiration bei *Salicornia herbacea*, sowie Untersuchungen uber den Einfluss der Salzaufnahme, auf die Wurzelatmung bei *Aster tripolium. Recueil des travaux botaniques néerlandais*, **36**, 559–657.

van Veen, J. (1955). *Dredge Drain Reclaim! The Art of a Nation*, 4th edn. The Hague: Martinus Nijhoff.

van Wingerden, W.K.R.E., Littel, A. & Boosma, J. J. (1981). Strategies and population dynamics of arthropod species from coastal plains and green beaches. In *Terrestrial and Freshwater Fauna of the Wadden Sea Area*, ed. C. J. Smit, J. den Hollander, W.K.R.E. van Wingerden & W.J. Wolff, pp. 101–25. Rotterdam: Balkema.

Venables, A. V. & Wilkins, D. A. (1978). Salt tolerance in pasture grasses. *New Phytologist*, **80**, 613–22.

Verger, F. (1968). *Marais et Wadden du Littoral Français*. Bordeaux: Biscaye.

Verhoeven, B. (1983). Geomorphology and soil of salt marshes. In *Flora and Vegetation of the Wadden Sea Islands and Coastal Areas*, ed. K. S. Dijkema & W. J. Wolff, pp. 26–37. Rotterdam: Balkema.

Vestergaard, P. (1978). Studies on vegetation and soil of coastal salt marshes in the Disko area. West Greenland. *Meddelelser om Grønland*, **204(2)**, 1–51.

(1982). Horizontal variability of some soil properties within homogeneous stands of coastal salt meadow vegetation. *Nordic Journal of Botany*, **2**, 343–51.

Vince, S. W. & Snow, A. A. (1984). Plant zonation in an Alaskan salt marsh. I. Distribution, abundance and environmental factors. *Journal of Ecology*, **12**, 651–67.

Vince, S. W., Valiela, I. & Teal, J. M. (1981). An experimental study of the structure of herbivorous insect communities in a salt marsh. *Ecology*, **62**, 1662–78.

Vivrette, N. V. & Muller, C. H. (1977). Mechanism of invasion and dominance of coastal grassland by *Mesembryanthemum crystallinum. Ecological Monographs*, **47**, 301–18.

Wagret, P. (1968). *Polderlands*. London: Methuen.

Wainwright, S. J. (1980). Plants in relation to salinity. *Advances in Botanical Research*, **8**, 221–61.

(1984). Adaptations of plants to flooding with salt water. In *Flooding and Plant Growth*, ed. T. T. Kozlowski, pp. 295–343. New York: Academic Press.

Waisel, Y. (1972). *Biology of Halophytes*. New York: Academic Press.

Waite, S. & Hutchings, M. J. (1978). The effects of sowing density, salinity and substrate upon the germination of seeds of *Plantago coronopus* L. *New Phytologist*, **81**, 341–8.

Waldren, S. (1982). *Frankenia laevis* L. in Mid Glamorgan. *Watsonia*, **14**, 185–6.

Walker, R. R., Sedgley, M., Blessing, M. A. & Douglas, T. J. (1984). Anatomy, ultrastructure and assimilate concentrations of roots of citrus genotypes differing in ability for salt exclusion. *Journal of Experimental Botany*, **35**, 1481–94.

Walker, R. R., Törökfalvy, E., Steele Scot, N. F. Kriedemann, P. E. (1981). An analysis of photosynthetic response to salt treatment in *Vitis vinifera. Australian Journal of Plant Physiology*, **8**, 359–74.

Wallentinus, H. G. (1973). Description of sea-shore meadow communities, with special reference to the *Juncetum gerardi*. The sea-shore meadows of the Tullgarnaset IV. *Svensk Botanisk Tidsskrift*, **67**, 401–22.

Wallentinus, H. G. & Jonson, L. (1972). Mapping vegetation of a Baltic sea-shore meadow by colour infra red photography. The sea-shore meadow of the Tullgarnaset. II. *Svensk Botanisk Tidsskrift*, **66**, 314–25.

Walter, H. (1977). Climate. In *Wet Coastal Ecosystems*, ed. V. J. Chapman, pp. 61–7. Amsterdam: Elsevier.

Walton, J. (1922). A Spitsbergen salt marsh: with observations on the ecological phenomena attendant on the emergence of land from the sea. *Journal of Ecology*, **10**, 109–21.

Warming, E. (1909). *Ecology of Plants. An Introduction to the Study of Plant Communities.* (English edition prepared by P. Groom & I.B. Balfour.). Oxford: Clarendon Press.

Warren, J. H. & Underwood, A. J. (1986). Effects of burrowing crabs on the topography of mangrove swamps in New South Wales. *Journal of Experimental Marine Biology and Ecology*, **102**, 223–35.

Warren, R. S. & Gould, A. R. (1982). Salt tolerance expressed as a cellular trait in suspension cultures developed from the halophytic grass *Distichlis spicata. Zeitschrift für Pflanzenphysiologie*, **107**, 347–56.

Watkinson, A. R. & Davy, A. J. (1985). Population biology of salt marsh and sand dune annuals. *Vegetatio*, **62**, 487–97.

Watkinson, A. R. & Harper, J. L. (1978). The demography of a sand dune annual: *Vulpia fasciculata*: I. The natural regulation of populations. *Journal of Ecology*, **66**, 15–33.

Watson, A. & Watson, S. J. (1984). The biting midge problem in Tweed Shire, New South Wales. *Wetlands (Australia)*, **4**, 73–8.

Webb, D. A. (1954). Is the classification of plant communities either possible or desirable? *Botanisk Tidsskrift*, **51**, 362–70.

Webb, L. J., Tracey, J. G. & Williams, W. T. (1976). The value of structural features in tropical forest typology. *Australian Journal of Ecology*, **1**, 3–28.

Wells, J. T. & Coleman, J. M. (1981). Periodic mudflat progradation, northeastern coast of South America: a hypothesis. *Journal of Sedimentary Petrology*, **51**, 1069–75.

West, R. C. (1977). Tidal salt-marsh and mangal formations of Middle and South America. In *Wet Coastal Ecosystems*, ed. V. J. Chapman, pp. 193–213. Amsterdam: Elsevier.

(1981). Nature and future of salt marshes in South America. In *XII International Botanical Congress Abstracts*, pp. 338. Sydney: IBC.

Westhoff, V. (1969). Langjahrige Beobachtungen an aussüssungs – dauerprobeflächen beweideter und unbeweideter Vegetation an der ehemaligen Zuiderzee. In *Experimentelle Pflanzensoziologie*, ed. R. Tüxen, pp. 246–55. The Hague: Dr Junk.

(1971). The dynamic structure of plant communities in relation to the objectives of conservation. In *The Scientific Management of Animal and Plant Communities for Conservation*, ed. E. Duffey & A. S. Watt, pp. 3–14. Oxford: Blackwell Scientific Publications.

(1985). Nature management in coastal areas of Western Europe. *Vegetatio*, **62**, 523–32.

Westhoff, V. & den Held, A. J. (1979). *Plantengemeenschappen in Nederland.* Zutphen: Thieme.

Westhoff, V. & Sykora, K.V. (1979). A study of the influence of desalination on the Juncetum gerardii. *Acta Botanica Neerlandica*, **28**, 505–12.

Westhoff, V. & van der Maarel, E. (1973). The Braun-Blanquet approach. In *Handbook of Vegetation Science V. Ordination and Classification of Vegetation*, ed. R. H. Whittaker, pp. 619–726. The Hague: Dr Junk.

Whinray, J. S. (1981). Barilla production and early soap making in Tasmania. In *Plants and Man in Australia*, ed. D.J. & S.G.M. Carr, pp. 281–96. Sydney: Academic Press.

Whitney, D. M., Chalmers, A. G., Haines, E. B., Hanson, R. B., Pomeroy, L. R. & Sherr, B. (1981). The cycles of nitrogen and phosphorus. In *The Ecology of a Salt Marsh*, ed. L. R. Pomeroy & R. G. Weigert, pp. 163–81. New York: Springer-Verlag.

Whittaker, R. H. (1973). *Handbook of Vegetation Science V. Ordination and Classification of Vegetation.* The Hague: Dr Junk.

(1975). *Communities and Ecosystems*, 2nd edn. New York: Macmillan.

Whitten, A. J. & Damanik, S. J. (1986). Mass defoliation of mangroves in Sumatra, Indonesia. *Biotropica*, **18**, 176.

Wiebe, W. J., Christian, R. R., Hansen, J. A., King, G., Sherr, B. & Skyring, G. (1981). Anaerobic respiration and fermentation. In *The Ecology of a Saltmarsh*, ed. L. R. Pomeroy & R. G. Weigert, pp. 137–59. New York: Springer-Verlag.

Wiegert, R. G. (1979). Ecological processes characteristic of coastal *Spartina* marshes of the south-eastern U.S.A. In *Ecological Processes in Coastal Environments*, ed. R. L. Jefferies & A. J. Davy, pp. 467–90. Oxford: Blackwell Scientific Publications.

Wiegert, R. G. & Pomeroy, L. R. (1981). The salt-marsh ecosystem: a synthesis. In *The Ecology of a Salt Marsh*, ed. L. R. Pomeroy & R. G. Wiegert, pp. 219–30. New York: Springer.

Wiehe, P.O. (1935). A quantitative study of the influence of tide upon populations of *Salicornia europea. Journal of Ecology*, **23**, 323–33.

Williams, G. & Hall, M. (1987). The loss of coastal grazing marshes in south and east England, with special references to east Essex, England. *Biological Conservation*, **39**, 243–53.

Williams, W. T. & Barber, D. A. (1961). The functional significance of aerenchyma in plants. In *Mechanisms in Biological Competition*, ed. F. L. Milthorpe, pp. 132–44. Cambridge: Cambridge University Press.

Winfrey, M. R. (1984). Microbial production of methane. In *Petroleum Microbiology*, ed. R. M. Atlas, pp. 153–212. New York: Macmillan.

Winter, E. (1982*a*). Salt tolerance of *Trifolium alexandrinum* L. II. Ion balance in relation to its salt tolerance. *Australian Journal of Plant Physiology*, **9**, 227–37.

(1982*b*). Salt tolerance of *Trifolium alexandrinum* L. III. Effects of salt on ultrastructure of phloem and xylem transfer cells in petioles and leaves. *Australian Journal of Plant Physiology*, **9**, 239–50.

Winter, E. & Läuchli, A. (1982). Salt tolerance of *Trifolium alexandrinum* L. I. Comparison of the salt response of *T. alexandrinum* and *T. pratense. Australian Journal of Plant Physiology*, **9**, 221–6.

Winter, E. & Preston, J. (1982). Salt tolerance of *Trifolium alexandrinum* L. IV. Ion measurements by X-ray microanalysis in unfixed, frozen hydrated leaf cells at various stages of salt treatment. *Australian Journal of Plant Physiology*, **9**, 251–9.

Winter, K. (1979). Photosynthetic and water relationship of higher plants in a saline environment. In *Ecological Processes in Coastal Environments*, ed. R. L. Jefferies & A. J. Davy, pp. 297–320. Oxford: Blackwell Scientific Publications.

Winter, K., Osmond, C. B. & Pate, J.S. (1981). Coping with salinity. In *The Biology of Australian Plants*, ed. J. S. Pate & A. J. McComb, pp. 88–113. Perth: University of W. A. Press.

Woodell, S. R. J. (1974). Anthill vegetation in a Norfolk saltmarsh. *Oecologia (Berlin)*, **16**, 221–5.

(1978). Directionality in bumblebees in relation to environmental factors. In *The Pollination of Flowers by Insects*, ed. A. J. Richards, pp. 31–9. London: Academic Press.

(1985). Salinity and seed germination patterns in coastal plants. *Vegetatio*, **61**, 223–9.

Woodroffe, C. D., Thom, B. G. & Chappell, J. (1985). Development of widespread mangrove swamps in mid-Holocene times in northern Australia. *Nature (London)*, **317**, 711–13.

Woodwell, G. M., Houghton, R. A., Hall, C. A. S., Whitney, D. E., Moll, R. A. & Juers, D. W. (1979). The Flax Pond Ecosystem Study: the annual metabolism and nutrient budgets of a salt marsh. In *Ecological Processes in Coastal Environments*, ed. R. L. Jefferies & A. J. Davy, pp. 491–511. Oxford: Blackwell Scientific Publications.

Woolhouse, H. W. (1981). Aspects of the carbon and energy requirements of

photosynthesis considered in relation to environmental constraints. In *Physiological Ecology. An Evolutionary Approach to Resource Use*, ed. C. R. Townsend & P. Calow, pp. 51–85. Oxford: Blackwell Scientific Publications.
Wyn Jones, R.G. (1980). An assessment of quaternary ammonium and related compounds as osmotic effectors in crop plants. In *Genetic Engineering of Osmoregulation*, ed. D. W. Rains, R. C. Valentine & A. Hollaender, pp. 155–70. New York: Plenum Press.
Wyn Jones, R. G. & Gorham, J. (1983). Osmoregulation. In *Physiological Plant Ecology. III. Responses to the Chemical and Biological Environment, Encyclopedia of Plant Physiology N.S.*, ed. O. L. Lange, P. S. Nobel, C. B. Osmond & H. Ziegler, pp. 35–58. Berlin: Springer-Verlag.
Wyn Jones, R. G. & Pollard, A. (1983). Proteins, enzymes and inorganic ions. In *Inorganic Plant Nutrition – Encyclopedia of Plant Physiology N.S. 15*, ed. A. Läuchli & R. L. Bieleski, pp. 528–62. Berlin: Springer-Verlag.
Wyn Jones, R. G. & Storey, R. (1978). Salt stress and comparative physiology in the Gramineae. IV. Comparison of salt stress in *Spartina* × *townsendii* and three barley cultivars. *Australian Journal of Plant Physiology*, 1978, **5**, 839–50.
(1981). Betaines. In *The Physiology and Biochemistry of Drought Resistance in Plants*, ed. L. G. Paleg & D. Aspinall, pp. 172–204. Sydney: Academic Press.
Wyn Jones, R. G., Storey, R., Leigh, R. A., Ahmad, N. & Pollard, A. (1977). A hypothesis on cytoplasmic osmoregulation. In *Regulation of Cell Membrane Activities in Plants*, ed. E. Marre & O. Ciferri, pp. 121–36. Amsterdam: North Holland.
Yancey, P. H., Clark, M. E.., Hand, S. C., Bowlus, R. D. & Somero, G. N. (1982). Living with water stress: evolution of osmolyte systems. *Science*, **217**, 1214–22.
Yapp, R. H., Johns, D. & Jones, O. T. (1917). The salt marshes of the Dovey estuary. Part II. The salt marshes (by R. H. Yapp & D. Johns). *Journal of Ecology*, **5**, 65–103.
Yeo, A. R. (1981). Salt tolerance in the halophyte *Suaeda maritima* (L.) Dum.: intracellular compartmentation of ions. *Journal of Experimental Botany*, **32**, 487–97.
(1983). Salinity resistance: physiologies and prices. *Physiologia Plantarum*, **58**, 214–22.
Yeo, A. R., Caporn, S. J. M. & Flowers, T. J. (1985). The effect of salinity upon photosynthesis in rice (*Oryza sativa* L.): gas exchange by individual leaves in relation to their salt content. *Journal of Experimental Botany*, **36**, 1240–8.
Yeo, A. R. & Flowers, T. J. (1980). Salt tolerance in the halophyte *Suaeda maritima* (L.) Dum: evaluation of the effect of salinity upon growth. *Journal of Experimental Botany*, **31**, 1171–83.
(1985). The absence of an effect of the Na/Ca ratio on sodium chloride uptake by rice (*Oryza sativa* L.). *New Phytologist*, **99**, 81–90.
(1986*a*). Salinity resistance in rice (*Oryza sativa* L.) and a pyramiding approach to breeding varieties for saline soils. *Australian Journal of Plant Physiology*, **13**, 161–73.
(1986*b*). Ion transport in *Suaeda maritima*: its relation to growth and implications for the pathway of radial transport of ions across the root. *Journal of Experimental Botany*, **37**, 143–59.
Yeo, A. R., Yeo, M. E. & Flowers, T. J. (1987). The contribution of an apoplastic pathway to sodium uptake by rice roots in saline conditions. *Journal of Experimental Botany*, **38**, 1141–53.
Yugovic, J. Z. (1984). The grey glasswort (*Halosaracia halocnemoides*) in coast Victoria and some implications for the Orange-bellied Parrot. *Victorian Naturalist*, **101**, 234–9.
Zahran, M. A. (1977). Africa A. Wet formations of the African Red Sea coast. In *Wet Coastal Ecosystems*, ed. V. J. Chapman, pp. 215–31. Amsterdam: Elsevier.
Zedler, J., Winfield, T. & Mauriello, D. (1978). Primary productivity in a So. California

estuary. In *Coastal Zone '78. Symposium on Technical, Environmental, Socioeconomic and Regulatory Aspects of Coastal Zone Management*, vol. **2**, pp. 649–62. New York: American Society of Civil Engineers.

Zonneveld, I. S. (1960). *De Brabantse Biesbosch. Een Studie van bodem en vegetatie van een Zoetwatergetijdendelta.* Wageningen: Pudoc.

Index